TRANSACTIONS
OF THE
INTERNATIONAL
ASTRONOMICAL UNION

VOLUME XVB – PROCEEDINGS

BENGT STRÖMGREN

PRESIDENT OF THE INTERNATIONAL ASTRONOMICAL UNION

1970–1973

INTERNATIONAL ASTRONOMICAL UNION
UNION ASTRONOMIQUE INTERNATIONALE

TRANSACTIONS OF THE IAU
VOLUME XVB

PROCEEDINGS OF THE FIFTEENTH GENERAL ASSEMBLY SYDNEY 1973

AND

EXTRAORDINARY GENERAL ASSEMBLY POLAND 1973

Edited by

G. CONTOPOULOS
General Secretary of the Union

A. JAPPEL
Executive Secretary

D. REIDEL PUBLISHING COMPANY
DORDRECHT-HOLLAND / BOSTON-U.S.A.
1974

Published on behalf of
the International Astronomical Union
by
D. Reidel Publishing Company
Dordrecht-Holland

Sold and distributed in the U.S.A., Canada and Mexico
by D. Reidel Publishing Company, Inc.
306 Dartmouth Street, Boston,
Mass. 02116, U.S.A.

Library of Congress Catalog Card Number 73–81827

ISBN-13: 978-94-010-2210-1 e-ISBN-13: 978-94-010-2208-8
DOI: 10.1007/978-94-010-2208-8

PREFACE

The three years since the Brighton General Assembly have been the most active period in the history of the Union. 33 IAU Symposia and Colloquia, the first Regional Meeting under the Auspices of the IAU, several co-sponsored Meetings and many other special projects. All this culminating with two General Assemblies in two opposite parts of the Earth, Australia and Poland. At the same time the membership of the Union rose to 3200, the number of Commissions to 40, the number of adhering countries to 47.

The present Volume gives a general picture of the Union's recent activity. It contains the report of the Executive Committee, the report of the General Assembly, including the Commissions, Meetings, a short report on the Extraordinary General Assembly and an Appendix with the Members and Commissions of the IAU and the approved names of Lunar and Martian features.

I take this opportunity to thank all our collaborators, members of the Executive Committee, Presidents of Commissions, Chairmen of Specific Projects, IAU Secretaries and all the Members of the IAU for their contribution to keep our Union a living body of active scientists and a big international family.

G. Contopoulos
General Secretary

CONTENTS

Part 3

Page No.

Reports of Meetings of Commissions
Comptes Rendus des Séances des Commissions

Commissions

Part 4

Page No.

Astronomer's Handbook

Part 5

Report of the Extraordinary General Assembly, Poland 1973

Appendix I

Appendix II

PART 1

REPORT OF THE EXECUTIVE COMMITTEE

1970–1972

EXECUTIVE COMMITTEE
1970–1973

PRESIDENT

Professor B. Strömgren, Observatoriet, Østervoldgade 3, Copenhagen K, Denmark

VICE-PRESIDENTS

Professor M. K. V. Bappu, Indian Institute of Astrophysics, Kodaikanal, India

Professor B. J. Bok, Steward Observatory, University of Arizona, Tucson, Arizona 85721, U.S.A.

Professor L. Gratton, Laboratorio di Astrofisica, Casella Postale 67, 00044 Frascati (Rome), Italy

Sir Bernard Lovell, Nuffield Radio Astronomy Laboratories, Jodrell Bank, Macclesfield, Cheshire, U.K.

Professor E. R. Mustel, Astronomical Council, USSR Academy of Sciences, Vavilov Str. 34, Moscow V-312, U.S.S.R.

Professor J. Sahade, Instituto de Astronomía y Física del Espacio, Casilla de Correo 67– Sucursal 28, Buenos Aires, Argentina

GENERAL SECRETARY

Professor C. de Jager, c/o Space Research Laboratory of the Astronomical Institute, 21 Beneluxlaan, Utrecht, The Netherlands

ASSISTANT GENERAL SECRETARY

Professor G. Contopoulos, Astronomical Department, University of Thessaloniki, Thessaloniki, Greece

Introduction

The present report covers the period of three years from 1 January 1970 to 31 December 1972. It follows closely the pattern as set by previous reports, especially that for the years 1967–1969, formally approved by the XIVth General Assembly, and published in volume XIVB of the *Transactions of the IAU.*

The period under report includes the first eight months of the year 1970 (until 27 August) which fell under the responsibility of the former Executive Committee, but it does not include the first eight months of the year 1973, still within the responsibility of the present Executive Committee. An oral report for the period from 1 January to 20 August 1973 will be presented at the XVth General Assembly.

Decisions of the Executive Committee related to the XIVth General Assembly and printed in *Transactions* volume XIVB or published in the *Information Bulletin* of the IAU will not be repeated here. Thus, the present report is actually restricted to formal matters and deals with facts necessary for permanent record. The reports on the activity of IAU Commissions, Inter-Union Commissions, Astronomical Services, etc., have been included in the Reports on Astronomy, 1973, to be published as volume XVA of the *Transactions.*

The part called Activities Report deals with the administration of the Union, meetings of the Executive Committee, IAU Membership, adherence of countries, Commission of the IAU, colloquia and symposia organized by the Union, IAU publications, and relations to other organizations.

The part called Report on IAU Finances gives a rounded off picture of the receipts and payments of the Union during the period under report, and presents an outline of the Union's financial problems in the years to come.

The report is supplemented by an appendix listing data on Adhering Organizations, IAU Members and IAU publications.

Activities Report

Since 1970, the International Astronomical Union has operated in accord with the provisions of the new Statutes and By-laws adopted by the XIVth General Assembly, and of the Working Rules, as approved at the 34th meeting of the Executive Committee.

The Executive Committee and the Officers responsible for the administration of the Union have had to cope with two conflicting trends: the expanding activity of the IAU, on the one hand, and permanent inflation, on the other.

The number of IAU Members increased, as did that of IAU Commissions and Adhering Organizations; relations with other international and national organizations became more involved; the number of scientific meetings sponsored by the Union doubled in comparison with the period 1967–1969; the number of IAU publications increased; two General Assemblies had to be prepared for 1973; a new type of scientific meetings held under the auspices of the IAU has been introduced, etc.

This difficult situation was met by adding a shorthand-typist to the existing staff of two over the periods of intensified work, and seeking external help when necessary. Moreover, the President, the General Secretary and the Assistant General Secretary benefited from the services of several secretaries and secretarial facilities in Copenhagen, Utrecht and Thessaloniki, generously provided by their respective home institutions. The Executive Committee realises that the staff of the Administrative Office is largely insufficient that, however, the finances of the Union, as they are, do not permit an improvement in this respect.

The expanded activity of the IAU had to be financed from budgeted funds whose value had dropped through inflation by more than 30% in the course of time. This prompted the Executive Committee to abandon the Union's policy of generosity, such as the free distribution of IAU Transactions to Members, and to propose to the General Assembly an increase of the unit of contribution payable by Adhering Organizations.

Executive Committee

The Executive Committee, in its former composition, held its 33rd meeting in Brighton during the XIVth General Assembly, having had sessions on 16, 17 and 25 August 1970. The meeting was presided over by the President of the Union Prof. Dr O. Heckmann. All members and both advisors were present.

The 34th meeting of the Executive Committee, in its present composition, was held in Brighton on 27 August 1970, under the presidency of Prof. B. Strömgren. Excused were the Vice-Presidents Sir Bernard Lovell and Prof. E. R. Mustel. The President invited Dr C. A. Ronan and Prof. A. B. Severny to attend the meeting in their stead. Further, the Executive Committee held the following meetings:

The 35th meeting on Ile d'Yeu, France, from 6 to 10 September 1971. Prof. Dr O. Heckmann, advisor, was excused.

The 36th meeting in Copenhagen, Denmark, from 28 August to 1 September 1972. Prof. B. J. Bok, Vice-President, was excused.

The Executive Committee adopted, at its 35th meeting, the following resolution: "The IAU recognizes the fact that in certain parts of the world the need may be felt to start new regional activities involving especially young professional astronomers. The IAU will be happy to assist in, and to stimulate, the early organization of such activities whenever there seems to be an occasion on which the IAU's assistance may be needed."

This resolution led to the organization of the First European Astronomical Meeting under the Auspices of the IAU in Athens, Greece, from 4 to 9 September 1972.

Officers' Meetings

The President of the IAU, the General Secretary accompanied by his staff, and the Assistant

General Secretary (Officers), met as follows: in Copenhagen, Denmark, on 27 October 1970, in Utrecht, The Netherlands, on 24 and 25 February 1971, in Thessaloniki, Greece, on 12 and 13 May 1971, in Copenhagen on 29 November 1971, in Utrecht on 30 and 31 January 1972, in Copenhagen on 9 May 1972, in Thessaloniki on 30 October 1972, in Paris, France, on 28 February 1973, and in Copenhagen on 8 June 1973.

The meetings were held to co-ordinate the Union's activities in its manifold fields of interest, to implement the decisions of the Executive Committee, and to outline future programmes.

Adhering Countries

Cuba and Uruguay have adhered to the Union since 1970 thus bringing the number of Adhering Countries to 46. The Republic of Korea has asked for information regarding admission to the Union.

The list of Adhering Countries and corresponding adhering bodies, together with the approximate number of Members in each country, is given in the appendix on pages 25–26.

Members of the IAU

The number of IAU Members was 1993 on 8 February 1970. A further 567 new Members were admitted at the XIVth General Assembly. On the other hand, the General Secretary had been informed of the decease of 46 Members and of the retirement of 5 Members, so that the Union had 2509 Members on 1 February 1973.

The list of Members whose decease has been notified since 28 February 1970 is given in the appendix on pages 26–27.

Commissions of the IAU

The *Transactions of the IAU*, volume XVA, subtitled *Reports on Astronomy 1973,* include the reports prepared by each of the 39 Presidents of Commissions of the Union, assisted by their Vice-Presidents, Organizing Committees, those responsible for the various Working Groups within the Commissions, and by the Commission members, themselves.

The XIVth General Assembly created Commissions No. 47 on Cosmology and No. 48 on High Energy Astrophysics, and approved the merger of Commissions Nos. 23 "Carte du Ciel" and 24 on Stellar Parallaxes and Proper Motions into a single Commission No. 24 on Photographic Astrometry, thus bringing the number of Commissions to 39.

In view of the unfortunate decease of Professor M. G. J. Minnaert, Dr. A Dollfus was appointed President ad interim of Commission No. 17 on the Moon, and Dr M. W. Feast became President ad interim of Commission No. 27 on Variable Stars after the resignation of Dr O. J. Eggen.

Between the General Assemblies, Commissions have co-opted a number of new members from among the Members of the Union. These co-options have been approved by the Executive Committee and published in the *Information Bulletin.*

The Presidents of Commissions will prepare proposals for the new composition of Commissions including co-opted members and new IAU Members to be admitted at the General Assembly.

Between the General Assemblies, Commissions organized several colloquia and other projects financially supported by the Union.

Colloquia of IAU Commissions
(held between 1 January 1970 and 31 December 1972)

IAU Colloquium No. 7 on Proper Motions, Minneapolis, Minnesota, U.S.A., 21–23 April 1970.

IAU Colloquium No. 8 on Techniques for Measurement of Fundamental Spectroscopic Data, London, U.K., 1–4 September 1970.

IAU Colloquium No. 9 on the IAU System of Astronomical Constants, Heidelberg, F.R.G., 12–14 August 1970.

IAU Colloquium No. 10 on the Gravitational *n*-Body Problem, Cambridge, U.K., 11–14 August 1970.

IAU Colloquium No. 11 on Automation in Optical Astrophysics, Edinburgh, U.K., 12–14 August 1970.

IAU Colloquium No. 12 "Physical Studies of Minor Planets", Tucson, Arizona, U.S.A., 8–11 March 1971.

IAU Colloquium No. 13 "Evolutionary and Physical Problems of Meteoroids", State University of New York, Albany, New York, U.S.A., 14–17 June 1971.

IAU Colloquium No. 14 "Ultraviolet and X-Ray Spectroscopy in Astrophysical and Laboratory Plasmas", Utrecht, The Netherlands, 22–26 August 1971.

IAU Colloquium No. 15 "New Directions and Frontiers in Variable Star Research", Bamberg, F.R.G., 31 August -3 September 1971.

IAU Colloquium No. 16 "Analytical Procedures for Eclipsing Binary Light Curves", Philadelphia, U.S.A., 8–11 September 1971.

IAU Colloquium No. 17 "Ages des Etoiles", Paris, France, 15–22 September 1972.

IAU Colloquium No. 18 "Orbital and Physical Parameters of Double Stars" Swarthmore College, Pennsylvania, U.S.A., 12–15 April 1972.

IAU Colloquium No. 19 "Stellar Chromospheres", Greenbelt, Maryland, U.S.A., 14–18 February 1972.

IAU Colloquium No. 20 "Meridian Astronomy", Copenhagen, Denmark, 25–28 September 1972.

IAU Colloquium No. 21 "Variable Stars in Globular Clusters and Related Systems", Richmond Hill, Canada, 29–31 August 1972.

IAU Colloquium No. 22 "Asteroïdes, Comètes, Matière Météorique", Nice, France, 4–6 April 1972.

IAU Colloquium No. 23 "Astronomical Polarimetry", Tucson, Arizona, 15–17 November 1972.

Specific Projects of Commissions

The list of Specific Projects of Commissions as approved by the XIVth General Assembly is given, together with the corresponding expenditure, in the report on the financial situation of the IAU (page 10). These projects are, as a rule, continuations of previous projects and thus need no further comment.

Symposia

IAU Symposia
(held between 1 January 1970 and 31 December 1972)

IAU Symposium No. 41 "New Techniques in Space Astronomy", Münich, F.R.G., 10–14 August 1970 (cosponsored by COSPAR).

IAU Symposium No. 42 "White Dwarfs", St Andrews, U.K., 10–14 August 1970.

IAU Symposium No. 43 "Solar Magnetic Fields", Paris, France, 31 August – 4 September 1970.

IAU Symposium No. 44 "External Galaxies and Quasi Stellar Sources", Uppsala, Sweden, 10–14 August 1970 (cosponsored by URSI).

IAU Symposium No. 45 "The Motion, Evolution of Orbits, and Origin of Comets", Leningrad, U.S.S.R., 4–11 August 1970.

IAU Symposium No. 46 "The Crab Nebula", Jodrell Bank, U.K., 5–7 August 1970 (cosponsored by COSPAR and URSI).

IAU Symposium No. 47 "The Moon", School of Physics, University of Newcastle upon Tyne, 22–26 March 1971.

IAU Symposium No. 48 "Rotation of the Earth", Morioka, Japan, 9–15 May 1971.

IAU Symposium No. 49 "Wolf-Rayet and High-Temperature Stars", Instituto de Astronomía y Física del Espacio, Buenos Aires, Argentina, 9–14 August 1971.

IAU Symposium No. 50 "Spectral Classification and Multicolour Photometry", Villa Carlos Paz near Córdoba, Argentina, 18–24 October 1971.

IAU Symposium No. 51 "Extended Atmospheres and Circumstellar Matter in Spectroscopic Binary Systems" (Struve Memorial), Parksville, Vancouver Island, B. C., Canada, 6–12 September 1972.

IAU Symposium No. 52 "Interstellar Dust and Related Topics", Albany, N. Y., U.S.A., 29 May to 2 June 1972.

IAU Symposium No. 53 "Physics of Dense Matter", Boulder, Colorado, U.S.A., 21–25 August 1972.

IAU Symposium No. 54 "Problems of Calibration of Absolute Magnitudes and Temperatures of Stars", Geneva, Switzerland, 12–15 September 1972.

IAU Symposium No. 55 "X- and Gamma-Ray Astronomy", Madrid, Spain, 11–13 May 1972

The following IAU Symposia will be held in conjunction with the XVth General Assembly and the Extraordinary General Assembly in commemoration of Copernicus:

IAU Symposium No. 56 "The Fine Structure of the Chromosphere", Surfers Paradise, Queensland, Australia, 3–7 September 1973.

IAU Symposium No. 57 "Coronal Disturbances", Surfers Paradise, Queensland, Australia, 7–11 September 1973.

IAU Symposium No. 58 "Formation and Dynamics of Galaxies", Canberra, Australia, 12–15 August 1973.

IAU Symposium No. 59 "Stellar Instability and Evolution", Canberra, Australia, 16–18 August 1973.

IAU Symposium No. 60 "Galactic Radio Astronomy", Maroochydore, Queensland, Australia, 3–7 September 1973.

IAU Symposium No. 61 "New Problems of Astrometry", Perth, Australia, 13–17 August 1973.

IAU Symposium No. 62 "The Stability of the Solar System and of Small Stellar Systems", Warsaw, Poland, 5–8 September 1973.

IAU Symposium No. 63 "Confrontation of Cosmological Theories with Observational Data", Cracow, Poland, 10–12 September 1973.

IAU Symposium No. 64 "Gravitational Radiation and Gravitational Collapse", Warsaw, Poland, 5–8 September 1973.

IAU Symposium No. 65 "Explorations of the Planetary System", Toruń, Poland, 5–8 September 1973.

IAU Symposium No. 66 "Late Stages of Stellar Evolution", Warsaw, Poland, 10–12 September 1973.

Symposia with IAU Participation

COSPAR/URSI/IAU/IAGA(IUGG) Symposium "1970 Solar Eclipse", Seattle, Washington, U.S.A., 18–21 June 1971.

COSPAR/IAU Symposium "High Angular Resolution Astronomical Observations from Space", Seattle, Washington, U.S.A., 24–26 June 1971.

SPARMO/IUGG/IAGA/IAU Symposium "Technology and Utilization of Stratospheric Balloons" Seattle, Washington, U.S.A., 26–28 June 1971.

IUHPS/IAU "Kepler Symposium", Austria and F.R.G., August 1971.

IAU/IAP/Meteorological Society/URSI Symposium "Planetology", Montreal, Canada, 22–29 August 1972.

URSI/COSPAR/IAU Symposium "Planetary Atmospheres and Surfaces", Madrid, Spain, 10–13 May 1972.

The IUPHS/IAU Symposium on the "Astronomy of Copernicus and its Background" will be held in Toruń, Poland, on 7 and 8 September 1973.

The IAU was, in addition, represented at numerous scientific meetings of ICSU organizations, as announced in the *IAU Information Bulletin*.

IAU Publications

Budgetary considerations led the Executive Committee to suspend the free distribution of IAU publications to Members. Thus only institutions of developing countries continue obtaining IAU Symposium volumes free of charge. Members of the Organizing Committees of Symposia receive each a complimentary copy of the corresponding Symposium volume. All IAU publications are being produced and distributed for the account and risk of the publisher, who pays the IAU a royalty.

The IAU continues financing the production and distribution of the *IAU Information Bulletin.*

Transactions of the IAU

The *Transactions of the XIVth General Assembly* were published in volumes A and B. Volume A, subtitled *Reports on Astronomy, 1970,* was distributed free of charge to IAU Members, Volume B, subtitled *Proceedings of the Fourteenth General Assembly, Brighton 1970,* was distributed free of charge to those who were Members of the IAU at the time of the XIVth General Assembly.

The Invited Discourses, proceedings of Joint Discussions and results of Special Meetings held during the XIVth General Assembly were published in the form of a symposium volume under the title *Highlights of Astronomy* (volume 2), as presented at the XIVth General Assembly of the IAU 1970.

Transactions volume A and B were financed from the budget of the IAU. The *Highlights of Astronomy* and all Symposium volumes went to the charge of the publisher.

The Transaction volumes sold as follows over the period under report:

vol. XIIIA	107 copies
vol. XIIIB	84 copies
vol. XIVA	489 copies
vol. XIVB	1.195 copies

IAU Symposia

The IAU Symposium volumes sold as follows over the period under report:

Highlights, vol.1 122 copies
Higlights, vol.2 364 copies

Symposium vol. No. (in number of copies)

32–181	36–868	40–777	45–550
33–72	37–905	41–661	46–701
34–92	38–1.021	42–715	47–320
35–91	39–292	43–694	48–397
		44–734	

IAU Information Bulletin

The *IAU Information Bulletin* was being prepared and edited by the General Secretary, and printed, published and distributed by D. Reidel Publishing Company, Dordrecht, Holland. It appeared twice a year and went, free of charge, to IAU Members, Consultants, members of International Science Writers Association, selected international organizations, and institutions on a special distribution list.

IAU Publication Policy

The deteriorating financial situation of the Union led the Executive Committee to measures of austerity which also reflect in the publication policy of the IAU. The free distribution of *Transactions* volumes XIVA and XIVB was a generosity which cost the IAU about $28.000 U.S. If the Union continued in this generosity it would have either to have recourse to its reserves, small as they are, or significantly curtail its scientific programme and thus fail to live up to its responsibilities. It was therefore resolved to discontinue the free distribution of Transaction volumes to Members.

A major and permanent concern of the Executive Committee is the rising prices of IAU publications. Efforts were and are being made to ameliorate the situation, but unfortunately no practicable solution could be found. Four outstanding publishing houses had been asked for conditions for a new publishing contract. Two of them declined outright, one offer was rather hypothetical, and the terms of the only positive response were not found more advantageous than those of the present arrangement. The Executive Committee therefore abided by the arrangements with the D. Reidel Publishing Company, but instructed the General Secretary closely to observe the book market for possibilities of reducing the prices of IAU publications.

Relations to other Organizations

IAU Representation

1. *International Council of Scientific Unions (ICSU).* The General Secretary represented, together with the former General Secretary Dr L. Perek, the Union at the XIIth meeting of the Executive Committee and 13th General Assembly of ICSU (Madrid, Spain, 24–29 September 1970). The General Secretary also represented the Union at the XIIIth meeting of the Executive Committeeof ICSU held in Ottawa, Canada, on 29 and 30 September 1971, and at the extraordinary General Assembly and related meetings (14th General Assembly, 1st Executive Board meeting) of ICSU in Helsinki, Finland, from 16 to 20 September 1972, where he was elected into the Executive Board as representative of International Unions.

2. *ICSU Organizations.* The Union participates in the activity of a number of Special and Scientific Committees, and Inter-Union Commissions, formed and sponsored by ICSU. For more details please consult pages 13 onwards of volume XIIC of the Transactions.

The following list only gives the present state of IAU representation in these organizations:

(*a*) Committee on Space Research (COSPAR): C. de Jager, President of COSPAR.
(*b*) ICSU Abstracting Board (IAB): J. B. Sykes.
(*c*) Federation of Astronomical and Geophysical Services (FAGS): B. Guinot, President, and G. A. Wilkins.
(*d*) Inter-Union Committee on Frequency Allocation for Radio Astronomy and Space Science (IUCAF): O. Hachenber; D. S. Heeschen, V. A. Sanamian, F. G. Smith.
(*e*) Inter-Union Commission on Solar Terrestrial Physics (IUCSTP – changed into a Special Commission to be re-named): Z. Švestka.
(*f*) Inter-Union Commission on Science Teaching (IUCST): Edith A. Müller.
(*g*) Committee for Data on Science and Technology (CODATA): G. A. Wilkins.
(*h*) Inter-Union Commission on Spectroscopy (IUCS): B. Edlén; G. Phillips; M. J. Seaton.
(*i*) Inter-Union Commission for Studies of the Moon (IUCM): A. Dollfus, Chairman, E. Anders.

Other Organizations

The Union is represented in the following organizations:

(*a*) La Fondation Internationale du Pic-du-Midi: A. Lallemand.
(*b*) Le Comité Consultatif pour la Définition de la Seconde (CCDS) du Bureau International des Poids et Mesures: Wm. Markowitz.
(*c*) Le Comité Consultatif pour la Définition du Mètre (CCDM) du Bureau International des Poids et Mesures: A. H. Cook.

(*d*) Le Comité Consultatif International des Radiocommunications (CCIR): F. G. Smith; H. M. Smith.
(*e*) European Physical Society (EPS): G. Contopoulos.
(*f*) Various Services of FAGS
Bureau International de l'Heure (BIH): H. M. Smith; G. M. R. Winkler (B. Guinot is Director of the Bureau).
International Polar Motion Service (IPMS): E. P. Fedorov; B. Guinot. The Chairman of the Scientific Council of IPMS is P. Melchior.
Quarterly Bulletin on Solar Activity (QBSA): Organizing Committee of IAU Commission 10
International Ursigrams and World Day Service (IUWDS): R. Michard.
Solar Particles and Radiation Monitoring Organization (SPARMO): C. W. Allen, C. de Jager, P. Simon.

Direct Relations with UNESCO

The IAU had the benefit of a direct contract with UNESCO, which provided for grants toward the Astronomy Courses held in Nairobi, Kenya, in August 1972. These courses proved a great success, and the Executive Committee wishes to express to UNESCO its thanks for the financial help.

Relations with ICSU

Over the period under report, the Union enjoyed the support of ICSU in all its efforts. In addition to the annual UNESCO subventions, transmitted to the Union by ICSU, this organism paid the Union, jointly with UNESCO, a further $13.000 U.S. The Executive Committee feels that it has to extend its special thanks to ICSU for the generous assistance given to the Union.

Report on IAU Finances

In accord with the policy adopted in 1961, the Executive Committee presents to the Members of the IAU a "Summarized Account of Receipts and Payments" for the period from 1 January 1970 to 31 December 1972. All accounts are in U.S. dollars converted from other currencies at the following ICSU conversion rates, treated as exact:

For the year 1970:

£ sterling = $2.40 U.S.

U.S. $ = 3,60 Dutch guilders = 5,55 French francs = 14,38 Czechoslovak crowns

For the year 1971:

U.S. $ = 3,25 Dutch guilders

For the year 1972:

U.S. $ = 3,24 Dutch guilders

The unstable situation on the money market, as reflected in the varying coversion rates, accounts for the disparities between the closing balances in 1969, 1970 and 1971, and the opening balances in 1970, 1971 and 1972 respectively:

Closing balance	Opening balance	Difference
1969 $55.274,47	1970 $54.929,49	– $ 344,98
1970 $46.444,98	1971 $48.020,49	$1.575,51
1971 $58.472,19	1972 $58.566,48	$ 94,29

The decreasing value of the U.S. dollar, and its eventual devaluation, depreciated the IAU assets on U.S. bank accounts by about 20 per cent.

Certain bank operations, such as transfers from one IAU account to the other, and reimbursements, were made at current rates of exchange and thus resulted in exchange differences.

The summarized account has been derived from 13 (7 in 1970, 3 each in 1971 and 1972) individual accounts of receipts and payments audited and certified as correct by the Union's professional accountants. These accounts will be available for examination by the Finance Committee at the XVth General Assembly.

Summarized Account of Receipts and Payments

The statement on pages xiii–xv is a simple account of receipts and payments collected for convenience under a few main headings. The following explanatory notes provide more detailed information on each item of the accounts.

SUMMARIZED ACCOUNT OF RECEIPTS AND PAYMENTS (IN U.S. DOLLARS)

Receipts

	U.S. dollars	U.S. dollars
1. Contributions from adhering countries		
–1970	54.161,24	
–1971	60.315,41	
–1972	59.805,43	
		174.282,08
2. Sales of IAU publications		
–1970	3.602,11	
–1971	9.514,86	
–1972	2.775,47	
		15.892,44
3. Interest on accounts		
–1970	1.331,16	
–1971	140,78	
–1972	81,67	
		1.553,61
4. Grant from UNESCO subvention to ICSU		
–1970	18.000,—	
–1971	18.000,—	
–1972	14.896,16	
		50.896,16
5. Contracts with UNESCO		
–1970	2.000,—	
–1971	2.000,—	
–1972	2.000,—	
		6.000,—
6. Other receipts		
–1970	43.457,71	
–1971	6.496,79	
–1972	2.480,16	
		52.434,66
	Total receipts	301.058,95

Bank accounts as on 31 December 1972:	
–Savings accounts	68.743,32
–Current accounts	69.050,34
Total balance on IAU Accounts as on 31 December 1972	137.793,66

SUMMARIZED ACCOUNT OF RECEIPTS AND PAYMENTS (IN U.S. DOLLARS)
FOR THE PERIOD 1970–1972

Payments

	U.S. dollars	U.S. dollars
1. Administrative office		
–1970	18.797,03	
–1971	20.396,96	
–1972	25.188,74	
		64.382,7
2. Contribution to ICSU		
–1970	1.052,29	
–1971	1.083,22	
–1972	1.507,88	
		3.643,39
3. Expenses of Commissions		
–1970	89,56	
–1971	—	
–1972	—	
		89,56
4. Specific Projects of Commissions approved by the General Assembly		
–1970	18.723,15	
–1971	9.320,64	
–1972	8.509,14	
		36.552,93
5. General Assembly		
–1970 (XIV GA)	21.786,04	
–1971 (XV GA)	52,—	
–1972 (XV GA)	916,44	
		22.754,48
6. IAU publications		
–1970	18.328,72	
–1971	23.864,36	
–1972	4.849,08	
		47.042,16
7. Meetings of Executive Committee		
–1970	—	
–1971	10.839,74	
–1972	6.204,21	
		17.043,95
8. Meetings of officers		
–1970	—	
–1971	2.666,94	
–1972	2.756,46	
		5.423,40
	to bring forward	196.932,60

	U.S. dollars	U.S. dollars
	brought forward	196.932,60
9. Organization of IAU symposia and colloquia, and participation in other scientific meetings		
–1970..	10.941,89	
–1971..	9.132,45	
–1972..	10.054,76	
		30.129,10
10. Inter-Union commissions, committees on the ICSU, etc.		
–1970..	2.300,—	
–1971..	3.000,—	
–1972..	4.299,30	
		9.599,30
11. Other expenses of representation		
–1970..	1.914,56	
–1971..	—	
–1972..	—	
		1.914,56
12. Projects authorized by the Executive Committee		
–1970..	—	
–1971..	200,—	
–1972..	1.578,45	
		1.778,45
13. Summer schools (partly under direct UNESCO Contract)		
–1970..	1.319,28	
–1971..	2.050,80	
–1972..	2.976,96	
		6.347,04
14. Reimbursement of loan from ICSU		
–1970..	1.000,—	
–1971..	3.000,—	
–1972..	2.000,—	
		6.000,—
15. Other expenses		
–1970..	34.784,21	
–1971..	409,03	
–1972..	713,61	
		35.906,85
	Total payments	288.607,90
Excess of receipts over payments		12.451.05
	Total	301.058,95

Bank Accounts as on 1 January 1970	
–Savings accounts ..	57.987,10
–Current accounts ..	54.929,49
Total balance on IAU Accounts as on 1 January 1970	112.916,59

Specific Projects of Commissions

Comm. No.	Project	Allocated in 1970 for 1971–1973 $	1970 $	Paid in 1971 $	1972 $	Total 1970/1972 $	Total 1971/1972 $	Balance for 1973 from allocation $
4	Information Bureau on Astronomical Ephemerides	600	–	200,—	200,—	400,—	400,—	200,—
6	IAU Telegram Bureau	1.600	533,34	533,33	533,33	1.600,—	1.066,66	533,34
16	Centre de Documentation sur les Planètes à Meudon	1.500	500,—	1.000,—	–	1.500,—	1.000,—	500,—
17	Special Working Groups of Comm. 17	1.000	–	333,33	333,—	666,33	666,33	333,67
20	Minor Planet Center, Cincinnati	1.500	400,—	500,—	–	900,—	500,—	1.000 – waived
27	Catalogue of Variable Stars	3.000	–	–	–	–	–	3.000,—
37	Card Catalogue of clusters and Associations	–	1.500,—	–	–	1.500,—		
38	Exchange of Astronomers	22.000	15.289,81	6.240,60	6.942,81	28.473,22	13.183,41	8.816,59
	Information Bulletin for the Southern Hemisphere	1.500	500,—	513,38	500,—	1.513,38	1.013,38	486,62
	Total	32.700	18.723,15	9.320,64	8.509,14	36.552,93	17.829,78	13.870,22

Receipts

1. *Contributions from Adhering Countries.* The XIVth General Assembly of the IAU, held in Brighton, resolved that the unit of contribution remain fixed at 900 gold francs until 1973. This unit of contribution equalled $293,94 U.S. at the time of the General Assembly. The parity of 900 gold francs = $293,94 became untenable with the deteriorating position of the U.S. dollar, so that the 1972 contributions had to be collected in gold francs. Unfortunately, not all Adhering Countries established the conversion rate of the gold franc into the currencies used for payment correctly, which led to irregularities in the case of 13 countries.

As on 31 December 1972, the IAU had 46 Adhering Countries paying annually a total of 221 units of contributions. Seventeen countries adhered in category 1, twelve in category 2, seven in category 3, two in category 4, three in category 5, four in category 7, and one in category 8.

The contributions of 32 Adhering Countries had passed through the bank before 31 December 1972, and one payment made in 1972 passed the accounts after 31 December 1972. Five Adhering Countries owe their dues to the Union since 1971, seven Adhering Countries failed to pay their contributions for 1972, and one country only paid 13/14 of its membership fee for that year.

2. *Sales of publications.* The receipts of $15.892,44 U.S. are by about $1.700 U.S. higher than anticipated in the budgets for 1970, 1971 and 1972. They consist of payments made by Academic Press Inc. (London) for sales of publications, royalties from the same publishers, royalties from Cambridge University Press, royalties from D. Reidel Publishing Company, and revenues from sales made by the Secretariat.

3. *Interest on accounts.* The interest of $1.190,50 U.S. earned on the savings account at the Amsterdam-Rotterdam Bank in 1970 has been transferred to the current account at that bank. For the years 1971 and 1972 it has been allowed to accumulate on the savings account. The same applies to the interest earned on the savings accounts at the Chemical Bank and the American Savings Bank for 1970 through 1972. $140,66, 140,78 and 81,67 were earned on the current account at the Amsterdam-Rotterdam Bank over 1970, 1971 and 1972 respectively. The total interest earned on the savings accounts at Amsterdam-Rotterdam Bank, Chemical Bank and American Savings Bank over 1970–1972 is $9.690,68 U.S.

4. *UNESCO subvention through ICSU.* Detailed accounts of the expenditure of this subvention are submitted to UNESCO (through ICSU) each year. The distribution of the subvention is left to the discretion of the IAU within the permitted UNESCO categories.

In 1970 and 1971 the normal UNESCO subvention of $14.000 U.S. was increased by $3.000 U.S. for each year, in 1972 by $1.000 U.S. The loss of $103,84 U.S. is accounted for by the difference between the ICSU conversion rate of the Dutch guilder and the actual bank rate, and by bank charges.

5. *UNESCO contracts.* The $2.000 U.S. in 1970 are the final payment by UNESCO towards the Young Astronomers' School held in Hyderabad, India, in 1969 (Contract UNESCO/SC/1492/69). The $2.000 U.S. in 1972 represent the final payment by UNESCO towards the Young Astronomers' School held in Cordoba, Argentina, in 1970 (Contract UNESCO/SC/1973/70), and the $2.000 U.S. in 1972 are a UNESCO grant for the Nairobi Astronomical Courses, 1972 (Contract UNESCO/SC/HP 250.727).

6. *Other receipts.* The entries under this heading largely misrepresent the actual receipts of the Union, as they include internal bank transfers, transient items, reimbursements, etc. The following is an analysis of the amounts of $43.457,71, $6.496,79 and $2.480,16 U.S. for the years 1970, 1971 and 1972 respectively:

Other receipts in 1970

	$U.S.
– reimbursement by ICSU of IAU subscription for 1969	881,12
– payment by Istanbul University for reason unknown	398,86
– reimbursement by ICSU of General Secretary's expenses with ICSU meetings	1.232,85
– reimbursement of unused grants for IAU Symposia and Colloquia	723,89
– reimbursement of unused Commission No. 38 grant	520,00
– rectification of erroneous debit	700,00
– reimbursement by LOC Brighton of young astronomers' registration fees	833,56
– reimbursement by LOC Brighton of 50% of expenditure for Preliminary Announcement	340,65
– General Secretary's personal deposit	257,13
– reimbursement of unused air-ticket coupon	32,00
– internal transfers	37.537,65
Total	43.457,71

Other receipts in 1971

– reimbursement by ICSU of IAU contribution paid in 1970	1.052,29
– reimbursement of unused parts of airtickets	25,76
– erroneous credits	14,08
– refund of unused amounts from EC 35 meeting	4.913,89
– reimbursement of unused part of additional contribution to IUCAF	380,92
– transfers to Petty-cash	109,85
Total	6.496,79

Other receipts in 1972

– reimbursement by ICSU of IAU contribution paid in 1971	1.084,82
– refund of unused amount advanced to 36th E.C. meeting	379,50
– rectification of erroneous debit	84,03
– erroneous credit	317,26
– refund of unused grants	248,66
– exchange-rate gain on cheque of $4.200 (35th E.C. meeting)	217,74
– tranfers to Petty-cash	148,15
Total	2.480,16

Recapitulation:	$ U.S.
Other expenses in 1970	43.457,71
1971	6.496,79
1972	2.480,16
Total	52.434,66

It is seen that the bulk of this sum is made up of transfers from one current account to another, of various reimbursements and erroneous credits which are balanced by corresponding payments and cannot therefore be considered as effective receipts.

Payments

The headings have been numbered to correspond to those of the budget approved by the XIVth General Assembly.

1. *Administrative Office.* The expenses of the Administrative Office include the salaries, social

security and pension payments of the staff of the IAU Secretariat and expenditure for a typewriter, stationery, etc. It should be noted that thanks to the generosity of the Czechoslovak Academy of Sciences, the Astronomical Institute in Utrecht, the Observatoriet in Cophenagen, and the University of Thessaloniki, the Union paid no rental, had the free use of all communication facilities, and only spent a negligible amount on stationery.

2. *Contributions to ICSU.* The annual contributions of the Union towards the administrative expenses of ICSU have been raised, from 1971 on, from 2 to 2,5 per cent of the income from membership fees for the year preceding payment.

3. *Specific Projects of Commissions.* Details of the allocations approved by the XIVth General Assembly are given in the report of the Finance Committee (*Transactions IAU*, vol. XIIIB for 1970; *Transactions IAU*, vol. XIVB for 1971–1973). Actual payments were as in the list on page 14 which requires only little comment.

The grant towards the publication of the Card Catalogue of Clusters and Associations is a balance from 1970 which is not covered by the allocations made in that year.

Commission No. 20 waived the grants attributed for 1972 and 1973 which represents a saving of $1.000 U.S. under this heading.

The balance as on 31 December 1972 consists of credits in favour of the different projects which may, but need not be spent in 1973.

4. *General Assembly expenses.* The amount of $21.786,04, as given in the summary account for 1970, had been withdrawn from the IAU current bank accounts in connection with the XIVth General Assembly in Brighton, 1970. This sum was used as follows:

Purpose	$U.S
(*a*) Travel and living expenses of members of the Executive Committee	6.729,62
(*b*) Travel expenses of Presidents of Commissions	4.170,60
(*c*) Travel grants to young astronomers	6.084,00
(*d*) IAU Staff and external help	2.931,16
(*e*) Officers meeting in 1970	1.824,07
(*f*) Local expenses (postage, telegrams, etc.)	46,59
Total	21.786,04

The costs of the XIVth General Assembly itself were $19.212,17 U.S., that is by about $1.400 U.S. more than in 1967.

The amounts of $52.00 U.S. in 1971 and $916, 44 in 1972 were paid in preparation for the General Assemblies in 1973.

5. *IAU Publications.* The IAU only financed the production of *Transactions* volumes XIVA and XIVB, and of the *Information Bulletin.* All other IAU publications were financed by the publisher. A considerable amount was spent on the purchase of IAU symposium volumes distributed free of charge to institutions of developing countries. The expenditure under this heading was as follows:

In 1970

	$U.S.
(*a*) D. Reidel Publishing Company for printing and distribution of *Transactions* Vol. XIVA	13.634,05
(*b*) D. Reidel for printing and distribution of Information Bulletin Nos. 23 and 24	2.851,41
(*c*) D. Reidel for IAU publications distributed free of charge to institutions of developing countries	1.793,26
(*d*) Contribution towards editorial expenses in connection with *Highlights of Astronomy 1967*	50,00
Total	18.328,72

In 1971

	$U.S.
(*a*) D. Reidel for additional charges in connection with printing and distribution of *Transactions* Vol. XIVA	1.124,88
(*b*) D. Reidel for free distribution of *Highlights of Astronomy 1970*	1.554,80
(*c*) D. Reidel for printing and distribution of *Transactions* Vol. XIVB and for reprints of alphabetical list, Statutes, etc.	13.738,72
(*d*) D. Reidel for free distribution of IAU Symposium volumes	5.524,22
(*e*) D. Reidel for printing and distribution of *Information Bulletin* Nos. 25 and 26	1.890,08
(*f*) D. Reidel for complimentary copies	31,66
Total	23.864,36

In 1972

	$U.S.
(*a*) D. Reidel for printing and distribution of Information Bulletin No. 27	1.728,26
(*b*) D. Reidel for free distribution of IAU publications	2.924,15
(*c*) D. Reidel for 5 copies of Symposium volume No. 46	87,30
(*d*) Refund to IAU Member of printing charges	109,37
Total	4.849,08

Recapitulation:		$U.S.
Costs of IAU publications and free distribution in	1970	18.328,72
	1971	23.864,36
	1972	4.849,08
	Total	47.042,16

6. *Meetings of the Executive Committee.* The costs of the 33rd and 34th meetings of the Executive Committee held in Brighton, 1970, are included under heading 4, General Assembly expenses.

The amount of $10.839,74, as costs of the 35th meeting of the Executive Committee, calls for explanation. A cheque of $4.200 sent to France for financing the meeting could not be collected owing to French monetary restrictions. Money in cash had to be taken to France from the current account at the Amsterdam-Rotterdam Bank. The refund of $4.200 U.S. for the unused cheque was credited to the Union with $4.229,08 U.S. as per new rate of exchange. Moreover, $684,81 were refunded to the account of the IAU as unused part of the money taken out in cash. Thus

$10.839,74 taken out from accounts
– $ 4.200,00 unused cheque
– $ 29,08 gain on conversion rate
– $ 684,81 refund in cash

$5.925,85 are the net expenses for the 35th meeting of the Executive Committee

The amount of $6.204,21 as costs of the 36th meeting of the Executive Committee gives the following net expenses if analysed:

$6.204,21 taken out from accounts
– $ 379,50 refunded in cash

$5.824,71 net expenses of the 36th meeting of the Executive Committee.

The figures of $5.925,85 U.S. for 1971 and $5.824,71 U.S. for 1972 compare well with the figures of $6.124,61 U.S. and $5.647,14 U.S. for 1968 and 1969 respectively.

7. *Meetings of Officers.* The ever more complicated administration of the Union requires that the President, General Secretary and Assistant General Secretary meet at more or less regular intervals to co-ordinate action. The bulk of the expenditure under this heading is for travel expenses of the

officers and staff, so that the total expenditure for the meetings increases with increasing distance between the participants. This should be borne in mind when preparing the budgets for the coming years.

8. *Symposia and Colloquia.* The payments were, as a rule, for travel grants towards participation in the meetings; organizational expenses incurred by the IAU were insignificant. The grants were allocated by the Organizing Committees of the meetings and had to be approved by the Assistant General Secretary. The payments were made by the Secretariat. The following table shows the payments in more detail:

Symposium No.	Subject, Place, Date	Payment in $U.S.
38	Spiral Structure of the Galactic System, Basel, Switzerland, 29 August – 4 September 1969	360,00*
41	New Techniques in Space Astronomy, Munich, F.R.G., 10–14 August 1970	1.080,00
42	White Dwarfs, St Andrews, U.K., 10–14 August 1970	900,00
43	Solar Magnetic Fields, Paris, France, 31 August – 4 September 1970	1.551,89
44	External Galaxies and Quasi Stellar Sources, Uppsala, Sweden, 10–14 August 1970	2.100,00
45	The Motion, Evolution of Orbits, and Origin of Comets, Leningrad, U.S.S.R., 4–11 August 1970	2.000,00
46	The Crab Nebula, Jodrell Bank, U.K., 5–7 August 1970	1.400,00
47	The Moon, Newcastle upon Tyne, U.K., 22–26 March 1971	1.661,54
48	Rotation of the Earth, Morioka, Japan, 9–15 May 1971	1.600,00
49	Wolf-Rayet and High-Temperature Stars, Buenos Aires, Argentina, 9–14 August 1971	1.500,00
50	Spectral Classification and Multicolour Photometry, Villa Carlos Paz near Cordoba, Argentina, 18–24 October 1970	1.337,00
51	Extended Atmospheres and Circumstellar Matter in Spectroscopic Binary Systems, Parksville, Vancouver Island, B. C., Canada, 6–12 September 1972	1.350,00
52	Interstellar Dust and Related Topics, Albany, N.Y., U.S.A., 29 May – 2 June 1972	1.500,00
53	Physics of Dense Matter, Boulder, Colorado, U.S.A., 21–25 August 1972	1.414,76
54	Problems of Calibration of Absolute Magnitudes and Temperatures of Stars, Geneva, Switzerland, 12–15 September 1972	1.370,00
55	X- and Gamma-Ray Astronomy, Madrid, Spain, 11–13 May 1972	1.500,00
	Total	22.625,19

Colloquium No.	Subject, Place, Date	Payment in $U.S.
5	Visual Double Stars, Nice, France, 8–10 September 1969	100,00**
7	Proper Motions, Minneapolis, Minnesota, U.S.A., 21–23 April 1970	— ***
8	Techniques for Measurement of Fundamental Spectroscopic Data, London, U.K., 1–4 September 1970	450,00
	To bring forward	550,00

* $ 300,00 U.S. are editorial expenses paid in 1970, and $ 60 U.S. represent a cheque made out in 1969, but collected in 1970.

** Additional grant paid in 1970.

*** Advance payment made in 1969.

Colloquium No.	Subject, Place, Date	Payments in $U.S.
	brought forward	550,00
9	The IAU System of Astronomical Constants, Heidelberg, F.R.G., 12–14 August 1970	500,00
10	The Gravitational *n*-Body Problem, Cambridge, U.K., 11–14 August 1970	400,00
11	Automation in Optical Astrophysics, Edinburgh, U.K., 12–14 August 1970	150,00
12	Physical Studies of Minor Planets, Tucson, Arizona, U.S.A., 8–11 March 1971	500,00
13	Evolutionary and Physical Problems of Meteoroids, Albany, N.Y., U.S.A., 14–17 June 1971	500,00
14	Ultraviolet and X-Ray Spectroscopy in Astrophysical and Laboratory Plasmas, Utrecht, The Netherlands, 24–26 August 1971	485,97
15	New Directions and Frontiers in Variable Star Research, Bamberg, F.R.G. 31 August – 3 September 1971	500,00
16	Analytical Procedures for Eclipsing Binary Light Curves, Philadelphia, U.S.A., 8–11 September 1971	500,00
17	Ages des Étoiles, Paris, France, 15–12 September 1972	500,00
18	Orbital and Physical Parameters of Double Stars, Swarthmore College, Pennsylvania, U.S.A., 12–15 April 1972	300,00
19	Stellar Chromospheres, Greenbelt, Maryland, U.S.A., 14–18 February 1972	470,00
20	Meridian Astronomy, Copenhagen, Denmark, 25–28 September 1972	500,00
21	Variable Stars in Globular Clusters and Related Systems, Richmond Hill, Canada, 29–31 August 1972	500,00
22	Asteroïdes, Comètes, Matière Météorique, Nice, France, 4–6 April 1972	350,00
23	Astronomical Polarimetry, Tucson, Arizona, U.S.A., 15–17 November 1972	300,00
	Total	7.005,97

IAU contribution towards participation in and/or organization of scientific meetings of other international organizations $497,94 U.S.

Recapitulation:	$U.S.
Symposia	22.625,19
Colloquia	7.005,97
Other meetings	497,94
Total	30.129,10

9. *Inter-Union Commissions.* The payments consisted of contributions to the Inter-Union bodies in question, and of expenses of IAU representatives at the meetings of ICSU AB and IUCM. They were as follows:

– Contribution to CIES (ICSU Committee on Science Teaching) 1970–1972	600,00
– Contribution to IUCAF 1970–1972	3.500,00*
– Contribution to IUCSTP 1970–1972	3.000,00
– Contribution to IAB (ICSU Abstracting Board) 1970–1972 and travel expenses of IAU representative to its meetings	839,30
To bring forward	79.39,30

* $ 380,92 were refunded as unused part of a grant of $ 500 U.S.

	$U.S.
brought forward	79.39,30
– Contribution to FAGS – 1972	1.000,00
– Travel expenses of IAU representative to IUCM meeting (Inter-Union Commission on the Moon) 1972	660,00**
Total	9.599,30

10. *Projects authorized by the Executive Committee.* The $200 U.S. paid in 1971 were a grant towards the publication of the Bibliography and Program Notes of Eclipsing Binaries (Commission No. 42) and $1.578,45 U.S. were expenses incurred, in 1972, with the preparation of the First European Astronomical Meeting held, under the auspices of the IAU, in Athens, September 1972 (initiated by the Executive Committee).

11. *Reimbursement of loan from ICSU.* The $6.000 U.S. under this heading represent instalments towards the reimbursement of a loan of $10.000 U.S. obtained from ICSU in 1967. The debt has now been cleared.

12. *Summer Schools (Now Young Astronomers' Schools).* The payments made under the terms of UNESCO contracts were as follows:

	$U.S.
Balance from contract UNESCO/SC/1492/69 (School in Hyderabad)	1.370,08
Balance from contract UNESCO/SC/1943/70 (Shcool in Cordoba)	2.000,00
Under the terms of contract UNESCO/SC/RP/250 727 (Courses in Nairobi)	2.000,00
Total	5.370,08
Travel and living expenses paid to the lecturer at Nairobi Courses from IAU funds	976,96
Grand total	6.347,04

Details of payments in respect of the contracts are given in the reports to UNESCO. They are available to the Finance Committee for examination.

13. *Other expenses.* Reference is made to the comment on heading 6 under Receipts. The analysis of the amounts of $34.784,21 U.S. for 1970, $409,03 U.S. for 1971 and $713,61 U.S. for 1972 is as follows:

Other expenses 1970

	$U.S.
– Transfer from Chemical Bank to Barclays Bank	10.000,00
– Transfer from Amsterdam-Rotterdam Bank to ČSOB	9.594,44
– Transfer from Amsterdam-Rotterdam Bank to Barclays Bank	277,78
– Transfer from Amsterdam-Rotterdam Bank to Petty Cash, Utrecht	27,78
– Transfer from Société Générale to Chemical Bank	4.631,22
– Transfer from Barclays Bank to Chemical Bank	1.159,00
– Transfer from ČSOB to Amsterdam-Rotterdam Bank	6.742,00
– Transfer from ČSOB to Petty-Cash, Prague	272,00
Internal transfers	32.704,22
Participation- and other expenses of General Secretary	1.619,69
Erroneous debit	350,00
Bank charges and exchange losses	110,30
Total for 1970	34.784,21

** Unused $ 15,66 U.S. refunded.

Other expensss 1971		$U.S.
Participation expenses of Assistant General Secretary		255,80
Transfer from Amsterdam-Rotterdam Bank to Petty Cash		109,85
Erroneous debits		14,08
Bank charges and losses		29,30
	Total for 1971	409,03

Other expenses in 1972		$U.S.
Participation expenses of Assistant General Secretary		442,70
Transfer from Amsterdam-Rotterdam Bank to Petty Cash		148,15
Erroneous debit		84,03
Bank charges		38,73
	Total for 1972	713,61

Recapitulation:

Other expenses in 1970	$34.784,21
1971	409,03
1972	713,61
Total	35.906,85

AUDIT FOR THE ACCOUNTS 1970–1972

Conformément à la mission que vous avez bien voulu nous confier, nous vous présentons le résultat de notre examen des comptes de l'Union Astronomique Internationale pour la période du 1 janvier 1970 au 31 décembre 1972.

L'examen, notamment, a porté sur les documents suivants:

- comptes des recettes et dépenses pour l'année 1970 vérifiés par nous-mêmes
- comptes des recettes et dépenses pour l'année 1971 vérifiés par nous-mêmes
- comptes des recettes et dépenses pour l'année 1972 vérifiés par nous-mêmes
- comptes des recettes et dépenses et situation des comptes bancaires pour la période de trois ans: 1970, 1971 et 1972.

Les comptes ont été comparés avec les cahiers de comptes, reçus et pièces comptables diverses.

Nous certifions que ces comptes sont en accord complet avec les informations et explications que nous avons demandées pour nous permettre d'effectuer notre examen.

Le compte, ci-inclus, des recettes et dépenses, résumant les comptes des années 1970, 1971 et 1972 est la représentation exacte des dépenses et recettes de l'Union Astronomique Internationale pour la période du premier janvier 1970 au trente et un décembre 1972.

A Paris, le 20 février 1973

Roger Bacle m.p.
H.E.C. – Expert-Comptable

CONCLUSION

The assets of $157.793,66 U.S., as on 31 December 1972, compare well with those of $113.261,57 U.S., as on 31 December 1969 ($112.916,59, as on 1 January 1970, due to the change of the conversion rate of the French franc): the balance in hand is $24.877,07 U.S.

The budget approved by the XIVth General Assembly for the years 1971–1973 provided for an expenditure of $225.000 U.S. balanced by an income of $225.000 U.S. This budget, compared with the actual receipts and payments of the years 1971 and 1972, gives the balance for the year 1973 as follows:

	Receipts in $U.S.	Payments in $U.S.
Budget 1971–1973, as approved by the XIVth General Assembly (UNESCO contracts not included)	225.000,00	225.000,00
Actual figures for 1971 and 1972 (adjusted and UNESCO contracts excluded)	– 174.248,73	– 153,520,37
	50.751,27	71.479,63
Excess of receipts over payments	20.728,36	
Total	71.479,63	71.479,63

Available for payments in 1973 – $20.728,36 U.S.

The budgetary estimates for 1973, with due regard to the first devaluation of the U.S. dollar, are as follows:

Budget 1973

Receipts in U.S. dollars

1. Contributions from Adhering Countries	91.200
2. Revenue from IAU publications	11.600
3. Interest on accounts	3.300
4. UNESCO subvention trhough ICSU	14.000
Total	120.100

Payments in U.S. dollars

1. Administrative Office	29.700
2. Contribution to ICSU	2.150
3. Commission expenses	300
4. Specific projects	15.400
5. General Assembly	49.500
6. IAU Publications	7.700
7. Free copies to developing countries	7.150
8. Executive Committee	0
9. Officers' meetings	3.850
10. Symposia and Colloquia	21.000
11. Inter-Union Commissions	4.500
12. Projects approved by the Executive Committee	800
13. Representation	500
14. Bank charges and losses	300
15. Young Astronomers' school	6.600
Total	149.450
Receipts	120,100
Excess of payments over receipts	29.350
Available from 1972 (about)	20.720
To be taken from reserves	8.630

Explanatory notes:

Receipts

1. The figure of $91.200 U.S. also includes $15.351,25 U.S. outstanding for 1972. All contributions not paid before the devaluation of the U.S. dollar have been augmented by 10 per cent.
2. D. Reidel Publishing Company owes the Union about $7.150 U.S. on royalties for the second half of 1971. From figures available on sales over 1972 it is safe to assume that $4.450 U.S. will be payable in 1973.
3. The sum of $3.000 U.S. represents interest earned in 1972. In 1973 they may be somewhat higher if the rate of interest does not drop.
4. UNESCO subvention through ICSU: The grant of $14.000 U.S. has already been allocated for 1973.

Payments

1. Administrative Office: $27.000 U.S. are the difference between the amounts budgeted for 1970–1973 and those actually paid. Raised by 10% they give $29,700 U.S.
2. Contribution to ICSU: has been set at 2.5% of the annual fees received from Adhering Countries. It had been paid before the devaluation of the U.S. dollar.
3. Commission expenses: are expected to be negligible
4. Specific Projects: see table on page 14. The balance for 1973 has been adjusted to account for the devaluation.
5. General Assembly: The estimate of $49.500 U.S. (adjusted) is rather vulnerable as it depends on a range of variable factors. The reserve of $10.000 U.S. at the disposal of the General Secretary should be considered as a reserve against underestimates (reserve approved by the XIVth General Assembly).
6. IAU Publications: The amount of $7.700 U.S. covers the production and distribution of 4 Information Bulletins ($1.650 U.S. each) and the printing of the Report of the Executive Committee.
7. Free copies to developing countries: Seven volumes of Symposium proceedings will have to be distributed free of charge in 1973 (about $990 U.S. each).
8. Executive Committee: The costs of the 37th and 38th meetings of the Executive Committee are included in those of the XVth General Assembly (heading 5).
9. Officers' meetings: The officers are likely to be farther apart after the General Assembly. This accounts for the increase of about $770 U.S. over the expenditure in 1972.
10. Symposia and Colloquia: 11 symposia ($1.775 U.S. each on the average) and 1 colloquium ($500 U.S.) will be held in 1973. $1.000 U.S. have been comitted as an IAU contribution to the Copernicus Symposium cosponsored by the Union.
11. Inter-Union Commissions: The contribution to CIES had to be increased by $200 for 1973. This accounts for the $4.500 U.S. instead of the $4.300 U.S. as in 1972.

12, 13, 14 Expenditure under these headings is expected to be negligible.

15. Young Astronomers' Schools: Nine Member Countries increased their categories of adherence in response to the General Secretary's appeal. The organization of a School for Young Astronomers in 1973 is therefore within the intentions of the XIVth General Assembly.

The Finance Committee, to convene during the XVth General Assembly, will be asked to endorse the budgetary estimate for 1973, as proposed above, and to approve a comprehensive budget for the years 1974–1976 to be submitted to the General Assembly.

APPENDIX

A. LIST OF COUNTRIES ADHERING TO THE UNION

The following is a list of the 46 countries that adhered to the IAU in December 1972, giving also the year of admission, the approximate number of Members, and the Adhering Organizations.

Country	Year	Members	Adhering Organization
Argentina	1927	16	Comité Nacional de Astronomía, La Plata.
Australia	1939	57	Australian National Committee for Astronomy, Sydney.
Austria	1955	15	Österreichische Akademie der Wissenschaften, Wien.
Belgium	1920	54	Académie Royale de Belgique, Bruxelles; Koninklijke Academie van België, Brussel.
Brazil	1961	9	Comissão Brasileira de Astronomia, Conselho Nacional de Pesquisas, Rio de Janeiro.
Bulgaria	1957	8	Bulgarian Academy of Sciences "7 November", Sofia.
Canada	1920	87	National Committee of Astronomy for Canada, Department of Geophysics and Astronomy, University of British Columbia, Vancouver 8, B. C.
Chile	1947	6	Universidad de Chile, Santiago de Chile.
Colombia	1967	1	Academia Colombiana de Ciencias Exactas, Físicas y Naturales, Bogotá.
Cuba	1970	0	Instituto de Astronomía de la Academia de Ciencias de Cuba, La Habana.
Czechoslovakia	1922	40	National Committee of Astronomy for Czechoslovakia, Czechoslovak Academy of Sciences, Praha.
Denmark	1922	17	Danish National Committee for Astronomy, det Kongelike Danske Videskabernes Selskab, København.
Finland	1948	9	Finnish Academy of Sciences and Letters, Helsinki.
France	1920	162	Comité National Français d'Astronomie, Académie des Sciences, Paris.
F.R.G.	1951	128	Rat Westdeutscher Sternwarten zu München.
G.D.R.	1951	32	Akademie der Wissenschaften der DDR, Berlin.
Greece	1920	18	Greek National Committee for Astronomy, Academy of Athens, Athens.
Hungary	1947	11	Hungarian Academy of Sciences, Budapest.
India	1948	47	Indian National Science Academy, New Delhi, Indian National Committee for the IAU.
Iran	1969	0	Tehran University, Tehran.
Ireland	1947	6	National Committee of Astronomy, Royal Irish Academy, Dublin.
Israel	1954	13	The Israel Academy of Sciences and Humanities, Jerusalem.
Italy	1921	76	Consiglio Nazionale delle Ricerche, Roma.
Japan	1920	106	National Committee of Astronomy for Japan, Science Council of Japan, Tokyo.
Korea (PDR)	1961	3	Academy of Sciences, Phyongyang.
Mexico	1921	13	Universidad Nacional de Mexico, Mexico D.F.
Netherlands	1922	58	Koninklijke Nederlandse Akademie van Wetenschappen, Amsterdam.
New Zealand	1964	5	Royal Society of New Zealand, Wellington.
Norway	1922	13	Det Norske Videnskaps-Akademi i Oslo, Oslo.

Country	Year	Members	Adhering Organization
Poland	1922	40	Polska Akademia Nauk, Warszawa.
Portugal	1924	12	Secção Portuguesa das Uniões Internacionais Astronómica, Geodésica e Geofísica, Lisboa.
Roumania	1928	17	Romanian National Committee of Astronomy, Bucarest.
South Africa	1938	13	Council for Scientific and Industrial Research, Pretoria.
Spain	1922	14	Comisión Nacional de Astronomía, Madrid.
Sweden	1925	41	Kungl. Vetenskaps-Akademien, Stockholm.
Switzerland	1923	20	Schweizerische Naturforschende Gesellshaft, Basel.
Taiwan	1959	9	Academia Sinica, Nankang, Taipei.
Turkey	1961	18	Türk Astronomi Dernegi, Istanbul.
U.A.R.	1925	9	National Committee of Astronomy, Helwan Observatory, Helwan near Cairo.
United Kingdom	1920	225	The Royal Society, London.
Uruguay	1970	0	Comité Nacional de Astronomia, Facultad de Ingeniería y Agrimensura, Montevideo.
U.S.A.	1920	712	The National Academy of Sciences, Washington D.C.
U.S.S.R.	1935	343	Academy of Sciences of the U.S.S.R., Moscow.
Vatican City State	1932	7	Pontificia Academia delle Scienze, Città del Vaticano.
Venezuela	1953	2	Academia de Ciencias Físicas, Matemáticas y Naturales, Caracas.
Yugoslavia	1935	14	Savez društava matematičara, fizičara i astronoma Jugoslavije, Zagreb.

B. LIST OF DECEASED MEMBERS

The Executive Committee regrets to record the loss by death of the following 58 Members of the Union (at the date of 1 May 1973):

Alexander A. F. O'D	29 January 1971
Alter G. A.	30 October 1972
Arbey L. M. J.	27 March 1972
Baker R. H.	23 June 1964
Baltá-Elias J.	3 January 1973
Banerji A. C.	in 1972
Barabashov N. P.	29 April 1971
Barros M. G. de Perreira	15 January 1971
Berkner L. V.	4 June 1967
Bernas R.	July 1970
Bowen I. S.	6 February 1973
Cameron R. C.	13 December 1972
Cox J. F. J. G.	20 October 1972
De Moraes A.	in 1970
Dick J.	22 March 1971
Dombrovskij V. A.	1 February 1972
Eckert W. J.	October 1971
Febrer J.	date unknown
Fesenkov V. G.	13 March 1972
Gamow G.	20 August 1968
Grandón R.	3 April 1971
Grigoryan K. A.	September 1971
Guigay G.	19 January 1971
Günther O.	8 January 1973
Harang L.	21 September 1970
Hargreaves F. J.	September 1970
Heiskanen V. A.	23 October 1971
Hellman C. D.	28 March 1973
Mrs Herzberg L. H.	3 June 1971
Humason M. L.	18 June 1972
Joy A. H.	18 April 1973
Kahrstedt A.	11 January 1971
Kao Ping-Tse	23 March 1970
König A.	in 1970
Kox H.	7 June 1971
Lindley W. M.	2 September 1972
Luplau Janssen C. E.	21 June 1971
Minnaert M. G. J.	26 October 1970
Narayan A. L.	6 March 1973
Nordström H.	28 July 1971
Nowacki H.	20 March 1972
Ölander V. R.	8 March 1973
Ramsey W. H.	1 January 1970
Rutllant F.	16 April 1971

Sarabai V. A.	30 December 1971	Väisälä Y.	21 July 1971
Sémirot P.	30 January 1972	Van Albada G. B.	18 December 1972
Sibahara R.	10 December 1971	Van Schewick H.	6 September 1972
Sterne T. E.	7 February 1970	Van Wijk U.	date unknown
Stewart J. Q.	19 March 1972	Vitkevich V. V.	29 January 1972
Stumpff K.	10 November 1970	Watts C. B.	17 July 1971
Tardi P.	5 August 1972	Wilkens A.	27 January 1968
Thomson J. H.	12 August 1969	Zanstra H.	3 October 1972
Tilles D.	30 March 1968	Zinner E.	30 August 1970

C. LIST OF IAU PUBLICATIONS

Transactions of the IAU, volume XIVA, pp. VIII+566, price $18,20 (Members $12,20)	Editor: C. de Jager. Published by D. Reidel Publishing Company, Dordrecht, Holland.
Transactions of the IAU, volume XIVB, pp. IX+378, price $14 (Members $9,35)	Editors: C. de Jager and A. Jappel. Published as *Transactions*, vol. XIVA.
Highlights of Astronomy as presented at the XIVth General Assembly of the IAU 1970, pp. XII+793, price $40 (Members $26,70)	Edited and published as *Transactions*, vol. XIVA.

IAU Symposium Volumes

No. 36. *Ultraviolet Stellar Spectra and Related Ground-Based Observations*, pp. XV + 361, price $19,80 (Members $13,20)	Edited by L. Houziaux and H. E. Butler. Published by D. Reidel Publishing Company, Dordrecht, Holland.
No. 37. *Non Solar X- and Gamma-Ray Astronomy*, pp. X+425, price $23,10 (Members $15,40)	Editor: L. Gratton. Published by D. Reidel's.
No. 38. *The Spiral Structure of our Galaxy*, pp. XIII+478, price $24,40 (Members $16,30)	Editors: W. Becker and G. Contopoulos. Published by D. Reidel's.
No. 39. *Interstellar Gas Dynamics*, pp. XXIV+ +388, price $25 (Members $16,70)	Editor: H. J. Habing. Published by D. Reidel's.
No. 40. *Planetary Atmospheres*, pp. XVIII+ 408, price $25 (Members $16,70)	Editors: C. Sagan, T. C. Owen, and H. J. Smith. Published by D. Reidel's.
No. 41. *New Techniques in Space Astronomy*, pp. XV+417, price $26,40 (Members $17,60)	Editors: F. Labuhn and R. Lüst. Published by D. Reidel's.
No. 42. *White Dwarfs*, pp. IX+164, price $11,75 (Members $8,95)	Edited by W. J. Luyten. Published by D. Reidel's.
No. 43. *Solar Magnetic Fields*, pp. XVI+782, price $46,20 (Members $30,80)	Editor: R. F. Howard. Published by D. Reidel's.
No. 44. *External Galaxies and Quasi-Stellar Sources*, pp. XVII+550, price $33 (Members $22)	Edited by D. E. Evans. Published by D. Reidel's.
No. 45. *The Motion, Evolution of Orbits, and Origin of Comets*, pp. XXII+521, price $39 (Members $26)	Editors: G. A. Chebotarev, H. J. Kazimirchak Polonskaya, and B. G. Marsden. Published by D. Reidel's.
No. 46. *The Crab Nebula*, pp. XX+470, price $25 (members $16,70)	Editors: R. D. Davies and F. G. Smith. Published by D. Reidel's.
No. 47. *The Moon*, pp. XVI+471, price $36 (Members $24)	Editors: S. K. Runcorn and H. C. Urey. Published by D. Reidel's.
No. 48. *Rotation of the Earth*, pp. XXII+244, price $19 (Members $13)	Editors: P. Melchior and S. Yumi. Published by D. Reidel's.

IAU Colloquium Volumes

No. 3. *Interstellar Dust*	M. Greenberg and H. Lambrecht. *Astronomische Nachrichten* **293**, Nos. 1–2, 1971.
No. 4. *Stellar Rotation*, price $19,80	A. Slettebak. D. Reidel Publishing Company, Dordrecht, Holland
No. 5. *La cohésion entre les différentes procédés de mesure d'étoiles doubles*	W. D. Heintz. *Astrophys. Space Sci.* **11**, 3–188, 1971
No. 6. *Mass Loss and Evolution in Close Binaries*, price $5	K. Gyldenkerne and R. M. West. Obtainable from Copenhagen University Fund, Københavns Universitet.
No. 7. *Proper Motions*	W. J. Luyten. University of Minnesota, Minneapolis. December 1970.
No. 8. *Techniques for the Measurement of Fundamental Spectroscopic Data*	*Quart. J. Roy. Astron. Soc.* **11**, 48–60, 1971. Report by B. W. Shore.
No. 9. *The IAU System of Astronomical Constants*	T. Lederle and B. Emerson. *Celest. Mech.* **4**, No. 2, 1971.
No. 10. *The Gravitanional n-Body Problem*. Price Dfl. 115,—	M. Lecar. D. Reidel Publishing Company, Dordrecht, Holland.
No. 11. *Automation in Optical Astrophysics*	H. Seddon and M. J. Smyth. *Publ. Roy. Obs. Edinburgh* **8**, 1–211, 1971
No. 12. *Physical Studies of Minor Planets*. Price $3 (paper cover)	T. Gehrels. NASA SP 267.
No. 13. *Evolutionary and Physical Problems of Meteoroids*	C. L. Hemenway and A. F. Cook. To be published as NASA Report.
No. 14. *Ultraviolet and X-Ray Spectroscopy of Astrophysical and Laboratory Plasmas*	A. Gabriel. *Space Science Reviews* **13**, 478–889.
No. 15. *New Directions and Frontiers in Variable Star Research*	W. Strohmeier. *Remeis-Sternwarte Bamberg, Astronomisches Institut der Universität Erlangen-Nürnberg* **9**, Nr. 100, 326, 1972

Publications

by, or on behalf of, IAU Commissions and by Astronomical Services

Commission No. 6. IAU (Telegram Bureau) *Circulars* Nos. 2197–2475	Issued by the IAU Central Bureau for Astronomical Telegrams.
Commission No. 10. *Cartes Synoptiques de la Chromosphère Solaire* (Volume V, fascicules 2, 3)	Prepared by Observatoire de Paris at Meudon under the direction of R. Michard.
Commission No. 16. *Newsletters* Nos. 7–10	Published by the President of Commission.
Commission No. 19. *Monthly Notes of the IPMS* Nos. 1(1970)–10(1972)	Published by International Polar Motion Service, Mizusawa-shi, Japan.
Commission No. 20. *Minor Planet Circulars* Nos. 3023–3406	Mimeographed circulars issued by the Minor Planet Center, Cincinnati.
Commission No. 24. *Astrographic Catalogues*	Information regarding the different sections of the Catalogue can be obtained from respective Observatories listed in *Transactions IAU* **XIVB**, 176, 177, 1971.
Commission No. 26. *Circulaire d'Information* Nos. 50–58.	President: J. Dommanget, Observatoire Royal de Belgique, Brussels, Belgium.
Commission No. 27. *Information Bulletin on Variable Stars* Nos. 397–752	Prepared by L. Detre, Konkoly Observatory, Budapest, Hungary.

Commission No. 31. *Annual Report of the BIH*	Published by the Bureau International de l'Heure, Paris, France.
Commission No. 35. *Circular Letters* up to No. 16	Editor: A. Massevich, Astronomical Council of the U.S.S.R. Academy of Sciences, Moscow, U.S.S.R.
Commission No. 41. *Bibliography of Books and Papers published in 1968–69 (1970) and 1969–71 (1972) on the History of Astronomy*	Editor: Z. K. Sokolovskaya, Institute of History of Sciences and Technology, Moscow, U.S.S.R.
Commission No. 41. *Information Circulars* 18–21	Published by the President of the Commission.
IAU Information Bulletin Nos. 24–29	Prepared by the General Secretary.

PART 2

WELCOMING CEREMONY

REPORT OF THE GENERAL ASSEMBLY

WELCOMING CEREMONY

Wednesday, 22 August at 20^h

The Welcoming Ceremony of the XVth General Assembly of the International Astronomical Union took place in the Sydney Town Hall, in the presence of a distinguished gathering of representatives of the Government of Australia, the Town of Sydney, the University of Sydney, the Australian Academy of Science, The National Organizing Committee, and other organizations.

Dr H.W. Wood, Chairman of the Local Organizing Committee, acted as chairman for the ceremony.

After a few words of welcome to Members and guests, the Chairman called upon His Excellency, the Honourable W.L. Morrison, Minister for Science, to address the Assembly.

ADDRESS BY THE MINISTER FOR SCIENCE, THE HONOURABLE W. L. MORRISON

'On behalf of the Government of Australia I am very pleased to welcome members of the International Astronomical Union to our country on the first occasion when an Assembly has been held in the Southern Hemisphere. I understand that the possibility of holding an Assembly here was first mentioned by a President of the Union in 1954 and that the moves which have led to your being here at present began in 1957.

The field of Astronomy has one of the longest and most continuous histories of international cooperation of which there are many examples before the foundation of the Union which was one of the earliest of the international scientific unions to be founded in 1919. The success of your Union is based on the promotion of cooperation and especially the formation of informal and personal relationships among astronomers.

This is not the first time by any means that Australia has been discovered astronomically. The main purpose of the voyage of James Cook was to observe the transit of Venus across the Sun from Tahiti in 1770 and incidentally this was a participation in an early example of international cooperation in science. Thus Cook's exploration of the east coast of Australia was incidental to a mission designed for astronomical studies.

The establishment of the penal colony in Sydney in 1788 has also an astronomical association. Second Lieutenant Dawes was to establish a station from which to observe the return of a periodic comet. Although he was unsuccessful in this, he played an important part in the exploration and survey of the Sydney area.

It is interesting that to finance Cook's voyage the Crown gave a Royal Grant of £4,000 while Sir Joseph Banks assisted by providing £10,000 from his own private resources. The balance between national and private funding for international science has long since been reversed.

Recognising the natural advantage that Australia has because of its location in the Southern Hemisphere, Australian Governments have given substantial support to astronomy both optical and radio. Our latest project, in cooperation with the Government of the United Kingdom, is the 3.8 metre Anglo-Australian telescope. I am very pleased with the Australian participation in this project and by this time next year the telescope will be formally opened and in operation.

In the funding of big sciences, such as astronomy, there are clear reasons for greater international cooperation and for development of facilities and the use of equipment on a multi-user basis. The Australian national objectives for astronomical research are directed towards continuing investigations in the areas of astronomy where we have demonstrated scientific excellence in the past or for which we have particular geographic advantage.

As a politician I must confess that there must be few votes to be gained from the pioneering work being done in Australia in chemistry of the Galaxy, but as an individual I am fascinated and intrigued by the significance of the new molecules discovered in interstellar space. The cooperation between the Division of Radiophysics of CSIRO and the Chemistry Department of Monash University in this work represents an interdisciplinary approach that is becoming so much a part of modern astronomy.

In Australia we have been interested in several active areas of astronomy, in quasars and their cosmological significance, in the nature of pulsars and in the physics of the Sun, including the basic nature of solar flares, high energy phenomena in the solar corona and in the behaviour and time changes of structures of all sizes in the solar atmosphere.

Mr President on behalf of the Government of Australia I wish the Assembly great success and after the words of the poet Milton shall now leave you "to model Heaven and calculate the stars; how they will wield the mighty frame, how build, unbuild, contrive".'

Dr Wood expressed his thanks to Minister Morrison, and invited the Right Honourable, the Lord Mayor of Sydney, Alderman David Griffin, C.B.E., to address the gathering.

ADDRESS BY THE LORD MAYOR OF SYDNEY, ALDERMAN D. GRIFFIN, C.B.E.

'Mr Chairman, Mr Minister, Ladies and Gentlemen,

It is a signal honour – perhaps in this company I should say an honour of the first magnitude – to welcome to Sydney for the first time members of the International Astronomical Union. Australia is no stranger to unions; already we have many – too many perhaps – and some may be asking why in heaven is the Lord Mayor welcoming still another. We welcome you, Mr President, partly because your union is closer to heaven than most, but mainly because yours is a most special union: You work not in your own interest, but in the interest of mankind. You strive not for better conditions, but for greater knowledge. Your studies throughout all ages of history reveal man at the highest point of his intellectual achievement. You are, it seems to me, the poets of science since you use your love romance to lead you to the truth. Sydney is honoured to be your host in this your first sally into the Southern Hemisphere.

Having thus established your credentials permit me to say a word about mine. The fact is that as Lord Mayor I am something of an old time astronomer myself. I am no heretic, I run no risk of being burnt at the stake, I hold fast to one demonstrably accurate astronomical proposition, namely that Sydney is the centre of the Earth. All other countries and peoples revolve around Sydney. The Lord Mayor is the centre of Sydney, therefore the Lord Mayor is at the centre of the centre of the Earth. It is an awfully pleasant place in which to be. Some minor figures have occasionally been ill-mannered enough to challenge this proposition. They are a doubting lot!

But perhaps I have over-simplified it a little. The Lord Mayor has other extraordinary qualitites. Let me, for example, explain his position in relation to energy and radiation. One starts with the simple proposition that a Lord Mayor gives off energy in the form of radiation – indeed I hope I am giving off some energy in the form of radiation at this very minute – yet, and here is the mystery, my mass is not diminishing. Einstein, by applying the Lorentz transformations, equated the relation between mass and radiation. Yet for all his genius he arrived at a wrong conclusion. His principle simply does not work when applied to a Lord Mayor. For Einstein postulated, if I remember correctly, that if a body gives off energy in the form of radiation its mass diminishes by E/c^2 where c is the velocity of light. True, Lord Mayors sometimes have to move pretty quickly but under no circumstances does their mass diminish. On the contrary, a Lord Mayor finds his mass increasing (He wishes it didn't) in direct proportion to the amount of radiation he gives off. Nevertheless, and with all its attendant risks, I am radiating to you all to-night the warmest of welcomes to Sydney!

I realise however, ladies and gentlemen, that you have come to Sydney not only to study Lord Mayors but to apply combined force of your intellects to even more important topics.

I hope however, that your programme will enable you to take some time off from your considerations of the wonders of outer space, to observe more closely the beauties and interest of a particular part of the planet Earth; that part lying at 151°12′ east of Greenwich and 33°52′ south latitude which the rest of the world knows as Sydney but which we in Sydney know as the most beautiful place on Earth.

But I must stop pretending. At heart the rest of us envy the astronomer. You have as your kingdom the wonder and mystery of the stars. By miracles beyond the range of normal human understanding you are able to project your thoughts to the most distant galaxies and beyond into the infinite night. Above all sciences astronomy – in the eyes of the lay population – stands supreme. For if there be a riddle of the Universe, if great truths touching the origin and purpose of life are to be discovered, it is surely an astronomer who will one day point the way. Was it not Gertrude Stein who on her death bed said to her faithful companion – 'Alice, what is the answer?' and on being told 'Gertrude, you know there is no answer' replied 'well, if there is no answer, what is the question?'? When one day we have the answer and having it will at last know the question, it will be an astronomer who, in all probability, will give us both. For all that you and your predecessors have revealed the city of Sydney honours you to-night.'

The Chairman thanked Alderman Griffin for his words and then invited Emeritus Professor W.M. O'Neil, Acting Vice-Chancellor, The University of Sydney, to take the floor.

ADDRESS BY THE ACTING VICE-CHANCELLOR OF THE UNIVERSITY OF SYDNEY, EMERITUS PROFESSOR W. M. O'NEIL

'On behalf of The University of Sydney I issue a very warm welcome to the participants in the Fifteenth General Assembly of the International Astronomical Union and I express the wish that their meetings will be comfortable, pleasant and highly rewarding.

International assemblies of this kind are important for several reasons. Though there is a full and fairly rapid exchange of information through the journals and even more rapid exchange through personal correspondence, it is important for scholars in a field to get together for oral exchanges of information, views and interpretation, especially where the material is preliminary and tentative. Another important reason is that an opportunity is provided for older workers to renew acquaintance with one another and for younger workers to made acquaintances which will be useful to them in their later work. It is much easier to write to someone we have met than to someone we have never. Finally international assemblies help to break down parochialism and nationalism: not that astronomy has ever suffered much from this sort of defect.

It seems to me appropriate that astronomers, the practitioners of the oldest empirical science, are holding their first meetings as an International body in Australia in the halls of The University of Sydney. First, it is the oldest of the Australian universities. Second, it is one of the few Australian universities with strong and diverse astronomical research programmes; perhaps as befits a young nation, our main programmes each involve a unique or novel installation. Third, I think Sydney was the first Australian university to have a Professor of Astronomy, a post occupied jointly from 1912 to 1926, with the post of New South Wales Government Astronomer. Fourth, though many Australian universities feature the Southern Cross in their coats of arms, Sydney has a reference to the constellations in its motto: *Sidere mens eadem mutato*, though the constellations be changed the spirit is the same, a reference to our hope to preserve northern intellectual interests, traditions and values under southern skies.

Let me say just a little about the age of astronomy, much greater, of course, than that of The University of Sydney and about its origins. It seems to have begun with star watching and recording for two main purposes. The first was quite practical, namely timekeeping. It was quite important to keep track of the days and to have some convention for deciding when they began and ended. The Moon with its phases marked out a convenient period of 29 or 30 days, and with its quarters subdivided this period into sets of about 7 days or so. Finally the Sun in its north-south movements in the sky marked out the round of the seasons so important for agrarian peoples. Twelve months measured by the Moon almost equalled one year measured by the Sun, and much observation, recording and thought had to be given to finding a method of reconciling the two periods. By the middle of the first millennium B.C. the Babylonians, the Greeks and possibly earlier the Chinese, found a workable mode of reconciliation in the form of the Metonic and the Callippic cycles. We have abandoned that form of reconciliation in favour of getting the year almost right and of being satisfied with vestigial or schematic months no longer tied to the phases of the Moon.

The second purpose was theoretical, made in the attempt to satisfy human curiosity and to find some regularity somewhere in the scheme of things so as to provide some sense of security. Life on Earth was obviously confused, capricious and uncertain. Husbands deserted wives, children rebelled against their parents, debtors failed to pay their debts, thieves robbed us of our property, the animals and plants sometimes failed to breed true, droughts or floods or insects sometimes destroyed the crops, our food sometimes spoiled and made us sick, other diseases molested us, even the lead pipes made of so obviously a suitable material for the delivery of water poisoned us, wages chased prices and in catching up forced up prices, lightning and earthquakes ruined our dwellings and so on and so on. Somewhere there must be some regularity, something that could be counted on. And there it was to be found if only we cared to look up, watch and record carefully and calculate. There were, of course, unexpected events in the skies such as eclipses and the appearance of comets, but even they proved amenable in the long run to prediction.

The events which at first sight seemed so regular proved on closer inspection to be less rational and less regular than may have been expected. Thus the Sumerians seem to have thought that the year ought to be 360 days, a good round sexagesimal number but the year proved not to be exactly even the 365 days which the Egyptians elected early in the second millennium B.C. Observation of a short run of months suggested that they were either 29 or 30 days in length, so 29.5 days would seem a good average for anyone unafraid of fractions. But who would have expected the average to be a silly number like 29.5305879 days, that the actual months varied between about 29.26 and 29.80 days and that the average was increasing by about 0.00000012 days or one hundredth of a second per century.

I can express only the greatest admiration for Naburiannu and Kidinnu, both Babylonians, and of the 5th and 4th centuries B.C. who estimated the mean lunar periods with errors of only two seconds, and half a second respectively and the year, sidereal I suspect in Naburiannu's case and tropical in Kidinnu's, with errors of only six minutes (too long) and eight minutes (too short) respectively. If Kidinnu did not distinguish the two solar periods, Hipparchos, perhaps the greatest Greek astronomer, did so in the 2nd century B.C. Further his estimate of the tropical year was only 5.5 minutes too long. Meanwhile the Greeks were working on the geometry underlying apparent celestial motions, major contributors being Eudoxos, Hipparchos and Ptolemaeus. If only Seleucus of Babylonia had had the geometrical talent of his near contemporary, Hipparchos, there would have been no need for us to have had to wait for Copernicus to generate a viable heliofocal planetary theory.

I have excused most astronomers from manifesting nationalism and parochialism. Not being one, perhaps I can be excused for manifesting both. One base used by Hipparchos to estimate the length of the tropical year was the difference in the angular separation of the bright star Spica or *alpha Virginis* from the Sun's Spring crossing point on the Celestial Equator, the First Point in Aries, as observed by him and as observed by Timocharis some 135 years before. My colleague Professor Hanbury Brown not long ago concluded some work with his Stellar Intensity Interferometer at Narrabri on this same Spica, a binary star. As a result he has developed an entirely new method of estimating stellar distances at ranges beyond reliable estimate by use of the parallax method. It may be used for stars which are not Cepheid variables though they must be binaries. I like to think that it would be nice to arrange a meeting between Hipparchos and Hanbury Brown, so that they could talk to each other about Spica and their observations of it. I must confess, however, that though Hipparchos would find Hanbury Brown a nice man whose Greek was deplorable, he would be very puzzled by his style of thought, his equipment and his discoveries. I must confess that as a non-astronomer, I feel more at ease reading about what Naburiannu, Kidinnu, Eudoxos, Hipparchos and Ptolemaeus were working on than I do in reading about quasars, pulsars, black holes, 21 cm frequencies, radiation pressures, electron streams and so on. Astronomy is a very old science but it has come a long way. May Astronomy go a little further as a result, direct or indirect, of your meetings in Sydney.'

Dr Wood expressed his thanks to Professor O'Neil for his address, and invited Professor Dr G. P. Wild, Australian Academy of Science, to speak.

ADDRESS BY PROF. DR G. P. WILD, AUSTRALIAN ACADEMY OF SCIENCE

'Mr Chairman, My Lord Mayor, Mr President, Mr Vice-Chancellor, Ladies and Gentlemen,

On behalf of the Australian Academy of Science, it is my very great pleasure to welcome you here on the occasion of the 15th General Assembly of the IAU.

I hope that in the coming days, besides meeting old and new friends, everyone of you derives new creative ideas and inspiration not only in the meeting room but perhaps more importantly from informal discussions while sipping coffee (or some other fluid). Welcome especially to the younger astronomers attending their first meeting; and perhaps at the other end of the scale we could extend a special welcome to the one astronomer with us who has been to every meeting since the Union began – that is, every astronomer's friend, Professor Jan Oort.

The President of our Academy, Sir Rutherford Robertson, would of course have been with us himself tonight had he not been out of the country, but he has specifically asked me to convey to you, Professor Strömgren, and to all participants his very good wishes and the sense of honour he feels that the General Assembly should be held in Australia. (Little did he know what a tussle it was.)

Academies of Science vary quite a lot from one country to another. In Australia's case the Academy is a body of scientists whose operations are quite independent of those of the government of the day, although it is of course dependent on the general goodwill of governments for financial support. One of the responsibilities decreed to the Academy by its Charter is 'to establish and maintain associations and relations between Australian scientists and the International Scientific Unions'. For the present General Assembly, therefore, the Academy has been the formal interface between the IAU and the host country. That is why I, as Foreign Secretary of the Academy, am on this platform tonight. (Not only tonight; in the words of the American mathematician Tom Lehrer, 'Come back tomorrow night, we're going to do fractions'.)

But tonight I promise not to keep you long at all. It is on occasions such as this that I do well to remember the famous story of Laplace, who once wrote profoundly and at very great length on a certain scientific topic. He finished by apologizing for writing so long a paper. 'Had I had time', he said, 'I would have written a short one'.

Mr Chairman, as I survey this gathering of people, one rather macabre thought occurs to me: perhaps never again am I likely to see such a strong concentration of people whose internal biological clocks are in such a state of complete disarray. (I hasten to say there is no outward evidence of it). To those of you who have travelled half way round the world to come and visit us, may I say thank you for being prepared to put up with this temporary inconvenience and assure you that in a few days all will be well. And thank you also for sacrificing part of your August for the equivalent of part of February. Although we are unable to supply you with summertime our orchestra will do its best with the next item by Vivaldi. Finally I hope those of you who find yourself so far south for the first time will find abiding compensation when you have the chance of getting away from the city lights to see our most treasured possession (the Lord Mayor excepted) – the southern sky: the Clouds of Magellan, the Southern Cross, and the sight of Sagittarius in the Zenith.

Ladies and Gentlemen, once again, welcome to the Southern Hemisphere, and welcome to Australia.'

The Chairman thanked Professor Wild for his words and called upon Professor B. Strömgren, President of the International Astronomical Union, to speak.

ADDRESS BY THE PRESIDENT OF THE INTERNATIONAL ASTRONOMICAL UNION, PROFESSOR B. STRÖMGREN

'Mr Lord Mayor, Dr Paul Wild, Mr Vice-Chancellor O'Neil, Dr Harley Wood, Ladies and Gentlemen,

The warm words of welcome to which we have just listened, spoken in the beautiful Sydney Town Hall, have been deeply appreciated, and on behalf of the International Astronomical Union I wish to express our sincere gratitude.

To you, Mr Lord Mayor, I wish to say that we are most happy that the Fifteenth General Assembly is taking place in your city. Already so much has been done for us to make our meeting a success, and we have been met with so much kindness and friendliness on the part of our hosts in Sydney that we feel very much at home. Most of us have come a long way to this General Assembly, and it has been a thrilling experience to get to know this fine city, with its many interesting landmarks and beautiful sights, with its magnificent harbour.

We are particularly grateful to you, Mr Lord Mayor Griffin, for the opportunity to hold the Inaugural Ceremony in the Sydney Town Hall and for the invitation to the beautiful reception that preceded it. I wish to thank you very much indeed.

Dr Paul Wild, I beg you to receive, as representative of the Australian Academy of Science, the sincere thanks of the International Astronomical Union. As one of the Host Organizations the Academy has played a highly appreciated active rôle in the preparations of the Fifteenth General Assembly. The great importance of your Academy in the scientific life of Australia is well known to us, and we are honoured and happy to hold our meeting under the auspices of the Australian Academy of Science. I wish to express warm thanks for your words of welcome, and I ask you to convey our gratitude to the Australian Academy of Science.

Dr O'Neil, I thank you most sincerely for your very kind words, and on behalf of the International Astronomical Union I wish to tell you how grateful we are to the University of Sydney for its most important contribution in support of the Fifteenth General Assembly of the IAU. We are all aware of the great burden put on a University that is a host to our General Assembly, and we wish to thank you for inviting us here and for providing such very excellent facilities for our meeting. It is a source of satisfaction to hold our General Assembly at a great University where Astronomy is so strongly represented and supported.

Dr Harley Wood, as Chairman of the Local Organizing Committee you have worked tirelessly and hard during the three years since our last General Assembly. We of the IAU owe you very much indeed, and I wish to tell you here how very grateful we are for all you have done for us. I should like to express to you and the members of the Local Organizing Committee special thanks from the Officers of the IAU. Through the whole period of preparation your most efficient help and fine cooperation have been invaluable and greatly appreciated.

In the life of the International Astronomical Union our Australian colleagues have played, and play a great rôle. Through many decades they have come regularly to the IAU meetings and contributed a most valuable scientific input. For astronomers in other parts of the world the opportunities to meet and have scientific discussions with colleagues from Australia have been most valuable.

And now a strong representation of the IAU community of astronomers is here for the General Assembly. This General Assembly serves many purposes, but a very important one is certainly to further strengthen the ties with our Australian colleagues.

We knew beforehand of the impressive facilities and the splendid achievements of astronomers in Australia. Yet, now that we are here, we appreciate all this even more.

The range of the astronomical activites in Australia is so striking. Let me mention astrometry first. The contributions to the undertaking of the Astrographic Catalogue are very extensive. Important work in this and other fields of Astrometry continues, as was quite clear at the recent IAU Symposium in Perth. Solar Astronomy is most strongly represented, and with the great development at Siding Spring contributions in observational optical astronomy will become even more important than has been the case up to now. The intensity interferometer has yielded remarkable results and Australia is making significant contributions in high-energy astrophysics and space astronomy.

And the Australian contributions in the whole range of radio astronomy that started right after World War II have developed into a magnificent and many-faceted effort of the greatest importance in Astronomy.

It is a wonderful chance for the participants in the Fifteenth General Assembly to get even better acquainted with the whole spectrum of these activities and to meet during these weeks for discussions in depth with Australian colleagues and friends.

We are most happy to be here, and on behalf of the International Astronomical Union I wish to say: Thank you so much for asking us to come.'

After Professor Strömgren's words of thanks, the Chairman formally declared the Welcoming Ceremony closed.

REPORT OF THE GENERAL ASSEMBLY (WITH RESOLUTIONS INCLUDED)

AGENDA

First Session

1. Formal opening of the General Assembly, by the President
2. Appointment of official interpreters
3. Report of the Executive Committee:
   (*a*) Discussion of the printed report
   (*b*) Report of decisions taken at meeting in Sydney, in particular of admission of new Adhering Countries and new IAU Members
4. Report by the General Secretary
5. Report by the President of the proposals for membership of the Executive Committee
6. Announcement of:
   (*a*) The names of representatives of Adhering Countries, empowered to vote on their behalf
   (*b*) The names of representatives to serve on the Nominating Committee
   (*c*) The names of Acting Presidents of Commissions
7. Appointment of the Finance Committee
8. Appointment of the Resolutions Committee
9. Suspension of the second sentence of Article 11(*a*) of the Statutes for the year 1973
10. Consideration of proposals for resolutions submitted by Adhering Organizations or National Committees of Astronomy. The proposals formally submitted before the statutory deadline are shortly as follows;
    (*a*) *(France):*
    (i) To change Article 11 of the Statutes to the effect that Adhering Countries which did not fully pay their annual contributions up to 31 December of the year preceding the General Assembly may participate in the voting on questions involving the budget of the Union with a number of votes by one greater than the number of category that corresponds to the payment actually made.
    (ii) To extend the interval between two General Assemblies of the IAU from 3 to 4 years.
    (iii) To organize regional astronomical meetings at intervals of 4 years or, with well defined topics, in years where there is no General Assembly of the IAU.
    (iv) To reduce, in co-operation with international and national organizations, air pollution by jet propelled aircraft, and the risks of airborne persons of being exposed to laser rays.
    (*b*) *(Yugoslavia):*
    (i) Not to increase the unit of contribution payable by Adhering Countries to the IAU.
    (ii) To organize young astronomers' competitions under the auspices of the IAU.
    (iii) To establish a centre for exchange and loan of temporarily unused astronomical instruments and accessories.
    The proposals will be submitted to the appropriate Commissions prior to the General Assembly.
11. Considerations of proposals for resolutions submitted by Commissions:
    (*a*) *(Commission 6)* To maintain the IAU subvention to the IAU Telegram Bureau at \$1600 U.S. per triennium.
    (*b*) *(Commission 7)* Creation of new IAU Commission on 'Astrodynamics'.

(*c*) *(Commission 8)* To endorse the proposals by *IAU Colloquium* No. 20 on 'Meridian Astronomy' for resolutions on new instrumentation, improvement of the FK4, the meridian catalogue, the SRS catalogue, etc.

(*d*) *(Commission 16)* To allocate $1500 U.S. to the Centre de documentation sur les planètes à Meudon for 1974-1976.

(*e*) *(Commission) 17)*

(i) To approve a list of fundamental lunar craters as basis for the co-ordinate system.

(ii) To include smaller topographic features in the Lunar Nomenclature.

(*f*) *(Commission 20)* To continue the subvention of $2000 U.S. to the Minor Planet Center, Cincinnati, for the period 1974–1976.

(*g*) *(Commission 30)*

(i) Sur l'adoption d'une formule unique pour l'expression de la Vitesse Radiale V_r en fonction du déplacement spectral relatif $z=\Delta\lambda/\lambda_0$ par tous les observateurs (Spectroscopie optique et Radio-Astronomie).

(ii) Sur la convention à adopter pour l'expression de λ_0 (valeur du vide ou de l'air) en vue du calcul de la vitesse radiale des Quasars.

(iii) Sur la necessité de déterminer par priorité les orbites de Binaires Spectroscopiques qui sont aussi Binaires à Eclipse en vue de l'obtention d'un plus grand nombre de Masses.

(*h*) *(Commission 31)* To recommend the adoption of the UTC system as the basis for the dissemination of Standard Time (Heure Normale) in all countries, to recommend definitions for TAI, TAI′ and UTC, to recommend a 'Modified Julian Date', and to recommend maximum limit of DUT1 and the introduction of leap seconds.

(*i*) *(Commission 35)* To consider the problem of overlapping international and large national symposia and colloquia on same or similar topics, and their possible coordination.

(*j*) *(Commission 42)* To change the name of IAU Commission 42 from 'Photometric Double Stars' to 'Close Binary Stars'.

Other proposals for resolutions have only been presented in draft form and will first be discussed by the Commissions. No proposals for resolutions have arrived from Inter-Union Commissions.

Second Session

12. Report of the Finance Committee
    (*a*) Accounts for 1970–1972
    (*b*) The revised budget for 1973
    (*c*) The budget for the ensuing period
13. The unit of subscription for 1974–1976. The Executive Committee will propose to the Finance Committee that the unit of subscription be increased by 25% from 900 gold francs to 1125 gold francs.
14. Appointment of the Special Nominating Committee
15. Proposals for resolutions presented by the Executive Committee
16. Proposals for resolutions presented by Commissions, subject to the recommendation by the Resolutions Committee (including admission of new Member Countries)
17. Announcement of new individual Members of the Union
18. *Commissions:*
    (*a*) Structural changes in Commissions
    (*b*) The election of Presidents and Vice-Presidents
    (*c*) The membership of Organizing Committees
19. *Place and date of the XVIth General Assembly:*
    Consideration of the French invitation to hold the XVIth General Assembly at Grenoble, France, in 1976
20. Election of a President, three Vice-Presidents, a General Secretary, and an Assistant General Secretary

21. Addresses by the retiring and newly elected Officers
22. Closing Ceremonies

At the first session of the General Assembly, to be held on 21 August, items 1 through 11 will be considered. The remaining items, and any that may have been adjourned from the first session, will be considered at the second session, to be held on 30 August.

FIRST SESSION

Held in the Great Hall of the University of Sydney, Sydney, N.S.W., Australia, on Tuesday, 21 *August* 1973 *at* 10^h30^m

Professor Bengt Strömgren, President, in the chair

The Government Astronomer, Dr H. W. Wood, Chairman of the Local Organizing Committee, welcomed the participants in the XVth General Assembly of the International Astronomical Union as follows:

'It makes us very happy to welcome to Australia so many colleagues who are well known to us through their work and who have become our dear friends when we have been fortunate enough to visit them in their own countries or to meet them at Assemblies of the Union and at meetings such as have been arranged in our part of the world for these few weeks.

On behalf of Australian astronomers I express the hope that you will enjoy your stay with us, that many old friendships will be renewed and new ones made, that our meetings will give momentum to progress of astronomy and, this may be tempting fate, that the weather will smile for us.

Tout ce que j'ai dit c'est: soyez les bienvenus! En avant l'astronomie! Et vive l'amitié!'

The President thanked Dr Wood in these words:

'Dr Harley Wood,

On behalf of the International Astronomical Union I wish to thank you sincerely for your warm words of welcome. I am happy to have the opportunity to tell you that we of the IAU have come here with high expectations. We greatly appreciate what you, as Chairman of the Local Organizing Committee, and the other members of that Committee have already done for us. Very many thanks.

The Inaugural Ceremony will take place in the Sydney Town Hall on Wednesday, 22 August, at 8. p.m. This means that today we shall go right on to the Agenda of the First Session of the General Assembly.'

1. *Formal Opening*. The President wished a hearty welcome to the Members of the Union, Invited Participants and their guests, representatives of Adhering Countries and National Committees of Astronomy, representatives of sister Unions and other organizations, and formally declared the first session of the fifteenth General Assembly of the Union open.

He welcomed in particular the official representatives of ICSU and other scientific Unions as follows:

ICSU	(International Council of Scientific Unions): Prof. C. de Jager
BIPM	(Bureau International des Poids et Mesures): Dr J. Terrien
COSPAR	(Committee on Space Research): Prof. J.W. Carver
FAGS	(Federation of Astronomical and Geophysical Services): Dr B. Guinot
IUCM	(Inter-Union Commission on Studies of the Moon): Dr A. Dollfus
IUCS	(Inter-Union Commission on Spectroscopy): Dr A. H. Cook
IUCST	(ICSU Committee on the Teaching of Science) Prof. Edith A. Müller
IUGG	(International Union of Geodesy and Geophysics): Prof. P. Melchior
IUPAP	(International Union of Pure and Applied Physics): Dr J. S. Dryden
URSI	(Union Radio Scientifique Internationale): Prof. W. N. Christiansen

The President then suggested that a message of thanks be sent to Sir Rutherford Robertson, President of the Australian Academy of Science, in appreciation of his kind words of welcome. This proposal was approved by acclamation.

The President continued by asking Members to stand while the General Secretary would read the names of Members who had died since the XIVth General Assembly, or whose death had not been known at the time of that meeting. The General Secretary then read the following list:

A. F. Alexander, G. A. Alter, L. M. J. Arbey, R. H. Baker, J. Baltà-Elias, A. C. Banerji, N. P. Barabashov, M. G. Barros de Perreira, L. V. Berkner, R. Bernas, I. S. Bowen, R. C. Cameron, J. F. J. G. Cox, A. De Moraes, J. Dick, V. A. Dombrovskij, W. J. Eckert, J. Febrer, V. G. Fesenkov, G. Gamov, K. A. Grigoryan, G. Guigay, O. Günther, L. Harang, F. J. Hargreaves, V. A. Heiskanen, C. D. Hellman, Mrs. L. H. Herzberg, A. Kahrstedt, M. Kamieński, Kao Ping-Tse, A. König, H. Kox, W. M. Lindley, C. E. Luplau Jansen, M. G. J. Minnaert, H. Nordström, H. Nowacki, V. R. Ölander, W. H. Ramsey, V. A. Sarabai, P. Sémirot, H. Shapley, R. Sibahara, T. E. Sterne, J. Q. Stewart, K. Stumpff, P. Tardi, J. H. Thomson, D. Tiles, Y. Väisälä, G. B. van Albada, U. van Schewick, U. van Wijk, V. V. Vitkevich, C. B. Watts, A. Wilkens, H. Zanstra, E. Zinner.

2. *Appointment of official interpreters.* The Assembly appointed by acclamation J.-C. Pecker as official interpreter.

3. *Report of the Executive Committee.* The President passed the word to the General Secretary who invited the Assembly to discuss the Report of the Executive Committee (see pp. 1–29), as printed and distributed at the statutory date, and said that he was prepared to answer any questions that might arise. There was no discussion, and the report was approved.

The General Secretary then informed the meeting that the Republic of Korea had applied for admission to the Union after the statutory deadline. This application requires, according to article 8(*b*) of the By-laws, the support of at least two thirds of the votes of Adhering Countries represented at the General Assembly and having the right to vote by virtue of Statute 11(*a*), to be placed on the Agenda for further discussion. The vote taken showed unanimity, with one abstention, for putting the application of the Republic of Korea on the Agenda.

4. *Report of the General Secretary.* The General Secretary then read the following addendum to the Report of the Executive Committee, covering the period January-August 1973:

'My supplementary report on the activities of the Union covers the period from 1 January to 20 August 1973 and also includes the finances of the IAU until 30 June 1973.

The President, General Secretary (with his staff) and the Assistant General Secretary met in Paris, France, on 28 February, and in Copenhagen, Denmark, on 8 June 1973 to discuss the most urgent problems of the Union, finances included, and to work out procedural guide-lines for the General Assemblies. It was decided to transfer the IAU savings on American accounts to an account in The Netherlands. The forward budgets for 1973 and 1974–1976 were revised in the light of the devaluation of the U.S. dollar, and new proposals were prepared for the consideration of the Executive Committee.

The work of the Executive Committee continued by correspondence. It prepared its Report to the General Assembly, approved the budgets drafted by the Officers for submission to the Finance Committee, resolved to ask the General Assembly for an increase of 25% of the unit of contribution, and released the Agenda for the XVth General Assembly, as prepared by the General Secretary, for distribution to Adhering Countries. The General Secretary attended part of the meeting of the Working Group for Lunar Nomenclature on May 21 to 23 in Konstanz, FRG, and represented the Union at the XVth COSPAR Meeting in Konstanz, F.R.G., 28 May–6 June.

The Reports on Astronomy, prepared by the Presidents of Commissions, are an outstanding achievement and I wish to thank all those who contributed to their success. The Presidents of Commissions participated in the preparations of the scientific programme of the General Assembly, sent in the Commissions proposals for new Presidents, Vice-Presidents, and Organizing Committees, made nominations for new Commission members, and presented motions for resolutions.

Only one scientific meeting was held in the span of 8 months I am reporting on. It was *Colloquium* No. 24 on 'Lunar Dynamics and Observational Coordinates', which took place in Houston, Texas, U.S.A. from 15 to 17 January and was cosponsored by COSPAR. However, 6 *Symposia* are being held in Australia simultaneously with this General Assembly and further 5 *Symposia* will be held in Poland in the framework of the Extraordinary General Assembly. The Union cosponsors another *Symposium* in Poland.

There is hardly a year in the annals of the Union so fertile in scientific meetings as the year 1973.

In addition to the Reports on Astronomy, which were published as *Transactions* **XVA** in June, there appeared 4 *Symposium* volumes this year: volume 49 on 'Wolf-Rayet and High-Temperature Stars', volume 51 on 'Extended Atmospheres and Circumstellar Matter in Spectroscopic Binary Systems', volume 54 on 'Problems of Calibration of Absolute Magnitudes and Temperature of Stars', and volume 55 on 'X- and Gamma-Ray Astronomy'. This brings the number of IAU Publications issued since the XIVth General Assembly to 17.

I wish to congratulate all those who contributed to these publications and especially to thank the Assistant General Secretary for his outstanding work, since the IAU publications were his responsibility. My thanks are also due to our publishers, the D. Reidel Publishing Company, whose task to bring out the volumes in time was not always the easiest.

The International School for Young Astronomers, held in Lembang, Indonesia, in July-August of this year, could be organized with local financial help, and thanks, mainly, to the generosity of those Adhering Countries that spontaneously increased their category of adherence. I wish to thank these Countries for their help.

The preparations of the observations of the solar eclipse of June 30, 1973, with its central track over North-Africa, needed intense coordination. An IAU party (Prof. Rigutti, Dr LaCount) visited several African countries to arrange for the local preparations, while also the IAU secretariat took an appreciable share in communicating with the governmental authorities of the various countries involved.

So much on the activities of the Union, which, as is seen, show an expanding tendency. Unfortunately, I cannot give you a favourable report on the finances of the Union. Our savings were, and still are, kept mainly in accounts of U.S. dollars whose value dropped through the devaluation of the U.S. currency and owing to general inflation to about 64% of their value in 1970. The situation will be examined by the Standing Finance Committee, to convene during this General Assembly. Clearly, the Union will have to choose a currency other than the U.S. dollar as basis for its budget and accounts.

In conclusion I wish to thank all those who contributed to the success of our activities during this period of office, and to express the hope that we shall find means and ways for stabilizing the finances of the IAU.'

5. Report by the President of the proposals for membership of the Executive Committee.

The President informed the Assembly of the rules adopted at the Moscow General Assembly in 1958 which provided for a Special Nominating Committee to advise the Executive Committee on nominations for the new Executive Committee. According to Article 12(*a*) of the By-laws in force the Special Nominating Committee submits its proposal to the General Assembly direct. The Special Nominating Committee appointed at the XIVth General Assembly in Brighton consisted of the present President, the former President, O. Heckmann, a Vice-President of the present Executive Committee, M. K. V. Bappu, and four IAU Members, G. H. Chebotarev, J. Delhaye, D. S. Heeschen, and A. Poveda.

The Special Nominating Committee made to the General Assembly the following proposals as to the composition of the next Executive Committee:

as President for the period 1973–1976: Prof. L. Goldberg, U.S.A.

as Vice-Presidents for the period 1973–1979:	Mr J. G. Bolton, Australia
	Prof. Ch. Fehrenbach, France
	Prof. W. Iwanowska, Poland
as General Secretary for the period 1973–1976:	Prof. G. Contopoulos, Greece
as Assistant General Secretary for the period 1973–1976:	Prof. Edith A. Müller, Switzerland

The full Executive Committee would thus consist of the above six together with:

the continuing Vice-Presidents:	Prof. B. J. Bok, U.S.A.
	Sir Bernard Lovell, U.K.
	Prof. E. R. Mustel, U.S.S.R.
and in an advisory capacity:	Prof. B. Strömgren, outgoing President, Denmark,
	Prof. C. de Jager, outgoing General Secretary, The Netherlands

In accord with Article 13(*c*) of the By-laws, the formal election would take place at the final session of the General Assembly.

6. *Announcements.* Called on by the President, the General Secretary read:
(*a*) the names of representatives of Adhering Countries, empowered to vote on their behalf;
(*b*) the names of representatives of Adhering Countries on the Nominating Committee.

Country	*(a) Official Representative*	*(b) Representative on Nominating Committee*
Argentina	J. Sahade	J. Sahade
Australia	H. W. Wood	J. H. Carver
Austria	not represented	not represented
Belgium	P. Swings	P. Ledoux
Brazil	L. Muniz Barreto	S. Ferraz Mello
Bulgaria	not represented	not represented
Canada	K. O. Wright	F. D. A. Hartwick
Chile	C. Anguita	C. Anguita
Colombia	E. Brieva	E. Brieva
Cuba	not represented	not represented
Czechoslovakia	V. Vanýsek	V. Bumba
Denmark	M. Rudkjøbing	K. Gyldenkerne
Finland	J. Tuominen	J. Tuominen
France	J.-C. Pecker	J. Kovalevsky
F.R.G.	M. Grewing	M. Grewing
G.D.R.	G. Ruben	G. Ruben
Greece	L. N. Mavridis	J. Argyrakos
Hungary	L. Detre	B. Szeidl
India	S. K. Trehan	G. Swarup
Iran	not represented	not represented
Ireland	P. K. Carrol	P. K. Carrol
Israel	Y. Ne'eman	S. Cuperman
Italy	M. G. Fracastoro	M. Hack
Japan	Y. Fujita	S. Miyamoto
Korea, D.P.R.	not represented	not represented
Mexico	P. Pişmiş	P. Pişmiş
Netherlands	J. Borgman	J. Borgman
New Zealand	F. M. Bateson	F. M. Bateson
Norway	E. Jensen	Ö. Elgaröy
Poland	W. Iwanowska	W. Iwanowska

Portugal	R. O. Vicente	R. O. Vicente
Roumania	not represented	not represented
South Africa	G. G. Cilliè	E. E. Baart
Spain	A. Orte	L. Quijano
Sweden	G. Larsson-Leander	T. Oja
Switzerland	M. Golay	U. Steinlin
Taiwan	not represented	not represented
Turkey	not represented	not represented
U.A.R.	A. S. Asaad	A. S. Asaad
U.K.	R. H. Stoy	R. J. Tayler
Uruguay	not represented	not represented
U.S.A.	D. S. Heeschen	G. Westerhout
U.S.S.R.	V. B. Nikonov	Yu. N. Parijskij
Vatical City State	P. J. Treanor	M. F. McCarthy
Venezuela	J. Stock	J. Stock
Yugoslavia	G. Teleki	G. Teleki

Note: In some cases National Committees designated substitute delegates. In order to simplify the above list the names of the substitute delegates have been omitted.

(c) *Acting Presidents.* The General Secretary announced that the Executive Committee had asked:

V. Szebehely to act for G. N. Duboshin as President of Commission No. 7 until the arrival of the latter

M. Golay to act for D. L. Crawford as President of Commission No. 25

A. H. Batten to act for R. Bouigue as President of Commission No. 30

D. MacRea to act for A. Reiz as President of Commission No. 38

A. Hayli to act for O. Gingerich as President of Commission No. 41

M. J. Rees to act for Ya. B. Zel'dovich as President of Commission No. 47, and for H. Friedman as President of Commission 48.

The acting Presidents were appointed for the duration of the General Assembly.

7. *Appointment of the Finance Committee.* In accord with Article 18(*a*) of the By-laws the General Assembly appointed the following Finance Committee consisting of one representative of each Adhering Country:

Argentina	J. Sahade	Japan	S. Miyamoto
Australia	A. G. Little	Korea, D.P.R.	not represented
Austria	not represented	Mexico	M. Peimbert-Sierra
Belgium	P. Ledoux	Netherlands	H. Habing
Brazil	P. Kaufmann	New Zealand	F. M. Bateson
Bulgaria	not represented	Norway	E. Jensen
Canada	J. L. Locke	Poland	W. Iwanowska
Chile	C. Anguita	Portugal	R. O. Vicente
Colombia	E. Brieva	Roumania	not represented
Cuba	not represented	South Africa	G. G. Cilliè
Czechoslovakia	L. Perek	Spain	L. Quijano
Denmark	H. E. Jørgensen	Sweden	P. O. Lindblad
Finland	J. Tuominen	Switzerland	B. Hauck
France	S. Débarbat	Taiwan	not represented
F.R.G.	W. Fricke	Turkey	not represented
G.D.R.	G. Ruben	U.A.R.	A. S. Asaad
Greece	L. N. Mavridis	U.K.	D. W. Stibbs
Hungary	L. Dezső	Uruguay	not represented

India	S. D. Sinwhal	U.S.A.	M. P. Savedoff
Iran	not represented	U.S.S.R.	G. N. Duboshin
Ireland	P. K. Carrol	Vatican City State	G. V. Coyne
Israel	Y. Ne'eman	Venezuela	J. Stock
Italy	G. Godoli	Yugoslavia	G. Teleki

Note: Substitutes designated by a few National Committees have been omitted in the above list.

8. *Appointment of the Resolutions Committee.* At the request of the President, the General Secretary informed the meeting that the Resolutions Committee is to advise the Executive Committee as regards the proposals for resolutions submitted for consideration by the General Assembly. The General Assembly appointed J.-C. Pecker and L. Perek to serve on the Resolutions Committee. The General Secretary and the Assistant General Secretary were to attend the meeting of the Resolutions Committee in an advisory capacity.

9. *Suspension of the second sentence of Article 11(a) of the Statutes for the year 1973.*
The President made the following announcement:
'Mainly owing to the uncertainties and variations in the rates of exchange, especially the instability of the U.S. dollar, and to administrative delays resulting from the necessary adjustments, no fewer than 13 (out of 46) Adhering Countries had not paid fully their annual contributions due to the end of 1972 when the Agenda for this General Assembly was being prepared. On the strict interpretation of Statute 11(*a*) these countries would not be able to participate in the voting at the General Assembly. The Executive Committee, anticipating this difficulty, placed on the Agenda a motion to suspend the second sentence of Statute 11(*a*) for the duration of the General Assembly. The effect would have been that all Adhering Countries could participate in the voting.

An examination of the books immediately before the General Assembly showed, however, that 12 out of the 13 Adhering Countries in debt had paid their contributions up to 31 December 1972 since the distribution of the Agenda, and may therefore participate in the voting at the General Assembly. This brings the number of Adhering Countries with voting right to 45, which is considered normal by the Executive Committee.

The Executive Committee accordingly wishes to withdraw the proposal to suspend the second sentence of Statute 11(*a*). The General Secretary has informed the representatives of the countries concerned of their rights to participate in the voting or not.

There is no motion before the General Assembly, but I invite comments.'

The withdrawal of the motion was approved without discussion.

10. *Consideration of proposals for resolutions submitted by Adhering Organizations or National Committees of Astronomy.* The proposals were read by the General Secretary as follows:

(*a*) *(France):*
- (i) To change Article 11 of the Statutes to the effect that Adhering Countries which, owing to irregularities in the rates of exchange, did not fully pay their annual contributions up to 31 December of the year preceding the General Assembly may participate in the voting on questions involving the budget of the Union with a number of votes by one greater than the number of category that corresponds to the payment actually made.
- (ii) To extend the interval between two General Assemblies of the IAU from 3 to 4 years.
- (iii) To organize regional astronomical meetings at intervals of 4 years or, with well defined topics, in years where there is no General Assembly of the IAU.
- (iv) To reduce, in co-operation with international and national organizations, air pollution by jet propelled aircraft, and the risks of airborne persons of being exposed to laser rays.

(*b*) *(Yugoslavia):*
- (i) Not to increase the unit of contribution payable by Adhering Countries to the IAU.
- (ii) To organize young astronomers' competitions under the auspices of the IAU.

(iii) To establish a centre for exchange and loan of temporarily unused astronomical instruments and accessories.

(*c*) (i) *(U.A.R.)* (sent in after the statutory deadline)

(i) To organize regional astronomical colloquia of 50 to 100 participants for astronomers from developing countries; to make available astronomical literature to astronomers and astronomical institutions in developing countries.

The General Secretary added that these proposals would first be discussed by the appropriate Commissions and then passed to the Resolutions Committee for further recommendation.

11. Consideration of proposals for resolutions submitted by Commissions. The General Secretary read the proposals as follows:

(*a*) *(Commission 6)* To maintain the IAU subvention to the IAU Telegram Bureau at $1600 U.S. per triennium.

(*b*) *(Commission 7)* Creation of new IAU Commission on 'Astrodynamics'.

(*c*) *(Commission 8)* To endorse the proposals by *IAU Colloquium* No. 20 on 'Meridian Astronomy' for resolutions on new instrumentation, improvement of the FK4, the meridian catalogue, the SRS catalogue, etc.

(*d*) *(Commission 16)* To allocate $1500 U.S. to the Centre de documentation sur les planètes à Meudon for 1974–1976.

(*e*) *(Commission 17)*

(i) To approve a list of fundamental lunar craters as basis for the co-ordinate system.

(ii) To include smaller topographic features in the Lunar Nomenclature.

(*f*) *(Commission 20)* To continue the subvention of $2000 U.S. to the Minor Planet Center, Cincinnati, for the period 1974–1976.

(*g*) *(Commission 30)*

(i) Sur l'adoption d'une formule unique pour l'expression de la Vitesse Radiale V_r en fonction du déplacement spectral relatif $z = \Delta\lambda/\lambda_0$ par tous les observateurs (Spectroscopie optique et Radio-Astronomie).

(ii) Sur la convention à adopter pour l'expression de λ_0 (valeur du vide ou de l'air) en vue du calcul de la vitesse radiale des Quasars.

(iii) Sur la necessité de déterminer par priorité les orbites de Binaires Spectroscopiquest qui son aussi Binaires à Eclipse en vue de l'obtention d'un plus grand nombre de Masses.

(*h*) *(Commission 31)* To recommend the adoption of the UTC system as the basis for the dissemination of Standard Time (Heure Normale) in all countries, to recommend definitions for TAI, TAI′ and UTC, to recommend a 'Modified Julian Date', and to recommend maximum limit of DUT1 and the introduction of leap seconds.

(*i*) *(Commission 35)* To consider the problem of overlapping international and large national symposia and colloquia on same or similar topics, and their possible coordination.

(*j*) *(Commission 42)* To change the name of IAU Commission 42 from 'Photometric Double Stars' to 'Close Binary Stars'.

Other proposals for resolutions had only been presented in draft form and would first be discussed by the interested Commissions.

The General Secretary reminded that the formal vote on resolutions would take place at the final session of the General Assembly, according to By-laws.

The General Secretary continued by announcing that all other items of the agenda would be deferred until the next session of the General Assembly.

The President then formally adjourned the meeting at 12^h00^m.

GENERAL ASSEMBLY

FINAL SESSION

Held in the Great Hall of the University of Sydney, N.S.W., Australia on Thursday, 30 August 1973 at $10^h\,00^m$

Professor B. Strömgren, President, in the chair

12 and 13. *Report of the Finance Committee and consideration of the unit of subscription.*
The Finance Committee, as appointed at the first session of the General Assembly, had to examine the accounts of the Union for the years 1970–1972 and to consider the budget of the IAU, as prepared by the Executive Committee, for the ensuing period. The President called upon the Chairman of the Finance Committee, Professor D. W. N. Stibbs, to present his report to the General Assembly. The Chairman of the Finance Committee read as follows:

REPORT OF THE FINANCE COMMITTEE TO THE GENERAL ASSEMBLY

1. The Finance Committee, as appointed at the First Session of the XVth General Assembly on 21st August 1973, met on that date and appointed D. W. N. Stibbs as Chairman. The General Secretary presented a Report on the finances of the Union and gave an outline of the work that the Committee was required to do in accordance with Article 18(*a*) of the By-laws. The Report was discussed in detail by the Committee and it appointed a Sub-Committee to examine the Account for the preceding period and the comprehensive budget for the following period as prepared by the Executive Committee. The Sub-Committee consisted of the Chairman ex officio, W. Fricke, P. Kaufmann, E. R. Mustel, G. Ruben, M. P. Savedoff with F. Edmondson ex officio as a member of the Executive Committee's Standing Committee on Finance; and C. de Jager, G. Contopoulos and A. Jappel ex officio as the General Secretary, Assistant General Secretary and Executive Secretary of the Union respectively. The Sub-Committee met on 23rd and 24th August, and the second meeting of the Finance Committee was held on 25th August.

2. The Sub-Committee examined the accounts for the years 1970–72 as summarized in the Narrative Report of the Executive Committee and verified by the Auditor to the Union, M. Roger Bacle of Paris. The very high standard at which the accounts had been kept was duly noted with appreciation by the Sub-Committee which gave its approval both on the basis of its inspection of the accounts and the supporting statement of the Auditor.

The Finance Committee recommends to the General Assembly that the accounts for 1970–72 be approved.

3. The Sub-Committee considered the Budget for 1973 as proposed in the Report of the Executive Committee. It was noted that it required a financial provision within the limit set by the XIVth General Assembly and that it represented activities entirely consistent with the policy of the Union. The Sub-Committee also examined the account for the first-half of the current financial year, found them to be in order and ascertained that the projected expenditure for the second-half of the year was within the financial provision.

The Finance Committee recommends to the General Assembly that the Budget for 1973 as proposed by the Executive Committee in its Report (vide p. 23) be accepted.

4. The Sub-Committee considered in detail the budgetary requirements for the period 1974–76 on the basis of two comprehensive budgets prepared by the Executive Committee, one based on the present unit of contribution and the other on a higher unit. It was recognized that, in order to pursue a programme of activities on the present level consistent with the function and policy of the Union without incurring a financial deficit, it would be necessary to increase the unit of contribution by about 25%. Without such an increase, the excess of payments over receipts would be about 56000 dollars at the June 1973 rates of exchange. The Sub-Committee noted that another form of increasing the Receipts, namely by introducing a fee for Membership of the Union, had not been well received. However, it felt that this possibility should be examined further, and it was suggested

that the Registration Fee at future General Assemblies should help to meet both local and IAU expenses in connection with the Assemblies. On the other hand, it was noted that the responses of Adhering Countries to the possibility of an increase in the unit of contribution has been encouraging, and the Sub-Committee considered that the proposed increase should be accepted for 1974–76.

The Finance Committee recommends to the General Assembly that the proposal of the Executive Committee (vide Agenda 13, p. 41) for the unit of contribution to be increased by 25 per cent from 900 gold francs to 1125 gold francs (1 gold franc=0.2903225 gm fine gold) be accepted for the purpose of the Budget for 1974–76, and that the Executive Committee should give active consideration to alternative ways in which the revenue of the Union might be increased, including those considered by the Sub-Committee.

5. The Finance Committee considered the Budget for 1974–76 prepared on the basis of the unit of contribution as 1125 gold francs and the present categories of membership, and it recommends for adoption by the General Assembly the following budget:

BUDGET FOR 1974–76

Unit of Contribution 1125 gold francs
($438,92 U.S. at 28 June 1973)

Receipts in U.S. dollars

1. Contributions from Adhering Countries	291000
2. Revenue from publications	16500
3. Interest on accounts	10300
4. UNESCO subvention through ICSU	36000
Total	353800
Payments	351760
Excess of receipts over payments	2040

List of projects under Payments: Item 4.2

	$U.S.
Comm. 4, Ephemerides	880
6, IAU Telegram Bureau	1980
16, Documentation, Meudon	1870
17, Special Working Groups	1870
20, Minor Planets	2200
27, Catalogue of Variables	3300
Inf. Bull. Southern Hemisphere	1650
Total	13750

Payments in U.S. dollars

1. Administrative Office	124300
2. Subscription to ICSU	5820
3. Commission expenses	1100
4. Projects of Commissions	
4.1. Exchange of Astronomers	29000
4.2. Other projects	13750
5. General Assembly	33000
6. IAU publications	12100
7. Symposium volumes for developing countries	15350
8. Executive Committee meetings	15400
9. Officers' meetings	9900
10. Symposia and Colloquia	39600
11. Inter-Union Commissions	14300
12. Projects authorized by the Executive Committee	3500
13. Representation	1650
14. Bank charges and exchange costs	990
15. Young Astronomers' schools	16500
16. Regional meetings	5500
17. Refund to reserves	10000
Total	351760

6. The Sub-Committee examined the question of the reference currency in which the financial affairs of the Union could be most effectively conducted, and it concluded on the basis of the stability and convertibility of currency that the Swiss franc, although subject to short-term fluctuations, would be the most favourable for the Union and its Adhering Countries.

The Finance Committee recommends to the General Assembly that the currency in which the affairs of the Union should be conducted should be the Swiss franc.

7. The Sub-Committee considered the investment of the funds of the Union at present on account in Dutch guilders and U.S. dollars. It was concluded on the basis of the available expert professional opinion that convertible debentures were the most attractive form of investment, and that a proposal from the Amsterdam–Rotterdam (AMRO) Bank for investment in convertible U.S. dollar

bonds and in a Netherlands Investment Trust offered the best investment prospects for the Union at the present time.

The Finance Committee recommends to the General Assembly that it should delegate to the Executive Committee the responsibility for the appropriate investment of the funds of the Union.

8. In conclusion, the Finance Committee would wish to express to the General Assembly its highest regard for the meticulous way in which the financial affairs of the Union have been conducted by the Secretariat and to thank them for their courtesy and helpfulness in making available all the documents and professional advice necessary for the Committee to effectively discharge its responsibilities.

D. W. N. STIBBS
Chairman, Finance Committee

The President, authorized by a show of hands, then formally asked the General Assembly to give approval to the report of the Finance Committee, explaining that voting would be by country, each country having a number of votes by one greater than the number of its category of adherence (Statute 11 (*b*)).

The recommendations of the Finance Committee were approved unanimously.

The President thanked Professor Stibbs, the members of the Finance Committee and the General Secretary for their valuable work.

14. *Special Nominating Committee.* The President referred to Article 12. of the By-laws which provides that proposals for elections to the President of the Union, six Vice-Presidents, the General Secretary, and the Assistant General Secretary are submitted to the General Assembly by the Special Nominating Committee which consists of the President and Past President of the Union, a member proposed by the retiring Executive Committee from Members not belonging to the Executive Committee, both immediately past and present, and four members selected by the Nominating Committee from among twelve Members proposed at the meeting of Presidents of Commissions.

The President then moved that the General Assembly elect the following Members into the Special Nominating Committee:

A. Blaauw (The Netherlands)
H. Elsässer (F.R.G.)
J.-C. Pecker (France)
L. Perek (Czechoslovakia)
A. B. Severny (U.S.S.R.)

This list was unanimously approved.

15 and 16. *Resolutions.* The General Secretary reported that the proposals for resolutions submitted to the General Assembly by the Executive Committee, the National Committees and by Commissions had eventually been passed on to the Resolutions Committee for consideration. Therefore, before passing to the vote on the resolutions, Professor J.-C. Pecker would present to the General Assembly the report of the Resolutions Committee.

The President then called upon Professor Pecker to take the floor.

REPORT BY THE PRESIDENT OF THE RESOLUTIONS COMMITTEE

Le Comité des Résolutions a examiné le texte des résolutions soumises par les Organisations Adhérentes, les Comités Nationaux, et les différentes Commissions de l'UAI. En sus des propositions incluses dans l'ordre du jour de l'Assemblée Générale (points 10 et 11), un grand nombre d'autres propositions ont été reçues et examinées.

A. Résolutions figurant au point 10 de l'ordre du jour.

Le Comité des Résolutions soumet au vote de l'Assemblée Générale:

1. la résolution 10a i sous la forme modifiée ci-après (Résolution No. 1).
2. la résolution 10a ii sous la forme modifiée ci-après (Résolution No. 2).
3. la résolution 10a iii sous la forme modifiée ci-après (Résolution No. 3).

Le Comité des Résolutions propose que la résolution 10a iv ne soit pas soumise au vote de l'Assemblée Générale, mais à la nouvelle Commission 50, si celle-ci est créée au point 18 de l'ordre du jour, pour considération, et sinon, au Comité Exécutif, pour étude.

Le Comité des Résolutions propose à l'Assemblée Générale de déférer la résolution 10b ii à la Commission 46, la résolution 10b iii à la Commission 9 pour considération pendant la periode 1973–1976, et d'abandonner la résolution 10b i, contradictoire avec les recommendations du Comité des Finances.

B. Résolutions issues de Comités Nationaux et d'Organisations Adhérentes et ne figurant pas à l'ordre du jour de l'Assemblée Générale.

Cinq telles résolutions ont été reçues après la date limite.

Le Comité des Résolutions recommande:

1. de déférer la résolution d'Israel concernant la libre circulation des savants au nouveau Comité Exécutif.
2. de déférer la résolution jointe de la France et de l'Australie concernant la désirabilité pour l'UAI de représenter tous les pays du monde actifs en astronomie, au nouveau Comité Exécutif.
3. de déférer la résolution des Pays-Bas concernant la création d'un Groupe de Travail sur l'Astronomie Infrarouge à la Commission 9, compte tenu du fait qu'un groupe similaire est formé au sein de cette Commission.
4. de déférer la résolution de la République Arabe Unie sur les colloques régionaux à la Commission 46 pour examen entre 1973 et 1976.
5. de déférer la résolution de la République Arabe Unie sur la dissémination de l'information dans les pays en voie de développement aux Commissions 5 et 46 pour examen entre 1973 et 1976.

C. Résolutions soumises par les Commissions et figurant au point 11 de l'ordre du jour de l'Assemblée Générale.

Le Comité des Résolutions propose à l'Assemblée Générale:

1. d'approuver la résolution 11a, complétée par le Comité des Finances.
2. d'abandonner la résolution 11b, compte tenu de la création par la Commission 7 d'un Groupe de Travail sur le Mouvement des Corps Célestes Artificiels.
3. d'approuver la résolution 11c sous sa forme complète, dont le texte intégral sera publié dans les Transactions de l'UAI, et qui sera couverte par la résolution No. 7.
4. d'approuver la résolution 11d complétée par le Comité des Finances.
5. d'adopter la résolution 11e sous sa forme définitive nouvelle mise au point par la Commission 17. Cette résolution sera couverte par la résolution No. 7.
6. d'approuver la résolution 11f, complétée par le Comité des Finances.
7. d'adopter la résolution 11g iii telle qu'elle a été modifiée par la Commission 30. Les résolutions 11 g i et ii ont été retirées de l'ordre du jour par la Commission 30.
8. d'adopter la résolution 11h, dont une partie est soumise à une vote séparée de l'Assemblée Générale (Résolution No. 4) et dont les autres parties sont couvertes par la résolution No. 7.
9. de ne pas voter sur la résolution 11i, compte tenu de l'action déjà entreprise par le Comité Exécutif.
10. d'adopter la résolution 11j, dans le cadre du point 18 de l'ordre du jour.

D. Résolutions provenant des diverses Commissions et ne figurant pas à l'ordre du jour de l'Assemblée Générale.

Certaines ont des implications financières et, pour simplification, le Comité des Résolutions les a affectées d'une lettre. Elles concernent les points suivants:

k (Commissions 4, 19, 31)	subside de 12000 $ pour le B.I.H. (par an).
l (Commission 42)	200 $ pour trois ans
m (Commission 4)	250 $ pour les éphémérides
n (Commission 27)	pour l'édition du catalogue des étoiles variables à Moscou
o (Commission 46)	1. 20000 $ pour les écoles de jeunes astronomes 2. 6000 $ pour les missions d'enseignement

Le Comité des Résolutions recommande l'adoption des résolutions *m* et *n* telles qu'elles ont été incorporées, avec des modifications, dans le rapport du Comité des Finances, et défère l'examen des résolutions *k*, *l* et *o* au Comité Exécutif qui devra prendre les décisions compatibles avec le budget de l'Union.

Certaines résolutions concernant le changement de nom des Commissions seront discutées sous le point 18 de l'ordre du jour.

Certaines résolutions, approuvées par les Commissions concernent des questions assez générales pour que le Comité des Résolutions propose à l'Assemblée Générale de voter séparément sur ces résolutions qui portent les numéros 4, 5, 6.

Les autres résolutions, fort nombreuses, soumises par les Commissions et qui feront l'objet d'une publication, in extenso, dans les Comptes Rendus de la XVe Assemblée Générale de l'UAI, dans le cadre des rapports des Commissions, ont été examinées individuellement par le Comité des Résolutions. Elles font l'objet d'une résolution plus générale, la résolution globale No. 7. Cette résolution globale couvre d'ailleurs des résolutions qui, bien que spécialisées, sont parfois d'une importance considérable; ainsi en est-il, par exemple, des résolutions sur la nomenclature martienne (Commission 16), les problèmes de la nomenclature lunaire (Commission 17), la définition des différents systèmes de définition du temps (Commission 31).

Le Comité des Résolutions recommande finalement que toute autre proposition qui aurait soumise à une commission, et approuvée par cette commission, soit le cas échéant couverte par la résolution globale No. 7.

J.-C. PECKER

Président de la Commission des Résolutions

RÉSOLUTION NO. 1

L'Assemblée Générale de l'Union Astronomique Internationale décide d'adopter la modification suivante des Statuts: A la fin des articles 11(*a*) et 11(*b*), la phrase suivante est ajoutée: 'Des exceptions à cette règle ne peuvent être envisagées que sur décision du Comité Exécutif.'

RÉSOLUTION NO. 2

L'Assemblée Générale de l'Union Astronomique Internationale demande au Comité Exécutif d'examiner la possibilité de modifier les Statuts, de telle sorte que l'intervalle de trois ans entre deux Assemblées Générales puisse être étendu à quatre ans, si telle est, après examen, l'opinion du Comité Exécutif.

RÉSOLUTION NO. 3

L'Assemblée Générale de l'Union Astronomique Internationale reconnaît l'intérêt scientifique des réunions régionales et recommande à leurs organisateurs de prendre les mesures appropriées en ce qui concerne les dates de ces réunions et leur programme, de telle sorte qu'elles ne constituent pas pour les pays membres une charge financière trop lourde.

RÉSOLUTION NO. 4

L'Assemblée Générale de l'Union Astronomique Internationale,

considérant

(*a*) qu'il existe des besoins de datation selon un système décimal dans des buts divers en liaison avec l'utilisation de signaux horaires et de codes horaires transmis radioélectriquement,

(*b*) qu'il est souhaitable de rapporter une telle datation à 0^h TU comme début de la journée plutôt qu'à 12^h TU,

(*c*) qu'un comptage continu des jours, le système des jours de la période julienne, a été depuis longtemps établi pour la datation en astronomie, chronologie, et dans les sciences connexes,

(*d*) qu'il est nécessaire d'éviter la prolifération de systèmes différents de datation,

(*e*) qu'une simple conversion du système classique des jours juliens, ci-dessus mentionné, en un comptage décimal moderne, serait avantageux,

(*f*) que le système de comptage actuel des jours juliens, qui se réfère à un midi moyen de Greenwich comme début du jour, doit être maintenu sans discontinuité, et

(*g*) qu'un système de 'jour julien modifié' satisfaisant aux exigences ci-dessus est déjà utilisé,

recommande

que le terme de 'jour julien modifié' et le symbole MJD soient utilisés seulement pour la quantité JD moins 2400000.5, MJD étant égal à 0 à la date du 17 novembre 1858, 0^h TU.

Note: Il est recommandé que cette Résolution soit communiquée à l'ISO et aux autres organismes intéressés.

RÉSOLUTION NO. 5

L'Assemblée Générale de l'Union Astronomique Internationale recommande, pour la détermination de z dans le spectre des objets fortement déplacés vers le rouge, l'utilisation des longueurs d'onde des raies dans le vide, lorsque la différence en λ_0 est significative.

RÉSOLUTION NO. 6

L'Assemblée Générale de l'Union Astronomique Internationale,

RESOLUTION NO. 1

The General Assembly of the International Astronomical Union resolves to adopt the following change of the Statutes: At the end of Articles 11(*a*) et 11(*b*), the following sentence is added: 'Exceptions from this rule cannot be considered unless decided by the Executive Committee.'

RESOLUTION NO. 2

The General Assembly of the International Astronomical Union requests the Executive Committee to examine the possibility of a change in the Statutes such as to allow the three years interval between two General Assemblies to be extended to a four-year interval, if such is, after study, the opinion of the Executive Committee.

RESOLUTION NO. 3

The General Assembly of the International Astronomical Union recognizes the scientific interest of regional meetings and recommends to their organizers to take the appropriate steps so that the dates of these meetings and their programme be such as not to constitute for the member countries an excessive financial burden.

RESOLUTION NO. 4

The General Assembly of the International Astronomical Union,

considering

(*a*) that there exist requirements for dating in a decimal sequence for various purposes in conjunction with the use of radio time signals and radio time codes,

(*b*) that it is desirable to refer such dating to 0^h UT as the beginning of a day instead of 12^h UT,

(*c*) that a decimal day count, the Julian Day count, has long been in established use for dating in astronomy, chronology, and related sciences,

(*d*) that it is necessary to avoid a proliferation of different dating systems,

(*e*) that a simple conversion from the above-mentioned conventional Julian Day count to a modern decimal day count will be advantageous,

(*f*) that the existing established Julian Day count which refers to a Greenwich Mean Noon as the beginning of the day needs to be maintained without discontinuity and,

(*g*) that a 'Modified Julian Day' count satisfying the above requirements is already in use,

recommends

that the term 'Modified Julian Date' and the symbol MJD be used only for the quantity JD minus 2400000.5, MJD being equal to 0 on 1858 November 17, 0^h UT.

Note: It is recommended that this Resolution be communicated to ISO and other concerned bodies.

RESOLUTION NO. 5

The General Assembly of the International Astronomical Union recommends, for the determination of z in the spectrum of objects of large redshifts, the use of vacuum wavelengths of lines, wherever the difference in λ_0 is significant.

RESOLUTION NO. 6

The General Assembly of the International Astronomical Union, *reconnaissant*

que la seconde du Système International peut être réalisée en pratique avec une précision meilleure que 2 sur 10^{13},

que les distances sur Terre et dans l'espace sont mesurées par le temps de parcours de rayonnement électromagnétique, et

que le Système des Constantes Astronomiques de l'UAI sera révisé dans le cours des trois années à venir,

considère

qu'une valeur unique de la vitesse de propagation du rayonnement électromagnétique devrait être utilisée lorsque l'on exprime en mètres les résultats des mesures de distance les plus précises, et

recommande

que, lorsque la valeur la plus précise de la vitesse de propagation du rayonnement électromagnétique dans le vide est nécessaire, la valeur proposée par le Comité Consultatif pour la Définition du Mètre à sa réunion de juin 1973, c'est-à-dire:

$$c = 299\,792\,458 \text{ mètres par seconde,}$$

soit utilisée,

que cette valeur de la vitesse de propagation du rayonnement électromagnétique soit incorporée dans la révision prochaine du Système des Constantes Astronomiques de l'UAI, et

que le Comité International des Poids et Mesures maintienne cette valeur dans toute éventuelle définition du Mètre.

RÉSOLUTION NO. 7

Proposée par le Comité des Résolutions

Sur les résolutions adoptées par les Commissions

Considérant qu'il lui est pratiquement impossible d'accorder une attention particulière à chaque résolution adoptée par chacune de ses diverses Commissions et ayant une confiance entière en ses Commissions, cette Assemblée Générale désire exprimer son approbation des Résolutions adoptées par ses différentes Commissions, recommande aux astronomes de les mettre en application dans la limite de leurs possibilités, et demande au Comité Exécutif de veiller, autant que possible, à la compatibilité de ces résolutions et à leur application.

recognizing

that the second of the Système International can be realized in practice with a precision of better than 2 parts in 10^{13},

that distances on Earth and in space are measured by times of travel of electromagnetic radiation, and

that the IAU System of Astronomical Constants will be revised within the next three years,

considers

that a common value of the speed of propagation of electromagnetic radiation should be used when expressing in metres the results of the most precise measurements of distance, and

recommends

that when the most precise value of the speed of propagation of electromagnetic radiation in vacuum is required, the value proposed by the Comité Consultatif pour la Définition du Mètre at their meeting in 1973 June, namely

$$c = 299\,792\,458 \text{ metres per second,}$$

should be used,

that this value of the speed of propagation of electromagnetic radiation should be incorporated in the next revision of the IAU System of Astronomical Constants, and

that the Comité International des Poids et Mesures should maintain this value in any re-definition of the metre.

RESOLUTION NO. 7

Proposed by the Resolutions Committee

On the resolutions adopted by Commissions

Considering the impracticability of giving individual attention to every resolution adopted by each of its Commissions, and having full confidence in its Commissions, this General Assembly wishes to give its endorsement to the Resolutions adopted by its individual Commissions, recommends that astronomers give effect to these Resolutions in so far as they are able, and asks the Executive Committee to insure, inasmuch as possible, the compatibility of the resolutions and to put them into effect.

The President then moved that the General Assembly approve the report of the Resolutions Committee and adopt resolutions 1 through 7, as recommended. This motion was carried by both representatives of Adhering Countries and Members.

17 (a) *Admission of new Member Country.* The General Secretary reported that the Executive Committee had carefully examined the application of the Republic of Korea for admission to the IAU, and had found that the development of astronomy in this country lives up to the standards required for IAU membership. He, therefore, moved that the General Assembly admit the Republic of Korea as a new Adhering Country to the Union. This motion was carried unanimously.

17 (b) *New Members of the Union.* The General Secretary announced that the Executive Committee had, on the proposal of Adhering Organizations and on the advice of the Nominating Committee, admitted 612 new Members to the Union, and deleted 10 names from its membership list. The names of new Members had been prominently displayed and the General Secretary did not propose to read them. He informed the General Assembly that these names would be incorporated in the alphabetical list of Members to appear in print, and that the new Members would be informed in writing of their election.

18. *Commissions.*

(*a*) The General Secretary submitted to the approval of the General Assembly the following proposal of the Executive Committee:

(i) that the name of IAU Commission No. 15 be henceforward 'Physics of Comets, Minor Planets and Meteorites'

(ii) that the name of IAU Commission No. 22 be henceforward 'Meteors and Interplanetary Dust'

(iii) that the name of IAU Commission No. 25 be henceforward 'Stellar Photometry and Polarimetry'

(iv) that the name of IAU Commission No. 42 be henceforward 'Close Binary Stars'

(v) that IAU Commission No. 43 'Astrophysical Plasmas and Magnetohydrodynamics' be abolished

(vi) that two new IAU Commissions be established, namely IAU Commission No. 49 'The Interplanetary Plasma and the Heliosphere', and IAU Commission No. 50 'Protection of Existing and Potential Observatory Sites'

This proposal was approved by the General Assembly.

(*b*) The General Secretary presented, on behalf of the Executive Committee, the following list of Presidents and Vice-Presidents of IAU Commissions for election by the General Assembly, subject to their willingness to serve:

Commission No.	*President*	*Vice-President*
4	R. L. Duncombe	V. K. Abalakin
5	J.-C. Pecker	W. D. Heintz
6	P. Simon	E. Roemer
7	P. J. Message	V. Szebehely
8	G. van Herk	R. H. Tucker
9	A. B. Meinel	J. Ring
10	K. O. Kiepenheuer	G. A. Newkirk
12	R. G. Giovanelli	M. K. V. Bappu
14	R. H. Garstang	W. Lochte-Holtgreven
15	A. H. Delsemme	N. B. Richter
16	C. H. Mayer	T. C. Owen
17	S. K. Runcorn	E. Anders and K. P. Florenskij
19	C. Sugava	R. O. Vicente

20	L. Kresák	B. G. Marsden
21	J. L. Weinberg	R. Dumond
22	B. A. Lindblad	I. Halliday
24	P. Lacroute	C. A. Murray
25	M. Golay	A. Hiltner
26	S. L. Lippincott	P. Muller
27	M. W. Feast	W. Wenzel
28	E. Holmberg	B. E. Markarian
29	A. A. Boyarchuk	M. Hack
30	R. F. Griffin	A. H. Batten
31	H. Enslin	D. A. Orte
33	L. Perek	F. J. Kerr
34	H. van Woerden	G. B. Field
35	L. Mestel	B. Paczynski
36	R. Cayrel	D. Mihalas
37	I. R. King	S. van den Bergh
38	P. M. Routly	D. A. MacRae
40	Yu. N. Parijskij	H. van der Laan
41	O. Gingerich	J. Dobrzycki
42	T. Herczeg	G. Larsson-Leander
44	A. Code	R. M. Bonnet
45	C. O. R. Jaschek	E. K. Kharadze
46	D. McNally	E. V. Kononovich and D. G. Wentzel
47	M. S. Longair	I. D. Novikov
48	M. J. Rees	I. S. Shklovski
49	E. N. Parker	W. I. Axford and A. Hewish
50	M. F. Walker	R. Cayrel

This list was approved by the General Assembly.

(*c*) The General Secretary reminded that, according to Article 20(*b*) of the By-laws, the members of the Organizing Committees of Commissions are appointed by the Commissions themselves subject to the approval by the Executive Committee. He presented to the General Assembly the list of Organizing Committees as approved by the Executive Committee on the advice of the Nominating Committee. The composition of the Organizing Committees of Commissions are given in the list of Commissions on pages 225–240.

19. *Place and date of the XVIth General Assembly.* The President announced that the Union had been favoured by the invitation of the French National Committee of Astronomy to hold the XVIth General Assembly in Grenoble, France, in the second half of August 1976. He then invited Professor Ch. Fehrenbach to address the Assembly. Professor Fehrenbach spoke, first in French, and then in English as follows:

Monsieur le Président de l'Union Astronomique Internationale, Mes chers Collègues,

C'est pour moi à la fois un grand honneur et un grand plaisir d'inviter l'Union Astronomique Internationale à tenir sa prochaine Assemblée Générale en France.

La dernière Assemblée Générale qui ait eu lieu dans notre pays est déjà ancienne. Elle s'est tenue en 1935 à Paris, et c'est avec plaisir que nous accueillerons de nouveau cette assemblée; mais nous vous proposons que la réunion n'ait pas lieu cette fois-ci à Paris, mais dans une de nos belles villes de province: Grenoble.

Les différentes autorités locales, le Maire, le Préfet du Département, le Président de l'Université, ont mis à notre disposition tous les moyens dont ils disposent pour permettre une organisation à la fois efficace et agréable de ce Congrès et nous serons accueillis avec plaisir.

La Ville de Grenoble, située au confluent de l'Isère et du Drac, n'est pas une très grande ville.

Elle a environ 200000 habitants, mais elle est extrêmement intéressante tant par son passé historique que par sa situation géographique et son développement moderne.

Bien que ville de plaine, elle est en effet située au coeur des Alpes et est entourée de beaux massifs montagneux. Cette situation privilégiée explique que Grenoble soit un centre touristique très attirant. Les Jeux Olympiques d'Hiver y ont eu lieu en 1968 et, de ce fait, la ville dispose d'un équipement hôtelier et de salles de réunions importantes et modernes.

La Ville de Grenoble a un passé prestigieux. Elle était déjà une bourgade habitée avant l'arrivée des Romains qui l'étendirent et la nommèrent Gratianapolis, dont le nom fut transformé en celui de Grenoble. Elle a toujours été un grand centre de passage, mais aussi un centre intellectuel renommé. Après Vienne sur le Rhône, Grenoble devint la capitale du Dauphiné, une province de France qui avait une place spéciale dans le royaume: le fils aîné du roi portait le titre de Dauphin.

La première Université y a été fondée dès 1339. Cette Université s'est beaucoup développée au cours des 30 ou 40 dernières années et aujourd'hui elle est une des Universités les plus modernes de France. Ses bâtiments, ses amphithéâtres s'étendent dans la plaine de l'Isère sur une surface de 183 hectares. Ses derniers bâtiments n'ont guère plus de deux ou trois ans d'âge. Ses multiples ressources rendront votre séjour agréable à Grenoble, bien que le mois d'Août risque d'y être un peu chaud.

Il n'y a pas d'observatoire à Grenoble, mais l'Observatoire de Haute-Provence est situé à 200 kilomètres et ce sera pour le Centre National de la Recherche Scientifique un très grand plaisir de vous accueillir dans ce centre astronomique français que j'ai l'honneur de diriger.

Il est facile d'atteindre Grenoble: la liaison avec Paris par train express est rapide. Grenoble possède un terrain d'aviation et, à l'époque de notre Assemblée Générale, le grand aérodrome international de Satolas, desservant à la fois les villes de Lyon et de Grenoble, sera en service. Les astronomes pressés pourront donc atterrir directement à Lyon-Grenoble.

Naturellement, nous organiserons pendant le Congrès des excursions plus ou moins importantes dans les très beaux massifs montagneux des Alpes Françaises.

C'est donc officiellement, au nom des Autorités Françaises, que j'invite l'Union Astronomique Internationale à tenir sa Seizième Assemblée Générale en Août 1976 à Grenoble.

J'ai l'honneur de préciser que:

(1) la France accueillera avec plaisir tous les Membres de l'Union Astronomique Internationale et les Participants Invités à l'Assemblée Générale qui seront désireux de participer à cette manifestation;

(2) la France offre un certain nombre de bourses de voyage à de jeunes astronomes. Il ne m'est pas possible de vous indiquer la somme ainsi offerte par le Gouvernement Français, mais nous pensons qu'elle sera suffisante pour permettre une centaine d'invitations de jeunes astronomes, et je précise que ces invitations, comme il est d'usage dans l'Union Astronomique Internationale, ne seront pas faites par la France, mais par l'Union Astronomique Internationale elle-même, avec les crédits que notre pays aura mis à sa disposition. D'autre part, le Directeur Général des Affaires Culturelles, Scientifiques et Techniques nous a fait connaître que le Ministère des Affaires Etrangères envisage également d'inviter un certain nombre d'astronomes, Présidents de Commissions ou Membres du Comité Exécutif. Ces dernières invitations supplémentaires seront faites par notre pays. Je puis aussi vous dire que les autorités de la Ville de Grenoble et du Département de l'Isère ont déjà fait voter un budget important pour permettre l'accueil de notre Assemblée Générale à Grenoble.

'Mr President, Members of the Union,

Allow me to renew, in English, the very warm invitation of the French authorities to hold our General Assembly at Grenoble in 1976.

The Town of Grenoble, built in a plain amidst the Alps, is a major University Centre, and all kind of suitable halls and rooms for our meetings and for accommodation will be at our disposal.

Moreover, due to the organization of the winter Olympic Games in 1968, many hotels have been built in which all of accommodation is available.

Grenoble, small as a town, is an important touristic centre from which the medium altitude

Alps, such as Chartreuse, Vercors and Belledonne, and the high mountains, such as Pelvoux or Mont Blanc are easy to reach.

Grenoble is an important road and railway junction, has an airport, and is therefore easily connected with the rest of the world.

Astronomy is not taught at the Grenoble University which is mainly known for physics, electronics, magnetism and nuclear research, but the Observatoire de Haute Provence, only 200 km south, and other observatories can be reached within a few hours.

It is my great pleasure to invite officially, on behalf of the French authorities, the IAU to hold the XVIth General Assembly at Grenoble.'

The motion of the President to hold the XVIth General Assembly of the IAU in Grenoble, France, in August 1976, was carried unanimously. The President thanked Professor Fehrenbach for his words.

20. *Election of the new Executive Committee.* The Chairman formally proposed that Professor Leo Goldberg be elected the new President of the IAU. This was approved by acclamation.

The President then reminded the Assembly of the names put forward for consideration at the First session, and formally proposed that Mr. J. G. Bolton, Prof. Ch. Fehrenbach, and Prof. W. Iwanowska be elected Vice-Presidents in place of Dr M. K. V. Bappu, Prof. L. Gratton and Dr J. Sahade. This was unanimously approved. Next, the President formally proposed that Prof. G. Contopoulos be elected General Secretary in place of Prof. C. de Jager, and Prof. Edith A. Müller Assistant General Secretary in place of Prof. G. Contopoulos. This was also unanimously approved.

The Chairman then invited Professor Goldberg, Mr Bolton, Professor Fehrenbach, Professor Iwanowska and Professor Müller to take seats on the platform.

21. *Addresses by the retiring and newly elected Officers.* The retiring President, Professor B. Strömgren, addressed the Assembly as flollows:

'And now I have the opportunity to thank all those within the IAU with whom I have collaborated during three years. I wish to thank all members of the Executive Committee for their cooperation and for their friendship. It has been a privilege to be a member of this group that worked so effectively and harmoniously for our Union. Special thanks go to Vainu Bappu, Livio Gratton and Jorge Sahade who now leave the Executive Committee after six years of dedicated work.

During the three-year period the Executive Committee has profited greatly from having the advice of the former President, Otto Heckmann, and of the former General Secretary, Luboš Perek. I feel most grateful to these two close friends of mine and of the members of the Executive Committee.

Anyone who participates in an IAU General Assembly will realise how very great is the contribution of the IAU General Secretaries toward the organization of the Assembly. And during the three-year period between assemblies they perform a number of tasks of the greatest importance to the Union. Kees de Jager and George Contopoulos have worked with great skill and enormous energy and always in a spirit of dedication to the IAU and its goals. Kees and George, I am particularly grateful to you for your cooperation, and for your friendship.

Kees de Jager is leaving the Executive Committee after six years of splendid work for our Union. Kees, I admire the way you have conducted the Office of the General Secretary, while at the same time you know and follow the research in your Space Institute in every last detail. I am very happy that we shall still have opportunities to discuss IAU matters together for another three years as advisers.

I propose a vote of thanks to Kees de Jager and George Contopoulos for all they have done for the IAU.

In their work, the General Secretaries have been most ably supported by the Executive Secretary, Arnošt Jappel, whose contributions have been greatly appreciated by all of us who have collaborated with him. I wish to add special thanks for the kind and effective help I have received from Arnošt Jappel on many occasions.

In his work, the Executive Secretary has again been most effectively helped by the Administrative Assistant J. Daňková. The IAU is greatly indebted to her for her fine work.

On the occasion of the Inaugural Ceremony there was an opportunity to thank our hosts for all they had done for us in their preparations for the General Assembly, and by making us feel so welcome and at home.

Today, as we are approaching the end of our Assembly, I wish again to express our very best thanks to the host organizations, the Australian Academy of Science, the University of Sydney, and the Local Organizing Committee, in particular its Chairman, Harley Wood, and its Secretary, A. J. Higgs.

The efforts of the host organizations before and during the General Assembly have contributed so very much toward making this a successful Assembly, and the International Astronomical Union is greatly indebted to the host organizations for all they have done for us. Very many thanks.'

Then the new President, Professor L. Goldberg spoke as follows.

'Professor Strömgren, Colleagues, Ladies and Gentlemen,

As you know, in the United Kingdom candidates stand for office, but in the United States they run. I can assure you that when Professor Strömgren asked me whether I would accept nomination as candidate for the Presidency of the IAU, my natural instinct was indeed to run. I am both touched by the honor associated with this lofty position and greatly intimidated by the responsibility it puts upon me. But more than anything I realize how much I owe the Union for the professional activities and personal friendships that I have so much enjoyed, and how fortunate I am to have been given an opportunity to repay at least a small part of the debt during the next three years. I am deeply grateful to the General Assembly for both the honor and the opportunity.

Among the many colleagues who have rewarded me with their friendship none has contributed more to astronomy and been more respected by his colleagues than you Bengt Strömgren, and I feel greatly honored to share this platform with you today, and to express my appreciation. Your classical paper on the chemical composition of the solar atmosphere and on the physical properties of the interstellar medium have provided inspiration for much of my own work, and that of my students, and your contributions to the theory of ionization in stellar interiors and to the observational determination of stellar ages have also been of fundamental importance. Professor Strömgren has practised internationalism as a way of life by dividing his working career almost equally on the two sides of the Atlantic Ocean and by helping astronomy to flourish in different parts of the world. As Director of Yerkes Observatory, he was a member of the committee that planned the establishment of the Kitt Peak National Observatory in Arizona and now, from Copenhagen, he is a leading figure in the development of the European Southern Observatory in Chile. He served the IAU well as General Secretary during the politically precarious and anxious years of 1948–52 when a visit by a President of the United States to either Moscow or Peking would have seemed no more realistic than a journey to the Moon.

And now, as President of the IAU, Bengt Strömgren has had the unprecedented and unenviable task of overseeing the preparations of two General Assemblies in rapid succession in cities of opposite sides of the earth. I cannot tell you how reassuring it is for me to know that both he and his immensely capable General Secreatry, Professor de Jager, will continue to serve as advisers to the President and Executive Committee for three more years. In fact, when I think of the combined wisdom and experience that they represent together with George Contopoulos, Edith Müller and Arnošt Jappel, and the Executive Committee, I realize that it will require a major effort on my part merely to create the impression that I am their leader.

Although I cannot foretell the future of astronomy during the next three years, I can safely assert that the opportunities for astronomical research and for international cooperation have never been as exciting and as challenging as they are at present. New technologies have exposed the entire electromagnetic spectrum to the view of highly sophisticated instrumentation and as a result there is almost literally no part of the spectrum in which theory and observation do not confront one another.

The explosion of research activity in astronomy during the past ten years has greatly enlarged the opportunities for international collaboration in ways that could not have been foreseen by the founders of the IAU. As examples, I can mention joint projects for the construction and operation of ground-based telescopes in remote locations, for the instrumenting and launching of rockets and satellites, for observations by very long baseline interferometry, and for the solution of important astronomical problems by coordinated observations of a single object in widely separated regions of the spectrum. It is neither necessary nor useful for the IAU to be formally concerned with every such international program, but I hope that the structure of the Union will continue to be flexible enough to accommodate any that need to make use of its capabilities and services.

Finally, although the Executive Committee faces many important and difficult tasks before the Sixteenth General Assembly meets in Grenoble, none is more urgent than finding some way to effect the return of our colleagues from China to the IAU. On this hopeful note, I will conclude and look forward to seeing many of you in Warsaw in a few days, and all of you in France in 1976.'

22. *Closing Ceremonies.* The Chairman called upon Mr H. M. Smith to propose a comprehensive vote of thanks. Mr Smith then spoke as follows:

'It is an honour and a privilege to be called upon to express, on behalf of the participants, our thanks to all who have assisted in the organization and the smooth running of this Fifteenth General Assembly of the International Astronomical Union.

This is the first occasion that a General Assembly has been held in the Southern Hemisphere, and it is a well-deserved tribute to the many valuable contributions made to astronomy by our Australian colleagues that this year we should meet in Sydney.

For some, it has been the first opportunity of seeing the Harbour, described in 1788 by an Englishman, Captain Arthur Phillip as the 'finest harbour in the world'. We have also seen the famous Harbour Bridge, the exciting and controversial Opera House, the well-preserved historic houses with their 'Sydney lace' balustrades and the modern tower blocks of this vibrant, cosmopolitan capital of the State of New South Wales.

The Assembly itself has given us the opportunity of catching-up with the latest developments in our rapidly advancing branch of science; of discussing new instruments, new techniques, new programmes, the results of new theoretical studies, investigations that have led to new discoveries and of formulating new policies to deal with new problems. The invited discourses have again provided the highlights of the Assembly, while the more informal unscheduled talks have aroused lively discussion. The Commission meetings continue to be the essence of the Assembly, and I have the impression that the meetings have been animated, sometimes controversial, and generally constructive. Joint discussions have been arranged for topics that cross the boundaries of the individual Commissions.

Our thanks are due to the General Secretary, Prof. Dr C. de Jager for the immense amount of work that he has done, the Assistant Secretary, Prof. Dr G. Contopoulos – we wish him success in his endeavours to find a satisfactory solution to the difficult problem of the publication of the Reports and Transactions – the long-suffering Executive Secretary, our good friend, Dr A. Jappel and to Mrs J. Daňkova, Mlle. G. Drouin and Miss B. Reijnen for the very efficient organization of the administrative and scientific activities. Locally they have been assisted by Mr F. M. Atkinson and Mr J. Hardy. I am sure that you will agree that our IAU Secretariat have once again achieved the impossible of fitting too many meetings into too short a time, with the overriding limitation that no participant should be required to be in more than two places at the same time.

We tender our sincere thanks to our host organizations: the Australian Academy of Science and the University of Sydney. We are most appreciative of their warm and generous welcome expressed at the Inaugural Ceremony by the Foreign Secretary of the Academy, Dr J. P. Wild, the Acting Vice-Chancellor of the University, Professor W. M. O'Neil and the Right Honourable the Lord Mayor of Sydney, Alderman David Griffin. We also express our thanks to the Minister for Science, the Honourable W. L. Morrison for attending this ceremony and for speaking to us and to the

Government of New South Wales for giving the reception for National Representatives and Presidents and Vice-Presidents of Commissions.

The University, whose excellent facilities have been so generously placed at our disposal, has as its motto the words 'Sidere Mens Eadem Mutato'. It is notoriously difficult to agree on the best translation of such mottoes. In a discussion last night one suggestion that emerged was 'The searching mind midst the changing truth'. It is open to question whether truth can change, or if it is only our conception of truth that changes. But during our stay at the University the searching minds of the members of the IAU have continued to probe into the mysteries of the Universe and we pursue the goal of the ultimate truth.

The Local Organizing Committee has performed wonders under the leadership of their genial chairman, Dr Harley Wood of Sydney Observatory. When one attends a smoothly-running and efficiently organized conference, one may not be aware of all the detailed planning that precedes the meetings and the hard work that goes on behind the scenes. Brighton is still fresh in my mind, and I know something of what is involved. All will wish to join with me in congratulating Harley on a truly magnificent effort. He has had the expert and untiring assistance of Mr A. J. Higgs as organizing secretary, Professor W. N. Christiansen who among other things took care of our musical delights, Mr A. G. Little who saw to the arrangements at the Great Hall, Mr. A. Watkinson who worked on the entertainment programme and Mr P. J. Kelly who dealt with many details in the arrangements. Let me tell them that the social functions and sightseeing trips have been much enjoyed and have provided welcome opportunities for relaxation; the traditional musical interludes at the Inaugural Ceremony were thoroughly appreciated. Our thanks would be incomplete without a mention of the charming and helpful staff of the Local Organizing Committee, Miss Sally Atkinson and her helpers, particularly Mrs Sue Gerdes, Mrs Elizabeth Ashard and Mrs Valerie Walker, who have obviously taken a delight in assisting us.

May I add a final word of thanks to the many people too numerous to mention by name who have willingly and cheerfully played their part in helping us during the past ten days or so, to the staff of the Mathematics Department of the University, who have attended to us in the Carslaw Building and to the staff of the University hostels where many participants have been accommodated. I am also sure you would like me to congratulate and thank those responsible for our daily newspaper, 'Astronomy 73'.

As one who was elected to membership of this Union 35 years ago, I greatly appreciate, as I am sure that you do, the opportunities of enjoying discussions with colleagues of long-standing and of becoming acquainted with new friends. I feel that this has not only been an interesting and profitable Assembly, but a very happy and friendly occasion. Will you please join with me in thanking all those whose hard work has made it possible.

On behalf of the ladies, Mrs J. M. Greenberg addressed the Assembly as follows:

Mr President, Ladies and Gentlemen,

It gives me particular pleasure to serve as spokesman for the more than 100 women who have been so graciously and lavishly entertained as guests of the Australian Ladies Committee of this XVth General Assembly of the IAU. To each and every one of you who has worked so hard – from all of us we say a most sincere and deeply felt thank you.

From the very first moment you put us at ease and made each of us feel personally cared for. You anticipated so many of our needs in so many little thoughtful ways and then, of course, you provided us with entertainment of such choice, variety and quality that it staggers our imagination to think of the hours, days and months spent in organizing and implementing so vast an undertaking.

We are aware that no one person can do the job alone, but rather, it represents the collective effort of a dedicated group working closely and tirelessly together.

Yet each time we complimented the women of the committee on the magnificent job they have done, we would invariably be told that it would not have been possible without a chairman like Elspeth Christiansen. And indeed, it seems that this gentle, soft-spoken woman we know, would tolerate no compromise that threatened the success or quality of this beautifully conceived program. And it occurred to me, perhaps a bit whimsically, that Mrs Christiansen may well be described much like the iron lace so unique to Sydney. For all its grace and delicacy, it has an intrinsic and enduring strength.

So once again, we compliment you and thank you all for introducing us to this most exquisitely beautiful part of your country. You have made this a memorable visit for us and we truly hope that one day we may have the opportunity to reciprocate.

Thank you.

The Chairman then called upon the retiring General Secretary, Professor C. de Jager, who spoke as follows:

'Dear Mr President, dear colleagues,

It is now many years since the General Secretaries of the IAU used to hold office during a long period: if we only remember that Fowler, Stratton and Oort held their posts during the first thirty years of the Union's life, then it is clear that some things have changed, since.

It is only three years ago that I took over this office from my friend and predecessor Luboš Perek. These three years, during which I was allowed to serve the Astronomical Community, were for me years of happiness, and at their conclusion I feel thankful to you all. The manifold relations with so many distinguished colleagues from all over our globe were a continual source of inspiration. My task was never felt as a burden, and I only hope, most sincerely, that I have been serving you well. To co-operate with you, President Bengt Strömgren, was a great experience, and the happy hours we spent with you and with Mrs Sigrid Strömgren in your hospitable home are kept in my memory as a costly treasure. Likewise the contacts with the members of the Executive Committee were always pleasant – it is rare indeed, to have the opportunity to regularly meet a group of eminent scientists, so diverse on one hand, and at the same time so homogeneous in their passion to our science.

Mr President, I just referred to Stratton, who was General Secretary for a long period. It is known that Stratton wrote all his letters himself, in long handwriting. Nowadays this would have been impossible, and without the able and diligent assistance of the Union's secretariat my work would just have been impossible.

I have greatly appreciated the co-operation with you, Dr Arnost Jappel and Mrs Jarka Dankova, both humanly as well as for the skilful and devoted way you handled the affairs of the Union. I, and my coworkers at the Astronomical Institute at Utrecht, will miss you greatly, for your personal qualities, for the friendship that has grown between us. At the same time I wish to thank you, Miss Beth Reynen, not only for having done so much for the Union, next to your task as head of the secretariat of the Utrecht Institute, but also for having been the one who thoughtfully determined and controlled my daily program – strictly, but always in a most charming way. It was you who thus enabled me still to do some scientific research, in the midst of my many obligations, to the Institute, to the Union and to other international organizations.

Not the least I thank my wife, Doetie, for her continuous moral aid, her stimulating remarks when things did not go as they should perhaps, and for allowing me to sacrifice so much of our family's time to the well-being of the Union.

And now the time has come for me to pass the torch to you, George Contopoulos. It was a good time we spent together, a time of fine and cordial co-operation during which it was partly my pleasant duty to prepare you for your new task in the chain of Secretaries of the Union. At this place I whish you all good luck, and as much pleasure in your function as I had. Thank you for the excellent spirit you showed and for the important contributions you made to the Union when you held the office of Assistant General Secretary.

Mr President, in terminating I should not conceal that it adds to my satisfaction that precisely this, so successful General Assembly, takes place on a continent where I had and have since acquired so many good friends and colleagues, in a country, once carrying the name New Holland, and where so many of my countrymen found a second home. I hope that astronomy will continue to flourish in this country.'

The President then invited the newly elected General Secretary, Professor G. Contopoulos, to address the Assembly. Professor Contopoulos spoke as follows:

'I also want to thank all of you, Mr President, Mr General Secretary, Members of the Executive Committee, Members of the IAU, and local organizers, for your co-operation.

As the number of IAU Members increases more and more, I rely more and more on this co-operation for the success of the Union, which I want to serve not only as a Scientific Organization, but also as a big family.

During the General Assembly many new friendships were formed. I am sure that these friendships will continue and develop, and that we will see each other again at the next General Assembly. Thus I wish you a happy return home and 'Au revoir à Grenoble'?

The Chairman then closed the meeting around noon with the following words:

'I thank you for your attendance, and I wish you all a happy return home. I hereby formally declare the fiftienth General Assembly of the International Astronomical Union closed.'

PART 3

REPORTS OF MEETINGS OF COMMISSIONS

COMPTES RENDUS DES SÉANCES DES COMMISSIONS

COMMISSION 4: EPHEMERIDES (EPHÉMÉRIDES)

Report of Meetings, 21 and 29 August 1973

PRESIDENT: J. Kovalevsky.
SECRETARY: B. Morando.

At the opening of the first session the chairman regretted that two prominent members of the Commission, Mr Sadler and Dr Duncombe were unable to attend. Greetings from the members of the Commission were conveyed to both by the Chairman.

Dr Morando was appointed as Secretary for both sessions of Commission 4.

Membership

Next president will be the past vice-president Dr Duncombe, the new vice-president Dr V. K. Abalakin. The Organizing Committee will include: J. Kovalevsky, T. Lederle, B. Morando, D. Mulholland, A. M. Sinzi, G. Wilkins.

New members: There is one retiring member Dr Sato. He will be replaced by Dr S. Aoki. Dr F. Gondolatsch has retired from the Astronomisches Rechen-Institut but, as he is still active, he was asked not to retire from the Commission. Dr Williams from JPL, proposed as member of the IAU by the U.S. National Committee, has been made a member of the Commission.

J. H. Lieske, C. Osterwinter and H. G. Walter who were already consulting members and were proposed members of the IAU, were made members of the Commission.

It was decided that there will be no consulting member of the Commission for the next three years.

Report of the Commission

In order to comply with the wish of the General Secretary of the Union, the Chairman of Commission 4 was not allowed to increase the length of the report. So it was that the reports of the directors of the almanac offices could not be included in full. Dr Wilkins remarked that the directors should have been asked to write shorter reports that could have been included in the final report.

The report of the commission as published in Volume 15A was formally adopted. The Commission expressed the wish that the publication of the commission reports be continued, and it is wished that the length of the report of Commission 4 should not be further reduced.

Report of the Working Group on Numerical Data

This group was set up at the last General Assembly in Brighton. The field of ephemerides is covered by the International Information Bureau on Astronomical Ephemerides. The group has not yet met as a group but will do so at the time of the present G. A.

Report of the BIIEA

Dr Morando, director of the International Information Bureau on Astronomical Ephemerides, (BIIEA) gave a very short report on the activity of the Bureau since the last General Assembly of the IAU when it was created. A more complete report was given at a public meeting of the Scientific Advisory Council of BIIEA on August 22nd.

A resolution was adopted asking an increased financial support of the BIIEA by the IAU. It was later decided by the general assembly that this support will be 880 U.S. dollars for the three years to come.

Reports and Future of Working Groups

The reports on the work of these working groups were the main topic of the IAU Joint Discussion N°1 and will be published in the Proceedings of this discussion. All three working groups considered that they have not reached final decisions. The commission decided that the three working groups should continue their work for the next three years. 1976 is a deadline for decisions to be taken if it is wished that they come into effect not later than 1980.

J. Kovalevsky conveyed R. L. Duncombe's invitation, supported by Dr Strand, to hold a symposium on astronomical constants in Washington in 1974. This invitation was gratefully accepted, but it was suggested that the attendance be restricted to very few active members of the working groups and the heads of the almanac offices who will have to apply the decisions. It was also agreed that the chairman of a WG may invite new people to the group if it is agreeable to the presidents of the commissions involved.

Priorities for Future Research

A long general discussion was held on this problem and the following summary, drafted for the second session was formally adopted.

'The goal of Commission 4 is to provide ephemerides to the users, with the required precision. This is a rather technical matter and involves essentially research in methods, generally related to Celestial Mechanics.

However, Commission 4 has a good knowledge of the needs expressed by the users and can therefore make some recommendation on the most urgent research.

These are the following. (The order is not indicative of any relative priority.)

- To obtain and adopt the best possible self consistent reference set of astronomical constants (angular constants, masses, time reference).
- To obtain much better ephemerides of natural satellites, in the first place, the Galilean satellites of Jupiter and the 7 first satellites of Saturn, for some will soon be visited by space probes.
- To improve in a general way the celestial system of reference and to bind it as well as possible to the terrestrial system.
- To obtain new planetary theories with the new system of masses.

The researches implied to meet these objectives are self-evident and the need of subsequent observations is to be stressed'.

Dr Morando suggested that a resolution be passed on the desirability of having good observations of the natural satellites of the planets. At the second session, the resolution prepared jointly by a small committee of Commissions 20 and 4 and already adopted by Commission 20 was unanimously adopted.

It reads as follows:

'Noting that the ephemerides of most of the natural satellites are based on observations made long ago, and that recent efforts to improve these ephemerides have been greatly hampered by the extreme paucity of recent observations;

Acknowledging that much higher accuracy in the ephemerides of these objects will be required in the near future, not only for earth-based studies, but also for the future exploration of the outer solar system by spacecrafts.

Commissions 4 and 20 urge that a reasonable amount of telescope time be made available for observations useful for the improvement of such ephemerides'.

Commission 4 also unanimously supported Commission's 8 resolution on the improvement of FK4, the addition of faint stars to the catalogue towards the building of FK5. It reads as follows:

'It is recommended that an improvement of FK4 and its extension to a fainter magnitude limit, resulting in a new fundamental catalogue, the FK5, be carried out at the Astronomisches Rechen-Institut, Heidelberg; that observatories throughout the world contribute to this project by providing basic observations, on punched cards if possible; and that all information pertinent to the formation of the FK5 be transmitted to the Astronomisches Rechen-Institut with the observations.'

Physical Ephemeris of Mars

A proposed resolution on new parameters for the physical ephemeris of Mars has been set up. It was explained by Dr de Vaucouleurs. The period of rotation has been improved by Ashbrook in 1953, and Mariner 9 has given better knowledge of the planet. Definitive landmarks being available on the surface of the planet, time has come for a new set of parameters to be used after 1979. (Ref. G. de Vaucouleurs *et al.*, *J. Geophys. Res.* **78**, No. 20, 4395, July 1973.)

This resolution reads as follows:

'Considering that during the past three years new determinations of the elements of Mars have indicated the need for a revision of the elements currently adopted in the physical ephemeris of Mars and that a new definition of the origin of the areographic longitudes has been proposed for the Mariner 9 cartographic products (G. de Vaucouleurs, M. E. Davies and F. M. Sturms, Jr., *J. Geophys. Res.*, in press),

Commission 4 recommends:

(1) that the new elements and new definition of the origin of the areographic longitudes given below be incorporated in the physical ephemeris of Mars after 1979 or as soon as deemed practicable in the judgement of the cognizant Directors of the National Ephemeris Offices, but not before the tie between the new and current systems has been firmly established by appropriate comparisons between the ground-based and Mariner 9 coordinate systems.

The new elements are as follows:

1. Celestial coordinates of North Pole of Mars (referred to the 1950.0 equinox).

$$\alpha_0 = 317^\circ\!.32 - 0^\circ\!.1011\, T$$
$$\delta_0 = +\, 52^\circ\!.68 - 0^\circ\!.0570\, T,$$

where the second terms are the Mars precession rates and T is measured in Julian centuries from JD 2433282.5 (Jan 0 1950.0) The corresponding inclination of the Mars equator to the Mars orbit is

$$I = 25^\circ\!.19969 + 0^\circ\!.01219T + 0^\circ\!.00006\, T^2.$$

2. Angle measured along the Mars equator from the ascending node on the mean 1950.0 Earth equator to the Mars *autumnal* equinox

$$\Delta_{50} = 42^\circ\!.93538 - 0^\circ\!.09040\, T - 0^\circ\!.00010\, T^2.$$

3. Angle from the Mars *vernal* equinox to the Mars prime meridian.

$$V = 148^\circ\!.24 + 350^\circ\!.892017\ (\mathrm{JD} - 2433282.5),$$

where the constant term is consistent with the new definition of the origin of areographic longitudes by the center of a small crater (Airy–0) and the rotation rate with a new determination of the rotation period (de Vaucouleurs *et al.*, *loc. cit.*). V_0 may require a small zero point correction when the relation between coordinates in the current ephemeris and in the Mariner 9 system has been finally established (Davies and de Vaucouleurs, work in progress).

(2) that the present investigations be continued in order to establish precisely the tie between the new and the current systems, and

(3) that a recomputation of the physical ephemeris based on these new elements be prepared for the years 1800 to 1979.'

It was remarked that $V_0 = 148^\circ\!.24$ might be a provisional value and that the usual quantity φ_0 is equal to $\Delta_{50}+180°$. It would be useful to compute the ephemeris back to the first observation of landmarks by Cassini in 1672.

The resolution was unanimously accepted after some remarks from Dr O'Handley, Dr Wilkins and Dr Mulholland from which it appeared that the change to be introduced will be less than 1/10th of a degree in longitude (5 km). It appeared also that the radar resolution is at most 10 km.

Future Changes in the National Ephemerides

Dr Kovalevsky stated how things stand. There are three users:

(1) Connaissance des Temps, American Ephemeris etc...users. They seem to be satisfied with what is done for them now.

(2) People who want to know roughly where the bodies are (amateurs, observers that wish to choose among predictable events etc...).

(3) People that need ephemerides very much in advance. For these people data like the Advanced proofs of the AE, or special ephemerides for a given body could be useful.

In a long letter, Dr. Everhart, explained that he would prefer that no precise values of α, δ be published, they should be good enough to settle the telescope. Dr Morando asked whether 1′ would be agreeable. Dr de Vaucouleurs replied that, as night assistants in observatories have generally a primary education, they should not have to do any calculations. Dr de Vaucouleurs thinks that a precision of 1″ would be desirable, which means that the amount of data to be published by the almanac offices would still be very bulky.

On the suggestion that computer printout sent to users would be good enough and would save printing, it was remarked that it would be just as expensive if not more. Dr Kovalevsky, in his conclusion of the discussion feels that more efforts in consulting users should be deployed.

New Definition of TAI

In view of the conclusions reached at the joint meetings of Commission 4 and 31, it was suggested that the new President should collect the feelings of the members of the commission on the desirability of changing the definition of TAI (or stepping it in the future) by an amount of exactly 32 seconds.

COMMISSION 5: DOCUMENTATION (DOCUMENTATION)

Report of Meeting, 24 August 1973

PRESIDENT: J. B. Sykes.
SECRETARY: F. Henn.

1. The *President's report* was approved on the motion of Fricke.

2. *Reports on Astronomy.* After discussing a letter from the General Secretary, the Commission carried unanimously the following resolution:

Commission 5 reiterates its resolution passed at the XIV General Assembly (*Transactions* **XIVB**, p. 86) concerning the desirability of continuing to publish the *Reports on Astronomy* in one volume. Commission 5 wishes in particular for its own report to continue to be published by the IAU.

It was regarded as understood that there would continue to be limitations on the length of reports, and means of obtaining the report of each Commission separately, and that the reports should be critical rather than merely enumerative.

3. *Composition of the Commission.* The President reported the deaths of four members (Alter, Heinemann, Luplau Janssen, and Zinner) and the resignation of Belorizky, Chandrasekhar, Sticker, and Velghe. P. Lantos (Meudon) was proposed as a new member subject to membership of the IAU, and Mrs C. A. Bain (Dudley Observatory) as a new consulting member.

4. *Officers of the Commission, 1973–76.* Ogorodnikov having declined the succession to the Presidency, Pecker was elected President and Heintz Vice-President, with an Organizing Committee of five other members: Henn, Kleczek, Shcherbina-Samojlova, Sykes and Wilkins. The name of Maxwell was later added to the Organizing Committee (see report of second joint meeting, below).

5. *Working Groups.* The question of a Working Group on Nomenclature, to coordinate nomenclature e.g. for galaxies, double stars, and formations on the Moon, was discussed. It was agreed to ascertain the conclusions of other Commissions discussing nomenclature problems.

The Working Group on Numerical Data, established by the Union in 1970, was in need of attachment to a Commission. It was agreed that this should become a Working Group of Commission 5 (Chairman: G. A. Wilkins), with the name 'Working Group on the Documentation, Exchange and Retrieval of Numerical Data'.

The Working Group on Universal Decimal Classification 52 (Astronomy) was continued.

Another Working Group was later added (see report of second joint meeting, below).

6. *Working Rules.* The President introduced some draft working rules of Commission 5, prepared in accordance with the wish of the Executive Committee. It was agreed that these should be discussed by the Organizing Committee.

7. *ICSU Abstracting Board.* The President reported his recent work in the Board. The text of a leaflet on secondary (abstracting and indexing) services prepared by Wilkins for wide distribution to astronomers and others was presented and generally approved. Similar leaflets already exist for services in physics and biology.

8. *Classification.* Some progress was reported concerning a joint committee with FID on the revision of UDC 52, but none towards the establishing of a common classification for use by secondary services. It was agreed that work on unification of classification schemes and development of multilingual thesauri deserves priority.

9. *Astronomical Dictionary.* The President reported that the *Astronomical Dictionary* by Kleczek (1961) had been revised and greatly enlarged to 15000 terms (91 sections) and eight languages. The Commission carried unanimously the following resolution:

Commission 5 has learnt with pleasure of the approaching completion in typescript of a revised edition of the *Astronomical Dictionary* by J. Kleczek, containing a greatly increased number of terms over the whole field of astronomy and space science and extended to include eight languages. Commission 5 urges institutions in Czechoslovakia and elsewhere to assist in the publication and distribution of this valuable work.

10. *Ancient books.* The President reported the completion of the catalogue of astronomical books of the fifteenth and sixteenth centuries in European observatory libraries.

11. *List of non-commercial publications.* The President reported the forthcoming third edition of the Utrecht bibliography.

12. *IAU Symposium and Colloquium proceedings.* It was noted that the Union publishes a list indicating the availability of these, supplemented by notes in the *Information Bulletin.*

Joint Meeting of Commissions 5, 9, 38 and 46

The meeting took place on 27 August under the chairmanship of J. B. Sykes. Its purpose was to continue consideration of the construction and publication of a world list of observatories, astronomers, instruments and programmes.

The chairman reported his failure to get adequate support from National Committees for Astronomy for the scheme discussed in 1970 (*Transactions* **XIVB**, p. 89).

After discussion of the forms in which information was already available, it was agreed that Heintz should seek to collate materials for the United States and to establish regional correspondents for similar work regarding other countries.

Joint Meeting of Commissions 5, 36, 38 and 46

The meeting took place on 29 August under the chairmanship of J. B. Sykes.

After discussion of the need to improve uniformity of editorial policy in astronomical journals, and reference to the existence of associations of editors in various other disciplines and geographical areas, it was resolved to form a Working Group of Commission 5 (Chairman: A. Maxwell), with the name of 'Working Group on Editorial Policy' and up to ten members. The terms of reference of the Working Group are:

(1) To investigate the possibilities of collaboration between editors of astronomical journals in such matters as style recommendations, relationships with secondary (abstracting and indexing) journals, publishers, authors, etc.

(2) To advise on the revision of the IAU Style Book.

After discussion of the economics of publication, the competition between quality and cheapness of production, and available information about methods of inexpensive publication, it was resolved to recommend to the Executive Committee the initiation or continuation of investigations with non-profit organizations such as Annual Reviews Inc., considering in particular: (1) the feasibility of establishing an independent non-profit IAU publishing house, (2) savings obtainable by the use of soft covers, cheaper paper, etc., (3) the cheap separate availability of reprints of proceedings of Commissions, Joint Meetings, etc.

COMMISSION 6: ASTRONOMICAL TELEGRAMS (TÉLÉGRAMMES ASTRONOMIQUES)

Report of Meeting, 23 August 1973

PRESIDENT: P. Simon.
SECRETARY: B. G. Marsden.

On opening the meeting, the President invited the Secretary to make a few remarks concerning the proposal for the new President of the Commission. The Secretary reported that the Vice-President, J. Hers, had decided that it would be inappropriate for him to take the normal course and succeed to the presidency. Elaborating on this point, the Secretary referred to the long and fruitful association of Mr Hers and three former directors of the Republic (formerly Union) Observatory, Johannesburg, with Commission 6. The makeshift relegation of the Republic Observatory to the status of an obscure and understaffed 'outstation' of the South African Astronomical Observatory had effectively removed Mr Hers from the astronomical scene. It should be stressed that Mr Hers' decision is entirely his own. Although no longer connected with any astronomical institution, Mr Hers continues to assist the work of the Commission in South Africa, and the Secretary was happy to note that he would remain a member of the Commission.

It was therefore decided that P. Simon would serve another term as President; E. Roemer was proposed as incoming Vice-President. No new members were proposed, but F. L. Whipple announced his retirement from the Commission: all present expressed their great appreciation of his services.

The President requested and received the approval of the meeting for his earlier recommendation to the Finance Committee that the Central Telegram Bureau receive a subvention of $1600 for the triennium 1974–76. [The recommendation was accepted, the amount of the subvention being increased to $1980.]

The question of the future of the reports of Commission 6 was discussed. The Secretary pointed out that for the past three years *annual* reports of the Central Bureau had appeared in the IAU *Information Bulletin*. There thus seemed to be no strong points in favor of publication of triennial reports in the IAU *Transactions*.

The Secretary then briefly described the new billing system for the IAU *Circulars*. It had been at the Prague meeting that the suggestion to put the system on to a calendrical basis was made. The Secretary was glad to report that, although the suggestion had taken almost six years to put into practice, such a system had in fact been in operation since 1 May 1973. The new system seemed to be working extremely satisfactorily, and it included the feature that all regular subscribers would be billed at the rate of 20¢ per issue, irrespective of geographical location. Subscribers in North America could receive the *Circulars* at the special rate of only 15¢ per issue (and for a specific number of *Circulars*), provided that they did not require invoices or reminder cards and would keep track themselves of the number of *Circulars* remaining in their subscriptions.

The Secretary announced that, since the conventions for the telegraphic code appeared to have stabilized, it would be appropriate to prepare a new edition of the 'code book', and he hoped that it would be possible to incorporate this in IAU *Trans.* **XVB.**

The Secretary then referred to his remarks at the Brighton meeting concerning the inappropriateness of the *Circulars* for the publication of many of the items transmitted to the Central Bureau. The situation was now even more extreme. During the twelve months preceding the Brighton meeting some 100 *Circulars* had been published: the figure for the twelve months preceding the Sydney meeting was closer to 140. The Commission agreed that the situation had now become unmanageable, and in the hope that the rate of publication of the *Circulars* would again be decreasing in the future, the Commission adopted the resolution given below.

Resolution

Commission 6 calls attention to the fact that the purpose of the IAU *Circulars* is to report and to stimulate timely observations of astronomical phenomena. Therefore, the Commission instructs the Director of the Central Bureau for Astronomical Telegrams to exercise discretion in editing material ıeceived to conform with this purpose.

IAU WORKING GROUP ON NUMERICAL DATA

Report of Meetings

CHAIRMAN/SECRETARY: G. A. Wilkins.

1. *Introduction*

The occasion of the General Assembly at Sydney provided the first opportunity for meetings of the Working Group on Numerical Data that had been formed as a consequence of the decision taken at the General Assembly in Brighton in 1970. Unfortunately, however, only about half of the members and a small proportion of the consultants were able to attend, and the open meeting organised by the Group was held at the same time as meetings of Commissions with considerable interests in numerical data activities. Even so, the Group was able to make useful progress in discussions on the nature of, and procedures for, future activities and was able to gather much additional information about available data files, as well as to learn about the needs and problems of other astronomers.

This report consists of two parts. Firstly, there is a summary account of the Open Meeting, which was intended to provide an opportunity for participants to learn about the past activities of the Group and to discuss current and future activities. Secondly, there is a record of the separate meetings of the Group to discuss administrative arrangements and how to implement the suggestions made at the Open Meeting. The survey of astronomical data activities that was drafted before the Assembly is being amended and extended as a result of the meetings and is being published separately.

2. *Open Meeting*

An Open Meeting on 'Numerical Data' was held during the afternoon of Tuesday, 1973 August 28, and was attended by about 30 persons.

The Chairman gave a brief account of the formation and initial activities of the Working Group (see first report in *Trans. IAU* **XVA**, 757–60, 1973). A draft survey of astronomical data activities had been prepared and distributed to all those who were associated with the Group. This contained brief details of the data activities of each of the Commissions of the IAU and of some other international organizations, and a listing by country and place of all known centres having substantial interests in the compilation and/or distribution of astronomical data. He invited additions, corrections and suggestions for improving this survey and drew attention to some of the questions about documentation, organization and policy that the Group was endeavouring to answer. He hoped that the meeting would provide an opportunity for general discussion on these questions, as well as for the presentation of brief reports on current activities.

Dr. Jung gave an account of the formation and activities of the Stellar Data Centre at Strasbourg, which depended on the active cooperation of several other European observatories. The Centre distributed an Information Bulletin about its holdings of data and met requests for the supply of data on magnetic tape. A particular effort was being made to produce a unified, homogeneous catalogue containing identifying, astrometric, astrophysical and bibliographic data for each star; so far about 400000 stars had been uniquely identified, but the associated data were not yet complete.

A. Underhill presented a report by J. Meade of the Laboratory for Optical Astronomy at the Goddard Space Flight Centre; pertinent data from many star catalogues were being combined and use was being made of the Strasbourg catalogue of identifications.

W. J. Luyten stated that he would soon have data for a further 400000 stars ready for publication and sought suggestions on the form that the catalogue should take so that it would be appropriate for inclusion in data centres.

T. Lederle described the work of the Astronomisches Rechen-Institut at Heidelberg on the compilation of astrometric data; many older catalogues still needed transcription to cards and tape; there were still difficulties in exchanging magnetic tapes.

B. Morando stated that over 200 institutions were on the mailing list for the information cards of the International Information Bureau for Astronomical Ephemerides at the Bureau des Longitudes, Paris; the cards give details of the availability, on cards or tape, of solar system ephemerides and catalogues of precise star positions.

C. W. Allen stated that a third edition of his compilation of *Astrophysical Quantities* would be published shortly, but he had now retired and was concerned to make arrangements for the continuation of the project – he hoped that one or two individuals, or an institution, would be willing to take over the responsibility. It was essential that the literature be continually monitored and work in preparation for the next edition should start immediately.

B. Hauck drew attention to some photometric and spectroscopic catalogues that have been issued recently, or are in preparation. The role of data centres had been discussed at the meeting of Commission 45; it is considered that a data centre should be active in the compilation of data and that journals should be prepared to publish details of the data available on magnetic tape but not printed in the journals. *R. H. Garstang* indicated that the main source of spectroscopic data was still the National Bureau of Standards (Washington, D.C.), but the Joint Institute for Laboratory Astrophysics (at Boulder, Colorado) had a data centre on collision processes and provided updated bibliographies every few years. He drew attention to the new *Journal of Physical and Chemical Reference Data*. The Chairman commented that a list of such journals, or other published sources of information about data activities, could be included in the Survey being prepared by the Group.

A. H. Batten stated that he is monitoring the literature in readiness for the publication of the seventh edition of the Lick Catalogue of orbital elements of spectroscopic binaries. Commission 42 considered that it is important that the catalogue should be published as a book, as well as being available on magnetic tape.

Y. Terzian stated that K. I. Kellerman had prepared a report on radioastronomical data and copies are available on request. R. H. Dixon maintains an updated version of his catalogue of radio sources and is prepared to supply copies on magnetic tape. A catalogue of pulsar data is maintained at Cornell University and computer listings are distributed at irregular intervals to about 100 active research workers; schematic representations of the data in graphical form are also distributed.

H. Tannenbaum described (through the Chairman) the availability of X-ray data from the Uhuru satellite; data are made available directly to Guest Investigators whose proposals for the analysis of the data have been approved by NASA but, in addition, accumulated data are available from the NASA National Data Centre under standard procedures.

3. *Group Meetings*

The Group met at 7 p.m. on August 23 and at 6.45 p.m. on August 28. The first meeting was primarily concerned with the arrangements for the open meeting and with the discussion on the suggestion that the Group should be associated with Commission 5 (Documentation). It had become apparent that the Group did not fit well into the structure of the IAU, but it did not appear to be appropriate for the Group to seek recognition as a separate Commission; the suggestion was adopted, subject to the agreement of Commission 5 and of the IAU Executive Committee to the appropriate extensions of the scope and membership of that Commission.

It was agreed that the survey of astronomical data activity should be published separately, and not in the Transactions of the Union; the idea that it be issued in an IAU Information Bulletin found general favour. Suggestions for amendments and additions were discussed, and the Chairman agreed to try to produce a revised copy by the end of October. The proposal that details of individual data files be made available on information cards, such as those of the Bureau on Astronomical Ephemerides, was discussed: Morando said that he did not have the resources to issue the large number of cards that would be involved, but he could issue some guide-line cards.

The Group felt that there was, as yet, no evidence of a need for a comprehensive astronomical data centre. There should be specialist data analysis centres whose staff are actively engaged on the use of the data, but there did not appear to be any need at present for international funding of such centres for astrophysical data.

It was agreed that it is desirable that the Group should meet between General Assemblies and it was provisionally agreed that meetings should be held in 1974 (perhaps at Herstmonceux) and in 1976 (perhaps at Strasbourg, just before the Assembly at Grenoble). Some suggestions for changes in the members and consultants of the Group were discussed; it was agreed that the Chairman should ask the new Presidents of Commissions to suggest possible consultants for their fields.

COMMISSION 7: CELESTIAL MECHANICS (MÉCANIQUE CÉLESTE)

Report of Meetings, 21 and 24 August 1973

PRESIDENT: G. N. Duboshin.
SECRETARY: S. Ferraz-Mello.

First Meeting

CHAIRMAN: V. Szebehely.

The meeting was scheduled as Administrative Session. The Commission discussed the proposal of creating a new IAU Commission on Astrodynamics. Since only 5 members of the Commission were present, no decision was taken and the Administrative Session adjourned.

Second Meeting

Scientific Reports

CHAIRMAN: G. N. Duboshin.

Speakers: G. N. Duboshin, 'The Lagrangian Solutions of the Three-Rigid-Bodies Problem'.

V. Szebehely, 'Recent Results in the Stellar Dynamical 3-Body Problem'.

S. Ferraz-Mello, 'Comparison Between Numerical Integrations and Analytical Theories for the Earth Shadowing Effects on Artificial Satellites'.

Third Meeting

Administrative Session

CHAIRMAN: G. N. Duboshin.

The Commission appointed the Organizing Committee for the next three years: Message (President), Szebehely (Vice-President), Aksenov, Duboshin, Brumberg, Davis, Deprit, Giacaglia, Hori, Kozai, Morando and Sehnal.

The Commission decided to create a Working Group on the Motion of Artificial Celestial Bodies, to be organized by Y. Kozai and under his chairmanship, instead of proposing a new IAU Commission on Astrodynamics.

The Commission elected Bec, Candy, Counselman, Demin, Everhart, Grebenikov, Jupp, Møller, Nahon, O'Handley, Sagnier, Shteins, Soultanov, Standish, Tapley and Tatevyan as new members of the Commission.

COMMISSION 8: POSITIONAL ASTRONOMY (ASTRONOMIE DE POSITION)

Report of Meetings, 22 and 23 August 1973

PRESIDENT: W. Fricke.
SECRETARY: W. Gliese.

I. REPORT OF THE COMMISSION

The President's Report of the Commission was discussed and approved without modification. In answer to an inquiry of the IAU General Secretary the Commission decided to request the continuation of the publication of the Reports of the Commission in the Transactions of the IAU.

II. OFFICERS OF THE COMMISSION

The Commission approved the following proposals for new officers: President: van Herk, Vice-President: Tucker, Organizing Committee: Anguita, Fricke, Klock, Lacroute, Nemiro, Schombert, Tavastsherna, Wood, Yasuda.

III. REVIEWS ON CURRENT PROBLEMS

The following reviews were presented:

(1) B. L. Klock: New Instrumentation and Techniques in Meridian Astronomy. This report was based on information provided by various observatories; it included recent progress made at Belgrade, Besançon, Bordeaux, Bucharest, Cagliari, Cape and Herstmonceux, Leiden, Mizusawa, Sao Paulo, Tokyo, and Washington.

(2) W. Fricke: Plans for the Improvement and Extension of the FK4. This paper will be published in the Proceedings of IAU Symposium No. 61, New Problems in Astrometry. The discussion in the Commission concentrated on the choice of objects to be included in the FK5. The new fundamental catalogue shall be extended to stars fainter than magnitude 7.5, it shall contain stars of all spectral types, and objects with radio emission. A supplement should also contain extragalactic radio sources in addition to stars. Contributors to the discussion were: Blaauw, Eichhorn-von Wurmb, Klock, Murray, Tucker, van Herk, and Walter. Fricke pointed out that the main criterion for the choice of fundamental stars will be the accuracy of available fundamental and differential observations to ensure very accurate positions and proper motions in a well defined system.

(3) Due to absence of M. S. Zverev who was invited to speak on 'Observations Relevant to the Improvement of the FK4 in the Southern Sky', W. Gliese summarized the information presented on this subject at IAU Symposium No. 61 'New Problems in Astrometry.'

(4) R. H. Tucker: Cape Astrolabe Catalogue Corrections. This is a report on results of observations carried out by D. V. Thomas from about 1965 to 1969 at the Cape.

(5) S. Débarbat presented a report prepared jointly with Davila, Grudler, and Sanchez on astrolabe observations of planets.

(6) G. Teleki reported on the activity of the Commission 8 – Study Group on Astronomical Refraction. The report was adopted by the Commission.

(7) Due to absence of J. L. Schombert, Chairman of the SRS Committee, the report on the SRS Programme was read by B. L. Klock. The status of the SRS Programme – 1 June 1973 is the following

Observatory	Zone	Commitment	Date started	Completed Obs'ns	Completed Red'ns	Final results received or expected
Abbadia	+ 5° to − 15°	1560 × 4	62.3	100%	100%	1968.8
Bordeaux	+ 5 to − 15	1560 × 4	62.5	100	100	1970.5
Bucharest	+ 5 to − 10	1176 × 4	62.6	100	97	1975.0
Nicolaiev	0 to − 20	5984 × 2	64.3	100	98	1974.0
San Fernando	− 10 to − 30	3709 × 4	63.3	100	100	1973.5
Tokyo	− 10 to − 30	3560 × 4	63.3	100	100	1972.7
USNO 6-in.	+ 5 to − 30	8706 × 2 1233 × 4	66.5	100	100	1973.3
Bergedorf (Bickley)	+ 5 to − 90	20495 × 4	67.9	100	95	1974.5
Cape	− 30 to − 40 − 40 to − 52 − 52 to − 64 − 64 to − 90	10082 × 4	61.3	100 100 80 —	97 100 75 —	1974.3 1966.0 1974.5 —
San Juan	− 40 to − 90	7190 × 2	69.5	99.4	100	1974.3
Santiago- Pulkovo	− 25 to − 47 − 47 to − 90	5992 × 4 5504 × 4	63.1	100 100	60 75	1974.0 1975.0
USNO 7-in. (El Leoncito)	+ 5 to − 20 − 20 to − 75 − 75 to − 75SP	7683 × 2 12121 × 4 1382 × 4	68.7	100	71	1975.0

(8) H. Yasuda: The PZT Star Programme, The status of the Northern PZT Programme – 1 May 1973 is the following

Observatory	Date started	Completed Obs'ns	Completed Red'ns
Abbadia	July 1971	30%	15%
Beograd	– –	0	0
Besançon	– –	–	–
Bucharest	Oct. 1971	46	0
Bordeaux	Jan. 1972	50	33
Copenhagen	April 1971	75	38
Engelhardt	April 1971	68	–
Herstmonceux	April 1972	52	0
Moscow	1967	100	75
San Fernando	Sep. 1973	0	0
Tokyo	July 1971	63	50
Uccle	– –	–	–
Washington	– –	0	0

IV. PRIORITIES

The 'Priorities' as suggested in the President's Report of the Commission were discussed and unanimously adopted.

V. RESOLUTIONS

The Commission has discussed the Resolutions submitted by IAU Colloquium No. 20 on Meridian Astronomy, held in Copenhagen, September 25–28, 1972. The President proposed a revision of some of these resolutions after consultation with members of the Commission from the Pulkovo Observatory and the U.S. Naval Observatory. Furthermore, members of the Commission proposed additional resolutions. As the result of the discussion the following resolutions were unanimously adopted:

(1) The recommendation made by Commission 8 at the XIV General Assembly concerning the development of new instrumentation and techniques is re-affirmed. In particular, the provision of a modern transit circle for the Tokyo Astronomical Observatory is strongly recommended. It is further recommended that a Study Group be established for horizontal meridian circles.

(2) It is recommended that an improvement of the FK4 and its extension to a fainter magnitude limit, resulting in a new fundamental catalogue, the FK5, be carried out at the Astronomisches Rechen-Institut, Heidelberg; that observatories throughout the world contribute to this project by providing basic observations, on punched cards if possible; and that all information pertinent to the formation of the FK5 be transmitted to the Astronomisches Rechen-Institut with the observations.

(3) It is recommended that meridian catalogues should include, in addition to the tabular magnitude on a recognised system, an indication of the screen or screens used for each star, so that the approximate magnitude at which the star appeared to the observer can be deduced.

(4) It is recommended that programmes of observations of fundamental stars, with meridian circles and astrolabes, should include members of the solar system.

(5) The recommendation made by Commission 8 at the XI General Assembly concerning the formation of catalogues of the Southern Reference Stars (in different zones) is amended in the sense that the individual observations of the whole southern sky be made available to the US Naval Observatory and that this observatory should send all basic data to the Pulkovo Observatory and, on request, to other institutions. The discussion of all observations and the compiling of the SRS catalogue for the whole southern sky may be made at the Pulkovo Observatory and at the US Naval Observatory. Concerning the best possible system of the SRS catalogue, consultations shall take place between the Astronomisches Rechen-Institut at Heidelberg, the Pulkovo Observatory and the US Naval Observatory.

(6) The recommendations made by Commission 8 at the XIV General Assembly concerning the inclusion of a number of faint stars up to the 9th magnitude (especially of FKSZ stars) into the fundamental catalogue is re-affirmed. The inclusion of FKSZ stars in fundamental observational programmes is recommended. Of special importance are absolute and differential meridian observations of the FKSZ stars together with FK4 stars in the southern sky (declinations from $+30°$ to $-90°$).

(7) It is recommended that the work of the Study Group on Astronomical Refraction that has been formed by Commission 8, should continue to be supported.

(8) Commission 8 re-affirms recommendations of the previous General Assemblies of the IAU on the following urgent programmes of meridian observations: 'Bright Stars' (BS), 'Double Stars' (DS), 'Latitude Stars' (LS) and 'Zodiacal Stars' (ZS). The Commission recommends also the new meridian programme: 'Reference Stars in the Areas with Galaxies of the Pulkovo Programme' (PS). The star lists for these programmes may be obtained from various observatories, as follows:

(BS), (DS), US Naval Observatory, Washington;

(LS), Sternberg Astronomical Institute in Moscow;

(ZS), Nikolajev Observatory, or, for a shorter list, Royal Greenwich Observatory, Herstmonceux.

(PS), Pulkovo Observatory.

Participation of the sourthern hemisphere observatories in these programmes is especially important.

(9) Commission 8 recommends the establishment of a Working Group charged with the following duties:

(i) To coordinate the various astrolabe programmes specially designed for the revision of the positions of the bright fundamental stars; and

(ii) To study the possibility of producing a homogeneous catalogue as a synthesis of existing astrolable catalogues, to be updated each time a fresh catalogue is published.

(10) Commission 8 takes the view that the provision of a stellar reference system with one star per square degree, reaching magnitude 9.5, with positions known to 0.01 arc second, would be a

major goal to reach in the future. The existence of such a reference system would have a spectacular influence on most of the astrometric programmes. Consequently the Commission recommends that detailed studies be made of all space and other astrometry projects that have the potentialities to attain this goal.

(11) The elaboration of a new international refraction table is needed. Therefore, Commission 8 concludes that the Soviet astronomers' proposals for revision of the Pulkovo Refraction Table and for active cooperation with the Commission 8 Study Group on Astronomical Refraction in this connection are very welcome, and should be accepted.

Resolutions of the IAU Symposium No. 61 'New Problems in Astrometry' endorsed by Commissions 8, 24, 33 and 40

1. This Symposium recognises the inadequacy of the existing optical astrometric data to meet modern requirements for precision positions and proper motions. It therefore urges that the great accuracy of radio positions and the potentiality of new optical and space techniques should be fully exploited. Both absolute and differential observations are necessary, and it is particularly important that observations in the southern hemisphere should not be allowed to lag behind the north as they do at present in both optical and radio astrometry.

2. This Symposium urges as the first priority the compilation by 1980 of an FK5 catalogue. The new reference frame must be extended to faint stars and extragalactic objects. This requires the vigorous continuation and, in fact, completion, especially in the Southern Hemisphere, of current programmes of photographic astrometry and, in addition, the development and exploitation of new photographic, radio and space techniques.

3. This Symposium notes the existence of the Working Party established by Commission 40 to draw up a list of suitable objects for positional calibration of radio observations and urges co-operation between it and Commissions 8 and 24 to ensure as complete an integration as possible of the radio and optical astrometric fundamental programmes.

Priorities

(1) The continuation of observations of positions of stars and members of the solar system with transit circles, meridian instruments, and astrolabes is of vital importance for progress in astronomical research, space research, geodesy, and geophysics.

(2) Observations relevant to (a) the improvement of the fundamental system of positions and proper motions, (b) the establishment of a reference system to the faintest possible magnitude limit, and (c) the derivation of proper motions in general and of special objects are of immediate importance. All investigations that contribute to these goals have to be encouraged.

(3) Developments of new instrumentation and techniques have to be encouraged which endeavor to improve the measurements of positions of stars and of members of the solar system.

COMMISSION 9: INSTRUMENTS AND TECHNIQUES (INSTRUMENTS ET TECHNIQUES)

Reports of Meetings, 21, 22 and 27 August 1973

PRESIDENT: V. B. Nikonov.
SECRETARY: H. C. Minnett.

Administrative Meetings, 21 and 27 August 1973

V. B. Nikonov announced the death of M. G. P. de Barros, I. S. Bowen and Y. Väisälä.

The proposed officers of the Commission were approved unanimously as follows: President: A. B. Meinel, Vice-President: J. Ring, Organizing Committee: B. Valniček, K. L. Hallam, I. M. Kopylov, W. C. Livingston, N. N. Mihelson, V. B. Nikonov, E. H. Richardson, M. F. Walker, G. Wlérick.

The following decisions were adopted.

The Working Group on 'Photoelectric Image Devices' should be organized as follows:

Chairman of Group: M. Walker; Members: Ables, A. Bijaoui, K. Ford, P. V. Shcheglov, G. Walker, E. Wampler, G. Wlérick; A Committee of Consultants: W. A. Baum, G. Kron, J. D. McGee, A. Lallemand, V. B. Nikonov.

To continue the work of the Working Group on 'Seeing and Site Testing' (Chairman: H. Brück; Members: H. Elsässer, K. O. Kiepenheuer, F. Roddier, J. Rösch, F. Saïssac, P. V. Shcheglov, J. Stock, M. Walker). The scope of its activities will be enlarged to cover all aspects of the problem, including: site testing, theory and practice of seeing and the application of interference methods. It was also resolved that the Executive Committee be informed that a Colloquium on Seeing and Site Testing would be desirable in 1975.

To continue the work of the following Working Groups established at the Brighton meeting of the Commission in 1970.

Working Group on 'Large Telescopes' (Chairman: J. Rösch; Members: I. M. Kopylov, A. B. Meinel, E. H. Richardson, G. M. Sisson, F. L. Whipple).

Working Group on 'Automation and Data Acquisition' (Chairman: L. Robinson; Members: E. Dennison, B. E. Lasker, N. N. Mihelson).

To establish a new Working Group on 'Infra-Red Techniques' (Chairman: J. Ring; Members: F. Low, L. S. Luud, P. Connes, R. V. Karandikar).

To enlarge the scope of the Working Group on 'Photographic Materials' to include all aspects of astronomical photography and to change its name to 'Working Group on Photographic Problems'. It was recommended that the Chairman of this Working Group should be W. Miller.

The Joint Working Group of Commissions 9 and 46 on the 'Exchange of Equipment' should be dissolved. The work carried out during the last term has shown that one or two people, acting as a centre of information, would now be adequate to perform its function. This work could be transferred to UNESCO or to a small centre in IAU.

Commission 9 supports the establishment of a new Commission on the 'Protection of Existing and Potential Observatory Sites', in the form of a Commission of the Executive Committee. M. Walker is proposed as the representative of Commission 9.

Commission 9 decided that publication of its Reports would be in the form of:

(a) A short report by the President indicating the main trends in astronomical techniques over the last few years.

(b) A longer report for circulation to all members of Commission 9.

Scientific Meeting, 27 August 1973

At the Scientific Meeting of the Commission devoted to the Problem of Large Telescopes the following papers were given:

A. B. Meinel: Introductory Talk.

A. B. Meinel and F. Whipple: Multiple-Mirror Telescope Project.

G. Odgers: Array Telescopes.

J. Rösch: Collecteurs de lumière composites.

Priorities

A very strong photon hunger is encountered in most cases when making detailed studies of galaxies, their nuclei, quasistellar objects, X-ray sources, pulsars and many other objects of great value to the understanding of the fundamental properties of the Universe.

To overcome this photon hunger a new generation of giant telescopes should be created and new types of observational equipment having maximal efficiency in recording single photons should be developed and put in operation.

These are the problems of highest priority in modern astronomical technique.

Construction of telescopes larger than 6–7 metres in diameter of classical design encounters practically insurmountable difficulties of both a technical and financial nature and new, realistic and economic ways of solving this problem are urgently needed.

The work on the new generation of large telescopes and, in particular, on the Multiple Mirror and Array Telescope projects is of the greatest value for the future of astronomy.

At the same time all aspects of the problem of how to reach the high potential resolving power of large telescopes, restricted in practice by the seeing, should be carefully considered. This should include not only the search for perfect places for observatories and the correct design of domes and telescopes, but also the use of such special interference methods as speckle interferometry.

A large telescope (diameter 3–4 metres) operating in space should also be created. It could make possible major breakthroughs in several fundamental areas not attainable with groundbased instruments. This is especially true in the photometry and spectroscopy of faint objects in the ultraviolet and infrared and in very high resolution imagery and astrometry.

In spite of remarkable success in developing very fast photographic emulsions in the last few years, photoelectric devices still possess the highest quantum efficiency. A great advantage of most of these devices is the possibility of applying photon counting methods, permitting digitization of the measurements.

New types of photoelectric devices developed recently, such as solid state photodiodes, digicons and microchannel, multianode photomultipliers, permit digitization of spectral observations and the observation of surface objects. Further perfection of photoelectric devices and the introduction of this modern technique into current astrophysical research is, as just mentioned, one of the most pressing problems in the near future.

However, for the studies of finest details of faintest galaxies and galactic nebulae, as well as for spectrographic work of high resolution and the accurate photometric observations of large numbers of very faint stars, electronic cameras are by far the best when compared either with common photography or with other image tube systems. The electronic cameras with large photocathodes developed in France offer new possibilities in galactic research by realizing precise and effective stellar photometry of the distant parts of the galaxy, especially if modern multicolour photometric systems are used. In this connection, further developments in operating large field electronic cameras and special instrumentation for automated measurements of electronographic stellar images are very much needed.

The growing role of infra-red research in modern astronomy requires further work on infra-red astronomical techniques, including further applications of Fourier spectroscopy for high resolution studies of spectra in the infra-red region.

One very serious problem must also be noted. It is immediately connected with the effective use of the astronomical techniques. This is the protection of present and future groundbased observing sites from light, air and radio pollution and from atmospheric disturbances produced by jet aviation. At the present time, this problem is becoming critical for some of the largest observatories in the world.

GROUPE DE TRAVAIL SUR L'EMPLOI EN ASTRONOMIE DES RÉCEPTEURS PHOTOÉLECTRIQUES D'IMAGES

Compte Rendu de la Séance, 22 Août 1973

PRÉSIDENT: G. Wlérick.
SECRÉTAIRE: F. Roddier.

1. *Résultats scientifiques*

Deux séances ont été consacrées à ces récepteurs le Mercredi 22 Août. Les quatre exposés de synthèse suivants ont été présentés:

H. Mazursky: The Mariner 9 Television system. Instrumentation and results.

G. Wlérick: L'emploi en Astronomie des récepteurs électronographiques.

L. B. Robinson et E. J. Wampler: Image Photon Counting for spectroscopy in Astronomy (présenté par Robinson).

J. L. Lowrance: Astronomical Television Development at Princeton University Observatory. Rickard, de l'ESO, a fait ensuite une courte communication pour montrer comment les récepteurs photoélectriques d'images sont utilisés dans un Observatoire de création récente.

Enfin J. D. McGee a tiré les conclusions de l'ensemble des exposés; il a souligné la nécessité de comparer de façon aussi quantitative que possible les performances des divers récepteurs pour l'étude d'un même objet; il a vivement insisté pour que les développements dans les observatoires se poursuivent au même titre que les développements industriels; il a remarqué que l'électronographie permet d'attaquer des problèmes de plus en plus difficiles et que l'utilisation de récepteurs linéaires en spectroscopie ouvre des possibilités nouvelles pour l'étude des astres faibles. Finalement il a prédit un bel avenir à ces récepteurs.

Tous ces exposés ont été suivis par une nombreuse assistance et largement discutés.

2. *Organisation*

Wlérick a indiqué qu'il a dirigé le noyau administratif du groupe pendant six ans (il avait succédé à Baum et McGee) et il a fait part de son souhait d'être remplacé dans cette tâche. L'organisation suivante à été adoptée:

Le 'Comité des Sages' est maintenu; ses membres sont: Baum, Kron, Lallemand, McGee, Nikonov.

Le Comité administratif comprend:

Président: M. Walker

Membres: Ables, Bijaoui, Carruthers, Ford, P. V. Shcheglov, G. Walker, Wampler, Wlérick.

C'est un devoir agréable de noter pour terminer, que la Commission 9 unanime a regretté l'absence de A. Lallemand et exprimé le souhait qu'il continue à la faire bénéficier de son concours.

COMMISSION 10: SOLAR ACTIVITY (ACTIVITÉ SOLAIRE)

Reports of Meetings

PRESIDENT: J. T. Jefferies.

Scientific Sessions

Three scientific sessions were held by Commission 10 during the IAU XV General Assembly. On 24 August, a session jointly with Commissions 12, 40 and 44 heard four reviews on 'Recent Advances in Solar Astronomy'. The topics of space observations, radio observations, groundbased optical, and a summary of future work were covered by L. W. Acton, D. J. McLean, A. Bruzek and G. Newkirk, respectively.

On 27 August, a joint session with Commissions 12, 29, and 36 considered 'Observational and Theoretical Objectives for the Next Decade in Solar and Stellar Atmospheres'. Stellar observations and theory were reviewed by L. Kuhi and P. Delache while J. M. Beckers and M. Kuperus (in a paper presented by J. Rosenberg) reviewed the observational and theoretical objectives of solar research.

Commission 10 organized a well-attended open session on 28 August at which Leo Goldberg presented exciting early results from the solar experiments on the SKYLAB manned space observatory.

Administrative Session

On 21 August, the Commission membership elected the officers and new Organizing Committee members proposed by the 1973 Organizing Committee. Their names are given below:

President: K. O. Kiepenheuer.

Vice President: G. A. Newkirk.

Members: L. W. Acton, V. Bumba, L. D. de Feiter, V. A. Krat, F. Moriyama, Paul Simon, J. T. Jefferies (Past President).

At the generous invitation of V. Bumba, it was decided to hold a Symposium on Solar and Stellar Activity in Prague, Czechoslovakia in August, 1975. The Organizing Committee of Commission 10 is to appoint a program committee charged to formulate a program sharply focused on the basic problems in solar activity.

The Working Group on Flare Sprays was reformed into a three-man Working Group on major prominence events, whose members are E. Tandberg-Hanssen, J. Parkinson and R. Stewart. This group is to evaluate the need for such a study, formulate the scientific objectives, and recommend the course of action. They are to report by mail to the membership of Commission 10, through its President, within one year.

The Commission requested the incoming Organizing Committee to establish a small group to recommend a standard nomenclature for solar features and to prepare an illustrated text.

It was resolved that the Organizing Committee should appoint a Working Group to consider the desirability of, and to formulate the scientific objectives of an 'International Study of the Active Sun' to cover the period of the next solar maximum. The Working Group is to define requirements for the success of such a study and is to report conclusions to the Commission 10 membership within one year.

Because of inadequate information available to the meeting, at the time, Commission 10 did not feel that action could be taken toward approval of a resolution put forward by the Committee of European Solar Radio Astronomers (CESRA) urging implementation of a Joint Interferometer Project.

The meeting heard reports from the Working Groups as outlines below.

Reports of Working Groups

STANDARDIZATION OF MAGNETIC FIELD MEASUREMENTS

Beckers presented the report of the Working Group on Standardization of Magnetic Field Measurements; the results are to be published in *Solar Physics.* The report of this Working Group was accepted with thanks. Disbandment of the Group was approved.

SOLAR ACTIVITY COOPERATIVE PROGRAMS (FORMER SUBCOMMISSION 10A)

International Survey Programs

P. Simon reported that he had received 25 replies from the questionnaire which had been distributed concerning the special problems of solar flare reporting. These problems covered patrol sites, apparatus, reports and final evaluation, a flare index, and high-resolutions data. Each of these was discussed in turn. It was decided to organize certain small groups to continue the study of the problems related to the Flare Index, Patrol Sites and Apparatus, and the Comparison and Reporting of Simultaneous Data.

MONSEE Program

H. Tanaka presented a brief report on MONSEE (Monitoring the Sun-Earth Environment), an inter-union group formed under the auspices of the Special Committee on Solar Terrestrial Physics (SCOSTEP) for the purpose of promoting activities in solar-terrestrial physics monitoring data exchange and associated matters. Tanaka also discussed improvements in the publication of solar ratio data in the QBSA for which he has accepted responsibility. His report was received with appreciation and approval.

Corona Monitoring

J. Rösch reported satisfactory normalization among the three observatories still doing routine coronal intensity monitoring in the green and red lines of Fe XIV and Fe X.

Sunspot Monitoring

M. Waldmeier reported on continuing activities in sunspot monitoring (about 30 observatories participated), and on the new format for QBSA.

Resolutions

The following resolutions were adopted for transmission to the Executive Council of the Union:

SPECIAL PROGRAMS DURING THE YEARS OF SOLAR MAXIMUM

The International Astronomical Union
noting that the next maximum in solar activity is expected to occur in 1979,
urges governmental and scientific agencies to implement special programs during the years of solar maximum to provide the data from space and ground which are necessary for studies of the active Sun.

ABSOLUTE CALIBRATION OF SOLAR RADIO FLUX MEASUREMENTS

The International Astronomical Union
considering

(a) that URSI Commission V noticed in 1966 important disagreements among absolute calibrations of microwave solar radio flux made at different stations;

(b) that most of the important disagreements were successfully removed before its General Assembly in 1972, and that the technique of making absolute calibrations has been established;

(c) that regular world-wide observations with an accuracy of a few parts in 100 are practicable only through a continuing international coordination;

(d) that IAU Commission 10 has long desired improvement of homogeneity in the field of solar activity survey;

expresses thanks to URSI Commission V for its successful work on radio flux calibrations; and

decides that Commission 10 takes over the responsibility for the continuation and future applications of the work done by URSI Commission V.

INTERGOVERNMENTAL ORGANIZATION FOR AN INTERNATIONAL SOLAR OBSERVATORY

The International Astronomical Union

being aware of the explorations made by astronomers from a number of IAU member countries during the last five years to find an excellent site for solar observations;

having been informed of the interim results during which some 40 prospective sites have been investigated, which have led to the identification of three very promising prospective sites that will be subjected to final testing with medium-sized solar telescopes during the years 1973 and 1974;

realizing the far-reaching scientific importance of solar observations that would be performed from a site with really excellent seeing;

realizing further that a truly international solar observatory can be established only as an intergovernmental organization;

urges the governments of the relevant member countries to examine the possibility of establishing such an organization, if the final exploration of the sites shows at least one of them to have excellent daytime seeing;

draws attention to the importance of such an observatory being accessible to visitors, not only from the member states but also from other countries; and

asks that measures to that end be examined when establishing the final structure of the Joint Organization for Solar Observations as an intergovernmental organization.

COMMISSION 12: RADIATION AND STRUCTURE OF THE SOLAR ATMOSPHERE (RADIATION ET STRUCTURE DE L'ATMOSPHÈRE SOLAIRE)

Business Meetings, 21 and 29 August 1973

PRESIDENT: R. G. Athay.

1. The Commission endorsed the following Organizing Committee:

President: R. G. Giovanelli.

Vice-President: M. K. V. Bappu.

Organizing Committee: J. M. Beckers, P. Delache, C. Jordan, G. M. Nikolsky, E. H. Schroter, R. G. Athay (ex-officio).

2. New members of the Commission were endorsed as follows: M. Altschuler, L. Acton, R. Bonnet, A. N. Cox, P. Delache, S. Dumont, I. Elliott, R. Falciani, A. Gabriel, L. House, J. M. Pasachoff, F. Roddier, K. R. Sivaraman, A. Wyller, C. Zwaan.

3. *Working Groups*

(a) *Sunspot Spectra.* Dr Pierce, Chairman of the Working Group, advised that a photographic map of the solar sunspot spectrum by Dr J. Harvey would be published within a few months, and that an atlas of sunspot spectra by Dr D. Hall had now been published. In general there were not enough groups engaged in sunspot spectra to warrant coordination of their work on an international basis, and he recommended that the Group be dissolved. The Commission carried this recommendation.

(b) *High-Resolution Atlas of the Solar Spectrum.* The first part of a high-resolution atlas had been prepared by Dr Delbouille, while Kitt Peak has prepared on microfile an atlas covering the region 2950 to 10000 Å for $\mu = 1$ and 0.2.

(c) *Solar Eclipses.* The Commission discussed a problem raised by the Chairman, Dr Rigutti, concerning aircraft diversion during eclipse times, and decided to endorse retention of Resolution No. 3 of Commission 12 in 1964.

The Commission also discussed the role of the Working Group and its relationship to groups such as NSF which were concerned with the collection of information relative to the organization of eclipse expeditions. It was decided that the Working Groups should be retained, but that the composition should be changed from time to time so as to include only those people concerned with any specific eclipse. Dr Newkirk was appointed chairman of the Working Group, his role being to act as coordinator with the Secretary-General of the IAU, and to find the best means of making arrangements for coordination for the various solar eclipses.

4. *Resolutions.* The following three resolutions were carried:

(a) The IAU wishes to express its appreciation to the National Science Foundation for its coordination of experiments during the last five total solar eclipses, and expresses the hope that this valuable service to the scientific community will be continued.

(b) The International Astronomical Union

being aware of the exploration made by astronomers from a number of IAU member countries during the last five years to find an excellent site for solar observations,

having been informed of the interim results during which some forty prospective sites have been investigated, which have led to the identification of three very promising prospective sites that will be subjected to final testing with medium-size solar telescopes during the years 1973 and 1974,

realizing the far-reaching scientific importance of solar observations that would be performed from a site with really excellent seeing,

realizing further that a truly international solar observatory can be established only as an intergovernmental organization, *urges* the governments of the relevant member countries to examine the possibility of establishing such an organization, if the final exploration of the sites shows at least one of them to have excellent day-time seeing,

draws attention to the importance of such an observatory being accessible to visitors, not only from the member states but also from other countries, and

asks that measures to that end be examined when establishing the final structure of the Joint Organization for Solar Observations as an intergovernmental organization.

(c) Commission 12 of the IAU strongly supports the Working Group on high resolution solar spectroscopy in the establishment of a center to bring together the observations and laboratory data for the publication of a definitive table of wavelengths, line strengths and identifications of solar Fraunhofer lines. Such work would provide the fundamental bases for the determination of solar and stellar abundances so important in an understanding of the Universe.

5. *Plans for the Future.* The following were suggested as possible items for future symposia or colloquia:

(a) 'The Chromospheric Network', to be held at Big Bear Observatory.

(b) Another Bildeburg conference, to be organized by Utrecht Observatory, taking into account the three-dimensional structure of the photosphere.

(c) 'A Model of the Chromosphere', to follow about two years after the completion of the ATM.

(d) 'The Fine Structure of the Corona'.

Scientific Sessions

The sessions of Commission 12 devoted to discussions of scientific topics were of an experimental nature of considerably different character than had been followed at preceding general assemblies. Each session was devoted to the discussion of an individual topic of current interest. The discussions were centered around one or two major review papers; most, or all, of the time being devoted to the presentation of the paper and discussion of it. This format provided ample time for both an extensive review and for short, unscheduled contributions from the audience. The experiment was successful in elevating the level of the scientific discussions and in providing a more effective role for the commission meetings.

A half day of discussion, on 25 August, was devoted to the topic 'Shifts and Asymmetries of Spectral Lines'. The focus in this session was on the complicated effects of velocity fields in the solar atmosphere on the shapes and locations of spectral lines. An extended review of these topics was prepared by J.-C. Pecker and C. Magnan.

A full day of discussion, planned jointly with Commissions 10 and 36, was held on 27 August. The topics for this discussion were 'The Next Decade of Research in Stellar Atmospheres' and 'The Next Decade of Research in the Solar Atmosphere'. L. Kuhi and P. Delache presented review papers, respectively, on observational and theoretical problems in Stellar Atmospheres. A review of problems in observational solar physics was presented by J. Beckers and a review of Problems in Theoretical Solar Physics was presented jointly by M. Kuperus and H. Rosenberg. The focus in these sessions was on the expected areas of development in observational capability as well as on those problems whose solutions seem to be particularly rewarding.

A second full day of discussion, planned jointly with Commission 36, was devoted to the topic 'Generation, Propagation and Dissipation of Waves in Stellar Atmospheres'. Extended review papers in these sessions, held on 29 August, were given by J. Leibacher and P. Souffrin. Although both speakers touched on all aspects of the problem, Leibacher placed main emphasis on wave generation and Souffrin placed main emphasis on propagation and dissipation. Both speakers led very constructive and extended discussions with a good deal of audience participation in the form of questions and clarifying comments.

It is planned to find a suitable means for collecting the various review papers presented in meetings of Commission 12 and in our joint meetings with Commissions 10 and 36 into a single volume that can be distributed to commission members and other interested astronomers.

COMMISSION 14: FUNDAMENTAL SPECTROSCOPIC DATA (DONNÉES SPECTROSCOPIQUES FONDAMENTALES)

Report of Meeting, 22 August 1973

PRESIDENT: A. H. Cook.
SECRETARY: J. G. Phillips.

Silent tribute was paid to two former members lost by death since the 1970 meeting: H. Barrell and M. C. J. Minnaert.

The Draft Report compiled from contributions by the five committees had been distributed to all members of the Commission prior to the meeting. The Draft Report was approved.

President Cook announced that the Commission was taking part in three joint meetings during the course of the General Assembly. One, with Commission 29, was to be on line identification in stellar spectra; a second, with Commission 40, was to discuss data for molecular radio astronomy, a third meeting, with Commissions 4, 31, and 40, was to be a discussion of recommendations of the Consultative Committee for the Definition of the Meter on realization of the Kr-86 wavelength standard of length, and on the value of the speed of propagation of light, the consequences of developments of absorption-stabilised laser sources of standard wavelengths.

Upon recommendation by the Organizing Committee, the following Officers and Committee Chairmen were approved for 1973–76:

President: R. H. Garstang.

Vice-president: W. Lochte-Holtgreven.

Organizing Committee: K. M. Baird, A. H. Cook, S. L. Mandel'shtam, C. Moore-Sitterly, J. G. Phillips, R. Tousey.

Chairmen of Committees:

1. Wavelengths: K. M. Baird.
2. Transition Probabilities: R. H. Garstang,
3. Cross Sections: E. Trefftz,
4. Atomic Structure: W. C. Martin,
5. Molecular Structure: R. W. Nicholls.

President Cook expressed the thanks of the Commission to the retiring members of the Organizing Committee and Committee Chairmen for their loyal service during the past three years.

Looking to the future work of the Commission, the President asked for consideration of the following Statement of Priorities:

'The priorities of work in the fields covered by Commission 14 are determined by the needs of other astronomers, for the aim of Commission 14 is to ensure that data are available by which astronomical observations may be interpreted in physical terms. Thus, there is a continuing need for precise measurements of standard wavelengths so that lines observed in astronomical spectra may be interpreted in terms of spectra of atoms or molecules studied in the laboratory. Until recently, only visible spectra were observed in astronomy, but it may be expected that in the next decade there will be a growing need for standard measurements of frequencies (or wavelengths) of lines in the radio, ultra-violet and X-ray regions of the spectrum. Transition probabilities and cross sections for various processes are needed to interpret physical conditions in astronomical sources, and here again, it is probable that there will be a growing demand for data relating to transitions at energies corresponding to radio, ultra-violet and X-ray spectra.'

In the subsequent general discussion of the above Statement of Priorities, Kessler pointed out that the expectations of the next decade are upon us already, while Martin commented that needs transcend the need for wavelength standards (the charge of Committee 1) and include the need for atomic analyses and wavelength measurements (Committee 4). In a letter read by the President,

Herzberg asks whether the present committee structure of the Commission is the most appropriate in view of these future needs, while in another letter, Praderie, Malaise and Lamers comment on the serious lack of transition probabilities for lines shortward of 2500 Å, which are of crucial importance for progress in ultra-violet stellar spectroscopy. This latter point was strongly supported by Garstang, who reminded the Commission that quasars shift such ultra-violet lines into the visible part of the spectrum. In the case of molecules, there is a serious lack of microwave identifications and transition probabilities, as well as of data on microwave collisional cross-sections. All these considerations were strongly supported by Seaton, who in a letter stressed the importance of extensions of data into the radio and X-ray regions. In a suggestion to the new Organizing Committee, President Cook urged that it consider within the next year its program for the next General Assembly, to stress these extended needs. In that connection, Kessler urged the Commission to seek more joint input from observational astronomers, possibly through more joint meetings.

On the basis of the above considerations, President Cook presented the following resolution, which was carried:

IMPORTANCE OF SPECTROSCOPIC DATA

The International Astronomical Union,

considering the great advances in astronomical observations throughout the electromagnetic spectrum from radio to X-ray frequencies, and

recognising the crucial importance of reliable numerical values of properties of atoms and molecules, many of which are at present lacking, if the full value and understanding is to be obtained from astronomical observations

urges all who provide financial support for astronomy to support also the measurement and calculation of relevant atomic and molecular properties.

A number of items of information were received by the President too late to be included in the Draft Report:

1. *From B. Edlén*

(a) *Precision Measurements of Stellar and Telluric Lines in the Region 6800–7400 Å. R. and R. Griffin* (*Monthly Notices Roy. Astron. Soc.* **162**, 255, 1973) give a list of wavelengths of 81 stellar and 159 telluric lines in the range 6841–7424 Å determined from high-dispersion spectrograms of Arcturus and Procyon. The random errors are estimated to 0.001 or 0.002 Å, which is a significant improvement over the solar spectrum data of H. D. Babcock and C. E. Moore (The Solar Spectrum, λ 6600 to λ 13495, Carnegie Inst. of Washington, 1947).

(b) *Near Infrared Wavenumber Standards.* A paper by *G. Guelachvili*, to appear in *Optics Communications*, reports wavenumbers in the (2, 0) vibration-rotation band of CO measured in absorption at Laboratoire Aime Cotton in reference to the Krypton 86 primary standard by means of a high information Fourier Interferometer operating in the vacuum. As compared to the best previous measurements the precision is increased by more than a factor of 10. Eight-figure wavenumbers are given for 51 lines from 4149 to 4337 cm^{-1}.

2. *Spectroscopic Results of Astrophysical Interest in the U.S.S.R. (1970–1972)*

V. Prokof'ev has submitted the following summary of results:

(A) *Spectra and Energy Levels.* Many experimental and theoretical investigations were devoted to the astrophysically important problem of the spectra of highly ionized atoms in the EUV and X-ray regions.

(a) Measures and new identifications have been made on a number of spectral lines in Al VII to Al XI (**1**), and S IX to S XIII (**2**). In addition, the spectra of K XIII to K XV and of Fe XVIII (**3**) have been obtained using focused laser radiation. Analyses are under way.

(b) Computation of energy levels are essential for interpretation of spectra of highly ionized atoms, where relativistic effects are pronounced. Several methods were used to compute energy levels for S IX to S XII (**4**); the isoelectronic sequence of Ne (S VII, Cl VIII, Sc XII to Cr XV) (**5**); resonance lines of ions important in solar corona and X-ray, such as He- and Li-like ions with $Z = 8, 10, 12, 16$ and 20, as well as Fe XXIV and Fe XXV (6); spectra of the sequence Ti I to Mn IV (7); the ions O II, Ne IV, Mg VI and Si VIII (8); helium-like ions ($Z \leqslant 6$) using a multiconfiguration approximation (9); autoionization levels from Li II through C V (10); the ions Ce III, Pr III, Pr IV, Nd X (11), and the configuration $1s^2 2s\, 2p^N$ ($N = 1 \ldots 6$) of atoms and ions (12).

Energy levels, wavelengths and oscillator strengths have been computed for the transition $3p^4$–$3p^3 4s$ for the sulphur sequence (S I to Cu XIV) (13); for the transition $2p^4$–$2p^3 3s$ in the spectra of Co XX, Ni XXI, and Cu XXII (14), and for the transition $2p^5$–$2p^4 3s$ of the isoelectronic sequence F I through Cu XXI (15). Energy levels have been computed in the configuration $2p^4 3d$ and $2p^4 4d$ for the isoelectronic sequence F I through Mg IV (16).

(B) *Oscillator Strengths and Transition Probabilities.* Several standard investigations were in progress of *f*-values and transition probabilities of atoms, ions, and molecules. Relative *f*-values for absorption lines in the Sm I spectra were measured using the hook method (17). Absorption line strengths in the spectrum of Procyon, yielded oscillator strengths of many lines of Fe II, Ti II, Cr II, V II, Zr II, La II, Sc II and G II, while the spectra of δ and π Sco as obtained by Morton and Spitzer were used to calculate transition probabilities of the $\lambda 1427$ and $\lambda 1577$ lines of C II. The lifetimes of levels $2p$ and $3p$ of Xe I and $6d(^4F)\, 7s^2P$ of Xe II have been measured (18). An experimental investigation of transition probabilities in autoionized series of Ca, Sr, and Be has been carried out (19) (20). The lifetime of the $2\ \Sigma_4{}^+$ level of the $CO_2{}^+$ ion has been measured as well as the intensities in the first positive band of active nitrogen (21).

In the area of theoretical calculations, perturbation theory has been invoked to calculate the asymptotic expansion of $1/z$ for the transition probabilities of resonance lines of atoms (22). Theoretical calculations have been carried out on oscillator strengths for excited states of helium-like systems (23); singlet and triplet transitions of Sr (24); Cd (25); Be and Ba (26); electric dipole transitions $1s^2 2s\, 2p^{N+1}$–$1s^2 2s^2 2p^N$ of Be, B, C, N, O and F (27); the transitions $3d^3 4p$–$3d^3 4s$ and $3d^4 4p$–$3d^5$ in Cr II (28, 29).

(C) *Collisions between Atoms, Molecules and Electrons.* Laboratory measures have been made of excitation cross-sections in singlet and triplet lines of He I by electron impact (30), as well as the excitation functions for resonance lines of Ca I and Ba I (31); Ba II and Sr II (32); N III (33); the absolute values of the ionization cross-sections by electron impact for Mg, Ca, Sr, and Ba (34), and the excitation cross-sections of lines of Xe II by impact of He and Xe in metastable states (35). The photoabsorption cross-section of barium vapor was measured near the ionization limit (36).

In the case of molecules, the formation of excited particles in He^+–CO collisions has been studied, as well as electron impact effective cross-sections of excitation of bands of CO and CO^+ (37), $CO_2{}^+$ (38), NO (39) and OH (40).

Theoretical calculations of excitation cross-sections have been carried out for the transition $2s^2 2p^2$–$2s\, 2p^3$ in C I, N II and O II, and for $2s^2 2p^2$–$2s^2 2p\, 3s$, $2s^2 2p\, 3p$ in C I (41). Semiempirical formulae have been developed for ionization cross-sections of atoms and molecules by electron impact (42), and other semi-empirical methods were invoked to calculate photoionization cross-sections of levels $3s$, $4s$, $5s$ and $3p$ of Na (43). Formulae have been derived for the calculation of the excitation cross-sections of atoms by electron impact (44); good agreement is found for the *S–P* dipole transition in H, He and Na and for the transition $4s^2$–$4p^2$ of Ca.

(D) *Other Studies.* The profile of the $2p_{10}$–$5d_5$ line of ^{86}Kr has been studied and corrections applied to its nominal wavenumber (45). The Stark-broadening of lines of the ions Si II and Si III has been calculated, and the resonance broadening of Lα has been studied experimentally (46). A general theory of Stark-broadening of hydrogen lines has been developed within the framework of the one-electron approximation, leading to a general analytic expression for the complete line contour (48).

3. *Report from the Centre for Research in Experimental Space Science-York University*

The following has been abstracted from a report submitted by *R. W. Nicholls*:

(A) *Intensity Measurements on Molecular Spectra.* Intensity measurements were completed on CO_2^+ spectra of importance in the spectrum of Venus (**49**), and a definitive set of band strengths have been obtained for the C_2 Swan system (**50**). Absolute absorption intensity measurements have been completed on the Cameron system of CO (**51**), the NO beta system (**52**), the O_2 Herzberg bands (**53**) and continua (**54**). The absolute absorption oscillator strengths of the O_2 Atmospheric Band system have been recently measured at high resolution, but not yet published. The 'hook' method of interferometric spectroscopy was used to measure band oscillator strengths for the NO Gamma system (**55**), and the NO Beta system (**56**); the potential precision of the method has been recently extended by about three orders of magnitude (**56, 57**). Shock tube techniques have been used to measure the absolute band strengths of the CN Red and Violet systems (**58, 59**), and a study has been completed of the intensity distribution in the B–X system of SO.

(B) *Wavelength Measurements and Structure Studies on Molecular Spectra.* A new study has been completed of the vibrational and rotational structure of the visible and near I–R ammonia spectrum (**60, 61**). The strongest feature at 6450 Å has been assigned to 5 ν_1, other vibrational assignments have been newly made and the 6450 Å band has been rotationally analyzed. A new vibrational analysis has been performed on shock-excited band systems of YO.

(C) *Identification Atlas of Molecular Spectra.* Atlases 8 and 9 on CN Red and CN Violet respectively have been issued (**62, 63**).

(D) *Synthetic Spectra.* The synthetic spectrum computer program SPECT III has been further refined to produce realistic emission, absorption and atmospheric transmission profiles.

(E) *Theoretical Studies.* A major study of line strength (Hönl-London) factors has been completed by Whiting (**64**), the prime result of which is a computer program for evaluation of any Hönl-London factor for any degree of coupling. The routine production of diatomic Franck-Condon factors has continued, and a series of Spectroscopic Reports to distribute them has been instituted (**65, 66, 67, 68, 69**). A new method has been developed for the computation of polyatomic Franck-Condon factors, and recent results have been published for CO_2^+ (**70**). The mathematical basis of the *r*-centroid approximation has been worked out (**71**); these ideas have been extended to polyatomic molecular transitions (**72**).

(F) *Atmospheric Absorption Coefficients.* A modification of the SPECT III computer program is being used to study realistic absorption coefficients of atmospheres (**73**).

(G) *Interpretative Work.* The recent discovery by Brewer *et al.* of an unusual absorption band at 30–60 km was identified as NH; spectroscopic and chemical evidence was put forward to explain the occurrence of stratospheric NH (**74**).

4. *From the Division of Quantum Metrology, National Physical Laboratory, Teddington, United Kingdom*

G. H. C. Freeman and *Vera P. Matthews* have used their vacuum Michelson Interferometer (**75**) for detailed studies of a few of the Hg^{198} UV lines from two r.f. standard lamps, at 1 and 3 torr (nominal) argon carrier gas pressure. The green line was used as a working internal standard. The procedure of interferometer scanning has been described elsewhere (**76**). Comparisons are made with previous work of Rowley (**77**) and Kaufman (**78**). From the visibility curves the line shapes and widths can be obtained. Their results indicate that for all the lines, except λ185 nm, the width at half intensity is approximately 10^{-6} of the wavelength, with those from the higher pressure lamp being 10% to 20% wider. The λ185 nm line is self reversed under all conditions. These results, together with temperature shift of the wavelength, have been discussed in detail.

5. *Additional Reports submitted at the Meeting*

(A) *K. L. Andrew* reported on the status of his research on argon. Interferometer measures were made of an argon tube cooled by liquid nitrogen at path lengths of 6, 30, 50 and 80 meters. Andrew presented a computer printout of results for 80 lines.

(B) *R. H. Garstang* reported that G. W. Wares has been using a shocktube method to obtain absolute transition probabilities of Mg I, Mg II, and the α-system of TiO, while M. Huber has employed the hook technique to derive transition probabilities for 107 lines of Fe I.

(C) A recent KPNO publication (Contrib. No. 559) describes the solar photospheric spectrum. 14624 lines are measured between 2918 Å and 9000 Å on their photographs with a probable error of ±0.0001 Å.

REFERENCES

1. Kononov Eh. Ya., Koshelev K. N: 1970, *Opt. i Spektrosk.* **29**, 217; Kononov Eh. Ya., Koshelev K. N., Ryabtsev A. N.: 1971, *Opt. i Spektrosk.* **30**, 996.
2. Podobedova L. I., Kononov Eh. Ya., Koshelev K. N.: 1971, *Opt. i Spektrosk.* **30**, 394.
3. Bojko V. A., Vojnov Yu. P., Gribkov V. A., Sklizkov G. V.: 1970, *Opt. i Spektrosk.* **29**, 1023.
4. Safronova U. I., Ivanova A. N., Rabinkina N. V.: 1971, *Opt. i Spektrosk.* **31**, 496.
5. Safronova U. I., Lezhava E. A.: 1972, *Opt. i Spektrosk.* **32**, 233.
6. Vajnshtejn L. A., Safronova U. I.: 1972, *Kratkie soobshcheniya po fiz.* No. 3, 40.
7. Bogdanovichjus P. O., Vizbarojte J. J.: 1971, Instit. Fis. i Matemat. Akad. Nauk Lit. S.S.R. VINITI No. 2805–71.
8. Kantseryavichyus A. Yu., Shadzhyuvene S. D.: 1971, *Lit. fiz. sbornik* **11**, 769.
9. Kavetskikh V. I., Kaminskas V. A.: 1971, *Lit. fiz. sbornik* **11**, 197.
10. Balachov V. V., Lipovetskij S. S., Pavlichenkov A. V., Palyudov A. N., Senashenko V. S.: 1971, *Vestn. Moskov. Univ. Fiz. i Astron.* **12**, 65.
11. Shadzhyuvene S. D., Surgajlene R. T., Ripskite P. V., Bogdanovichyus P. A., Karaziya R. I.: 1970, *Lit. fiz. sbornik* **10**, 873.
12. Kuplyauskis Z. I., Lazauskas V. M., Yutsis A. P.: 1971, *Lit. fiz. sbornik* **11**, 191.
13. Gruzdev P. F.: 1970, *Opt. i Spectrosk.* **28**, 615.
14. Gruzdev P. F.: 1971, *Zh. prikladnoj spektrosk.* **14**, 1100.
15. Gruzdev P. F.: 1971, *Opt. i Spektrosk.* **30**, 525.
16. Gruzdev P. F., Loginov A. V.: 1972, *Opt. i Spektrosk.* **33**, 3.
17. Komarovskij V. A., Penkin N. P., Nikiforova G. P.: 1971, *Opt. i Spektrosk.* **29**, 220.
18. Karimov R. G., Klimkin V. M.: 1971, *Izv. VUZ, Fiz.*, No. 3, 24.
19. Kozlov M. G., Startsev G. P.: 1971, *Opt. i Spektrosk.* **30**, 575.
20. Basiev T. T., Malakhov Yu. I., Tsys' S. N.: 1971, *Opt. i Spektrosk.* **30**, 421.
21. Pavlovskaja E. N., Yakovleva A. V.: 1971, *Opt. i Spektrosk.* **30**, 365.
22. Khristenko S. V., Vetchinkin S. I.: 1971, *Opt. i Spektrosk.* **31**, 503.
23. Safronova U. I., Ivanova A. N., Kharitonova V. N.: 1970, *Opt. i Spektrosk.* **28**, 585.
24. Zilitis V. A.: 1970, *Opt. i Spektrosk.* **29**, 821.
25. Zilitis V. A.: 1972, *Opt. i Spektrosk.* **31**, 161.
26. Zilitis V. A.: 1970, *Opt. i Spektrosk. Uchenye zapiski Latv. Univers. Riga*, 15.
27. Nashlenas Eh. P., Kaminskas V. A., Yutsis A. P.: 1971, *Lit. fiz. sbornik* **11**, 203.
28. Kuzmitskite L. L., Ushpalis K. K.: 1971, *Lit. fiz. sbornik* **10**, 201.
29. Kuzmitskite L. L., Ushpalis K. K., Dembavskajte A. V.: 1970, *Lit. fiz. sbornik* **10**, 213.
30. Smirnov Yu. M., Sharonov Yu. D.: 1971 *Opt i Spektrosk.* **30**, 1001.
31. Aleksakhin I. S., Zapesochnyj I. P., Garta I. I., Starodub V. P.: 1971, *Opt. I Spektrosk.* **30**, 1178.
32. Bejtman I. Ya., Vajnshtejn L. A., Dronov A. P., Kudzyavtsev E. M.: 1970, *Zh. Eksperim. Teor. Fiz.*, **59**, 1991.
33. Smirnov Yu. M., Sharonov Yu. D.: 1972, *Opt. i Spektrosk.* **32**, 624.
34. Vajnshtejn L. A., Ochkur V. I., Rakhovskij V. I., Stepanov A. M.: 1971, *Zh. Eksperim. Teor. Fiz.* **61**, 511.
35. Bochkova O. P., Tolmachov Yu. A.: 1972, *Opt. i Spektrosk.* **32**, 827
36. Kozlov M. G., Startsev G. P.: 1970, *Opt. i Spektrosk.* **28**, 1217.
37. Koppe V. T., Koval' A. G., Danilevskij N. P., Popova L. I., Fizgeer B. M.: 1971, *Zh. Eksperim. Teor. Fiz.* **61**, 906.
38. Koval' A. G., Koppe V. T., Danilevskij N. P., Popova L. I.: 1972, *Opt. i Spektrosk.* **32**, 1037.
39. Povch M. M., Skubenich V. V., Zapesochnyi I. P.: 1972, *Opt. i Spektrosk.* **32**, 1044.
40. Sushanin I. V., Kishko S. M.: 1971, *Opt. i Spektrosk.* **30**, 580.
41. Poshyunajte N. P., Lyash A. V., Bolotin A. B.: 1970, *Opt. i Spektrosk.* **29**, 799.
42. Krasavtsev L. M., Krinberg I. A.: 1971, *Issledovanija po geomagn. aeron. fiz. Solntsa*, No. 16, 182.

43. Zilitis V. A.: 1970, *Opt. i Spektrosk.* **29**, 617.
44. Vajnshtein L. A., Vinogradov A. V.: 1970, *J. Phys.* **B3**, 1090.
45. Batarchukova N. R., Kartashev A. I., Ptitsina E. A.:1970, *Trudy metrolog. Inst. S.S.S.R.*, No. 114 (174), 5.
46. Skidan V. V., Shrejder E. Ya.: 1971, *Opt. i Spektrosk.* **28**, 627.
47. Kobzeev G. A.: 1971, *Opt. i Spektrosk.* **30**, 199.
48. Lisitsa V. S., Sholin G. V.: 1971, *Zh. Eksperim. Teor. Fiz.* **61**, 912.
49. McCallum J. C., Nicholls R. W.: 1972, *J. Phys.* **B5**, 1417.
50. Danylewych L.: Ph. D. thesis, York University.
51. Hasson V., Nicholls R. W.: 1971, *J. Phys.* **B4**, 681.
52. Hasson V., Nicholls R. W.: 1971, *J. Phys.* **B4**, 1769.
53. Hasson V., Nicholls R. W.: 1971, *J. Phys.* **B4**, 1778.
54. Hasson V., Nicholls R. W.: 1971, *J. Phys.* **B4**, 1789.
55. Farmer A. J. D., Hasson V., Nicholls R. W.: 1972, *J. Quant. Spectr. Rad. Transfer* **12**, 627.
56. Farmer A. J. D., Hasson V., Nicholls R. W.: 1972, *J. Quant. Spectr. Rad. Transfer* **12**, 635.
Hasson V., Farmer A. J. D., Nicholls R. W.: 1972, *J. Phys.* **B5**, 1241.
57. Hasson V., Farmer A. J. D., Nicholls R. W., Anketell J.: 1972, *J. Phys.* **B5**, 1248.
58. Arnold J. O., Nicholls R. W.: 1972, *J. Quant. Spectr. Rad. Transfer* **12**, 1435.
59. Arnold J. O., Nicholls R. W.: 1972, *J. Quant. Spectr. Rad. Transfer* (in press).
60. McBride J. O. P., Nicholls R. W.: 1972, *J. Phys.* **B5**, 408.
61. McBride J. O. P., Nicholls R. W.: 1972, *Canadian J. Phys.* **50**, 93.
62. Brocklehurst B., Seel R. M., Hébert G. R., Nicholls R. W.: 1971, CRESS, York University.
63. Brocklehurst B., Seel R. M., Hébert G. R., Nicholls R. W.: 1972, CRESS, York University.
64. Whiting E. E., Paterson J. A., Kovacs I., Nicholls R. W.: 1972, *J. Mol. Spectr.*
65. McCallum J. C., Jarmain W. R., Nicholls R. W.: 1970, CRESS Spectr. Report No. 1.
66. McCallum J. C., Nicholls R. W.: 1971, CRESS Spectr. Report No. 2.
67. McCallum J. C., Jarmain W. R., Nicholls R. W.: 1972, CRESS Spectr. Report No. 3.
68. McCallum J. C., Jarmain W. R., Nicholls R. W.: 1972, CRESS Spectr. Report No. 4.
69. McCallum J. C., Jarmain W. R., Nicholls R. W.: 1972, CRESS Spectr. Report No. 5.
70. McCallum J. C., Nicholls R. W.: 1971, *J. Phys.* **B4**, 1096.
71. Nicholls R. W.: 1972, *Chem. Phys. Letters* (in press).
72. McCallum J. C., Nicholls R. W.: 1971, *J. Phys.* **B4**, 1096.
73. Cann H. W. P., Nicholls R. W.: 1972, *J. Atmos. Sci.* (in press).
74. Nicholls R. W.: 1972, *Nature* (in press).
75. Freeman G. H. C.: 1972, *Optics Commun.* **4**, 66. 4th E.G.A.S. (Amsterdam) abstracts, paper 25.
76. Terrien J.: 1967, *J. Physique* **28**, C2-3 to C2-10.
Terrien J.: 1954, National Physical Laboratory Symposium No. 11 on Interferometry pp. 437–456.
Rowley W. R. C., Hamon J.: 1963, *Rev. Optique* **10**, 519.
77. Rowley W. R. C.: 1962, Comité consultatif pour la définition du mètre (CCDM). Comité international des poids et mesures, Paris, Gauthier-Villars.
78. Kaufman V.: 1962, *J. Opt. Soc. Am.* **52**, 866.

Joint Meeting of Commissions 14 and 40, 25 August 1973

CHAIRMEN: D. S. Heeschen, A. H. Cook.

SPECTROSCOPIC DATA FOR THE INTERPRETATION OF RADIO FREQUENCY RADIATION FROM MOLECULES

The purpose of the joint meeting of the two Commissions was to explore what data were needed for the interpretation of astronomical observations of radiation from molecules and to see how far the needs were met by existing data or by laboratory or theoretical work at present under way. It appeared that a great deal more information from experimental and theoretical sources was required.

Dr B. J. Robinson discussed some current problems of identification of molecules which required more experimental determination of transition frequencies, pointing out that the recent high rate of discovery of new molecules was in part a consequence of the many frequencies that had been measured by spectroscopists, but that most simple molecules for which frequencies were known had been found or looked for. An outstanding problem was that of lines observed in astronomical sources which could not at present be attributed to any molecules; the main ones had frequencies of 89.189, 90.664 and 85.4 GHz, the first two being prominent in a number of sources and the last in Sgr B2. Although the corresponding molecules were not so far identified, some indications of their properties could be obtained from comparisons of maps of the emissions with those from known molecules. A second problem was that of transient molecules which, with brief lives in laboratory apparatus were difficult to study, but which might be quite stable in the conditions of the interstellar medium. An associated problem was that of molecules with very low vapour pressure, again difficult to study in the laboratory. Thus no molecules incorporating the relatively abundant magnesium had been detected, possibly because the vapour pressure is too low to enable transition frequencies to be measured in the laboratory. A final problem is that of frequencies of astronomical sources which are inconsistent with laboratory measurements.

Dr B. E. Turner said that the use of interstellar molecules as probes of physical conditions and processes in interstellar clouds is hampered by the lack of basic laboratory data. Two basic areas of research are involved: the study of the excitation of interstellar molecules and a determination of their abundances and ambient physical conditions; and a study of formation and destruction of these molecules.

The excitation of interstellar molecules is poorly understood because the nature of the interaction of molecules with radiation and with neutral particles is largely unknown. Regarding neutral particles (H_2 primarily), we need to know both the selection rules for collisional interaction and the total cross sections. Only if the interaction is pure dipole-dipole in nature will the excitation temperature of the molecules be a well behaved monotonically increasing function of the particle density, varying from the background radiation temperature at low densities to the kinetic temperature at high densities. If higher moment-induced moment interactions are important (leading to collisional selection rules $\Delta J = \pm 2, \pm 3, \ldots$) then the excitation temperature of the lower rotational levels can greatly exceed the kinetic temperature, or it can be negative, for even the simplest diatomic molecules. Abundances and physical conditions can be derived in these cases only if the detailed nature of these interactions is known. The laboratory technique of double resonance spectroscopy, recently applied to NH_3 and CH_3OH, seems best able to attack these problems, but much more detailed experiments are needed on these and other molecules.

Regarding the interaction of interstellar molecules with radiation, much more information is needed on the precise positions of potential curves for electronically excited states, and on oscillator strengths for transitions from these states to the ground state. Rates and wavelengths for photodissociation and photoionization depend on these parameters, as do the rates for certain possible formation mechanisms such as indirect radiative association by inverse predissociation. In some cases, interstellar molecules may even be pumped by interaction with radiation involving these states. In the infrared, the problem is solely with uncertain line strengths for vibrational transitions, but these are crucial since radiation at these wavelengths (2–350 μ) is likely to affect significantly the excitation of molecules that are observed to occur preferentially near strong infrared sources. The 2.8 μ vibrational transition of OH has played a prominent role in theories of the anomalous excitation of interstellar OH, although its strength is highly uncertain.

Interstellar molecules are destroyed primarily by photodissociation (in clouds of low density) and by adsorption onto dust grain surfaces and possibly dissociation by cosmic rays (in clouds of high density). A good start has been made on the photochemistry of several of the more complex interstellar molecules, but little has been done on diatomic species, which, because they are observed primarily in low-density clouds, are the species most likely affected by radiative destruction. Questions of adsorption rates of various molecules on surfaces are almost completely unanswered, although in dense clouds these rates are important in determining relative molecular abundances. Various 'rules' of surface chemistry need to be tested, such as that physical adsorption but not chemiadsorption can proceed on contaminated surfaces, that chemisorption but not physical adsorption requires appreciable activation energy, and that physical adsorption is very similar in mechanism to the condensation of a vapor on the surface of its own liquid. Virtually nothing is known of the dissociation and ionization rates of molecules by interaction with high-energy particles.

Interstellar molecules are presently thought to be formed by several processes including radiative association, neutral gas exchange reactions, ion-molecule reactions, surface reactions, and possibly even under LTE conditions, for example in pre-solar nebulae. Radiative association, which is relevant primarily to diatomic molecules observed optically in low-density clouds, has highly uncertain rates which have undergone many drastic revisions recently. The most important, and uncertain, rates, involve direct radiative association of CH^+, from which CH, CN and CO are formed by exchange reactions. Other molecules, such as OH, CN, CO, NO, C_2 may form by inverse predissociation but rates for these processes are equally uncertain.

Ion-molecule reactions appear best able to explain observed abundances and patterns of interstellar molecules in both low and high density clouds. Among types of ion-molecules reactions that need further study are those of the form $A^+ + H_2 \rightarrow AH^+ + H$ (particularly $A \equiv O$, OH, CN), and those reactions involving C^+ with a large variety of small molecules (NH, CH, H_2O, NH_2, NH_3, HCN, H_2CO and others). All gas-phase reactions, whether initiated by radiative association or ion-molecule reactions, also involve neutral exchange reactions, and many of these, of particular importance, are unstudied such as $CH + X \rightarrow CX + H$ ($X \equiv N$, C, O) and similar reactions involving OH and CO.

The problems of formation of molecules on surfaces are too numerous to cite here. In particular, the products formed depend strongly on whether the surfaces are saturated (i.e. contain a layer of relatively unreactive molecules such as H_2O, NH_3, CH_4) or not. In conjunction with a strong theoretical approach, well-designed laboratory experiments might be able to suggest whether interstellar grain surfaces are saturated or not, and hence offer a qualitative idea of what types of molecules may preferentially be formed, and whether the rates are adequate. Such progress will probably also require definite observational information on the presence or absence of sources of UV radiation inside dense molecular clouds, the temperature, and the density of both gas and grains. Present experiments on surface catalysis have not adequately simulated the expected interstellar conditions, either with regard to the surfaces used, or in the choice of initial gas phase species. A severe difficulty, deduced theoretically, seems to be that of desorbing molecules from the surfaces at very low temperatures, once they have formed. Experiments on desorption by UV and infrared radiation, as well as energetic particles, are greatly needed. Experiments on isotope separation and the separation of ortho-para forms of molecules such as NH_3, H_2, and H_2CO by adsorption at low temperatures are also required to determine whether some of the isotope and ortho-para anomalies observed in interstellar molecules are indicative of a direct role played by interstellar grains.

Dr D. R. Johnson described work of the microwave laboratory at the National Bureau of Standards in three different areas of activity that directly concern molecular radio astronomy. The determination of the molecular origin of unidentified interstellar lines is an exceedingly difficult but important area of laboratory activity. Recent experiments leading to the assignment of an unidentified interstellar line at 84521 MHz to methanol (CH_3OH) were described. This example also served to illustrate the need for detailed spectral maps of all known interstellar molecules. The NBS laboratory has initiated a review series which is intended to provide a complete and critical review of the literature for all known interstellar molecules. Predicted transition frequencies complete with 95% confidence limits will be provided for the entire frequency range accessible to the telescope. Additional information on hyperfine splittings, line strengths and energy levels will also be included. Reviews of this type have been completed for formamide (NH_2 CHO), formaldehyde (H_2CO), thioformaldehyde (H_2CS), methylenimine (CH_2NH), methanol (CH_3OH), water (H_2O), hydrogen sulfide (H_2S), carbonyl sulfide (OCS), hydrogen cyanide (HCN) and sulfur monoxide (SO). Reviews are in progress for most of the other known interstellar molecules. Another equally important facet of the NBS laboratory is the generation, detection and spectral mapping of new molecules which potentially play an important role in interstellar chemistry. Thioformaldehyde (H_2CS) and methylenimine (CH_2NH) served as examples to illustrate how studies of new molecules proceed from ideas on a blackboard to telescope detection in the interstellar medium. Current laboratory experiments were discussed in which the possibility of generating and detecting short-lived tautomers of known interstellar molecules are being explored.

Dr R. H. Garstang described recent work by Oka who had used a method of double resonance spectroscopy to study collision cross-sections for various types of transition. He found that in pure gases, collisions taking place between similar molecules, the change of the rotational quantum number was greater than 1 in HCN, formaldehyde and ketene but in other cases followed dipole rules. In collisions of NH_3 with rare gases, especially helium, the dipole selection rules for rotational quantum number and parity are not followed closely, and in the collisions of NH_3 with H_2 there is a substantial number in which the change of rotational quantum number is 2. Finally Garstang argued that in the interstellar medium, radiative transitions are not important compared with collisions and that while collisions with electrons are very effective in changing the energy of a molecule, there are not enough electrons in the interstellar medium for them to be important.

All speakers, as well as contributors to the discussion, emphasised the need for more experimental and theoretical work to obtain the data needed to interpret the astronomical observations.

Joint Meeting of Commissions 4, 14, 31 and 40, 27 August 1973

CHAIRMAN: A. H. Cook.

THE RE-DEFINITION OF THE METRE AND THE VALUE OF THE SPEED OF LIGHT

The meeting was convened by *A. H. Cook*, the representative of the IAU on the Consultative Committee for the Definition of the Metre, to explain to members of the IAU the background to recommendations recently made by the Committee, to consider how the recommendations affected astronomical interests and to discuss the response that the IAU should make to the recommendations. He explained that the meeting of the Committee held in 1973 June was in response to a request from the IAU for guidance on the effect that recent developments in the measurements of frequencies of laser radiation would have in astronomy.

Shortly after the helium-neon laser was devised it was shown that the wavelength of the radiation could be controlled by varying the separation of the mirrors in a servo-system so that the wavelength was that at which the gain was a maximum and subsequently was that of the Lamb dip. Much higher stability results if the reference wavelength is that of the maximum absorption in gas in an external cell, for example, an absorption line of methane for the 3.39 μm line of the helium neon laser and one of a number of lines in the iodine molecule for the 633 nm line. By using saturated absorption the effective Doppler width is very greatly reduced and the stability of the resulting wavelength is very high. A second important development is that lasers giving radiations over a wide range of frequencies oscillating continuously at high power are now available, especially those using HCN, water, carbon dioxide and helium-neon; the frequencies range from 300 m to 633 nm. Lastly efficient mixers and harmonic generators operate over almost the whole of the range, making use of point contacts, the metal-insulator-metal diode, and the Josephson junction. The fact that laser radiations of high stability and high power can be mixed in such devices and their harmonics compared means that it is possible to lock the frequency of an HCN maser at 1000 GHz to the output of a klystron at 30 GHz and so to the caesium standard frequency (9 GHz) and on the other hand to compare frequencies up through a chain to those of the carbon dioxide laser at 10 μm (3×10^3 Hz) and of the helium-neon laser at 3.39 μm (10^4 Hz). Some eight standards laboratories are now engaged on this work and the related measurements of wavelengths of laser radiations in terms of the wavelength of the standard orange line of krypton-86.

At the meeting of the Consultative Committee, which is a committee of the International Committee of Weights and Measures, members reported on techniques of stabilising lasers, on comparisons of wavelengths of stabilised lasers with krypton-86 standard and on values of the speed of light derived from measurements of the frequency and wavelength of the same transition. The conclusions of the Committee were embodied in the recommendations that are reproduced in the Appendix to this Report.

Having explained the recommendations of the Committee, Cook said that in his view the IAU was no longer concerned, as it had been in the past, with the definition of the Metre and with the primary standards of wavelength, for the accuracy that could now be achieved was far beyond anything required in astronomy, but that astronomers, who measure distances, as do geodesists, in terms of the times of travel of electromagnetic radiation, may be concerned to use the new value of the speed of light proposed by the Consultative Committee.

The discussion was opened by *J. Terrien,* the Director of the Bureau International des Poids et Mesures, who said that in his view it was possible and desirable for the metre to be redefined so that in effect the value of the speed of light recommended by the Consultative Committee is adopted as a constant, the metre being derived as the distance travelled by light in a specified time. G. Winkler, P. L. Bender, D. J. Mulholland, G. A. Wilkins, J. Kovalevsky and others took part in the discussion and discussed the effect that the adoption of a new value of the speed of light would have upon the branches of astronomy with which they were concerned. It was agreed that it was desirable to recommend that the new value should be used in future in preference to that in the System of Astronomical Constants, and a resolution was passed by the meeting and subsequently by the General Assembly of the Union. (See Resolution No. 6.)

COMMISSION 15: PHYSICAL STUDY OF COMETS, MINOR PLANETS AND METEORITES (L'ÉTUDE PHYSIQUE DES COMÈTES, DES PETITES PLANÈTES ET DES MÉTÉORITES)

Report of Meetings, 22, 24, 25 and 28 August 1973

PRESIDENT: V. Vanýsek.
SECRETARY: J. Rahe.

22 August 1973

I. SCIENTIFIC PRESENTATIONS

The following communications were presented during the meeting of the Commission, in order of their presentation:

J. Rahe, V. Vanýsek: Report on Planned Observing Programs of Comet Kohoutek 1973f.
J. M. Greenberg: Skylab Comet Program.
J. C. Brandt: Remarks on NASA Comet Program and Comet Observatory.
A. H. Delsemme: Brightness Laws and Photometric Profiles of Neutral Coma.
V. Vanýsek, J. Rahe: Lifetime of Parent Particles in Cometary Atmospheres.
L. Biermann: Chemical Processes in Comets.

II. WORKING GROUPS

Three working groups were established for discussions about observing programs for Comet Kohoutek 1973f, and for the preparation of Commission Resolutions concerning Early Space Missions to Comets and Asteroids, and Priorities in Cometary Research.

24 August 1973

I. ADMINISTRATIVE MATTERS

During the meeting the following administrative matters were discussed:

As new officers of the Commission were proposed and unanimously nominated by the Commission for approval by the Executive-Committee of the Union: President: A. H. Delsemme, Vice-President: N. B. Richter.

As members of the Organizing Committee were elected by the Commission: E. Anders, C. Arpigny, L. Biermann, O. V. Dobrovolsky, B. D. Donn, T. Gehrels, J. Rahe, E. Roemer, V. Vanýsek, F. L. Whipple.

As new members of the Commission were approved: J. E. Blamont, M. P. Candy, Z. Ceplecha, V. I. Cherednichenko, A. A. Demenko (Mrs), A. E. Douglas, T. Gehrels, J. M. Greenberg, W.F. Huebner, V. P. Konopleva (Mrs), C. F. Lillie, S. P. Maran, B. G. Marsden, C. D. McCord, N.F. Ness, J. A. O'Keefe, P. Proisy, G. W. Wetherhill.

Commission 15 was renamed *Physical Study of Comets, Minor Planets and Meteorites.*

II. RESOLUTIONS

The following Resolutions were discussed and approved by the Commission. They were submitted for approval and approved by the XVth General Assembly of the IAU (Resolution No. 7).

1. *Resolution concerning Early Space Missions to Comets and Asteroids*

Whereas Comets, unlike planets, probably still preserve the original composition of early condensation in the solar system, possibly including complex carbon compounds;

Whereas the interaction of a cometary atmosphere with the solar wind constitutes a natural plasma laboratory impossible to simulate;

Whereas Comet observations can provide information about the solar wind in inaccessible regions of space after calibration by space probes; and

Whereas most Asteroids represent matter aggregated in different regions of the primordial solar nebula than the planets or the comets.

Therefore the XVth General Assembly of the International Astronomical Union strongly recommends Early Space Missions to Comets and to Asteroids in continuing programs.

2. *Resolution concerning Priorities in Cometary Research*

The transient and unpredictable appearance of most comets prevents cometary astronomers from scheduling time on large telescopes.

In order to overcome the deficiency in cometary data, the XVth General Assembly of the International Astronomical Union strongly recommends the priority allocation of time for cometary observations on larger telescopes.

25 August 1973

SCIENTIFIC PRESENTATIONS

The following communications were presented during the meeting of the Commission, in order of their presentation:

F. L. Whipple: Space Mission to Comets.

B. D. Donn: A Proposal to Space Mission to Comets.

B. G. Marsden: Nongravitational Effects on Comets.

After these presentations, Commissions 15 and 20 held a joint meeting (see Report of Commission 20).

28 August 1973

SCIENTIFIC PRESENTATIONS

The following communications were presented during the meeting of the Commission, in order of their presentation:

Z. Sekanina: Solid Matter in Cometary Tails.

L. F. Burlaga, J. Rahe, B. Donn, M. Neugebauer: Solar Wind Interaction with Comet Bennett.

Z. Kviz: Cometary Dust as Cloud Seeding Agents.

COMMISSION 16: PHYSICAL STUDY OF PLANETS AND SATELLITES (ÉTUDE PHYSIQUE DES PLANÈTES ET DES SATELLITES)

Report of Meetings, 23, 24 and 28 August 1973

PRESIDENT: G. H. Pettengill.

Three regular half-day sessions of the Commission were held with topics as listed below. In addition, the Commission sponsored (together with Commission 17) a Joint Discussion on the Origin of the Moon and Satellites, which took place on 29 August.

General Business Session, 23 August 1973

The Working Group on Martian Nomenclature, chaired by G. de Vaucouleurs, presented its final draft report and recommendations. These will be published in full in a forthcoming issue of the journal *Icarus*, since they are too lengthy for inclusion here. The draft was approved by the Commission members present. An abbreviated version describing the general scheme of the recommendations is given below, followed by two adopted resolutions concerned with coordinate systems.

RECOMMENDATIONS OF THE WORKING GROUP ON MARTIAN NOMENCLATURE

1. *Designation of Provinces*

Instead of the original concept of 'provinces' defined by classical albedo features, which proved unworkable in practice because of the variability of many of these markings and the frequent indefiniteness of their boundaries, we propose to divide the surface of the planet into 30 geometric areas. The equatorial areas between $+30°$ and $-30°$ latitude are covered by 16 maps in Mercator projection measuring 45° in longitude by 30° in latitude. The intermediate latitudes between $\pm 30°$ and $\pm 65°$ are covered by 12 maps in Lambert conformal projection each measuring 60° in longitude by 35° in latitude; the polar regions between $\pm 65°$ and $\pm 90°$ are covered by two maps in polar stereographic projection.

Each chart is designated by the name of a prominent classical albedo feature within its confines and by a 3-letter abbreviation which will be used as a prefix to the crater designation scheme described below (Figure 1).

2. *Crater Designations*

Within each chart all craters larger than approximately 20 km are assigned a two-letter designation from Aa to Za in which the first letter is alphabetically in order of increasing longitude from east to west and the second in order of increasing latitude from south to north. A total of approximately 6000 craters have been so designated and are marked on the charts of the *Atlas of Topographic Maps* prepared at the U.S. Geological Survey (available to Commission 16 members on request). In a few heavily cratered areas the capacity of the two-letter scheme is exceeded and a third letter is added to provide a designation for all craters larger than the 20 km limit. It is recommended that Martian craters be identified by the chart prefix followed by the letter designation (e.g. SYR Aa).

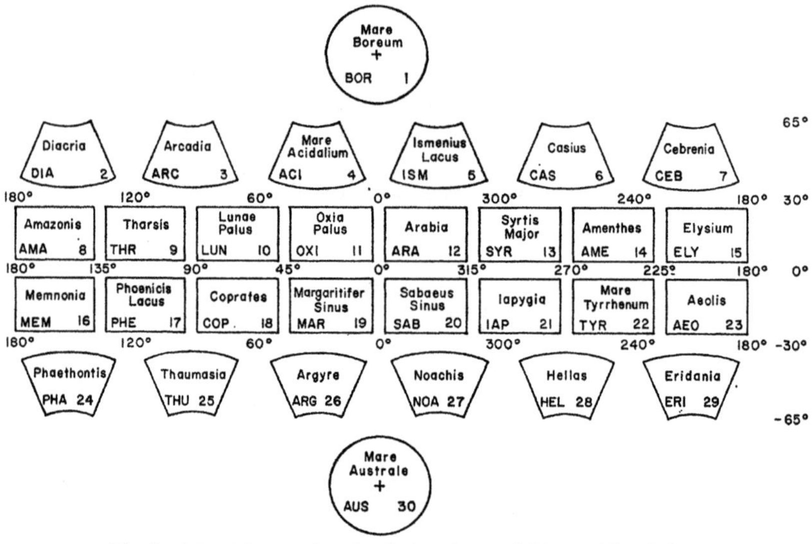

Fig. 1. Mars Map quadrangles designations and 3-letter abbreviations.

3. *Named Craters*

In addition to the letter scheme, a selection of the largest craters, generally larger than 100 km in diameter, receive names of prominent, deceased individuals who have contributed either to the scientific study of Mars, or the lore of the planet, or to basic discoveries of significance to the exploration of Mars or the interpretation of its phenomena. After extensive discussions a consensus list of about 180 names was selected.

4. *Other Topographic Features*

Thirteen classes of topographic features are designated below. Features in twelve of these classes receive specific names adopted from nearby classical albedo features taken from the Schiaparelli or Antoniadi maps. Usually following the name is the class of feature, e.g. Olympus Mons. One class of features, the valleys, is treated differently and named after 'MARS' in various languages.

Name of Class	*Description of Feature*
Catena	chain of craters
Chasma	canyon
Dorsum (Dorsa)	ridge(s)
Fossa (Fossae)	long, narrow, straight or curved valley(s)
Labyrinthus	intersecting valley complex
Mensa (Mensae)	mesa; flat-topped elevation(s)
Mons (Montes)	mountain(s)
Patera	shallow complex crater with scalloped edges
Planitia	low plain
Planum	plateau or high plain
Tholus	small domical mountain or hill
Vallis (Valles)	sinuous valley(s)
Vastitas	widespread lowlands

5. *Craters on Satellites*

Craters on Phobos are named after those involved with the discovery, dynamics, or properties of the Martian satellites and who are not already commemorated on Mars. Craters on Deimos are named after authors of literary or artistic allusions to the Martian satellites.

RESOLUTION ON THE PHYSICAL EPHEMERIS OF MARS

(Adopted by Commissions 4 and 16)

Considering that recent new determinations of the rotational elements of Mars indicate the need for a revision of the elements currently adopted in the physical ephemeris of Mars, and that a new approach to the definition of the origin of areographic longitudes appears useful (G. de Vaucouleurs, M. E. Davies and F. M. Sturms, Jr, *J. Geophys. Res.* **78**, 4395, 1973), Commissions 4 and 16 recommend

(1) that new elements and a new definition of the origin of the areographic longitudes, as given below, be incorporated in the physical ephemeris of Mars after 1979 or as soon as deemed practicable in the judgment of the cognizant Directors of the National Ephemeris Offices, but not before the tie between the new and current systems has been firmly established by appropriate comparisons between the ground-based and Mariner 9 coordinate systems.

The new elements are as follows:

(a) The R.A. and D of the North Pole of Mars (referred to the standard equinox and equator of 1950.0) are given by

$$\alpha_0 = 317.^\circ 32 - 0.^\circ 1011\,T$$
$$\delta_0 = +\,52.^\circ 68 - 0.^\circ 0570\,T,$$

where T is measured in Julian centuries of 36525 days from JD 2433282.5.

(b) The angle measured in the direct sense along the Mars equator from its ascending node on the standard equator of 1950.0 to the zero point of areocentric RA is given by

$$\phi_0 = 222.^\circ 93538 - 0.^\circ 09040\,T - 0.^\circ 000010\,T^2.$$

(c) The areocentric RA of the prime meridian of Mars is given by

$$V = V_0 + 350.^\circ 892017\,(\mathrm{JD} - 2433282.5).$$

The Julian Date is reckoned in ephemeris time. The provisionally adopted value of V_0 is 148.°24. The rotation rate results from a new analysis of the observations of the past century (de Vaucouleurs *et al., loc. cit.*).

The following notes on the preceding statement of the elements are intended for explanatory purposes only.

(i) the linear terms in the expression for α_0 and δ_0 correspond to a precession of the axis of rotation about the pole of the orbit at the rate of 0.°0837 per century.

(ii) the corresponding inclination of the equator to the orbital plane of Mars is given by

$$I = 25.^\circ 19969 + 0.^\circ 01219\,T + 0.^\circ 000006\,T^2$$

(iii) the zero point of areocentric right ascension corresponds to the vernal equinox (i.e. the ascending node of the orbital plane on the equator of Mars).

(iv) the adopted prime meridian is intended to be such that the center of the small crater Airy-0 has zero areographic longitude.

(2) that the present investigations be continued in order to establish precisely the tie between the new and the current systems, and

(3) that a recomputation of the physical ephemeris based on these new elements be prepared for the years 1800 to 1979.

RESOLUTION ON SATELLITE COORDINATE SYSTEMS

For all satellites except the Moon, the origin of longitude is defined by the sub-planetary intersection of the satellite's equator and the plane containing the center of the satellite, the planet and the Sun at the first superior heliocentric conjunction of the satellite after 1950.0. The conventions for the direction of increasing longitude and for the choice of north pole shall be the same as adopted for the planets at the IAU 14th General Assembly in 1970.

WORKING GROUP FOR PLANETARY AND SATELLITE NOMENCLATURE

The accelerating rate of planetary exploration by spacecraft has created the need for a more rapid procedure for naming surface features than is possible in the present approach, tied to triennial IAU General Assemblies. To this end, the IAU Executive Committee has created an Inter-Commission Working Group for Planetary and Satellite Nomenclature, with considerable power to act in the name of the Committee between Assemblies. The Chairman of the Group is Peter Millman; other members include the presidents of Commissions 16 and 17 and the Heads of 5 Panels. The latter are charged with the responsibility for naming features on the Moon, Mercury, Venus, Mars and the Galilean satellites of Jupiter. Recommendations are originated in the Panels, passed to the main Group for consideration and possible revision, then sent to the Executive Committee for approval. The actions of this group in the preceding 3 years will be reviewed, of course, at each General Assembly.

SCIENTIFIC PRIORITIES FOR PLANETARY RESEARCH

Commission 16 unanimously voted in this session to refrain from the setting of scientific priorities. It was felt that a list of such priorities would serve no useful purpose and might, in fact, lead to embarrassment if misunderstood by national funding agencies.

Spacecraft Results and Plans, 24 August 1973

Titles of presentations:

C. Sagan: Variable Features and Climatic Cycles on Mars.

H. Masursky: Fixed Surface Features and Geology of the Martian Surface.

B. A. Smith: Experimental Objectives of the U.S. Mariner Venus/Mercury and Mariner Jupiter/Saturn Missions.

J. S. Hall: Report on the Lowell Observatory.

Scientific Papers, 28 August 1973

Titles of presentations:

D. B. Campbell: Topography and Other Radar Results for Mercury and Venus.

S. K. Runcorn: The Shape and Internal Dynamics of Mars.

J. R. Dickel: Jupiter's Radiation Belts and Upper Atmosphere.

A. Dollfus: Cartography of the Galilean Satellites.

S. K. Runcorn: The Rotation of Jupiter.

A. F. Cook: The Rings of Saturn.

C. Sagan: The Atmosphere and Surface of Titan.

COMMISSION 17: THE MOON (LA LUNE)

Report of Meetings

PRESIDENT: A. Dollfus.
SECRETARY: J. A. O'Keefe.

During the IAU General Assembly of Sydney, Commission 17 'The Moon' held meetings for each of its 3 Working Groups, an Administrative Session, a Scientific Session and a Joint Discussion with Commission 16 on 'The Origin of the Moon and Satellites'. The Secretary for the session was Dr J. A. O'Keefe. The minutes of these sessions are available at the Office of the retiring president, A. Dollfus – Observatoire de Paris, 92190 Meudon, France.

WORKING GROUP: FIGURE, MOUVEMENTS ET POSITIONS OBSERVÉES DE LA LUNE

Scientific and Working Session, 21 August 1973

The Chairman of the WG, Dr Th. Weimer, having had unforeseen plane delays, the meeting was chaired by the President of the Commission, A. Dollfus.

The Scientific part of the session prepared by Th. Weimer was a series of 5 invited review papers summarizing the major problems and latest results in the field:

L. V. Morrison: The Current Program of Occultations.

L. A. Schimerman: The Apollo Metric Camera Selenodetic Program.

P. N. Bender: Selenodetic Coordinates of the Appollo Retroreflectors.

J. G. Williams: Determination of the Lunar Libration from Laser Range Measurements.

C. Counselman: Precision Selenodetic and Lunar Libration through Very Long Base-line Interferometry.

Each presentation was discussed at a rather high level, due to the presence of the appropriate experts in the audience.

The working session which followed was centered on the problem of the list of craters which should be standard points for all future lunar geodetic work. One of the assignments of the WG coordinated by Th. Weimer was the analysis of the problem and the preparation of a proposal already discussed by experts. A sub-group has been appointed with Weimer, Moutsoulas and Gavrilov as members to study the resolution. In Dr Weimer's absence, Dr Moutsoulas presented the results to the whole WG. The careful discussion which followed suggested some additional detailed improvements. The final resolution was unanimously adopted by vote to be submitted to the Administrative Session of the whole Commission 17.

WORKING GROUP: GEOLOGY AND GEOPHYSICS OF THE MOON

Working Session, 22 August 1973

The Chairman of the WG, G. Fielder, being unable to attend, had asked to be replaced in the chair by A. Dollfus. The topic to be discussed was the problem of scientific priorities for the study of the Moon during the next decade. Invited by ICSU to try and prospect a balance between the different aspects of their scientific needs in research, the different International Scientific Unions belonging to ICSU initiated action within their committees, commissions and working groups.

Within this framework, both the Inter-Union Commission for Lunar Studies (IUCM–CIUL) and the IAU Commission 17 'The Moon' undertook the difficult task of trying to identify research studies on the Moon in the context of present and future scientific effort, to define its impact on mankind, to sum up the state of the art, knowledge already acquired, the major gaps remaining to be filled. The minimum level of effort to be made during the next decade was estimated; a strategy for developing lunar science was discussed. The Chairman had collected by correspondence and discussions the views expressed by a large fraction of the Commission members interested. He also analysed the work done by IUCM in its conference of November 1972 in Paris; IUCM involves lunar studies through its different disciplines: astronomy, geophysics, geology, space research, radio science.

The Chairman summarized for the WG the major points of agreement between views expressed by individuals and the draft proposal issued by IUCM. A long discussion followed during which new views were exchanged, and new balances between the emphasis on different priorities were suggested. Finally, the President A. Dollfus and the Secretary J. O'Keefe were invited to prepare a draft of a document summing up the discussion to be submitted to the Administrative Session of the whole Commission 17.

WORKING GROUP: LUNAR NOMENCLATURE

Chairman: D. H. Menzel, Members: B. Levin, H. Masursky.

Prof. Levin was unable to attend.

The Working Group held two open sessions on 22 August afternoon and 25 August morning. The attendance was around 15–20 members.

The WG had been entrusted with the responsibility of defining the system of nomenclature for the craters and topographic features at smaller scale than the telescopic resolving power and to apply it on the new detailed lunar cartographic program. In order to meet this assignment, the WG had particularly intensive activity during the past three years, to arrive at a proposal reaching a general consensus. After several meetings in 1971 to explore the field, it held formal sessions on 17–18 May 1972 in Madrid, on 27 November 1972 in Paris, on 9–10 March 1972 in Houston, on 21–23 May 1973 in Constance, plus working sessions at the IUCM General Assemblies of 15 November 1971 and 28 November 1972.

The Chairman, D. H. Menzel, summarized the work done, presented the scheme proposed for the designation of small lunar features and opened the proposal to discussion. A 2-hours discussion followed in which at least 10 participants expressed comments, remarks, first on the general scheme, then on more precise points. The Chairman ended the session by proposing that although the general spirit of the first draft remained unchanged, he would prepare for the second session of the WG an improved version of the draft, accounting for the new balance of consensus on details resulting from the advices of the participants.

At the second session, the new draft of the resolution was presented to the audience. Again, about 10 members participated actively in the improvement of certain aspects, although clearly the consensus on the general principles was already reached at the beginning of the discussion. A final proposal was then agreed upon to be presented for adoption at the General Assembly of the Commission, and adopted with unanimity of the voting members, as follows:

Resolution on Lunar Nomenclature

"*Commission 17 recommends* for adoption the following resolutions concerning lunar nomenclature.

(1) The abundant data secured by space vehicles have made possible new programs of lunar mapping, which require a more precise and detailed system of nomenclature than the current system originated by Mädler. The revised system employs a geometrical grid of Regions and Provinces.

(2) In the new system, parallels of latitude and meridian arcs divide the lunar surface into 144

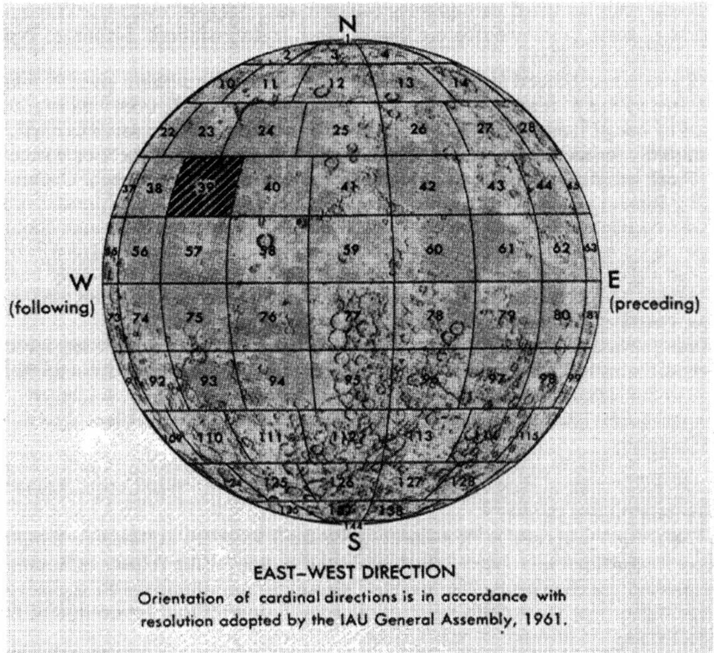

Figure 1

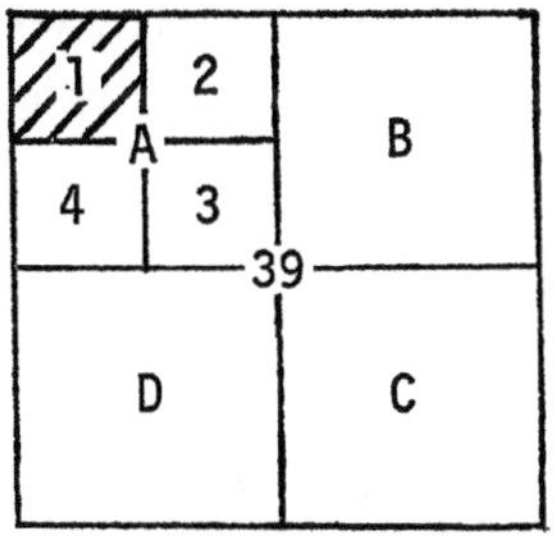

Figure 2

Regions, as in the NASA Lunar Aeronautical Charts (LAC), on the scale 1:1000000, Figure 1. (For details see Robert Carder, *The Moon*, IAU Symposium 14, p. 17). Each region is subdivided into sixteen Provinces, as in the NASA Lunar Topograhic Orthophotomaps (LTO), on the scale 1:250000. Each Province carries a code designation, consisting of its LAC Region number, followed by a letter A, B, C, or D and a number 1, 2, 3, or 4. For example, Figure 2 illustrates the coding for Province 39 A 1.

(3) Consistent with standard cartographic practice, each Region and each Province will also bear the name of some prominent feature (preferably a crater) within it. Names of Provinces will not duplicate those of Regions.

(4) A grid system may be used for further identification of features within a given Province. Abscissae are indicated by one of the capital letters A–Z (I and O are omitted), from west to east. Ordinates are indicated by one of the lower-case letters a–z (i and o omitted), from south to north. This letter system is primarily for identification and reference and will not generally be used on the maps.

(5) The Greek letters previously indicating lunar elevations will be dropped. Craters previously designated by Roman capital letters will progressively be assigned new and distinctive names. To facilitate cross correlation between the current and the new system, the old names should be listed on the larger-scale maps. At an appropriate time a complete cross-reference table will be prepared.

(6) Elevated regions will generally require new designations. Ridges, previously unnamed, will be called Dorsum, plural Dorsa. Rimae and Rima systems will receive new and more appropriate designations. Crater chains will be designated Catena, such as Catena Davy.

(7) Whereas previously assigned lunar names have been, with few exceptions, those of distinguished deceased scientists, future assignments may also contain the names of distinguished, deceased contributors to human culture and knowledge, such as writers, painters, musicians, etc., chosen on an international basis. Exclused, however, are political, military, and religious figures, as well as modern philosophers.

(8) Very small features requiring identification for some special reason may be assigned, with IAU approval, first names, male or female, from an international list. Excluded, however, are names having more than three syllables.

(9) For operational purposes, such as designating areas specially explored or sampled, Apollo missions have assigned names to certain features in the vicinity of the respective landing sites. The list* was corrected and revised by the Working Group in consultation with the astronauts and Principal Investigators of the Apollo missions. The IAU requests that future lists of this type be submitted to the IAU in advance of the mission.

(10) The complete corrected list of the adopted names is published in Appendix I. The list contains the name of M. Minnaert, who died shortly after the 1970 General Assembly. It contains also the name of Abbot (C. G.), who is still living.** It was assigned to honor Abbot's 100th birthday."

ADMINISTRATIVE SESSION OF THE WHOLE COMMISSION 17

The meeting was called to order on August 28 morning by A. Dollfus, President. He presented, in French, the following agenda:

(1) Resolution on fundamental craters.
(2) Recommendation on scientific priorities.
(3) Resolution on lunar nomenclature.
(4) Composition of Commission for 1973–76.
(5) Structure of Working Groups.

The Agenda was agreed upon.

Under Item (1) of the Agenda, the following Resolution was presented by Th. Weimer, read by G. O'Keefe, and accepted after a short discussion with unanimity.

Resolution on Selenographic Reference Points

Commission 17 recommends:

"(1) That in selenographic catalogues as many reference points as possible from the 211 listed in Appendix I should be included, so that direct comparisons between the corresponding networks can

* Copies are available at the office of Dr H. Menzel.
** C. G. Abbot died after the XVth General Assembly.

be made; thus the establishment of a common selenographic control system could be ultimately achieved;

(2) That, in the catalogues, reference to the constants used for the reductions be given."

Under Item (2), A. Dollfus presented the proposed priorities on the study of the Moon. He noted, however, that this document is only for use at the level of Commission 17 and IUCM, since IAU decided that it appeared not feasible to reply to ICSU's request for a list of scientific priorities between the different aspects of research within Astronomy. The draft was then read. A number of language corrections were proposed to the members of the Commission and were adopted. The text was finally voted and still slightly re-worked later on by the President, to care for the new views expressed in September–October 1973 after the General Assembly, by members and experts unable to attend the Sydney Conference. The final wording of the Priorities in the study of the Moon is given at the end of the Report of Commission 17.

Under Item (3) of the Agenda, the resolution on lunar nomenclature which improved during the last sessions of the WG was read by D. H. Menzel. The discussion went on with some changes in wording. D. H. Menzel noted that most of the name changes replaced y by ij in Russian names. A large unnamed crater near the South Pole was assigned 'Minnaert'. Finally, the resolution on lunar nomenclature was adopted unaminously.

For Item (4), composition of the Commission for 1973–1976, President A. Dollfus proposed for adoption the new board as follows.

President: Runcorn, S. K.

Vice-Presidents: Anders, E., Florenskij, K. P.

Organizing Committee: Dollfus, A., Dzhapiashvili, V. P., Koziel, K., Kuiper, G. P., Mulholland, J. D., O'Keefe, J., Sonett, C. P.

Ex-officio-members: President IAU Commission 16: Mayer, C. H.; Chairman COSPAR Working Group 7 'Space related sudies of the Moon and Planets': Kuzmin, A. D.; Vice-Chairman COSPAR Working Group 7: Sagan, C.

Consultant: Chairman of IAU–COSPAR Working Group on laser study of the Moon: Alley, C. O.

The new list of members as appended in the 'Proceedings of the IAU General Assembly', and the list of consultants, will be circulated by the new President.

The new President, S. K. Runcorn was then invited to take the Chair and to explain his views about the future activity of the Commission. He stated that he plans to continue the 3 existing Working Groups.

After some discussion, the title of Working Group 1 was changed to: *Figure and Rotation of the Moon*. Under this title, however, the occultation work will still be included; S. K. Runcorn stated that he includes occultations under this title on account of their contribution to the figure of the Moon. He proposed to retain Th. Weimer as Chairman of this group. The permanent members will be Gavrilov, Gurstein, Moutsoulas, Mulholland. A number of experts will be added later by the Chairman Th. Weimer.

Under Working Group 2, 'Geology and Geophysics of the Moon', S. K. Runcorn proposed to drop the prefix 'Geo' since it refers to the Earth. After discussion, the following title was decided on: *Physics, Chemistry and Geology of the Moon*. J. O'Keefe was appointed new Chairman of this group with members Dollfus, El Baz, Levin, Sonett, Urey. The WG has also the possibility to appoint consultants.

For the problem of lunar nomenclature, the frame will be slightly modified. The chair was again turned to A. Dollfus. He noted that during the next coming years, the Moon, Mars, Venus, Mercury and the galilean satellites of Jupiter will all involve nomenclature problems. The experience gained in past years shows that a certain level of organized coordination is needed between these different aspects of nomenclature. Moreover, the speed of response of a heavy body such as an IAU Working Group reporting to an IAU Commission and then to the General Assembly meeting every three years is too slow for the modern needs of space research work. Hence, a new organisation is being prepared which will allow the appropriate IAU Commissions (17, 16) and IUCM to report directly to the IAU.

The new organisation will be a Working Group for 'Planetary System Nomenclature' with the following structure: (1) there being several Task Groups, one for each celestial body (the Moon, Mars, Venus, Mercury... etc.). The lunar Task Group is traditionally nominated by the Executive Committee, the others by the Executive Committee on proposal of Commission 16. (2) the WG as such, consists of the Chairmen of the Task Groups, and in addition, the Presidents of Commissions 16 and 17 and a representative of IUCM. It is headed by a President nominated by the Executive Committee, not necessarily from among members of the WG.

In this new scheme, the responsibility of lunar nomenclature will no longer be the exclusive concern of Commission 17. The question of delegating this authority to the Executive Committee was voted on and approved unanimously.

The membership of the new 'Task Group' for lunar nomenclature will be: D. H. Menzel (Chairman), A. Dollfus (Vice-Chairman), F. El Baz, K. Florenskij, H. Masursky (for IUCM) and S. K. Runcorn (ex-officio). This composition decided by the Executive Committee was agreed upon by the Commission.

The new President S. K. Runcorn thanked the retiring President A. Dollfus for his many years on the Commission and adjourned the meeting.

Scientific Sessions

On 28 August 1973 afternoon, four invited review papers were presented to summarize the knowledge recently acquired on the following basic aspects of lunar research:

A. T. Strickland: U.S. Cartographic and Selenodetic Lunar Programs.

J. D. Mulholland: Lunar Dynamics and Observational Coordinates Systems

H. Masursky: Geology of the Lunar Crust.

S. K. Runcorn: Interior of the Moon.

On 29 August morning and afternoon, a Joint Discussion between Commissions 16 and 17 was held.

Priorities in the Study of the Moon (1973)

"The *International Astronomical Union* and the *Inter-Union Commission for Lunar Studies* having, at the end of 1973, reviewed and discussed the scientific priorities to be recommended for the continuation of the study of the Moon, realize that:

– The underlying justification for lunar studies is the light which it can cast on the origin and evolution of the Solar System as a whole.

– Extensive programmes of investigation and exploration have been developed over the past few years, including: six manned lunar missions; the operation of instrumentation on the lunar surface, aboard both fixed and mobile automatic stations; remote sensing by lunar orbiters; ground-based optical, radio, radar and laser techniques; and the analysis of lunar samples returned to Earth. Coordination of many of these programmes has been particularly well developed.

– As a result of this intensive effort, very basic progress has been made in elucidating fundamental problems such as the origin of the Moon, the ages of various surface features, the effects of meteoroidal bombardment, the Moon's thermal history, the formation of a thick lithosphere, processes of melting and magmatic differentiation, etc.

– Because of gaps which remain, both in the data presently available and in our theoretical understanding, a number of important questions about the nature of the Moon cannot yet be answered.

– In connection with these studies of the Moon, research on other Solar System bodies is likewise being directed toward laying bare the overall structure and history of the Solar System.

This research includes coordinated fly-by, orbiting or landing missions for all planets out to and including Saturn, as well as the study of some of their satellites; it will be accomplished in parallel with ground-based operations and theoretical studies. Work on asteroids, comets, meteorites and the interplanetary medium will be intensified.

– These studies will permit major advances in our knowledge of planetary processes, and will enhance the value of the lunar work; at the same time, they will be dependent on lunar studies for clarification of many points in the overall history of the Solar System, such as accretion processes, cosmochemistry, bombardment histories, etc.

From these considerations it is clear that the study of the Moon should be pursued at an intense level during the next coming decade. The *International Astronomical Union* and the *Inter-Union Commission for Lunar Studies* recommend the following further objectives for the study of the Moon.

(1) Data concerning the shape, orbital motion and dynamics of the Moon need to be improved. For this purpose both a continued extensive ground-based observational programme and high-precision tracking of lunar orbiters are required.

(2) A programme of in-situ lunar surface studies should be continued. The importance of remotely controlled roving vehicles is stressed; these offer the possibility of reaching lunar terrains of different types, of returning samples and of carrying sophisticated geophysical, chemical and imaging instrumentation.

(3) Physical and chemical parameters relating to the lunar surface and interior, having been evaluated in detail at landing sites and to a lesser degree but more widespread extent by orbiting vehicles, should now be extended to achieve whole-Moon coverage and to include new techniques. This task could be undertaken by high-inclination orbiting vehicles, supported by ground-based research.

(4) Increased activity should be fostered at an international level and on a pluridisciplinary basis, both with regard to the interpretation of available lunar data and to the mutual interaction with results from other Solar System objects. Special importance should be attached to the safeguarding, processing and wide dissemination of the body of material acquired.

It is anticipated that, with this basic minimum programme for the Moon, combined with the simultaneous studies of the other bodies of the Solar System, the fundamental problem of the origin of the Solar System could be greatly clarified during the next decade."

COMMISSION 19: ROTATION OF THE EARTH
(ROTATION DE LA TERRE)

Report of Meeting, 24 August 1973

PRESIDENT: H. M. Smith.
SECRETARY: H. J. Abraham.

First Session

The President opened the meeting and called on W. Markowitz to speak on the Independence of the Definitive ILS Polar Motion from Star Catalogues. Markowitz described how Carnera had obtained the coordinates of the Pole separately for each group for each month from the equations $x = \Sigma a_j \Delta\varphi_j$; $y = \Sigma b_j \Delta\varphi_j$; $\Sigma a_j = \Sigma b_j = 0$. Consequently Carnera's definitive declinations and definitive x and y are independent of the initial catalogue. This had been demonstrated by the method of Control Latitudes and by comparison with Fichera's results which are based on the accurate catalogue of Melchior and Dejaiffe. It follows that no change in accuracy in the determination of the periodic or secular motions would result from a change in the conventional latitudes. The CIO should be retained for continuity.

In a further paper describing the ILS and BIH Secular Motions of the Pole for 1962 to 1970, Markowitz showed that for this period the secular motion from the BIH results, based on numerous stations, was in general agreement with that from the five ILS stations, namely a motion approximately along the 135°W meridian. H. J. Abraham added that the rate of increase in the period of the librations that are shown by the ILS data is remarkably constant.

C. A. Murray was asked by the President to present two proposals by R. d'E Atkinson for discussion. The first proposal was that the zero point of right ascension on the true equator of date, used in published ephemerides including the Day Numbers shall be defined by the intersection of the true equator of date with the great circle containing the true pole of date and the mean equinox of date. The second proposal was that the luni-solar nutation used in the ephemerides, including the Day Numbers, shall correspond to that of the Earth's pole of figure and not, as at present, to that of the instantaneous pole of rotation. T. Lederle considered that these proposals should be considered in conjunction with others of which notice had been given. He had strong objections to the first proposal. The Chairman suggested that no decisions should be taken at this stage and that the proposals should be fully considered before the next Assembly.

R. O. Vicente was invited by the President to speak about certain observations obtained in Rio de Janeiro. Vicente explained that these had been made from 1924 to 1931 with an instrument that was similar to the ILS instruments and that they were now being reduced. P. J. Melchior offered to obtain corrections for the star positions.

The President then asked S. K. Runcorn to lead a discussion on the Irregular Rotation of the Earth. Runcorn first distinguished the secular changes due to lunar tidal friction, and the irregular changes which could not be ascribed to surface effects. He referred to Vestine's evidence that the Earth's electromagnetic field was drifting by about 0°2 per year, and then discussed the dynamo theory and the consequences of having to conserve momentum. There must have been exchanges of material or changes in the moment of momentum. Urey's suggestion that iron might be finding its way from the mantle to the core could explain a progressive change but the irregular slowing down and speeding up must be due to variations in coupling. It appeared that sufficiently strong currents could occur in the lower mantle as silicates would be semi-conducting at such high temperatures. The torque required to change the length of the day could be calculated. Similar impulsive torques about other axes would cause polar displacements, and such shifts were in fact observed. The polar wobble could arise from shifts of mass, which would displace the mean pole from the axis of figure. He suggested that it could also be caused by impulsive torques that would change the radius of the polhode but leave the mean pole unchanged. Examples of this could be seen from Guinot's analysis of the Chandler wobble. There was a possibility of fluctuations being caused through differences in the ellipticities and precessions of the mantle and the core, plus electromagnetic coupling. N. Sekiguchi remarked that the so-called irregular changes were not so irregular. The fluctuations had followed a sawtooth pattern that repeated almost every 30 years, except between 1920 and 1940, during which period the polar motion had been irregular. He ascribed this to anomalous action of the core. In reply to questions from Markowitz, Runcorn added that energy would be transferred through changes of field. It was important to know whether changes in the length of the day were sudden or gradual. Brouwer had fitted parabolic arcs although this implied sudden changes in torque.

L. V. Morrison was invited by the President to present a paper on Decade Fluctuations in the Rate of Rotation of the Earth. Morrison said that he had used 25000 occultation observations, obtained between 1955.5 and 1972, in fitting the coefficients in the expression for the Moon's mean longitude to the atomic (TAI) time scale. He had then used this expression for the Moon to analyse occultations before 1955.5. He had obtained $-20''T^2$ where Spencer Jones had used $-11''.22\ T^2$ in expressing the acceleration in the lunar mean longitude. The results then indicated that since about A.D. 1600 the length of the day had fluctuated irregularly with semi-amplitude about 4ms and had increased steadily by 1.5 ms/cy. Ancient data gave 2 ms/cy over a time base of 2000 years, but conservation of angular momentum in the Earth–Moon system implied 4 ms/cy. S. K. Runcorn recalled that Spencer Jones had shown that lunar effects were distinguishable but solar effects could not be isolated from the effects of internal changes. C. A. Murray said in regard to the secular acceleration term that in 1957 he had obtained $-22''T^2$ from the magnitudes of ancient eclipses, which were independent of position on the Earth's surface and hence were independent of the Earth's rotation. J. D. Mulholland added that compatible values had been obtained by Oesterwinter and Cohen ($-19''T^2 \pm 3$) and by Van Flandern. The study by Oesterwinter and Cohen might be regarded as a logical equivalent of the work by Spencer Jones but with the advantages of a more comprehensive internally consistent model and of over a decade of data based on an atomic time scale. The currently available lunar theory was adequate for reliable determinations of the secular acceleration.

B. Guinot was asked by the President to speak on his determination of the Chandler wobble. Guinot explained that he had carefully minimized the effects of changes in the catalogue or the observatories. He had found amplitude changes of 0''.1 that occurred in less than 1 year. A change occurred about every 5 years and this strongly supported Runcorn's hypothesis. No correlation with lunar data had been found.

W. Markowitz was then invited by the President to speak on Changes in the Speed of Rotation of the Earth. Markowitz referred to Brouwer's use of parabolic arcs and remarked that many other curves could be fitted. He discussed a diagram that showed variations in the length of the day from 1955 to 1970 and noted that Brouwer's and Morrison's data had indicated changes every 10 or 15 years but that recently there had been none. Melchior and Runcorn considered that the choice of parabolae was subjective and presupposed a theory. Winkler remarked that the variations were reminiscent of those in atomic clock comparisons. Markowitz then added that K. Lambeck had found that wind effects caused variations with a period of one to two weeks. Runcorn commented that the suggested change of G in Dirac's hypothesis, and the suggested change in the moment of inertia in Urey's hypothesis, would lead to opposite effects. Markowitz went on to say that the probable errors in the length of day variations were very small. He said also there was no correlation with earthquakes. The Chairman remarked that investigations by O'Hora confirmed Markowitz's views. Winkler asked whether ocean currents could account for the changes but Runcorn replied that there were many orders of magnitude in the uncertainty as to the influence of such currents.

The President then thanked the speakers for their papers and declared the session closed.

Second Session

The President opened the second session which was devoted to administrative matters. He asked the Director of the BIH to present his report. B. Guinot said that, following preliminary studies, the polar motion derived by the Dahlgren Polar Monitoring Service from Doppler measurements on artificial satellites was now included in the current work of the BIH. The DPMS data had reduced the standard deviation of the 5-day values of x and y respectively from 0ʺ011 to 0ʺ008 and from 0ʺ010 to 0ʺ008. The solution based on classical data alone would still be published. The use of Doppler measurements from stations not in the DPMS net showed satisfactory results, and a long-term programme was being inaugurated. A study of the stability of the classical instruments had emphasized the quality of PZTs and the ZTL 180 zenith telescopes. The systematic error in the proper motions of the PZT stars relative to the FK4 appeared to be not significant. The precision of the coordinates of the pole and of UT1 computed from the 80 participating observatories (with classical instruments) was improving. The possibility of further improvement was illustrated by the seasonal variation in precision. The BIH had begun to operate a rapid service in May 1971 under contract to JPL, thus giving x, y and UT1–UTC on a weekly basis with a delay of 3 days. During critical periods daily results were provided with 1 day delay. The results generally agreed with the definitive results to within 0ʺ03 for x and y, and 0ˢ003 for UT1. Winkler referred to differences between the DPMS and the classical data, and explained that the Doppler observations are affected by ionospheric conditions. Constant harmonics were assumed but seasonal effects must occur. Moreover, there are large effects due to earth tides. The President thanked Guinot for his report.

The President then asked the Director of the IPMS to present his report. S. Yumi said that more than 50 stations and observatories co-operate with the IPMS. Few are in the southern hemisphere, however. The preliminary coordinates of the pole are published in the Monthly Notes with a delay of two months, and the detailed results are published annually. Urgent problems stem from star and station errors, the definition of mean latitude, and discrepancies between the coordinates derived from time and latitude data. It was a matter of particular concern that the ILS results for 1949.0 to 1962.0, which Cecchini had compiled, were still not published. With regard to PZTs, the second at Mizusawa was completed in 1972, there would soon be one at Kitab, one was to be installed at Cagliari, Sardinia, and it was intended to install one at Gaithersburg and another at Ukiah. The Punta-Indio PZT was in operation but co-operation with Mount Stromlo had been temporarily interrupted. There was interest in installing a PZT at Tjililitan near Jakarta. The Wellington astrolabe could be used if support were provided. Brazilian astronomers had plans for an astrolabe at Macapá ($\varphi = 0°0'0''$) to co-operate with Quito, and for a PZT at about $\varphi = -8°$. It had not been possible so far to establish co-operation with the Tientsin ZTL 180 or the Shanghai astrolabe and transit instrument. Many observatories were interested in laser ranging, VLBI and Doppler observations, the results of which should be used in studies of the polar motion. Melchior stated that the IUGG had provided funds for Cecchini's results to be published.

The President thanked Yumi for his report and asked him to present the report of the Working Group on the Pole Coordinates, of which he is President. Yumi explained that the purpose of the Working Group was to organize international co-operation and promote agreement. This referred particularly to collecting and reducing the original data to an appropriate frame of reference and to obtain the best coordinates of the pole. The Working Group favoured the forwarding of all ILS data to Mizusawa, conversion to machine-readable form, and the use of the Melchior catalogue and of constants referred to the present IAU system. It had agreed also on the methods of obtaining the apparent places of the stars and the coordinates of the Sun. Data to be punched would cover 70 years and require about 2 million cards. On behalf of the Commission the President thanked Yumi for this report.

It was agreed by the meeting that the names of the following members should be submitted to the Executive Committee for future office.

President: C. Sugawa

Vice President: R. O. Vicente

Organizing Committee: H. J. Abraham, P. L. Bender, M. H. Cohen, K. Lambeck, V. Milovanovic, P. Pâquet, H. M. Smith, Ya. S. Yatskiv, S. Yumi (ex officio).

Resolutions recognizing the importance of VLBI and laser ranging to the Moon and artificial satellites were moved by Markowitz and seconded by Melchior. They were strongly supported by Bender and were passed unanimously.

In conclusion the President expressed his appreciation of the work by members of the Commission, and a vote of thanks to the President was carried with acclamation.

Resolution Adopted by Commission 19:

whereas new methods have been developed for determining the angular position of the Earth by long baseline radio interferometry and by laser range measurements to the Moon and to artificial satellites; and

whereas the accuracy expected from these techniques for determining polar motion, UT1, and crustal movements is very high; and

whereas more accurate determination of these quantities is expected to give valuable information on earthquake mechanisms, on the transfer of angular momentum between the atmosphere and the Earth, and on interaction between the core and the mantle; and

whereas the importance of some of this information is great for practical applications as well as for scientific purposes;

it is therefore recommended that: preliminary multi-national and global networks of stations to determine polar motion and Earth rotation by these techniques be established; and

that these global networks be operated for a long enough time to permit both an evaluation of the accuracies achievable by these techniques and a comparison of the results with those obtained by other methods; and

that additional technical developments be undertaken to improve further the accuracy of these methods.

Joint Session of Commission 19 and 31, 24 August 1973

CHAIRMAN: H. M. Smith.
SECRETARY: H. Enslin.

The Chairman opened the session devoted to new developments of conventional instruments for the determination of UT and Polar Motion.

S. Yumi described the new PZT at Mizusawa which has been in operation since May 1972. It has been installed in an underground room; temperature and humidity are kept constant. The plate carriage is driven by a stepping motor; photo-electric timing is employed. A cylindrical dew-cap goes up from the lens into the open air. A problem still to be solved arises from air mass movements in the cylinder in winter when the outside temperature is below that of the instrument.

The chairman noted that the Herstmonceux PZT has also been equipped with a stepping motor. The accuracy achieved in the time coordinate is inferior to that obtained in the latitude coordinate: a moiré fringe timing unit is being fitted in an endeavour to improve the timing accuracy.

G. Winkler reported on the new 65 cm/13 m PZT of the U.S. Naval Observatory. Special features are: four-element apochromatic lens, field of view of about one square degree, variable length shutter timing mechanism, facility for automatic plate changing, pneumatic device for rotating the head. Great care will be taken to avoid thermal disturbances, especially by keeping the instrument during the day below the expected night temperature. All heat-producing equipment will be removed some hundred feet from the instrument. It will be operated by an IBM Data Acquisition System and stars down to 11m will be observed. The lens is corrected for 5000–6500 Å with a filter for cut off at 5000 Å. Tidal variations of the plumb line will be checked by a mercury tiltmeter. At some later date, photo-electric observation might be considered: it promises no higher accuracy than the use of photographic plates, but it would eliminate the plate measuring.

Discussion arose on the problem of tiltmeter drift, following remarks by S. Iijima. All speakers agreed on the importance of geological and underground features. Winkler said that the tiltmeter at the U.S. Naval Observatory drifts in rain because it has been installed on a slope. He pointed out that by cross-correlation of tiltmeter and astronomical observations the tiltmeter drift can be determined. J. Terrien referred to new ideas on measuring mercury levels. The chairman said that the use of a tiltmeter is being considered at the Royal Greenwich Observatory.

S. Débarbat outlined the photo-electric astrolabe being studied at the Paris Observatory by G. Billaud. It is basically similar in construction to the full aperture astrolabe APP but a mirror will be used instead of the lens. For photo-electric observing, a fixed grid is used in conjunction with a photomultiplier tube which collects the modulated light from a star image passing the successive slits of the grid (photon-counting technique according to Høg's principle is employed). It will probably be possible to observe all FK4 stars and a reduction of errors is expected from impersonal observations as well as from several instrumental improvements. By changing the optical square, observations can be made at different zenith distances, e.g. for deducing absolute declinations.

Technical details of the astrolabe were discussed by Guinot and others. Winkler mentioned that experiments with automatic registration have been made by D. Monger with one of the two Richmond astrolabes. Two methods have been tested. A photo-electric camera has been used in one case, a photographic technique in the other case.

C. Costain proposed that the work of the PZT's and astrolabes which form chains should be encouraged. A resolution was drafted by W. Markowitz and adopted by the Commission. (Resolution attached.)

The Chairman closed the session.

Joint Resolution by Commissions 19 and 31

Considering that

Optical observations made with chains of PZT's and astrolabes which observe the same stars provide basic data for studies of the secular motion of the pole and continental drift, and that the establishment of such chains has been recommedend by the IAU,

Resolution

Urges that such existing chains, for example, Calgary – Herstmonceux and Punta-Indio – Mt. Stromlo, shall be fully supported by the administrative agencies concerned.

Proposal to the Executive Committee for a Motion by Commissions 4, 19 and 31

Commissions 4, 19 and 31 invite attention to the increasing work load of the BIH for the international coordination of time of high precision, which is needed now for satellite and deep space tracking and for radio astronomy (e.g. pulsars, VLBI, lunar laser ranging and precise coordinates). The BIH is the *only* agency which provides this important service for astronomy (as well as for other sciences).

Therefore, Commissions 4, 19 and 31 urgently request that a subsidy of $12000 per year be granted to the BIH.

23 August 1973

NEW TECHNIQUES FOR DETERMINATION OF UT1 AND PV

CHAIRMAN: H. M. Smith.
SECRETARY: L. V. Morrison.

C. Counselman (MIT) spoke on the results obtained from wide-band Very Long Base Interferometry. The technique consisted of measuring the time delay between widely spaced antennae of wave-fronts from extragalactic sources whose relative positions were known with an accuracy of about $0\overset{''}{.}01$. These data together with an atomic clock pulse were recorded on magnetic tape at each antenna and later analysed using the clock pulses to establish the coherence of the signals. The difference in clock rates was eliminated by observing four different sources in each set of observations. The present accuracy of the measured time delay, expressed in angular resolution, was of the order of 0".05, but it was likely that this would be improved by a factor of 10. A comparison of four results for the x-component of polar motion agreed with the BIH values to about 2 metres, and the UT1 results agreed with the USNO PZT values to about 5 ms. It was found that the N–S base line varied by about 2 metres. P. Bender asked if the comparisons had been made with the mean BIH 5-day values. B. Guinot advised using the definitive values in the BIH Annual Reports. N. Matsunami enquired about the stability of the masers, and Counselman said that clock jumps could be easily identified. In reply to a question by L. Morrison, Counselman stated that 17 sources were used, but that there are more with sufficiently accurate positions. I. Soltau asked if Earth tides could explain the 2 metre variation: P. Melchior said that they are no more than ½ metre.

B. Elsmore (Mullard Radio Astronomy Observatory, Cambridge, U.K.) reported on results obtained using a linked E–W interferometer comprising two aerials separated by 2 km. A radio source of known position was tracked continuously from -6^h to $+6^h$ in hour angle. The measured phase difference gave UT1 directly, as the measurements were independent of the instantaneous position of the pole. The accuracy of UT1 realized from one passage of a radio source was ± 4.6 ms, which is to be compared with the value of ± 4.0 ms for a single night's observation using a PZT. A limitation of the method is the irregularity of water vapour in the troposphere. B. Guinot regarded an accuracy of ± 4.0 ms as too optimistic for the results of one night using a PZT.

C. Oesterwinter (U.S. Naval Weapons Laboratory) reported on the results of polar motion for 1972 obtained by R. Anderle using doppler observations of artificial satellites. The standard error for the determination of the position of the pole based on 5 days of observations was as low as 20 cm during periods when data from 19 stations on 3 satellites were used. The x and y components of polar motion were in good agreement with the BIH values. P. Melchior (Royal Observatory, Belgium) reported on the work of P. Pâquet and R. Dejaiffe who had intercompared the results from doppler and astrolabe observations, both made at the Royal Observatory. They had carried out an analysis deducing the internal and external standard deviations due to both methods. In order to reduce the standard deviation in longitude for the doppler results to that of the astrolabe, about 5% of the satellite passes had to be rejected. A comparison of the doppler latitude results with BIH values for 1972 showed a small systematic shift in phase and amplitude. B. Guinot remarked that the phase-amplitude difference was not surprising; the results were subject to annual fluctuations. P. Melchior agreed and suggested that in the case of doppler measurements, annual variation in the troposphere should be investigated. W. Markowitz asked if the results quoted for the coordinates were astronomical or geodetic. Melchior said that as derived from the doppler measurements, they were geodetic, using the NWL Earth model. J. Mulholland agreed that there are reasons for transforming station coordinates determined geocentrically into some geodetic system, but that for global work the geocentric values are the most useful. He asked that these also be published. Bender accepted the drift in excess of 50 cm/yr earlier reported by Oesterwinter as possible for the Japanese and Philippine stations, but not for the U.S. ones. He suggested that they might be due to short term phenomena, possibly originating in the ionosphere.

K. Johnston (U.S. Naval Research Laboratory) reported on the determination of UT1 using VLBI measurements of molecular H_2O sources at 22.2356 Hz. Since the sources have very narrow intrinsic bandwidths, only the fringe rate was measured. For the strongest sources, the fringe rate residuals were about 2.5 mHz, and using a 4000 km E–W base line it would be possible to determine UT1 with an accuracy of about 2 ms. G. Winkler asked how the maser clocks were synchronized and suggested that some external means was required because they drift with time. Johnston replied that they were simply measuring a difference in transit time.

P. Bender (National Bureau of Standards and University of Colorado) reported on the use of Lunar Laser Ranging (LLR) to determine UT1 and PV. The 5-day mean BIH values were accurate to about ± 40 cm ($\pm 0".03$) for PV: with about 6 stations distributed in both hemispheres observing at hour angles -3^h, 0^h and $+3^h$ LLR could attain an accuracy of a few centimetres for PV after the removal of lunar ephemeris errors. From a 9- to 10- hour observing period on a particular day the residual from the McDonald observations had a RMS of about 11 cm which demonstrated that there were no short term unknowns in the lunar ephemeris. In future it was hoped that LLR stations would be set up in other countries, particularly in the southern hemisphere, which would undertake programs consisting of range measurements three times daily to at least one reflector in order to detect changes in the Earth's rotation which occur at intervals down to one day. Chairman: Before leaving the meeting, I would like to say that there seems to be a need to coordinate the results on UT1 and PV coming from the various techniques that we have heard about this afternoon. I now invite Professor Melchior to take the Chair. P. Brosche enquired if the solid Earth tides of the order of 20 cm would affect the measurements. Melchior pointed out that only the radial component was involved; and that this is dependent on the Love number h. Bender said that the effect was less than 1 cm. H. Eichhorn referred to the distance of the Moon being measured to about 15 cm, but wondered if the analytical theory was equally accurate. Mulholland said it was over a short period and added that the limitation comes in the angular position of the Moon in the sky and here the accuracy of the ephemerides is at present comparable to the ILE, that is, about 0″75. This was achieved by the synthetic fit of a numerical integration to the ILE.

L. Morrison said that by using occultations this could be improved to 0″4. Bender drew attention to the limitation of the libration theory but said that with the extension to include third and fourth-degree terms in the lunar gravity field, fitting of data with a residual RMS of about 2 m has been achieved: the inclination and node corrections are required. In reply to a further question by Morrison, Mulholland said that no estimates had been obtained for the corrections to the inclination and node, but the problem was being investigated with the co-operation of the USNO.

D. Smith (Goddard Space Flight Center) reported on results obtained from doppler ranging to artificial satellites. After adjustment for the harmonics of the Earth's field, a residual fit with RMS of 50 cm had been achieved for single passes over 2 to 3 hours. This had later been improved to 30 cm. Tidal deformation had been removed by taking k_2 as 0.25. The variation in latitude had been compared with BIH data and the residual RMS was of the order of 0″03. There were plans to launch a high-altitude laser reflector and here the limitation would be due to variations in the Earth's albedo pressure. The aim was to have a daily measure of PV to within a few centimetres and UT1 to 100 μs.

There then followed a general discussion on the possibility of future use of the new techniques in determining PV and UT1 on a regular and routine basis. In particular, it was considered doubtful whether VLBI could provide a routine service as the technique relied on the co-operation of the existing network of NASA stations and these were at present subject to financial constraints.

The Chairman then closed the session.

COMMISSION 20: POSITIONS AND MOTIONS OF MINOR PLANETS, COMETS AND SATELLITES (POSITIONS ET MOUVEMENTS DES PETITES PLANÈTES, DES COMÈTES ET DES SATELLITES)

Report of Meetings, 24, 25 and 27 August 1973

PRESIDENT: F. K. Edmondson.
SECRETARY: B. G. Marsden.

First Session

The President opened the meeting by calling upon the Secretary to read notices of the deaths of two members since the last meeting: Yrjö Väisälä (1891–1971) and Michael Kamieński (1879–1973). The meeting stood in silence for a few moments in their memory.

A survey indicated that only about half of the members present had seen the report of the Commission, and discussion of the report was therefore postponed until the final session.

The meeting confirmed the proposals of L. Kresák for incoming President and B. G. Marsden for incoming Vice-President. The following members retired from the Commission: S. Böhme, G. M. Clemence, Y. Kozai and I. I. Shapiro. The following new members were elected: K. Aksnes, Yu. V. Batrakov, S. Ferraz-Mello, R. S. Harrington, A. R. Klemola, L. Kohoutek, B. A. Lindblad (co-opted since the 1970 meeting), R. E. McCrosky, B. Mintz Blanco, B. Morando, K. Tomita and J. G. Williams. The following were elected to the Organizing Committee for the next triennium: F. K. Edmondson, P. Herget, E. I. Kazimirchak-Polonskaya, B. Morando, W. H. Robertson, E. Roemer and J. Schubart. E. Roemer agreed to continue as Chairman of the Working Group on the Orbits and Ephemerides of Comets, and the other members of the Working Group were to be: M. P. Candy, E. I. Kazimirchak-Polonskaya, L. Kresák, B. G. Marsden, G. Sitarski and K. Tomita.

There was then some discussion on the desirability of appointing a Working Group devoted to the motions of satellites. Although the study of satellites is indicated in the name of the Commission, satellites have habitually been de-emphasized in relation to minor planets and comets. On the other hand, the reality of space missions to the outer planets had now made it very necessary to overhaul the theories of the motions of the satellites. The meeting agreed that the appointment of a Working Group on the Orbits and Ephemerides of Satellites would be very appropriate, and the following members were proposed for the group: B. Morando (Chairman), K. Aksnes, Yu. V. Batrakov, S. Ferraz-Mello, J. Kovalevsky, J. L. Sagnier and G. A. Wilkins; J. D. Mulholland and J. Lieske were co-opted as consulting members.

The possible co-sponsorship of two IAU Colloquia was then considered:

(1) 'An International Conference on the Natural Satellites of the Planets', to be held at Cornell University, Ithaca, New York, in June 1974. The President read sections of a letter received from J. A. Burns, the initial organizer of the conference. K. Aksnes was appointed the Commission's official representative on the Organizing Committee.

(2) 'Study of Comets', to be held at Goddard Space Flight Center, Greenbelt, Maryland, in October or November 1974. B. Donn was prepared to talk about this proposed colloquium at the third meeting of the Commission.

The Commission approved the earlier action by the President in asking the Finance Committee for continued support of the Minor Planet Center in the amount of $2000 for the triennium 1974–76. (The Finance Committee increased the amount to $2200.)

The Secretary read a resolution, prepared with the assistance of B. Morando and J. D. Mulholland, concerning the need for more intensive observations of the natural satellites of the outer planets, and after a brief discussion the resolution was adopted by the Commission. (See Resolution No. 3 at the end of this report.)

Z. Sekanina proposed that observers of positions of comets should be encouraged to supply information about the position angles and lengths of cometary tails. E. Roemer suggested that a more general proposal might be desirable, and further action was deferred until the final session.

The President mentioned an enquiry received from M. B. Protitch about the resolution passed at the General Assembly in Brighton concerning the difficulty of obtaining good quality photographic plates in reasonable quantities. No additional information was available on this subject.

In response to a question relayed from the Assistant General Secretary concerning the need for commissions to adopt 'rules', F. L. Whipple proposed that Commission 20 adopt a rule that there be no rules. This was seconded by R. E. McCrosky and carried unanimously.

There was then some discussion on the suggestion by the Assistant General Secretary concerning future commission reports. After some discussion by the President, the Secretary, B. Morando, F. L. Whipple, E. Roemer and others, it was agreed that Commission 20 should take a strong position in favor of continued publication of the reports in the IAU *Transactions*. It was also suggested that the Executive Committee should make an effort to ensure that copies of the report of a commission would be available to members of that commission in advance of its first meeting.

Finally, the President briefly reported on the results of the joint meeting to discuss the future of the solar system commissions and on the subsequent deliberation by the Executive Committee. It was indicated that there would be little change in the activities of Commission 20, except that the physical studies of minor planets would be transferred to Commission 15.

Second Session

This session, again chaired by the President, was devoted to scientific papers on minor planets. J. G. Williams gave an intensive review on the importance of continuing to observe minor planets and suggested priorities for future observing programs. B. G. Marsden gave an account of a joint investigation by C. M. Bardwell of the Minor Planet Center and himself on the 'critical list'. E. Everhart then discussed the possibility of the existence of 'horseshoe orbits' in the Sun–Earth system and suggested that appropriate observations near the quadrature points could perhaps reveal objects moving in such orbits.

Third Session

This session was a joint meeting with Commissions 15 and 22 and was chaired by A. H. Delsemme, Vice-President of Commission 15.

E. Roemer brought up the question of the future of 'Comet Notes', and it was agreed that *Icarus* would be a suitable home for them if special arrangements could be made to cut down to no more than three months the time between submission of the manuscript and publication. The *Bulletin of the American Astronomical Society* was considered as an alternative if page charges could be waived.

B. Donn described the proposed conference on 'Study of Comets'.

Papers by F. L. Whipple and B. Donn then considered the desirability of launching space missions to comets; and B. G. Marsden gave an account of recent work by Z. Sekanina and himself on the nongravitational forces that affect the motions of comets.

Fourth Session

This session was organized by Commission 20's Working Group on Comets and was chaired by E. Roemer, B. Donn, E. Everhart and A. H. Delsemme all presented papers on the 'Origin of Comets', the last-named author restricting himself to the short-period comets. Z. Sekanina then discussed 'New Empirical Data on Distant Comets'; and T. Gehrels and C. Sagan reported on proposals for high-resolution imaging of cometary nuclei from spacecraft.

Fifth Session

Held under the chairmanship of B. Morando, this session was devoted to the study of orbits and positional observations of the natural satellites of the outer planets. The following papers were presented: J. Kovalevsky, 'The Present Status of the Theories of the Motions of the Satellites of the Planets'; G. A. Wilkins, 'Current Work in the United Kingdom on the Motions of the Satellites'; K. Aksnes, 'On the Need to Improve Satellite Ephemerides'; and S. Ferraz-Mello, 'An Analysis of Photographic Series of Observations of Jupiter's Galilean Satellites'.

Sixth Session

The final administrative session was held under the chairmanship of the President. M. P. Candy gave a brief description of the minor-planet observing program at the Perth Observatory, pointing out that the program is conducted in a manner somewhat similar to the suggestions by Williams and Marsden at the second session.

In response to the invitation for criticism of the Commission's report in IAU *Trans.* XVA, E. Everhart indicated that lines 11–13 on page 557 should read: "(1) Everhart reports that he has papers in press on 'Horseshoe and Trojan Orbits Associated with Jupiter and Saturn'; 'Examination of Several Ideas on Comet Origins'; 'Earth Horseshoe Orbits'; and 'A New Method of Integrating Orbits'." The report was then accepted.

It was agreed that Commission 20 should support Commission 15 in the proposal by B. Donn for a conference on 'Study of Comets'. F. L. Whipple, E. Roemer and B. G. Marsden would represent Commission 20 on the Organizing Committee.

A rewording of Z. Sekanina's proposal, deferred from the first session, was discussed by E. Roemer, E. Everhart, F. L. Whipple and B. G. Marsden and adopted by the Commission (see Resolution No. 2 below). The Secretary then brought up a proposal, by J. G. Williams and himself concerning minor planets (Resolution No. 1), and after some discussion this too was adopted.

Resolutions

(1) Noting that the statistics of the minor planets are complete only to a mean opposition magnitude of about 15, and that many of the families of minor planets therefore contain very few known members;

recognizing that certain classes of unusual minor planets (Apollo and Amor objects, objects with high orbital inclinations, objects whose motions are commensurable with that of Jupiter, etc.) appear to be underrepresented among the numbered objects and in minor-planet surveys;

Commission 20 urges observers to search for and to report unusual objects.

The Commission points out that it is still scientifically useful to discover and follow up main-belt objects.

(2) Observers of positions of comets are urged also to note and report physical characteristics whenever possible. Of particular interest are:

Brightness, distinguishing between 'total' (m_1) and 'nuclear' (m_2) magnitudes;

Dimensions and any distinctive properties of the coma, such as asymmetries or other unusual brightness distribution;

Length, orientation and structural characteristics of the tail, if any.

Notice of unusual activity should be communicated by telegram or airmail letter to the IAU Central Telegram Bureau.

(3) Noting that the ephemerides of most of the natural satellites are based on observations made

long ago, and that efforts to improve the ephemerides have been hampered by the paucity of recent observations;

acknowledging that much higher accuracy in the ephemerides of these objects will soon be required, particularly for the exploration of the outer solar system by spacecraft;

Commission 20 urges that a reasonable amount of telescope time be made available for observations useful for the improvement of satellite ephemerides. (This resolution was endorsed by Commission 4.)

COMMISSION 21: LIGHT OF THE NIGHT SKY (LUMINESCENCE DU CIEL)

Report of Meeting, 24 August 1973

PRESIDENT: H. Elsässer.

Business Meeting

It was agreed that the Commission, in cooperation with Commissions 15 and 22, and the Dust Panel of COSPAR, should sponsor a colloquium on 'Interplanetary dust and the zodiacal light' which shall take place in Heidelberg in 1974 or 1975. New results of the space experiments being in orbit or being in preparation at the present time will represent one of the main topics of this colloquium.

With regard to a letter of Dr Weill on the problem of man-induced light and to the plans of the IAU-Executive Committee to form a new commission devoted to the protection of observatory sites the resolution was adopted:

'Commission 21 supports the creation of an IAU-working group on light pollution problems and is willing to assist in this work. Dr Weill has proposed a colloquium on this subject in Paris and we fully support this.'

Scientific Meeting

The following papers were presented:

R. Dumont and F. Sánchez-Martinez: From Zodiacal Light Photo-Polarimetry Around the Ecliptic to Phase Functions and Polarization Curves of Interplanetary Scatterer.

H. Elsässer: New results on the Zodiacal Light from Rocket- and Balloon-Measurements.

R. D. Wolstencroft: Recent work in Hawaii on Polarimetry of the Night Sky.

H. Tanabe: Integrated Star Light Derived from Star Counts.

COMMISSION 22: METEORS AND INTERPLANETARY DUST (MÉTÉORES ET LA POUSSIÈRE INTERPLANÉTAIRE)

Report of Meeting

PRESIDENT: R. E. McCrosky.
SECRETARY: W. G. Elford.

The business meeting was called to order at 15.15 on 24 August 1973. Nine members and two guests were present. The President was in the chair. He reported with regret the deaths of Dr Guigay of France and Professor Fesenkov of the USSR.

REPORT

The new format of the report has been generally favorably received. Two adverse comments were noted. The President pointed out that the discontinuance of the draft report procedure had made possible more up-to-date reporting, but as a consequence there was no way of reaching a consensus in the case of any disagreement. It was agreed that the new procedure was to be preferred. The current report was accepted.

In reply to a question from the Executive regarding future reports, it was agreed that;

(1) Reports are of scientific value, are used for reference purposes by members and should be continued in their present format.

(2) The Executive seek a less expensive form of publication.

NEW OFFICERS

The Executive Committee announced the appointments of B. A. Lindblad as the new President and I. Halliday as the new Vice President. The slate for the new Organizing Committee was accepted. The members are: A. F. Cook, W. G. Elford, T. R. Kaiser, L. Kresak, B. J. Levin, R. E. McCrosky and F. F. Verniani.

NEW MEMBERS

The following were welcomed as new members of Commission 22: B. L. Kashcheev, V. N. Lebedinets, J. A. M. McConnell, R. G. Roosen, Z. Sekanina, A. E. Simonenko and G. W. Wetherill.

CONSULTANTS

Recommendations for consultants to the Commission are to be forwarded to the new President within three months.

REORGANIZATION

The proposals on reorganization of Commissions 15, 16, 20, 21, and 22 as presented in the interim report of the General Secretary were approved unanimously. It was further agreed that the General Secretary be informed that the topics of meteorite craters and tektites should be retained within Commission 22.

COMMITTEES

It was agreed that, as a consequence of the reorganization of Commissions 15 and 22, and the specific inclusion of the words 'interplanetary dust' in the title of Commission 22, the two existing committees be abolished.

ZLIM CONFERENCE

The Commission agreed to the proposal to co-sponsor, with Commission 21, a conference on the 'Zodiacal Light and Interplanetary Medium' to be held in Heidelberg in 1974 or 1975.

METEORITE NOMENCLATURE

It was agreed to support the establishment of a committee on meteorite nomenclature as proposed by the Meteoritical Society.

TEPEXITL CRATER

Dr Paris Pişmiş has supplied the Commission with preliminary information on a newly recognized meteorite crater in Puebla, Mexico. The structure is 1180 m in diameter, 65 to 92 m in depth and the rim is from 18 to 35 m high. This was proposed as a probable meteorite crater by Lucrecia Manpome in 1972. A full report will be published in *Bol. Obs. Tonantzintla y Tacubaya*. The Commission congratulates Mrs Manpome on this exciting discovery and recommends that investigation be continued.

A committee on Radar Observations of Meteor Flux and Radiants, and Anomalies at the Base of the Thermosphere was established to investigate the value of and the optimum methods for making world wide observations. The committee consists of: Keay (Chairman), Elford, Kaiser, Lindblad, McIntosh, Verniani.

COMMISSION 24: PHOTOGRAPHIC ASTROMETRY (ASTROMÉTRIE PHOTOGRAPHIQUE)

Report of Meetings, 24 and 29 August 1973

PRESIDENT: S. Vasilevskis.
SECRETARY: C. A. Murray.

After welcoming members to the meetings of the Commission, the President reported on changes in the list of members. Six members, Cox, Günther, König, Semirot, Stearns and Wilkens, had died, eight had resigned and 17 new members had been admitted. The current membership was now 57.

The President announced that the following had been nominated as officers and Organizing Committee for the ensuing three years, for approval by the General Assembly:

President: P. Lacroute.

Vice-President: C. A. Murray.

Organizing Committee: A. N. Deutsch, W. Dieckvoss, K. Aa. Strand, H. W. Wood, A. Blaauw, W. Gliese, W. van Altena

The President then said that the Executive Committee of the Union had asked Commission Presidents to review the scientific objects and the rules of conduct of their Commissions, and to indicate broad scientific priorities for future work within their particular fields. In the discussion it was agreed that the Commission did not need a set of rigid rules, but that nominations for membership should be made solely by members of the Commission on the basis of the individual candidate's contribution to parallax and proper motion work. The new name of the Commission did not give due weight to modern developments in photo-electric and space techniques for astrometric work, but nevertheless it was generally felt that it would be unwise to seek to change the name of the Commission again. The objectives of the Commission remained the measurement of stellar parallaxes and proper motions by any suitable techniques.

The Executive Committee had also called for views on the publication of *Reports on Astronomy* in Volume A of the *Transactions* of the Union. In discussion it was generally agreed that the volume as a whole, by bringing together reports on different branches of the subject, provided a very useful reference on the state of astronomy, and should be continued in its present form.

The President then referred to the draft report of the Working Group on Nomenclature of Faint Stars which had been set up at the Brighton meeting under the Chairmanship of *Murray*.

The Working Group had proposed a scheme based on rectangular co-cordinates on Palomar Sky Survey prints for the northern hemisphere, with extension to the south when the current surveys had been completed and published. The report had been circulated to all members, and had received a majority approval in a postal ballot. However, there had been two objections and counterproposals, by *Thomas* and by *Giclas* and the President regretted that he had acted incorrectly in putting the original draft report to the vote. Accordingly he proposed that the Working Group continue to act, in consultation with the Stellar Data Centre at Strasbourg, to see whether a practicable scheme could be implemented. In discussion, *Luyten* urged that whenever possible, established identifications such as BD or CoD numbers should continue to be used; editors of journals should insist on precise statements from authors on their sources of nomenclature for stars. *Murray* remarked that the southern sky charts, to be based on surveys with the ESO and U. K. Schmidt telescopes would have position fiducial marks printed on them to facilitate identifications by means of rectangular co-ordinates, as proposed by the Working Group.

Gliese opened the first scientific session with a review of the need for trigonometric parallaxes in the southern hemisphere. While 45% of the stars in the two Jenkins Catalogues had negative declinations, the distribution in declination of stars with the most accurate parallaxes (p.e. ≤ 0″.006) was:

δ+90°	+30°	0°	−30°	−90°
%	38	42	17	3

This emphasized the great disparity between the two hemispheres. Since 1963 only the 24 in. refractor at the Cape had contributed to parallaxes in the south. Urgent problems requiring parallax observations in the southern hemisphere included the frequency of nearby M-dwarfs, particularly in the region of the south galactic pole. He welcomed the plans for measurement of parallaxes in this region announced by *Murray* and *Hoffleit* at the recent Perth Symposium.

Stock said that he was anxious to undertake astrometric programmes at the new observatory in Venezuela (latitude +8° 48′). His equipment would include a 1 m *f*/3 Schmidt, with objective prism, a 1 m *f*/20 coudé reflector, a long focus refractor with aperture 65 cm and focal length 10 m and a double astrograph with aperture 51 cm and focal length 3.75 m.

Stoy then spoke about the present status and future needs of proper motions in the southern hemisphere. He drew attention particularly to the large amount of material at present available in published catalogues extending from the early work of Gill, the Carte-du-Ciel catalogues and the AGK1, to the extensive photographic surveys by Yale and the Cape observatories, all of which should be combined into a general catalogue; this could then be used with the modern Cape and Sydney surveys, which will be reduced to the FK4 system by means of the extensive SRS programme, to give a much improved system of fundamental proper motions in the south.

In the second scientific session, *van Altena* described the various methods of reducing relative trigonometric parallaxes to an absolute system. Current trends in parallax work were toward higher internal accuracy and a reduction in the size of systematic differences between observatories; for example the difference between the system of the new Yerkes parallaxes and that of the 61-in. USNO reflector was only 0″.0006 with a standard error of ±0″.0025. He described a new method of reducing parallaxes to absolute, which was based on a revised kinematic model for secular parallaxes, but gave some weight to the relative proper motion distribution of the reference stars in each particular field.

Murray then reviewed the various methods of deriving absolute proper motions. He pointed out examples of apparent inconsistencies of about 0″.01 p.a. between AGK3 and FK4, and AGK3 and the kinematic reduction method, and suggested that an extensive comparison between AGK3 and the FK4 system, through the GC, should be carried out. He also emphasized the apparent discrepancy between the Lick and Pulkovo results for the rotation in right ascension, but pointed out that systematic errors of a few microns in either or both surveys could account for it; it was an urgent matter to resolve the discrepancy. *Dieckvoss* warned that the zero point of AGK3 motions in a limited field would have a mean error of about 0″.003; also, a detailed comparison between AGK3 and GC was being carried out at Bergedorf by de Vegt.

Luyten summarised IAU Symposium No. 54 on 'Problems of Calibration of Absolute Magnitudes and Temperatures of Stars' which was held at Geneva on 1972, September 12–15, and *Hoffleit* gave a short account of the recent Symposium No. 61 on 'New Problems in Astrometry' which was held in Perth immediately prior to the General Assembly.

After general discussion the following statement of priorities was approved.

Astrometry provides data essential for the solution of such basic problems as the determination of the scale of the Universe, and thus for studies of the luminosities, distribution and kinematics of celestial objects; experience has shown that useful astrometric data can only be acquired by long, sustained series of observations. Commission 24 therefore sets as its most important priorities:

(1) The vigorous continuation of existing trigonometric parallax programmes in the northern hemisphere and a major expansion in the southern hemisphere, with an extension to faint magnitudes over the whole sky.

(2) Consolidation of the zone catalogues over the whole sky into a definitive general catalogue of positions and proper motions relative to the fundamental system.

(3) Determination of proper motions relative to extragalactic objects over the whole sphere in order to relate the fundamental system to an inertial frame and to provide absolute data for the study of kinematics in the Galaxy.

The following resolutions were proposed and carried unanimously; a preliminary text of Resolution No. 1 had previously been endorsed by a joint meeting of Commissions 8, 24, 33 and 40 which was held on 25 August.

1. Commission 24, taking note of Resolutions No. 1 and 2 of IAU Symposium No. 61, emphasises

(a) the extreme importance to fundamental and stellar astronomy of the Yale-Columbia programme for the determination of proper motions relative to extra-galactic objects in the southern hemisphere; it therefore urges that the full programme of first epoch photography in two colours should be completed as soon as possible, and arrangements made to ensure that second epoch photography should start within the next five years.

(b) the potential value of new optical, radio and space techniques for astrometry, and supports programmes for the measurement of proper motions and parallaxes which may be devised for the Large Space Telescope and other space projects.

2. Commission 24 urges that regular and continuing allocation of observing time on suitable large telescopes be made for programmes of trigonometric parallaxes and proper motions of faint stars.

3. Commission 24, noting the extremely favourable position close to the Equator of the new observatory in Venezuela and the urgent need for basic astrometric data in the southern sky, recommends as highly desirable fields of work

(a) trigonometric parallax observations, especially of stars in the southern hemisphere

(b) positional observations in the equatorial zone between declinations $+30°$ and $-30°$, with specially high priority for the portion between 0° and $-30°$.

4. Commission 24, noting the key role played by astrometric data in the solution of many of the most important problems of astrophysics, urges that the methods and achievements of astrometry should be regarded as an integral part of the education of young astronomers.

5. Commission 24 notes with admiration the speed, accuracy and adaptability of the automatic plate scanning and measuring machine developed for the Palomar Schmidt Proper Motion Survey and its great potential for other astronomical problems. It therefore urges that this machine be kept in commission and made available to qualified investigators after the termination of the present project.

COMMISSION 25: STELLAR PHOTOMETRY AND POLARIMETRY (PHOTOMÉTRIE ET POLARIMÉTRIE STELLAIRES)

Report of Meeting, 23 August 1973

PRESIDENT: D. L. Crawford.

The members of the Commission have decided to add polarimetry to the objectives of Commission 25. These members suggest also to change the name of the Commission and recommend 'Stellar Photometry and Polarimetry'.

The members of the Commission support the proposal of Dr T. Gehrels concerning definition and convention in polarimetry.

The question of 'priorities' has been discussed and the need of precise multicolour photometry in a very limited number of celestial locations but up to the 20th magnitude has been underligned. Only stars with at least rough spectral data must be measured.

On behalf of the Commission, Dr A. R. Hyland must prepare a project for a photometric standard in infra-red.

The following papers have been presented:

B. Hauck: Activity of the European Data Center (Strasbourg) in the Field of Photometry.

B. Hauck: Photometric Catalogues.

T. Gehrels: Astronomical Polarimetry.

T. Gehrels: Conventions and Definitions of Terms in Polarimetry.

B. J. Bok: Catalogue of Photometric Sequences.

A. R. Hyland and J. A. Thomas: IR Photometry Standards.

G. Goy: UBV Diagram with 30000 Stars.

COMMISSION 26: DOUBLE STARS (ÉTOILES DOUBLES)

Comptes Rendus des Séances

PRÉSIDENT: J. Dommanget.
SECRÉTAIRE: O. G. Franz.

Première Séance

Avant de passer à l'ordre du jour, l'assemblée, sous la présidence de J. Dommanget, désigne le secrétaire de séance. Le choix se porte sur O. G. Franz, qui accepte d'en assumer la charge.

La Commission respecte ensuite quelques instants de silence pour honorer la mémoire de deux de ses membres: C. E. Luplau-Janssen et G. B. van Albada, ainsi que d'un ancien membre: J. Dick, tous trois récemment décédés.

Le Président signale que la Commission a dû enregistrer avec grand regret, l'arrêt de l'inlassable activité de trois de ses meilleurs observateurs: W. H. van den Bos (72000 mesures), P. Baize (40000 mesures) et W. S. Finsen (15000 mesures dont 6000 mesures interférométriques de couples particulièrement difficiles. Il leur exprime une fois de plus, au nom de la Commission, sa plus profonde gratitude pour le travail accompli.

Etant donné la faible participation que l'on pouvait craindre depuis longtemps à la présente Assemblée Générale et l'impossibilité de prendre aucune décision valable dans une telle situation, le Président a décidé d'interroger tous les membres de la Commission par une lettre circulaire du 13 juin 1973, sur les principaux points prévus alors à l'ordre du jour de la Commission. Sur ses 38 membres, 15 membres absents aujourd'hui ont répondu, ce qui, avec les 7 membres présents à la séance, permet d'assurer une majorité de 22 votes valables. Il fut précisé par ailleurs que l'absence de réponse serait considérée comme une approbation implicite des décisions qui seraient prises sur la base des votes effectifs. A la demande du Président, l'assemblée approuve le processus.

Le Président signale aussi avoir reçu de la plupart des quinze membres absents ayant répondu à la lettre-circulaire, des souhaits de fructueux travaux pour la Commission. Il les en remercie vivement.

1. Changement de nom de la commission no. 42 (photometric binaries)

Ce point de l'ordre du jour est examiné en tout premier lieu afin de permettre au Président de la Commission No. 42, M. Plavec, de présenter son point de vue et de participer à l'éventuelle discussion qui s'en suivra, avant d'assister à d'autres réunions.

Le Président signale que la nouvelle dénomination proposée pour cette commission (Close Binaries) a suscité la réticence unanime des 15 membres cités plus haut, cette dénomination ayant une signification différente pour les observateurs d'étoiles doubles visuelles. M. Plavec démontre la nécessité absolue pour sa Commission d'abandonner l'ancienne dénomination, laquelle ne recouvre plus son activité actuelle, puisqu'elle porte sur les couples réellement serrés, qu'ils soient observables photométriquement par suite d'une forte inclinaison orbitale ou qu'ils soient seulement observables spectroscopiquement. Après un échange de vues entre les membres présents, la Commission décide de ne pas formuler d'opposition à cette proposition de changement de nom, mais de faire connaître au Secrétaire Général, ses objections et ses regrets devant l'expression choisie.

M. Plavec remercie la Commission pour sa compréhension.

2. Composition de la commission

Sur proposition du Président et en application de l'article 20b, chapitre III des 'Directives' de l'Union, sont considérés comme démissionnaires: B. Cester, O. J. Eggen et G. P. Kuiper, ces collègues n'ayant pas répondu à plusieurs lettres les invitant à faire connaître leur désir de poursuivre leur activité au sein de la Commission. Toutefois, dans le cas de B. Cester, sur proposition de M. G. Fracastoro, le statu quo sera provisoirement observé.

Le Président fait part de la proposition du Comité Organisateur sortant de considérer comme nouveaux membres d'une part: A. A. Kiselev et P. S. The, membres de l'Union et d'autre part: R. S. Harrington, P. J. Morel, R. E. Nather, G. A. Starikova et R. L. Walker, pour autant que ces chercheurs soient élus membres de l'Union sur recommandation de leurs comités nationaux respectifs.

Ces propositions sont adoptées.

La Commission choisit ensuite son nouveau Comité d'Organisation dont la composition sera soumise au Comité Exécutif de l'Union pour approbation. La proposition est issue de concertations entre les membres du Comité sortant, visant à une représentation aussi variée que possible des diverses spécialités existant au sein de la Commission, tout en respectant un minimum de continuité dans sa gestion.

W. Heintz ayant, pour des raisons personnelles à caractère scientifique, décliné l'offre de faire partie du Comité d'Organisation, sont proposés: A. H. Batten, P. C. Couteau, A. N. Deutsch, J. Dommanget, O. G. Franz, S. L. Lippincott, P. Muller et K. Aa. Strand.

Pour la Présidence et la Vice-Présidence, le Comité d'Organisation sortant propose respectivement: Miss S. L. Lippincott et P. Muller.

Ces propositions sont chacune adoptées malgré deux votes défavorables que leurs auteurs ont justifiés auprès du Président par une opposition au mode d'élection du Bureau et du Comité d'Organisation, insuffisamment démocratique à leur avis.

L'un d'eux, F. Holden a adressé à ce sujet une lettre-circulaire (14 juillet 1973) à tous les membres de la Commission, les priant de faire connaître au Président avant l'Assemblée Générale, leur éventuelle approbation d'une réforme consistant dans des élections au scrutin secret ('written ballot'). Seules 5 réponses ont été reçues: elles comportent une abstention, deux réponses favorables et deux autres entièrement opposées à la proposition. Une des réponses favorables (U. Güntzel-Lingner) comporte par ailleurs une proposition d'amendement.

Le Président estime qu'un large débat sur cette importante question est souhaitable et que, pour lui donner une solution précisément aussi démocratique qu'on la réclame, la présence d'un grand nombre de membres est exigé. Aussi, propose-t-il de renvoyer cette question aux séances de la Commission lors de la prochaine Assemblée Générale et de la traiter, comme le conçoit P. Muller, dans le cadre d'un règlement d'ordre intérieur. Indépendamment d'ailleurs, le Secrétaire Général vient de prier les Commissions d'établir un tel règlement conformément à l'article 22 du 'Règlement' de l'Union.

Comme bien d'autres présidents de commissions, J. Dommanget estime toutefois qu'il faudra éviter l'établissement de règles trop strictes, mais bien plutôt rechercher la mise au point de principes de fonctionnement démocratiques. De telles règles risqueraient de compliquer sinon de bloquer le fonctionnement de la Commission.

Le Président transmettra à tous les membres de la Commission, une copie des documents reçus à ce sujet du Secrétaire Général. Comme le souhaitent K. Aa. Strand et A. H. Batten, la mise au point d'un projet à discuter à la prochaine Assemblée Générale, sera laissée aux soins du nouveau Comité d'Organisation et du nouveau Bureau.

L'ensemble de ce processus est approuvé par les membres présents.

3. Report on astronomy

Le rapport établi par le Président sur la base des informations reçues des membres de la Commission et publié dans les *Transactions* **XVA**, est adopté sans modification, compte tenu des votes des membres absents.

4. Priorités en astronomie

Le texte publié sur les priorités dans le même volume, à la suite du Rapport signalé au point précédent, est également approuvé sans modification et dans les mêmes conditions de vote.

5. Prochain colloque

Par la lecture d'une lettre du Prof. A. Poveda, Directeur de l'*Observatorio Astronómico* de l'*Universidad Nacional Autonoma de Mexico*, le Président fait part de l'invitation de celui-ci à tenir le prochain colloque sur les étoiles doubles et multiples, à Mexico City en octobre 1974.

La Commission approuve l'idée d'un tel colloque mais estime cependant préférable d'en reporter la date à l'automne 1975. Elle est reconnaissante au Prof. A. Poveda pour son invitation et l'en remercie vivement.

6. Publication des mesures

Le Président rend compte de sa négociation avec les éditeurs de *Astronomy and Astrophysics* destinée à résoudre les problèmes que la publication des mesures avait fait surgir à une certaine époque. Il a été convenu que celles-ci seraient généralement publiées—ainsi que la plupart des orbites—dans les Supplement Series, pour des raisons d'ordre pratique et budgétaire et qu'il serait tenu compte des recommandations de la commission en cette matière. L'accord conclu avec les éditeurs de la revue a fait l'objet d'une publication dans la Circulaire d'Information No. 58—annexe (nov. 1972) de la Commission, éditée par P. Muller. Il prévoit également un groupement, dans la mesure du possible, des séries de mesures et des calculs d'orbites dans un minimum de fascicules.

De tels accords pourraient être passés avec d'autres grandes revues astronomiques où sont publiées des mesures d'étoiles doubles et des orbites, afin d'assurer la régularité de leur publication mais aussi le respect des recommandations de la Commission (rappelées dans la circulaire citée).

La Commission approuve cette suggestion. J. Dommanget est chargé de poursuivre sa tâche auprès de *Astronomy and Astrophysics*, cette revue ayant demandé que la Commission désigne officiellement son représentant. Le nouveau Président, S. L. Lippincott pourrait jouer un rôle identique auprès des grandes revues astronomiques américaines.

7. Recommandations (voir *Transactions* XVA, p. 309)

(a) *Publication des Mesures*

Voir ci-dessus: 6.

(b) *Révision de l'Index Catalogue*

La Commission marque un vif intérêt au projet de C. E. Worley d'une révision de l'Index Catalogue, consistant dans l'adjonction des couples nouveaux et dans la recherche systématique des erreurs qu'il contient. P. C. Couteau estime particulièrement satisfaisant, le délai de cinq ans proposé pour l'achèvement du travail, le survey qu'il poursuit en collaboration avec P. Muller devant alors toucher à sa fin.

Comme proposé dans une récente lettre par C. E. Worley, la Commission décide de ne pas entamer dès maintenant une discussion approfondie des modalités d'exécution de ce travail, mais de procéder par lettre-circulaire, pour connaître les suggestions de chacun.

Le Président félicite C. E. Worley pour le travail déjà accompli (la recherche des erreurs et leur correction a porté sur les $\frac{3}{4}$ du travail et a nécessité le traitement de quelque 300000 cartes). Il remercie le U.S. Naval Observatory et son directeur, K. Aa. Strand de mettre à la disposition de C. E. Worley, le personnel et les moyens nécessaires à sa tâche.

(c) *Diffusion des Publications*

La Commission adopte le voeu de J. Dommanget de voir chacun de ses membres adresser, dans la mesure du possible, des tirés-à-part de ses travaux à tous ses collègues de la Commission.

(d) *Terminologie Relative à la Parallaxe Dynamique*

Parmi les réponses reçues des membres absents au sujet de cette recommandation, certaines expériment des remarques ou objections qu'il y aurait lieu de discuter devant une assemblée plus vaste. Aussi, sur proposition de l'auteur même de la recommandation, la Commission décide d'en reporter l'examen à la prochaine Assemblée Générale

J. Dommanget profite toutefois de l'occasion pour rappeler que sa proposition porte principalement sur la distinction essentielle à faire entre les parallaxes dynamiques orbitales et non-orbitales, sous peine de voir le crédit accordé aux parallaxes dynamiques, très sérieusement affecté. Trop d'astronomes ne connaissent pas cette distinction et portent dès lors un jugement regrettable sur la qualité des parallaxes orbitales car ils les confondent avec celles données par exemple par H. N. Russell et C. E. Moore dans le tableau 53 de 'The Masses of the Stars'. Ces dernières sont en effect non-orbitales, c'est-à-dire calculées sur une base statistique tenant compte seulement du mouvement relatif des composantes en un point de la trajectoire, en l'absence d'orbite connue.

Seconde Séance

8. Situation dans l'hémisphère sud (voir: *Transactions* XVA, p. 310)

Le Président rend compte des contacts suivis qu'il a eus tant avec le *Council for Scientific and Industrial Research* d'Afrique du Sud qu'avec les autorités de la *University of Michigan* (U.S.A.) concernant l'arrêt des programmes d'observation d'étoiles doubles respectivement au Republic Observatory à Johannesbourg et au Lamont-Hussey Observatory à Bloemfontein.

Comme aucun représentant officiel de chacune de ces deux institutions n'est présent à Sydney, sa proposition de constituer une délégation chargée d'un échange de vues avec eux, perd tout son sens et est donc retirée, malgré l'appui qu'elle avait trouvé auprès des membres absents ayant fait connaître leur position à ce sujet.

Par ailleurs, la proposition de F. Holden de constituer une sous-commission chargée d'étudier le problème de l'observation dans l'hémisphère sud et de prendre toute disposition nécessaire pour mettre en route de nouveaux programmes, n'a pas rencontré l'approbation des membres de la Commission. Seules cinq réponses ont été reçues à la lettre de F. Holden du 14 juillet 1973, parmi lesquelles deux seulement sont favorables. La Commission estime donc que son activité dans ce domaine ne réclame pas la mise en place d'une telle sous-commission.

Le Président souligne les appels du Comité d'Organistion de la Commission auprès de tous les organismes astronomiques de l'hémisphère sud pour qu'ils prennent conscience de l'importance de l'observation des couples visuels.

Au sujet de l'Afrique du Sud, il ressort d'informations reçues tant du Directeur de l'Observatoire Sud-Africain, Sir R. Woolley, que du Président du CSIR, le Dr F. J. Hewitt et confirmées au cours de la présente Assemblée Générale par le Prof. G. G. Cillie, chef de la délégation sud-africaine, que la station de Johannesbourg sera maintenue en fonctionnement pour l'instant. G. Knipe serait prêt à y reprendre les observations de binaires visuelles dès la fin de l'année en cours.

Par ailleurs, M. Klerk envisagerait l'observation de couples visuels au Bosscha Observatory à Lembang (Indonésie) ce qui ramenerait un grand réfracteur au service des binaires visuelles, comme le souhaite son Directeur, B. Hidajat.

P. R. Hurly a fait savoir qu'il dispose à l'Université du Cap, d'un "area scanner' construit par Warner et Nather et qu'il compte entreprendre un programme photométrique d'observation des couples visuels en UBV.

A l'Observatoire de Cordoba, G. M. Iannini s'occupe de la mise au point d'un équipement devant lui permettre l'observation photographique de couples visuels à l'aide du télescope de 154 cm de Bosque Alegre.

R. R. de Freitas Mourão précise son intérêt pour l'échange d'astronomes entre observatoires de manière à profiter au maximum des meilleures conditions climatiques dans chacun d'eux.

Aucune information précise ne peut malheureusement être donnée sur la destination du réfracteur du Lamont-Hussey Observatory situé près de Bloemfontein.

9. Questions scientifiques

La séance se termine par les trois exposés suivants:

A. H. Batten: Orbit and Period of β 1163 (en cours d'impression dans: *Journal of the Royal Astronomical Society of Canada*);

A. M. Sinzi: Measurement of Double Stars by Occultations with Small Telescopes (en cours de publication dans: *Report of Hydrographic Researches, Tokyo, No. 8, 1974*);

W. J. Luyten: Double Stars with Common Proper Motion (voir: *Proper Motion Survey with the 48 in. Schmidt Telescope, No. 25 et 29, University of Minnesota, Minneapolis*).

Erratum
à
'Reports on Astronomy'

p. 298, deuxième ligne à partir du bas de la page:
lire: 'réflecteurs' au lieu de: 'réfracteurs'

COMMISSION 27: VARIABLE STARS (ÉTOILES VARIABLES)

Report of Meetings

PRESIDENT AD INTERIM: M. W. Feast.

Administrative Session, 21 August 1973

The President announced with regret the death of Dr A. C. Banerji, a member of the Commission. He also recalled the recent deaths of two of the most distinguished workers in the field of variable stars during the present century, Dr A. H. Joy and Dr H. Shapley (both at one time members of the Commission). The composition of the new Organizing Committee for submission to the IAU Executive was approved. A list of new members of the Commission was approved. Plans for a symposium in Moscow in 1974 on 'Variables as a Key to the Evolution of Stars and Stellar Systems' and a colloquium in Budapest in 1975 on 'Multiple Periodic Variable Stars' were announced. The President asked for views on the future of the Report on Variable Stars published in Transactions. In view of the large amount of work this Report entailed, he felt it only worth continuing if it was fairly extensively used and he had some doubts as to whether this was the case. None of the members present appeared to have strong views for or against continuing the Report and it was agreed that the incoming Organizing Committee should decide, after consultation with Commission members and the IAU Executive, whether to continue the Reports in the present or another form or whether to drop them entirely. The request from the Executive that each Commission compile a list of priorities and expected developments in its field was discussed. The meeting concurred with the view of the President (*Trans.* IAU XVA, p. 313) that such a list could easily be misused and asked him to urge the Executive not to repeat this request in future years. There was some discussion of a circular letter from the Moscow Variable Star Bureau asking for suggestions for changes in the *General Catalogue of Variable Stars*. It was announced that Mrs H. Sawyer Hogg's third catalogue of variable stars in globular clusters was now in the hands of the printer and that she was already preparing for the fourth edition which she envisaged publishing within the next five years. A motion that the annual subvention from the IAU to the Moscow Variable Star Bureau for their work on the *General Catalogue of Variable Stars* and related programmes be continued for the next three years, was proposed by Dr D. H. McNamara, seconded and carried unanimously. The difficulty experienced by observers in correctly identifying faint variable stars was raised. It was agreed that everything should be done to provide good maps (many existing maps are inadequate and for many variables there are no maps at all). In some cases variables will only be positively identified by rediscovering them. A small informal group (Rodgers, Kunkel, Feast) was set up to look into this problem. They later co-opted Miss Hoffleit into the group and it was agreed to work in close collaboration with the Moscow Bureau. The Commission voted for the continuation of four Working Groups under the following chairmen: Mrs H. Sawyer Hogg (Variables in Globular Clusters), Prof. P. F. Chugainov (Flare Stars), Dr F. Zwicky (Supernovae), Dr S. C. B. Gascoigne (Variables in the Magellanic Clouds). The possible need for Commission rules (suggested by the IAU Executive) was referred to the incoming Organizing Committee. However it was felt that the Commission functioned well without formal rules other than the by-laws of the IAU.

Scientific Meeting, 25 August 1973

The meeting began with a talk by Dr A. D. Code on observations of variable stars from outside the atmosphere. After a break the meeting continued as an informal discussion group organized by

Dr Lesh and Dr Aizenman on β CMa stars. A letter from Dr Van Hoof asking for collaboration in a programme on ν^2 Pup was discussed at this meeting.

Other Meetings

Commission 27 was responsible for a large part of the organization of the Joint Meeting (Commissions 27, 35, 40, 42, 44, 48) on *Variable XR Sources* which occupied a whole day (23 August 1973). Speakers were:

Tananbaum, Parkinson, Boyd, Petersen: XR Observations.

Hutchings, Garrison, Kemp, Vidal, Wilson: Optical Observations.

Wade, Miley: Radio Observations.

Kundu, Rees: Theory

Commission 27 was responsible for much of the organization of a Joint Meeting (Commissions 25, 27, 34, 37, 40) on *Very Compact H_{II} regions* (which included some work on Variable OH/IR sources), (27 August 1973, half day). Speakers were:

Habing, Mezger, R. D. Davies, Palmer, Zuckerman, Baars, Hoglund, Donn, Kahn.

Commission 27 was represented on the organizing of the following Joint Discussions:

22 August 1973: The Outer Layers of Novae and Supernovae.

24 August 1973: Stellar Infrared Spectroscopy.

27 August 1973: Kinematics and Ages of Stars in the Solar Neighbourhood.

28 August 1973: Very Short Time Scale Phenomena.

COMMISSION 28: GALAXIES (GALAXIES)

Report of Meetings, 21, 22, 24 and 29 August 1973

PRESIDENT: E. M. Burbidge.
ACTING SECRETARY: G. O. Abell.

There were four sessions, of which the first two dealt mainly with business matters, the third (a double session) dealt with residual business matters but was mainly devoted to science, and the fourth mainly to reports from observatories on work in progress.

Business Meeting, 21 August 1973

G. O. Abell was appointed Acting Secretary, in the absence of M. S. Roberts.
Members stood in silence for one minute *in memoriam* of M. L. Humason and H. Shapley.
The President noted with regret the resignation from the Commission, and from the Union, of A. R. Sandage.

NEW OFFICERS AND MEMBERS

The Commission unanimously approved the election by the Organizing Committee of the new President, E. B. Holmberg, and Vice-President, B. E. Markarian. The President proposed and the Commission unanimously approved, the list of names proposed for the new Organizing Committee. The number of Committee members has been increased to provide better coverage both of countries and of topics. It was agreed that the Secretary should be selected by the new President.

Several proposals for new members of the Commission had been received, some from the individuals themselves and others from National Committees. The Organizing Committee had screened the list, and presented the names of those they regarded qualified to the Commission for consideration. Several new names were added by members present. The amended slate was then approved unanimously by the Commission members present.

WORKING GROUPS

Since the Chairman of the Working Group on Galaxy Photometry and Spectrophotometry, J. L. Sérsic, was unable to be present, G. de Vaucouleurs agreed to chair its meeting. Similarly, Ch. Fehrenbach agreed to replace the Chairman of the Working Group on the Magellanic Clouds, A. D. Thackeray, who could not be present.

P. Pişmiş proposed that a new Working Group on Internal Motions in Galaxies be formed, and agreed to hold an exploratory meeting to see if there was sufficient interest.

COMMISSION RULES

A discussion was held on the suggestion by the Secretariat that new rules for the organization and conduct of Commissions be formulated and adopted. Since there already exist general rules, contained in the IAU Statutes and By-Laws, the opinion of the Commission was against the formulation of further rules, at least for the present. The members voted unanimously, on a motion of W. G. Tifft seconded by G. de Vaucouleurs, to submit the list of suggestions provided by the Secretariat to the new Organizing Committee for consideration. The Committee would present recommendations, if any, to the Commission at its business meeting during the next General Assembly of the Union.

Business and Scientific Meeting, 22 August 1973

REST WAVELENGTHS

The President put before the Commission a two-part Resolution proposed to Commission 30. D. S. Heeschen reported that, following discussion by Commissions 30 and 40, it had been agreed that the first part of the proposed Resolution should be withdrawn. The second part concerned whether to define the rest wavelength, λ_0, used in determining redshifts as the vacuum or air value. The question arises only for large redshifts.

DECISION TAKEN

It was agreed that, in cases where the difference between vacuum and air values made a significant difference in redshift determinations (e.g. in large redshifts exhibited by narrow absorption lines in quasi-stellar objects (QSOs)) vacuum values of λ_0 should be used.

NOMENCLATURE OF FAINT GALAXIES AND QSOS

A proposal had been received by V. C. Rubin that, to avoid confusion, a standard nomenclature be adopted for the naming of faint galaxies and QSOs. An *ad hoc* Working Group of six, consisting of members of Commission 40 as well as Commission 28, was set up under the Chairmanship of J. G. Bolton to consider the matter and formulate a recommendation.

SCIENTIFIC MATTERS

Since most of the Commission members present had attended the preceding IAU Symposium No 58 in Canberra, on 'The Formation and Dynamics of Galaxies', no summary of this was presented. The meeting concluded with the presentation of two short reports:

W. G. Tifft: On Redshift Uncertainties, Intercomparisons, and the Need for Standards.

G. Burbidge: Description of Lick Observations on a Double QSO, by E. J. Wampler *et al.*

Scientific and Residual Business Meeting, 24 August 1973

Prior to the scientific meeting, two residual business matters were settled.

NOMENCLATURE OF FAINT GALAXIES AND OTHER OBJECTS

J. G. Bolton's *ad hoc* Working Group presented its recommendation, which was unanimously approved by the Commission as follows:

RECOMMENDATION

Commission 28 recommends that a co-ordinate numbering system be adopted for all future catalogues of radio sources, X-ray sources, QSOs, galaxies, clusters of galaxies, etc. For general purpose catalogues of the order of 10^5 objects the present 8-digit Parkes system is recommended. The first four digits give the hours and minutes of R. A., the fifth the sign of the declination and the remaining three the degrees and truncated tenths of a degree of declination for the mean equator and equinox of 1950. For special purposes such as occur in dense clusters of galaxies, it is recommended that the eight-digit system be extended as necessary.

REFERENCE FRAME FOR RADIAL VELOCITIES

Some discussion was held on confusion arising when workers refer radial velocities to different standards of rest. It was unanimously agreed that in the future workers should publish observed values of velocity relative to the Sun, and not to the local standard of rest, the galactic centre, or any other reference frame.

SCIENTIFIC PAPERS

The following short papers or comments were presented.

J. Heidmann: Neutral Hydrogen in Haro Galaxies.
H. E. Bond: Galaxies in the Variable-Star Literature.
W. Wlérick: BL Lacertae.
H. van Woerden: Work at Westerbork on M101, M81, and NGC 3718 (50 cm continuum).
G. Sancisi: Radio Observations of NGC 5383 and NGC 891.
W. G. Tifft: A Revision of the NGC.
J. H. Oort: Westerbork Results on NGC 4258.

The scientific session concluded with a discussion of anomalous redshifts. A session of IAU Symposium 58 had been devoted to presentation of the orthodox and unorthodox views, by W. L. W. Sargent and H. C. Arp respectively, and members wished to explore the unorthodox view further. The following short papers were presented:

H. C. Arp: Summary of Observations of Anomalies.
W. G. Tifft: Banded Structures in Redshift-Magnitude Diagram for Clusters.
G. de Vaucouleurs: Virgo Cluster Redshifts.
J.-C. Pecker: New Observational Data, and a Theory of Abnormal Redshifts.

Scientific Meeting, 29 August 1973

The following reports of work in progress were presented:

G. Alcaino: Extragalactic work at ESO, Chile.
G. de Vaucouleurs: Some Extragalactic Programs at McDonald Observatory.
M.-H. Ulrich: A Complete Sample of Radio Galaxies Identified with Ellipticals.
B. and D. Wills: QSO Spectroscopy at McDonald Observatory.
F. Bertola: Rotation Curves of Early-Type Galaxies.
H. Murdoch: Radio Source Program at Molonglo.

Reports of Working Groups

G. de Vaucouleurs, C. Fehrenbach and P. Pişmiş submitted the following reports:

GALAXY PHOTOMETRY AND SPECTROPHOTOMETRY

The group convened on August 28 with G. de Vaucouleurs acting at the request of the Chairman, J. L. Sérsic, who was unable to attend the Sydney Assembly.

ELECTION OF NEW CHAIRMAN (1973–76)

As it has been the practice to rotate the chairmanship of the Group, nominations from the floor were requested. Bertola and de Vaucouleurs declined to be nominated. Dr H. D. Ables, U. S. Naval Observatory, Flagstaff Station, Flagstaff, Arizona was nominated and duly elected unanimously (in absentia) by the 8 members of the group present at the meeting.

Since the WG is open to anyone interested or active in galaxy photometry, no formal election of new members was held; those present not previously on the mailing list will be added automatically.

SCIENTIFIC PAPERS

G. de Vaucouleurs reviewed recent and current programs in galaxy photometry and spectrophotometry at Austin and McDonald Observatory (see report of Commission 28 and of WG).

G. Alcaino reported on his program of UBV photometry of 40 southern galaxies of all types selected from Sérsic's Atlas.

A. E. Whitford described progress of his scanner observations of the bulges of M31, M81 (Sb), M33, M51 (Sc) and of the Milky Way. Our Galaxy appears to be a strong-lined Sb spiral.

M. Walker presented preliminary results of spectracon observations with the Lick 120-in. of 3C120 and NGC6044 showing faint outer extensions of these galaxies.

F. Bertola listed objects under study at Asiago: NGC 128, 3627, 4125, 4258, 4697 for which detailed photometry is in progress.

J. Heidmann indicated how combinations of optical and 21-cm parameters can be used to derive extragalactic distances and made a plea for observations of diameters at the 25.0 or 26.5 (B) mag./(arc sec)^2 levels.

LES NUAGES DE MAGELLAN

Le groupe a tenu une séance le 28 Août.

1. Le rapport du Dr Thackeray est approuvé mais le complément ci joint (1)–(8) est ajouté; par suite d'une erreur postale, ce rapport est arrivé avec trois mois de retard.

2. Une discussion sur le renouvellement du WG s'engage. M. Westerlund est proposé pour le présider. Les membres de l'ancien groupe pourraient être complétés par Van den Bergh, Reddish, Davies, Toomre, Freeman et Graham. Le nouveau président prendra contact avec le président de la Commission 28 et les membres proposés.

3. Une séance scientifique a lieu où sont données les communications suivantes:

Alcaino: Photometry of globular clusters in MC.

Davies: Magellanic streams.

Fehrenbach: Supergéantes anormales de type A à F dans le GNM.

Nandy: Observation of LMC by the S2/68 sky telescope satellite of ESRO.

Le rapport du Groupe de Marseille-Haute Provence sur les Nuages de Magellan a été retardé plus de trois mois par la poste et le Dr Thackeray m'a demandé d'ajouter exceptionnellement la partie de ce rapport qui n'est pas analysée.

(1) Le catalogue de base de 1970 a été complété par un nouveau catalogue (1972) qui contient 88 étoiles du GNM et 253 étoiles galactiques (Fehrenbach, Duflot). Une troisième liste de 25 étoiles GNM est en préparation.

(2) Une liste de 1832 VR d'étoiles obtenues au P.O. dans la direction du GNM dont 398 étoiles du Nuage est sous presse (A et A).

(3) Des études de la rotation (Prévot) et une étude des populations stellaires ont été faites (Brunet) (sous presse A et A).

(4) N. Carozzi a publié en 1971 deux listes d'étoiles à Grande Vitesse dans l'aile du PNM.

(5) Des études de l'absorption dans la direction du GNM ont été publiées (1971, 72, Fehrenbach, Brunet, Prévot).

(6) Des études d'étoiles à GVR galactiques ont été publiées (Fehrenbach, Prévot).

(7) Des études d'astres spéciaux ont été publiées—l'étoile S22 à Fe II (Fehrenbach 1971) dispersion de 20 Å mm^{-1}.

(8) La comparaison des catalogues de Sanduleak et Fehrenbach montre l'existence dans le champ du GNM d'une trentaine d'étoiles A à F à fortes raies d'hydrogène et forte absorption UV.

Sanduleak leur attribue la classe de luminosité II à III. Duflot et Fehrenbach estiment que ce sont des étoiles du GNM de magnitude − 6, 2 à spectre anormal.

REPORT ON THE FORMATION OF A WORKING GROUP ON INTERNAL MOTIONS IN GALAXIES

Twelve astronomers—six optical and six radio—were present at the meeting scheduled on August 27 under the Chairmanship of P. Pişmiş to discuss the formation of a Working Group on the velocity fields in galaxies. It was unanimously agreed that a coordination between the various types of work done on velocity fields was very desirable as it would serve to integrate data obtained by different methods and at different regions of the spectrum; in other words, a collaboration between the radio and optical astronomers working in the field should be carried out in a concerted manner.

It was therefore agreed to form the group under the name of 'Working Group on the Internal Motions of Galaxies' and some discussion was carried out to delineate the functions of the group.

The scope of this Working Group, as far as it is possible to determine at the moment, will be to compile the pertinent velocity data already existing and in the process of being obtained. The WG will recommend or rather appeal to workers in the field to include in their programs certain galaxies in order to attain a complete coverage in all the different regions of the spectrum; these methods might be: absorption-line velocities, H II region-velocities (spectroscopic, Fabry Perot or by radio recombination lines), 21 cm H I line velocities, molecular line velocities, etc.

All those active and interested in this project are invited to membership of the WG. The working group will be coordinated by a Chairman. The first Chairman, unanimously approved by the Commission, will be P. Pişmiş.

COMMISSION 29: STELLAR SPECTRA (SPECTRES STELLAIRES)

Report of Meetings, 22, 23, 25 and 28 August 1973

PRESIDENT: Y. Fujita.

First Session, 22 August 1973

The Draft Report was accepted. The proposed members for the Organizing Committee were approved. New members and resignations of members were also approved.

The President reported about the activity of members who communicated with him since Circular No. 5 had been sent to each member on June 1, 1973.

The President asked the Chairmen of the Working Groups in attendance to report. D. C. Morton suggested L. N. Houziaux as his successor in the Working Group on 'Ultraviolet Stellar Spectra and related Ground-based Observations'. R. Herman is willing to continue as Chairman of the Working Group on 'Be Stars'. G. Cayrel de Strobel reported on the Working Group on 'Line Intensity Standards' as follows: The intercomparison of photometric calibration of stellar spectra, using Kienle's step filters has come to an end with negative results; e.g., calibration errors are not the major cause of divergence between equivalent widths measured with different equipments. Discrepancies of the order of 15% to 30% are still common, and as yet no clue exists as to what values are the best ones, an entirely new coordinated program should be undertaken to give a sort of metrologic tool to spectroscopists. As proposed earlier by Griffin, some spectra obtained in a double pass photoelectric spectrometer should be provided. The light of the source should obviously become available, both for northern and southern observatories. The best proposal could well be the integrated solar spectrum, which is also an invaluable comparison standard for stellar abundances. The solar spectrum can be obtained in stellar observatories either from diffusers of blue sky, the Moon, or, from stellar like sources as the minor planets supplying a large range of exposure times for comparison.

Copies of the memorandum to the members of Commissions 27, 29, 33, 37, and 42, issued from the Moscow Bureau of Variable Stars in June, 1973, were distributed to members and the President invited the members to discuss it.

The proposal by M. Plavec for a symposium on Be and shell stars possibly in 1975 was supported. The proposal by L. H. Aller in connection with the 'Publication of Observational Data in Complex Stellar Spectra' was discussed and it was decided to re-discuss it in the Second Session on 28 August. Aller emphasized that the proper thing to do would be to publish the actual tracings. As spectrum synthesis methods become more and more popular, so becomes the necessity of presenting high-resolution spectral intensity details and not just equivalent-widths.

Then the scientific session was continued with three invited speakers:

D. C. Morton: Ultraviolet Spectra of Bright O and B Stars by the OAO-3 Satellite.

S. B. Parsons: An Ultraviolet Stellar Spectroscopy Experiment on NASA Skylab.

J. Smoliński, J. L. Climenhaga, and K. O. Wright (read by Climenhaga): Blanketing Effect as related to Luminosity and Turbulence in Late-type Stars.

Second Session, 28 August 1973

The proposal by the President for the addition of a new member, Dr S. B. Parsons was accepted.

The proposal by Aller in connection with the 'Publication of Observational Data in Complex Stellar Spectra' was again discussed and the following two recommendations were accepted:

(1) Commission 29 recommends to astronomers who have data on the spectra of stars on mag-

netic tape, that they fully document these data and make copies of the data available for other astronomers. Individual announcements about the availability of data should be made in each paper on stellar spectra.

(2) Commission 29 recommends to editors that they permit publishing of lists of line identifications and equivalent widths as well as profiles of lines in stellar spectra, for these data are fundamental to the study of stellar atmospheres. We urge the continued publication of atomic and molecular data on spectra of atoms, ions and molecules.

Then, the following statement was approved: Commission 29 fully endorses the proposal of the Commission 12 Working Group on the Solar Spectrum to prepare a fourth revision of the Rowland Table.

In discussing the future of the Commission the activity of each member was stressed, though new members are welcome.

The priorities for future research in the field of our Commission were shortly discussed.

Then followed a meeting of the Working Group on Be Stars, under the Chairmanship of Cayrel de Strobel. The following two papers were presented:

T. Kogure: Hydrogen Spectrum of Be Stars.

J. P. Maillard: Spectrum of Be Star γ Cas for 1 to 2.5 μm.

Joint Meetings

Two Joint Meetings were organized by the President with the cooperation of A. H. Cook, President of Commission 14, and R. N. Thomas, President of Commission 36:

1. Joint meeting on 'line identification in stellar spectra for atoms and molecules'

(Commissions 14 and 29), August 23

Chairman: A. H. Cook.

C. M. Sitterly: Atomic Spectra—The Future Outlook.

J. E. Ross and L. H. Aller: Spectrum Synthesis and Spectral Line Identifications.

Chairman: L. H. Aller.

J. G. Phillips: Survey of Molecules in Astronomical Sources and of Recent Pertinent Laboratory Work.

Y. Fujita: Identification of Spectra in Violet and Infrared Regions of Carbon Stars.

2. Joint meeting on 'spectroscopy of peculiar A-type stars'

(Commissions 29 and 36), August 25

Chairman: R. N. Thomas.

M. Hack: Surface Composition and Physical Structure of the Ap Stars.

D. S. Leckrone: The Ultraviolet Flux Envelopes of Ap Stars.

W. Buscombe: Variations in the Line Spectrum of A Supergiants.

Chairman: M. Hack.

E. R. Mustel: Investigation of Ap Stars in U.S.S.R.

M. Smith: Abundance Studies of Am Stars across the HR Diagram.

Short comments by C. Meggesier and J. Dachs followed.

COMMISSION 30: RADIAL VELOCITIES (VITESSES RADIALES)

Report of Meetings, 21 and 23 August 1973

ACTING PRESIDENT: A. H. Batten.
SECRETARY: A. G. Davis Philip.

(1) No objections were raised to the printed report of the Commission. A formal motion that the list of 21 faint stars listed in the report be adopted as new standard-velocity stars was carried unanimously. The thanks of the Commission were expressed to Drs Heard and Fehrenbach for their work on these new standards. Thanks were also expressed to Dr D. S. Evans and Drs. Abt and Biggs for their work on catalogues and bibliographies of radial velocities, respectively.

(2) The three resolutions in the name of the Commission on the General Assembly agenda (items 11g (i)–(iii)) were discussed. After discussions between the Acting President and the Secretary of Commission 40 the two Commissions agreed to set up a joint working group to study (i) and (ii) to report back to them at the XVI General Assembly. Meanwhile these two resolutions were to be withdrawn from the agenda of the General Assembly. The Commission appointed Batten (Chairman), Fehrenbach, and de Vaucouleurs as its representatives on the working group. The resolution (iii) was affirmed by the Commission. In addition, the Commission unanimously welcomed the proposal of Commission 42 to enlarge its scope to include spectroscopic binaries.

(3) After discussion the following officers and Organizing Committee were adopted, subject to approval by the General Assembly and Executive Committee, as appropriate.

President: R. F. Griffin, Vice-President: A. H. Batten, Organizing Committee: R. Bouigue, Mme M. Duflot, H. Gollnow, J. F. Heard, L. Oetken, A. G. Davis Philip, Mrs V. C. Rubin.

Four new members were elected: Y. Georgelin, A. Pedoussaut, K. Serkowski, and R. V. Willstrop. The Commission reaffirmed its desire to include radio astronomers in its membership.

(4) The question of working rules for the Commission was referred to the incoming Organizing Committee. The Commission stressed that IAU Reports were very useful, and should continue to be published in substantially their present form. The determination of priorities in the field of radial velocities was delegated to a committee consisting of Batten, Fehrenbach, and de Vaucouleurs. They reported as follows to the General Secretary.

Radial velocities are fundamental data essential to studies of:

(i) galactic structure and dynamics,

(ii) stellar masses,

(iii) structure of stellar atmospheres (including those of Cepheid variables),

(iv) velocity dispersions in star clusters, galaxies, and clusters of galaxies,

(v) anomalous red shifts in the spectra of stars (e.g. *W–R* stars) and of galaxies. (Are the red shifts Doppler effects?).

Therefore the Commission urges the importance of obtaining large numbers of radial velocities of the accuracy required for the resolution of these problems. It considers the provision of adequate telescope time and modern instrumentation for the measurement of radial velocities to be of the highest priority.

(5) The second meeting of the Commission was largely devoted to a scientific discussion on 'Methods of Obtaining and Measuring Radial Velocities'. The following communications were given:

Ch. Fehrenbach: Introductory Lecture.

G. de Vaucouleurs: Measurements of Radial Velocities and Velocity Dispersion in Galaxies by High-Resolution Photoelectric Spectrophotometry.

K. C. Freeman: A New Technique for Measuring Velocity Dispersions in Stellar Systems.

The meeting included a discussion on some aspects of the design of radial-velocity scanning spectrometers, with contributions by K. O. Wright and N. J. Rumsey.

COMMISSION 31: TIME (L'HEURE)

Report of Meetings

PRESIDENT: G. M. R. Winkler.

First Session, 21 August 1973

The session was opened by President G. M. R. Winkler.

IA

The first topic on the agenda was consideration of procedures to be used in maintaining an active membership list.

Markowitz and Smith both felt that the original wording was vague. The rescinding of membership should also extend to the consulting members to this Commission. The action of this Commission should be sent to all current members and failure to respond would automatically cause them to be dropped from the membership.

This matter will be resolved at the meeting on 29 August.

IB

The President next put forth the candidates for office:

Proposal for Offices (1973–1976)

President: H. Enslin.

Vice-President: A. Orte.

The candidates for the Organizing Committee will be put forth at the session of 27 August. The President informed the Commission that it was the wish of the General Secretary of the Union that the Organizing Committee consist of no more than 10 members and that individuals should serve no more than 2 consecutive terms. A consensus of many members indicated that continuity was important and therefore $\frac{1}{2}$ or 5 new members should be placed on the Committee and should serve 2 terms and the 5 remaining members serve 1 term and thereafter the rotation would be established. Both Markowitz and Smith indicated they would offer to leave the Organizing Committee. A new list will be presented on 27 August.

The proposed President and Vice-President were approved by acclamation for transmittal to the Executive Committee.

IC. CONSULTANTS TO THE COMMISSION

The President stated the importance of Consultants to this Commission, especially from Standards Laboratories. It was noted in discussion that this list should fall under the same scrutiny as full membership.

The new Consultants put forward by the President were Barnes, J. A., Costain, C. C., Morgan, P., Muller, I. I., Penny, C. J. A. (Miss), and Steele, J. McA.

IIa. REVIEW OF DRAFT REPORT AND DOCUMENTS

Because of the draft reports not being readily available, the topic was deferred to a later session. Copies were made available by the President.

IIb. PRIORITIES IN TIME

(1) The BIH must receive more financial support.

(2) Details of 1 second steps in UTC and DUT 1 dissemination should be resolved.

(3) The questions regarding Ephemeris Time must be clarified.

(4) The derivation of TAI should include more contributors as links with other time scales become available.

(5) Close cooperation between the scientific and treaty organizations, CIPM, CCIR, IUGG, URSI, and IAU, must assure that the present evolution in worldwide standardization in the provision and use of precise time can continue in an orderly way.

These priorities will be voted on by the Commission on 29 August 1973. Remarks and additional items were requested by the President.

IIIa. DRAFT RESOLUTIONS

The following draft resolutions were placed before the Commission. The result of this meeting concluded with the establishment of a drafting committee to revise draft resolutions to a form acceptable to all members. This committee consisted of Markowitz (Chairman), Becker, Enslin, Sinzi, and Wilkins.

Draft Resolution No. 1. Adoption of UTC as basis of standard time in all countries.

Draft Resolution No. 2. TAI decimal counting.

Draft Resolution No. 3. Modified Julian Date

Draft Resolution No. 5. Redefinition of ET with reference to TAI.

Draft Resolution No. 4 and No. 6. Maximum limit of TU1-UTC.

Draft Resolution No. 7. This resolution was deferred until after the joint meeting on UTC.

The remainder were individually discussed at later sessions.

IVa. REPORT ON CCDS MATTERS BY W. MARKOWITZ

The CCDS met on 6 and 7 July 1972 to discuss the definition of an international atomic time scale for daily use. It was agreed that the problems involved should be studied by interested agencies, in particular, the IAU.

It was noted that the almost universal acceptance of UTC makes it a suitable basis for a time system for common use.

Agencies concerned with comparisons of atomic clocks were urged to send their results to the BIH, in the form which it may request.

The President passed out a paper 'Note on Ephemeris Time (ET), Relativity and the Problem of Uniform Time in Astronomy', by G. M. R. Winkler.

Second Session, 24 August 1973

Jointly with Commission 19 (see Commission 19 Report).

Third Session, 24 August 1973

Dr *Winkler* opened the session.

Reports by the directors of the BIH and BIPM were presented along with a report on the URSI meeting by Dr Winkler.

Dr *A. M. Sinzi*, Director of the Japanese Hydrographic Office, was approved for membership in Commission 31.

I. REPORT OF BIH DIRECTOR

Dr *B. Guinot* presented the Bureau International De L'Heure (BIH) report. A complete text with figures was handed out to the assembled participants. 'Report to Commission 19 and 31, 15th IAU General Assembly, Sydney, 21–30 August 1973'.

On the subject of International Atomic Time (TAI) it was noted that the XIVth General Conference of Weights and Measures, formally recognized the BIH atomic time as TAI. Problems resulting from this rôle were noted with regard to (1) Uniformity and accuracy of TAI since it is the mean of the frequencies of a small group of commerical cesium standards. There are currently 7 contributing laboratories and the stability for 1 year is of the order of 1×10^{-13}. (2) Traveling clock service provided by the USNO made frequent checks of LORAN-C propagation delay. The 10-day averages of propagation delay showed a scattering of a few 0.1 μs. The apparent propagation delay of the composite LORAN-C link between America and Europe has a periodic nature with an amplitude of 0.2 μs over periods of 1 to 2 months. There appears a positive correlation between observatories/laboratories in North America. However, these fluctuations are of the order of magnitude of 0.2 μs to 0.4 μs and indicated the precision of LORAN-C reception.

With respect to implementation of Universal Coordinated Time (UTC) two facts were noted: (1) The step of 0.3 μs, in coordinated time made on January 1, 1972, solved the problem of homogeneity of the time data between the laboratories and the BIH. (2) Announcement of a leap second is made at least 8 weeks in advance. Because of this the difference between UT1-UTC exceeded the tolerance of 0.7 μs before 31 December, 1972 when the leap second was made. The Director of BIH requests more precise guidelines for future introduction of these jumps.

In regard to Time Signals, it was noted that time signals are emitted within ± 0.2 ms from UTC and this is much smaller than the CCIR tolerance and the propagation errors. The synchronization of the time signals has been practically achieved.

Matters relating to Universal Time and Coordinates of the Pole will be presented at a later session.

Dr *Guinot* reported that as a result of favorable action by the BIPM (Bureau International des Poids et Mesures) an assistant, responsible to work solely on TAI, will begin work on 1 September 1973. It was noted that increased responsibilities with respect to scientific research have resulted in significant increases in costs which need to be met. H. Smith strongly urged additional support and he and W. Markowitz will assist H. Enslin in providing a succinct resolution to the IAU finance committee requesting additional support because of increasing responsibilities to the astronomical community.

Discussion:

Dr Winkler presented additional facts regarding the calibrations and operation of the LORAN-C chain across the Atlantic. Also indicated was the fact that time synchronization across the Pacific is still a problem. Dr Morgan discussed the satellite link between the USNO and Australia. Dr Guinot also brought up in the course of the discussion a proposal to have guest workers from various laboratories around the world spend a residency period in Paris. This would essentially alleviate the manpower problem which Dr Guinot has, and also provide a broad dissemination of the problem in operation of the BIH. Additional discussion ensued as to the possible frequency adjustments and second jumps in the TAI Timekeeping System.

II. REPORT OF BIPM DIRECTOR

Dr *J. Terrien* presented the report of the BIPM. The CGPM held its meeting in October, 1971, at which time a resolution was adopted which defined the basis of TAI. CIPM was asked to conclude the necessary arrangements with the BIH for implementation of the TAI.

The CCDS met in July, 1972 with Prof. Markowitz as the representative of the IAU. The following four actions were taken at this meeting. (1) Laboratories and observatories of cooperating countries were asked to give the BIH scientific support with more detailed information. (2) TAI should be adopted as a symbolic notation in every language for International Atomic Time. (3) That scales of time either be counted in decimals of a second or with normal designations of days, hours, minutes, and seconds. It was referred to the IAU for a recommendation. (4) Should the CCDS take official action to make civil time the same as UTC? This question was discussed and it was referred back for further studies with other appropriate international organizations.

CCDM met in June of 1973. Discussion were undertaken at this particular meeting with respect to the velocity of light and its value $c = 299\,792\,458$ m s^{-1}. A conclusion was to keep this value regardless of the definition of the meter and the definition of the second.

III. REPORT ON THE URSI MEETING IN WARSAW, POLAND

Dr *Winkler* presented the results of the URSI meeting. They were (1) the recommendation to adopt UTC, (2) recommendation for the definition of a meter, and (3) a resolution with respect to mutual interference of time signals. With respect to the latter, the time signals cannot be received in certain areas of the world. The various frequencies used in various countries and the principles under which they operate were briefly discussed with respect to CCIR, CHU in Canada, Russian Broadcasts and in the United States. The conclusion of this discussion seemed to be that the action of Commission 31 should be one of encouraging the CCIR to continue work in improvements in the dissemination of time signals.

Cook of the UK continued the discussion with respect to the definition of the meter and proposed that there needs to be a distinct view presented by the IAU.

Fourth Session, 25 August 1973

This session was opened by Dr Kovalevsky, President of Commission 4. It was a joint session of Commissions 31 and 4

This joint session had as its principal topic a discussion of 'UTC'.

H. Smith was asked to present the CCIR report. As the Chairman of the working party, it has been his responsibility to make sure that the UTC system has been implemented successfully. It is recognized that the UTC system is a compromise system. It also is a uniform system for meeting the general requirements for time. Questions were then posed: (1) Should any changes be made now in the UTC system? H. Smith recommended maintenance of status quo unless absolutely necessary. (2) Question of the leap second and the excess of departure above 0.7 s. He asked the question as to what might be preferred dates for change and stated that as far as he was aware, no serious inconvenience was to result from excess departures of the time limit. Referring to the draft resolution No. 6 by G. F. Sadler, having to do with the departure of UTC from UT1, he made the observation that if the earth continues to behave in its rate of rotation as past experience has shown, then perhaps two changes per year are inadequate. Possibly one solution is to allow a tolerance of 0.9 s. and also the possibility that the number of preferred dates go from say the current two to possibly four. Some combination of both or an adoption of one might be superior to the current resolution. Mr Smith indicated that there was a need for clear indication of the position of Commissions 31 and 4.

Dr *Barnes* stated that there were no complaints at the National Bureau of Standards with respect to the excess of 0.7 s. Dr Mulholland stated that the change of tolerance needs a resolution, but the number of dates does not. Dr Iijima proposed that the tolerance be set at 1.0 s. Dr *Winkler* indicated that a major difficulty would exist in the change required in the code format if the excess time deviation became larger than 0.95 s. *H. Schlegel* of the US said that the 0.7 s must be rigrously interpreted and any adjustment that should be made should be rigorously defined as hereto. Dr *Sinzi* indicated a study which they had conducted in Japan which concluded that the need for high accuracy in the sense of a one second tolerance seems not to have been demonstrated. Dr *Mulholland* subsequently indicated that if 20% of the polled people in this study felt requirements on the value not to be less than 0.7, that these people should not be ignored.

At this time Dr *Winkler* presented the revised draft of resolution 1, to be found below. Resolution 1 was approved with two abstentions in the voting.

A discussion as to the name to be attached to UTC and TAI followed. Nine attendants favored a designation of international atomic time immediately; 14 indicated a desire to change the designation at a later date. With respect to universal time coordinated, 19 attendants indicated that it should be retained as the current designation of time. As a result, draft resolution 7 which contained a proposal for UTC to be changed to ICT (International Coordinated Time) submitted by Dr *Belocerkovskij* and supplemented by Drs. *A.* and *N. Stoyko* was defeated 16 against to 1 for, with no abstentions. Draft resolutions 4 and 6 dealt with the tolerance to be allowed with respect to the leap seconds. Dr *Morrison* of the Royal Greenwich Observatory presented a discussion on the rate of rotation of the earth and anticipated requirements both in leap seconds and in the number of dates. His conclusion was that if a recommendation were presented which allowed four unique dates and a tolerance limit of 0.9, that this modified resolution would satisfy requirements for many years. As a result of much discussion, draft resolutions 4 and 6 were referred to a drafting committee consisting of Winkler, Guinot and Morrison to be reworked and presented at the session on the 29th of August.

Fifth Session, 27 August 1973

The session was opened by *G. Winkler*, President. The president proposed a slate for the new Organizing Committee. This slate, which took the new requirements for composition and tenure of the Organizing Committee into account, was approved unanimously.

Dr *Guinot* presented comments on 'Advances in Timekeeping at the BIH'. Dr *J. Barnes*, NBS, reported on the status of the 'Primary Frequency Standards'. After complete evaluation and account of the individual frequency standard biases, it appears that the three standards at PTB, NRC, and NBS are within about 2×10^{-13}. TAI appears to be high in frequency by about 1×10^{-12}.

Dr *Morgan* gave a report on timekeeping in Australia. Clocks at various establishments are now intercompared by TV with a composite scale computed by the Mt. Stromlo (Dept. of National Mapping) Time Service. Australia is tied to the rest of the world by portable clocks (with USNO) and also recently by satellite (Doppler tracking of TRANSIT satellites). Experiments with Timation satellite established a link to the USNO with a scatter from day to day of about 0.2 μs.

R. Keating reported on the relativity experiments with clocks which he conducted together with *J. Hafele*. The discussion centered on the question of claimed accuracy of, and confidence in, the experimental results (internal precision $\pm$ 10 ns, estimated external accuracy $\pm$ 30 ns). The difficulty of estimating disturbing factors as for instance the Earth's magnetic field, temperature shocks, physical motion, etc., was discussed. A repetition of such experiments has been suggested by several.

Dr *Becker* reported that an analysis of TAI and its contributors in reference to his laboratory primary frequency standard clearly shows tendencies of time scales to decrease their frequency. TAI decreased by 1.4×10^{-13} per year. This is suggested as typical for the commercial frequency standards used. The USNO time scale proved to have the best long term stability, whereas PTB was best for 100 day intervals.

Sixth Session, 29 August 1973

G. Winkler, President, opened the session. The question of membership retention of inactive members was resolved with a recommendation to the President to take the necessary steps to reduce the membership periodically to include only active members and consultants. Consultation with the Organizing Committee is advisable in doubtful cases.

After much discussion the resolutions and "Priorities" given below were adopted in final form.

Resolutions

RESOLUTION NO. 1 BY COMISSIONS 4 AND 31

(Adopted 25 August 1973)

Considering:

(a) that a single worldwide coordinated clock time scale based upon the SI second is desirable,

(b) that the Coordinated Universal Time system (UTC) makes the International Atomic Time (TAI), which is based on the SI second, generally available, and

(c) that the UTC system provides mean solar time directly to a precision that is needed for navigation and surveying,

Recommend:

the adoption of the UTC system as the basis for the dissemination of standard time (heure normale) in all countries.

RESOLUTION NO. 2 BY COMMISSIONS 4 AND 31

(Adopted 29 August 1973)

Considering:

(a) that TAI equalled UT2 (nominally) when UT2 was 1958 January 1^d, $0^h\ 0^m\ 0^s$,

(b) the necessity of maintaining all conventional calendarial and time designations,

(c) the desirability of bringing into accordance ET and TAI, which differ by about 32s;

Recommend:

(1) that a change be introduced in the TAI as soon as practicable so that TAI (new) equals TAI (old) plus 32 seconds precisely,

(2) that DTA (DAT) be defined by TAI minus UTC, where DTA is an integral number of seconds,

(3) that the CCDS be invited to take the necessary action.

Explanation:

(1) TAI is not affected by leap seconds but UTC is.

(2) The value of DTA is determined and announced by the BIH.

(3) UTC is the basis of standard time, the time in common (civil) use, as disseminated by radio time signals.

RESOLUTION NO. 3 BY COMMISSIONS 4 AND 31

(Adopted 29 August 1973)

This was adopted by the General Assembly specifically because of its wide importance (Resolution No. 4).

RESOLUTION NO. 4 BY COMMISSIONS 4 AND 31

(Adopted 29 August 1973)

Considering:

(a) that present procedures governing the insertion of leap seconds have been variously interpreted and,

(b) that if the present trend of the rotation of the earth continues, it will become impossible to maintain UTC within the present limits by the insertion of leap seconds on two preferred dates only and,

(c) that it is important for many users to have UT1 minus UTC remain within a *fixed* limit;

Recommend:

(1) that the *maximum limit* of UT1 minus UTC be set at ±0.950 seconds,

(2) that the maximum deviation of UT1 from UTC plus DUT1 be ±0.100 s,

(3) that, when necessary, leap seconds may be introduced at the end of any month, but that first preference be given to the end of June and December, and second preference be given to the end of March and September.

RESOLUTION NO. 5 BY COMMISSIONS 4, 14, 31 AND 40

This was adopted by the General Assembly specifically (Resolution No. 6).

Applications of Time – Priorities

There is still a very large number of users who need UT1 and who depend on the time signals entirely. There is also a considerable group of users who need UT1 with greatest precision (Satellite Geodesy, deep space tracking). There is a growing number of observatories who need to point their telescope to within 1′ or 2′ (for optical retroreflector satellite tracking, Pulsar studies, etc.). They are very concerned about DUT1.

The need for precise clock time has increased even more dramatically. Pulsar time of arrival observations can be done with microsecond precision. Lunar optical ranging measures travel times with 10^{-10} s resolution. Indeed there is no doubt that in a large number of scientific observations the superior accuracy of time measurements, compared to any other quantity, will become useful.

In view of these ever growing applications, not only in pure science, the following priorities in our work can be recommended:

(1) The BIH must receive more support, financial and otherwise.

(2) Details of the 1 second steps in UTC and of the DUT1 dissemination should be resolved.

(3) The questions regarding Ephemeris Time must be clarified.

(4) The derivation of TAI should include more contributors as links with other time scales become available.

(5) Close cooperation between the scientific and treaty organizations, CIPM, CCIR, IUGG, URSI and IAU, must assure that the present evolution in worldwide standardization in the provision and use of precise time can continue in an orderly way.

Special Session, 29 August 1973

APPLICATIONS OF PRECISE TIME IN ASTRONOMY

This session was organized and chaired by the Vice President, H. Enslin.

G. Winkler gave a report on present capabilities of various electronic systems to disseminate time. Loran-C in its area of groundwave coverage is the most useful system which offers a precision of better than a microsecond. Television is being used increasingly as a local and intermediate distance link. Satellites, looking very promising, have not yet been used for routine operations. They are still experimental.

L. V. Morrison reviewed a comparison of ET2 with AT over the period 1955.5 to 1972.0. It was found that the value of the tidal deceleration in mean longitude of the Moon should be corrected by $-10'' T^2$. The speaker recommended redefinition of ET with reference to TAI.

J. D. Mulholland presented a paper on 'The Use of Precise Clock Time in Laser Ranging'. At a level of ±15 cm range uncertainty as now achieved at McDonald Observatory, the availability of precise epoch and frequency are crucial to the observing strategy, and also to data recognition and applications.

K. I. Kellermann spoke about 'Very Long Baseline Radio Interferometry'. Starting-synchronization of the clocks at the two stations is generally obtained from Loran-C. Final synchronization of about 0.1 μs as often required can be achieved after the observation by comparing the tape records with different time delays until maximum correlation is found.

R. Wielebinski gave an account of 'Applications of Precise Time in Pulsar Research'. Time of arrival observations can be done now with sub-microsecond precision. This precision over periods of months constitutes the most demanding requirement for a uniform clock time including the necessary reductions to a coordinate time. The "corrections" involved are critical for the reduction of observations (see Hunt, G. C.: 1971, *Monthly Notices Roy. Astron. Soc.* **153**, 119).

H. F. Fliegel gave a review on the utilization of time in deep space tracking. In order to obtain accurate orbit determinations, tracking station clocks must be synchronized with each other, and their relationship to UT1, TAI, and ET must be known precisely. Use of the BIH rapid service for UT1 and the pole coordinates has greatly improved the precision of tracking.

COMMISSION 33: STRUCTURE AND DYNAMICS OF THE GALACTIC SYSTEM (STRUCTURE ET DYNAMIQUE DU SYSTÈME GALACTIQUE)

Report of Meetings, 22, 25, 27 and 28 August 1973

PRESIDENT: S. W. McCuskey.
SECRETARIES: K. C. Freeman, A. G. D. Philip.

22 August 1973

SPIRAL GALACTIC STRUCTURE

Joint Meeting with Commissions 34, and 40.
Invited speakers were:
Th. Schmidt-Kaler: Introductory Remarks
K. C. Freeman: Report on Some Highlights from Symposium No. 58 *Formation and Dynamics of Galaxies* (Canberra).
Th. Schmidt-Kaler: General Review and Optical Observations.
K. Rohlfs (read by Prof. O. Hachenberg): Radio-Astronomical Observations.
W. W. Roberts: Theoretical Aspects of Galactic Spiral Structure.

Considerable discussion followed, including brief papers by Wielebinski (Continuum Observations of M31); van Woerden (H-Line Observations of M101); Simonson (A Density-Wave Map of the Galactic Spiral Structure); Piddington (Galactic Spiral Arms and Evolution); and Mark (An Amplifier of Galactic Density Waves). Two papers read by title were: A Simple Model of a Spiral Wave at the Center of a Galaxy by E. Athanassoula-Georgala, and Periodic Orbits in Barred Spiral Galaxies by M. Michalodemitrakis.

25 August 1973

A. JOINT MEETING WITH COMMISSIONS 8 AND 24

B. J. Bok: Report on IAU Symposium No. 61 *New Problems of Astrometry* (Perth).

This comprehensive report was followed by comments from W. Fricke, President of Commission 8 and S. Vasilevskis, President of Commission 24. Resolution No. 1 of IAU Symposium No. 61, concerning needs and priorities in the fields of astrometry was strongly endorsed by Commission 33. Note was taken also, but no action was deemed appropriate, on a resolution to be proposed by Commission 24 regarding the importance of the Yale-Columbia program for determination of proper motions in the southern hemisphere.

B. BUSINESS MEETING OF COMMISSION 33

(1) The new officers and Organizing Committee members for the period 1973–1976, proposed by the present O.C., were duly elected. (2) Guide-lines and rules of operation for the Commission were discussed. The concensus of opinion was that freedom of action and a minimum of rules were essential. Application of the simple procedures given in *IAU Trans.* **XIVB**, p. 200 were deemed sufficient. (3) A lengthy discussion regarding the presentation to the IAU and ICSU of a list of 'priorities' for the work of the Commission was followed by a unanimous vote that no such list should be formulated. (4) It was agreed by acclamation that a wide-reaching and detailed report, such as those prepared by the Commission in 1970 and 1973, is a most desirable feature and should be continued.

(5) A memorandum from The Moscow Variable Star Bureau concerning the GCVS, its format and inclusion of new material was called to the attention of the Commission. Copies of the memorandum were available to members but lack of time prevented discussion.

27 August 1973

Joint Discussion III. *Kinematics and Ages of Stars Near the Sun.* (Included in the *Highlights of Astronomy* 3.)

28 August 1973

THE GALACTIC DISTANCE SCALE

Invited speakers were:

S. V. M. Clube: The Galactic Distance Scale.

W. F. van Altena: The Distance of the Hyades Cluster.

I. Iben, Jr.: The Cepheid Zero-Point.

The invited speakers presented evidence that the distance R_0 from the Sun to the galactic center is likely to be nearer 7 kpc than the 10 kpc now usually assumed. In the discussion period Oort summarized the new results by Plaut on the distribution of RR Lyr stars, which indicate $R_0 \sim 7.8$ kpc. Feast pointed out that the number-apparent magnitude distribution of these stars would indicate a value R_0 nearer 8.0 kpc. Bok mentioned that $\langle M_v \rangle = +0.6$ mag. for the RR Lyr stars, obtained by Graham for stars in the Magellanic Clouds, would also be of importance here. Van den Bergh reported on the large number of very faint stars, presumably in the galactic bulge, now being observed with the 200 in. telescope. These are probably main sequence stars and may lead to another estimate of R_0.

We record here our appreciation to Dr L. Perek and his committee for the excellent program of the Joint Discussion; to Dr Th. Schmidt-Kaler for organizing the program on Spiral Galactic Structure; and to Dr S. V. M. Clube for the program on the Galactic Distance Scale.

COMMISSION 34: INTERSTELLAR MATTER AND PLANETARY NEBULAE (MATIÈRE INTERSTELLAIRE ET NÉBULEUSES PLANÉTAIRES)

Report of Meetings, 22, 23, 24, 27, 28 and 29 August 1973

PRESIDENT: F. D. Kahn

22 August 1973

This was a Joint Meeting with Commissions 33 and 40, largely devoted to a discussion of galactic spiral structure.

23 August 1973

L. Biermann gave an invited lecture on the 'Interaction of the Interstellar Medium with the Solar System'.

There followed a brief business meeting, at which the members of the Commission

(i) approved the proposals for the next President, Vice-President and Organizing Committee;

(ii) considered various ways of reducing the cost of Union publications. Several suggestions were made and were later passed on to the General Secretary;

(iii) decided that they could not make any recommendations to the General Secretary as to the scientific priorities in range of interests of the Commission.

After the business meeting contributed papers were read by B. D. Donn, C. Heiles, P. G. Mezger, K. Nandy and T. Terzian.

24 August 1973

The theme of this meeting was 'Observational Data'. The invited speakers were *D. C. Morton*, on the results from the Copernicus Satellite, *B. J. Robinson*, on interstellar molecules, *J. E. Baldwin*, on interstellar electrons, *J. M. Greenberg*, on dust and *R. D. Davies*, on the interstellar magnetic field. *R. L. F. Boyd* also spoke, on X-ray absorption data.

27 August 1973

The morning meeting dealt with 'Extreme Conditions in Interstellar Space'. Invited contributions were presented by *J. H. Oort*, on the galactic centre, *H. van Woerden*, on high velocity clouds, *D. P. Cox*, on the effects of supernova explosions, and *M. J. Rees*, on interaction with very low frequency radiation.

The afternoon meeting was held jointly with Commissions 25, 27, 37 and 40. It dealt with galactic infra-red and microwave sources, with particular emphasis on cocoon stars.

28 August 1973

A meeting was held jointly with Commission 35, on the subject of 'Fragmentation and Collapse'.

29 August 1973

The theme of this meeting was 'The Role of Instabilities in Interstellar Space'. Invited speakers were *L. Mestel*, on MHD instabilities, *D. G. Wentzel*, on plasma instabilities, *E. R. Capriotti*, on radiation pressure induced instabilities, *D. W. Goldsmith*, on thermal instabilities and *F. D. Kahn*, on Rayleigh-Taylor instabilities.

COMMISSION 35: STELLAR CONSTITUTION (CONSTITUTION DES ÉTOILES)

Report of Meetings

PRESIDENT: E. Schatzman.

A. Administrative Session, August 22, 1973

(a) Election of the new Organizing Committee, and President and Vice-President.

(b) Examination of the list of proposed new members of the Commission and new proposals.

(c) Draft reports.

The following motion was carried:

'Commission 35 feels that conditions have developed in such a way as to make future draft reports of the Commission undesirable. If however, the IAU decision is to retain draft reports, Commission 35 would wish to present a brief report on trends in the subject of stellar constitution'.

(d) Commission Rules.

The following motion was carried:

'Commission 35 intends to carry out its activities responsibly, fairly and with maximum scientific effectiveness, but prefers to operate without formal rules other than those prescribed by the By-laws of the IAU.'

(e) Consultants.

It was agreed that many physicists who are not IAU Members make substantial contribution to the subjects studied by the Commission, and that the Commission President should arrange for some of them to be Consultant.

B. Scientific Meetings

The following topics have been discussed:

(1) The solar neutrinos.

P. Demarque gave a report mainly based on his work on the effect of fast rotating cores on evolution and neutrino fluxes. *P. Ledoux* discussed the possibility of local instabilities near the center of the Sun and their effect on overall stability. An extensive discussion followed. *W. F. Huebner* gave informations on the opacity library, to which access will be possible in 1974.

(2) Transport problems.

G. Michaud gave a report on microscopic transport (diffusion). *J. P. Zahn* gave a report on macroscopic transport (overshooting, transport in radiative zone: Goldreich Schubert Fricke instability and shear flow instability (linear and non linear regions)) (based on work by Gough, Latour, Spiegel, Toomre and Zahn). *E. E. Salpeter* presented some results on convective overshoot and *A. V. Maeder* on evolutionary sequences with mixing in a small zone above the convective core.

(3) Informal discussion of the limits of the mixing lengths theory.

(4) Star formation and cloud fragmentation: (Joint Meeting with Commission 34).

Review paper by *T. Arny* (read *in absentia* by E. Schatzman).

Invited paper by *R. B. Larson*, centered on numerical methods in hydrodynamical studies of fragmentation (including a short invited paper by *Quirk* on the condensation of fast rotating discs).

Three other short papers were presented by *D. Mc Nally*: Interstellar gas cloud collapse; *Savedoff* and *Taff* (read by Savedoff): Coalescence model; *Appenzeller:* Massive protostars.

COMMISSION 36: THE THEORY OF STELLAR ATMOSPHERES (LA THÉORIE DES ATMOSPHÈRES STELLAIRES)

Report of Meetings

PRESIDENT: R. N. Thomas.

1. One business meeting of the Commission was held:
The following officers were elected:
President: R. Cayrel.
Vice-President: D. Mihalas.
Organizing Committee: E. Avrett, K. Hunger, V. L. Khoklova, J. Hutchings, B. E. J. Pagel, M. Vardya and R. N. Thomas.
In addition, it was agreed that the President might co-opt one or two additional members of the Organizing Committee, one of which might be the Secretary of the Commission. (This has been done by Cayrel. F. Praderie has been co-opted as Secretary to the Organizing Committee; S. Rucinski has been co-opted as Commission member.)

2. A full-day symposium on *the Next Decade of Research in Stellar and Solar Atmospheres* was held. The four invited speakers were:
L. Kuhi: Stellar Observational Work.
P. Delache: Stellar Theory.
J. Beckers: Solar Observational Work.
H. Rosenberg and M. Kuperus: Solar Theory.
The rest of the day was devoted to informal discussion.
The entire symposium was tape-recorded; it is currently under debate what to do with the tapes. This was jointly organized by Commissions 10, 12 and 36.

3. A full-day symposium on *Waves in Stellar and Solar Atmospheres* was held. The two invited speakers were:
J. Leibacher: Wave-Generation.
P. Souffrin: Wave Propagation and Dissipation.
The rest of the day was devoted to informal discussion. The symposium was not tape-recorded. Jointly organized by Commissions 12 and 36.

4. A half-day symposium was held on *Ap Stars*, jointly organized by Commissions 29 and 36 (see the Report of Commission 29).

5. Commission 36 went firmly on record *against* publishing a bibliography of the work of the Commission in terms of reference numbers, it wishes title and author. Commission 36 makes no suggested changes in its rules over what presently exists.

COMMISSION 37: STAR CLUSTERS AND ASSOCIATIONS (AMAS STELLAIRES ET ASSOCIATIONS)

Report of Meetings, 21, 24, 27 and 29 August 1973

PRESIDENT: G. Larsson-Leander.
SECRETARY: R. Wielen.

A. Administrative Session

The first meeting of the Commission, on August 21, was devoted to administrative matters. The following items of the agenda were discussed:

(1) The Commission Report. The Report for the past triennium was prepared in due time and submitted to the General Secretary for printing. In order to give the Commission members opportunity to read the Report before publication, and thus possibility to suggest minor changes and corrections, the type-script had been mimeographed and copies distributed to the Commission members and to all other colleagues who had forwarded progress reports to the President. This procedure was much appreciated by the Commission, and the Report in its present form, as a critical review, was considered very useful. The Commission especially stressed the value of the extensive tabulations of current work, which have been a regular feature of the Commission 37 Reports. The Commission was strongly in favour of having all Reports of the various Commissions published in book form, as has been done up to now, and the President was urged to act accordingly when discussing the question with the General Secretary.

(2) Election of Commission officers. Proposals for new officers had been made in a circular letter issued by the President, after consulting the Organizing Committee. The proposed names were approved by the Commission. The President expressed his thanks to the colleagues now leaving the Organizing Committee.

(3) New Commission members. Sixteen colleagues had been proposed to, or had applied for membership in the Commission. Most of these names had been considered and approved by the Organizing Committee and almost all had been announced by the President in a previous circular letter. The Commission approved of all the candidates proposed as new members.

(4) Commission rules. The Commission opposed to the idea of establishing strict rules, but agreed that certain guide lines might be useful. It was decided that the length of service in the Organizing Committee should in general be limited to three periods (of three years each), exceptions being made, if necessary, for those elected as Presidents and Vice-Presidents. For the sake of continuity, a past President should remain within the Organizing Committee also during the subsequent three-year period. Further, it was emphasized that proposals for new members of the Commission should be decided upon by the Commission itself.

(5) Atlas of globular clusters. The President announced that Dr G. Alcaino is preparing an atlas of galactic globular clusters, giving identification charts, stellar magnitudes, and other cluster data.

(6) Priorities in research. The question of assigning research priorities within the fields of the various Commissions had been raised by the General Secretary at a meeting with Commission Presidents. The matter was discussed by the Commission, and a number of important research topics were suggested and listed on the blackboard. It was found impossible, however, to assign priorities to the various topics. On the proposal of the President he was given the task to arrange the material and to prepare a draft to be distributed for the consideration of Commission members and other interested colleagues present in Sydney. The suggested additions and improvements to the draft were taken into account for the final version, given at the end of this Report.

(7) Circular letter from the Moscow Variable Star Bureau. The contents of the circular letter were discussed, but as the letter mainly concerned the field of Commission 27 no formal action was taken.

There being no further business items the meeting was adjourned.

B. Scientific Sessions

The meeting on August 24 was mainly devoted to the topic of Stellar Rings. After an extensive introductory review, 'The Stellar Rings', given by J. Isserstedt, two critical papers were read in abstract form:

N. V. Vidal and K. Å. Bern: Stellar Rings and the Structure of the Galaxy (presented by Bern),

T. A. Uranova: To the Problem of the Stellar Rings (read by the President).

A large number of people participated in the ensuing general discussion.

After this the following two papers were presented:

F. J. Kerr: Neutral Hydrogen Observations of Globular Clusters,

G. Alcaino: An Atlas of Galactic Globular Clusters.

In connection with his paper Dr Alcaino exhibited for inspection pre-print copies of the *Atlas*.

The meeting on August 27 was a joint meeting with Commissions 25, 27, 34 and 40, on 'Galactic Infrared and Microwave Sources'. The full report is given by Commisison 27.

The August 29 meeting, on 'Stellar Content of Associations and Clusters', was held jointly with Commission 45. The full report of the meeting is given by Commission 45.

Importance of Research on Star Clusters and Associations

Besides being of great interest *per se*, as stellar systems of various size, the clusters and associations are of great importance for many branches of astronomy and astrophysics. They provide information on star formation and stellar evolution, and they constitute almost ideal objects for testing theories in these fields. The correlation between the ages, abundances, and kinematics of clusters provides an essential clue to the early evolution of our Galaxy, and in the globular clusters we have the oldest known objects in the Universe. Further, the clusters and associations are especially useful for calibration of various astrophysical relationships, needed as fundamental data both in theoretical studies of stellar structure and in galactic research.

If the clusters and associations are to meet these expectations and demands, all kind of such objects have to be studied intensively, in order to segregate physical members from field stars. All possible methods should be used, but of particular value are spectral classifications, radial-velocity measurements, and proper-motion studies. To further the solution of the calibration problem, a list of especially useful objects might be prepared. Special regard has to be given to the chemical abundances, frequency of binaries, and the stellar-rotation velocities. Intermediate-band photometry and spectral studies of individual stars are needed. The faint ends of nearby clusters should be investigated more closely.

Extremely young clusters and other very young aggregates should be studied, because of their importance for the theories of star formation and the early stellar evolution. Similarly, the stars above the horizontal branch in globular clusters are of special interest for the late evolution.

A satisfactory classification scheme for globular clusters is needed, as well as mass and distance determinations for such clusters, using various methods.

The kinematical ages of associations should be more thoroughly investigated, and the reality of moving stellar groups should be checked. In the field of cluster dynamics much work remains to be done; thus, for instance, the time scale of dynamic evolution is rather uncertain.

The above is a summary of a discussion at the Commission 37 meeting on August 21, 1973. Many more important research items could be added. The Commission finds it impossible to assign priorities by number to the various problems outlined, as they are all of great importance and cover different areas of a large field. The question of priority must be decided upon by the individual researcher himself.

COMMISSION 38: EXCHANGE OF ASTRONOMERS
(ÉCHANGE DES ASTRONOMES)

Report of Meeting

ACTING PRESIDENT: D. A. MacRae.
SECRETARY: F. B. Wood.

Commission 38 held its business meeting on the 21st August, 1973. Present were J. Sahade, K. Strand, J. P. Wild, F. B. Wood, Secretary, D. A. MacRae, presiding, in the absence of Drs Reiz and Routly.

The sad losses of Drs Minnaert and Bowen were noted with regret. These two members of the IAU had served Commission 38 loyally and energetically.

The draft report as prepared by the President was approved in principle.

P. M. Routly was unanimously elected president for the coming triennium.

J. P. Wild declined nomination as vice-president; D. A. MacRae was nominated and unanimously elected.

The following were nominated to serve (with the president and vice-president as ex officio members) as the Organizing Committee for the interval 1973–1976: J. Delhaye, G. S. Khromov, A. Reiz, J. Sahade, F. G. Smith and F. B. Wood.

Dr *Sahade* was requested to report to the Executive Committee of the Union the difficulties caused in the work of this Commission in particular by the devaluation of the U.S. dollar, and to urge an appropriate increase in the funds allotted to the Commission.

A letter from *Dr Reiz* concerning the grant allocation procedure was read. The Commission agreed with Dr Reiz's conclusion that there was no need for introducing a more formal procedure than currently existing for the efficient operation of the Commission.

F. B. Wood was delegated to represent the Commission at the joint meetings with Commissions 5, 9 and 46 on the World List of Observatories and D. A. MacRae to represent Commission 38 in the joint meetings with 5, 36 and 46 on Publication Economies and Policies.

COMMISSION 40: RADIO ASTRONOMY
(RADIO ASTRONOMIE)

Report of Meetings, 21 and 29 August 1973

PRESIDENT: D. S. Heeschen.
SECRETARY: W. E. Howard.

Business Meetings

The following matters were discussed at the two business meetings of Commission 40:

1. THE REPORT OF THE COMMISSION

Opinions were divided concerning the scientific value of Commission Reports in *Transactions*, relative to the cost to produce and distribute it. Many members were of the opinion that the size limit imposed by the Executive Committee restricted the Report to a bibliography of only limited usefulness. Others felt that summaries, even if brief, are valuable. There was no clear consensus that Commission 40's recent practice should be changed.

2. MEMBERSHIP

A subcommittee consisting of R. D. Davies (chairman) and W. E. Howard received and reviewed nominations for membership. Thirty-five new members were elected by the Commission.

3. COORDINATION OF FLUX DENSITY CALIBRATIONS OF STANDARD SOURCES

The Working Group set up at Brighton, Shakeshaft and Kellermann, has undertaken an analysis of the flux density scales in various catalogues as a function of the observing frequency. The results on scale corrections and a list of secondary calibrators will be published shortly. The Working Group is being continued.

4. FUNDAMENTAL POSITIONAL SYSTEMS FOR RADIO ASTRONOMY

A working Group consisting of Ryle (convener), Burke, Christiansen, Cohen, Gent, H. Palmer and Raimond has suggested that a group of 21 radio sources be used as fundamental calibration sources for positions. Desirable characteristics of this group include: (1) appreciable flux density, (2) well distributed in the sky, (3) relatively flat spectra, (4) small diameter, and (5) no low frequency cut-off. In addition, a larger number of sources is suggested as secondary calibrators. These sources may be found in the following references:

Fundamental Calibrators

Ryle, M. *et al.*: 1973, *Monthly Notices Roy. Astron. Soc.*, in press.
Tucker, R. *et al.*: 1973, *Monthly Notices Roy. Astron. Soc.*, in press.

Secondary Calibrators

Adgie, R. *et al.*: 1972, *Monthly Notices Roy. Astron. Soc.* **159**, 233.
Browne, W. *et al.*: 1973, *Nature* **244**, 146.
Hunstead, R.: 1972, *Monthly Notices Roy. Astron. Soc.* **157**, 367.

Brosche, P. *et al.*: 1973, *Astrophys. J.* **183**, 803.
Kapahi, V. *et al.*: 1973, *Astron. J.*, in press.
Swarup, G. *et al.*: 1971, *Astrophys. Letters* **9**, 53.
Joshi, M. *et al.*: 1973, *Astron. J.*, in press.

The Working Group pointed out that calibration sources are still needed for the southern hemisphere.

The report was adopted and the Working Group dissolved.

5. The unit of flux density

Kerr proposed the following resolution that was passed by the Commission: 'RESOLVED, that the name 'Jansky', abbreviated 'Jy', be adopted as the unit of flux density in radio astronomy and that this unit, equal to 10^{-26} W m^{-2} Hz^{-1}, be incorporated into the international system of physical units.'

6. Standard equinox for astronomical catalogues

Bolton proposed the following resolution that was passed by the Commission, after discussions with Commission 8: 'RESOLVED, that the equinox of 1950.0 should remain the standard equinox for astronomical catalogues.'

7. Optical and radio definitions of radio velocity

Optical astronomers have customarily defined radial velocity as $cz = c\Delta\lambda/\lambda_0$. Radio astronomers usually express radial velocity as $c\Delta f/f_0$. For vacuum ultraviolet lines redshifted into the visual, some spectroscopists have quoted a $z = (\lambda_{air} - \lambda_{0,vac})/\lambda_{0,vac}$. Differences between these various definitions become appreciable for velocities above 1000 km s^{-1}. A joint working group convened by Commission 30 and involving Commission 40 members M. Roberts and B. Clark will be established to attempt to unify the definitions.

8. Polarization definitions

A working Group chaired by Westerhout was convened to discuss the definition of polarization brightness temperatures used in the description of polarized extended objects and the galactic background. The following resolution was adopted by Commissions 25 and 40: 'RESOLVED, that the frame of reference for the Stokes parameters is that of Right Ascension and Declination with the position angle of electric-vector maximum, θ, starting from North and increasing through East. Elliptical polarization is defined in conformity with the definitions of the Institute of Electrical and Electronics Engineers (IEEE Standard 211, 1969). This means that the polarization of incoming radiation, for which the position angle, θ, of the electric vector, measured at a fixed point in space, increases with time, is described as right-handed and positive.'

9. Newsletter on unidentified melecular lines or negative searches

D. Johnson and *B. Donn* proposed a means for reporting and documenting new unidentified radio lines and negative-result searches for molecular lines. Forms with a tentative format were presented that will be sent to active observers who can summarize past and future line search results. These summaries will be circulated among the observers or published occasionally in the existing literature.

10. Interference and frequency allocations

F. G. Smith summarized the worldwide status of frequency allocations for the radio astronomy service that resulted from the World Administrative Radio Conference (WARC) in 1971, and presented the new system of correspondents now in use by the Inter-Union Commission for the Allocation of Frequencies for Radio Astronomy and Space Science (IUCAF). Smith asked for advice on whether allocated frequencies in the range 10–15 MHz would be desirable and asked for advice on priorities for future protection for newly discovered molecular lines. Howard summarized the broadcast satellite problem at the 2690 MHz band edge, noting that the satellite, expected to become operational in 1974, will now contain a filter that should reduce interference to below the levels recommended for the 11-cm radio astronomy band by Report 224 of the CCIR.

11. Research priorities

The Commission went on record as opposing any suggestion to attach priorities to research programs in radio astronomy.

12. Other resolutions

Commission 40 endorsed the resolutions passed at the *IAU Symposium* **61** in Perth concerned with astrometry.

13. Election of the organizing committee

The President presented the following list of nominations for Officers and Organizing Committee of the Commission which was approved without further nominations:

President: Yu. N. Parijskij

Vice President: H. van der Laan

Organizing Committee: Blum, Heeschen, Kuzmin, McLean, Menon, Mezger, H. Palmer, Robinson, Zuckerman.

Scientific Meetings

A Joint Discussion on Very Short Time Scale Phenomena was held on August 28, organized by F. G. Smith and is described in *Highlights of Astronomy* **3**.

The Commission participated in the following joint scientific meetings, with the other Commissions indicated: *Galactic Spiral Structure* (33, 34, 40), *X-Ray Sources, Black Holes and Radio Stars in Binary Systems* (27, 35, 40, 42, 44, 48), *Solar Work* (10, 40, 44), *Molecular Radio Astronomy* (14, 40), *Very Compact H II Regions* (25, 27, 34, 37, 40), and the *Definition of the Meter* (4, 14, 31, 40).

One scientific session devoted to extragalactic radio astronomy was held on August 28, organized by Shakeshaft.

COMMISSION 42: CLOSE BINARY STARS
(ÉTOILES BINAIRES SERRÉES)

Report of Meetings, 22, 23, 28 and 29 August 1973

PRESIDENT: M. Plavec.
SECRETARY: I. Jurkevich.

Business Meetings

The business meetings were held on August 22 and 28, and this summary covers both of them. A. H. Batten remembered briefly the deceased IAU members O. Günther, A. H. Joy and H. Shapley. Draft Reports could not be discussed since to most members they had not been accessible. The President promised to distribute the reprints to all members as soon as the reprints are available. The Commission was very strongly in favor of continuing the style and publication of the Reports on Astronomy in spite of the high price of the volume. It is recommended to future Presidents to consider the possibility of distributing a mimeographed form of the Commission report to members before the General Assembly.

The name of the Commission was discussed, also with other Commissions (26 and 30). It was decided to change the name to Close Binary Stars, thereby indicating that the field of the Commission will from now on include eclipsing binaries, spectroscopic binaries, and all other types of binary stars that are close enough in space to invoke serious interaction between the components. Visual double stars as well as astrometric binaries continue to be the domain of Commission 26, as before.

The Organizing Committee presented the following slate for election: President: T. Herczeg, Vice-President: G. Larsson-Leander, Organizing Committee: A. H. Batten, M. Fracastoro, K. Gyldenkerne, M. Kitamura, D. M. Popper, J. Smak, F. B. Wood, M. Plavec (Past President). K. O. Wright proposed to add two more names to improve the geographical distribution and increase the representation of young astronomers. The Organizing Committee then proposed to add E. P. J. van den Heuvel and B. Warner, and in this form the list of new officers was approved by the Commission.

The Commission elected 18 new members, partly from young astronomers who just became members of the IAU, partly from among astronomers working in the broader field covered by the new name of the Commission. Future procedures were discussed for election or co-option of new members. It was decided unanimously that if a National Committee proposes a new Commission member, this proposal will be considered only provided satisfactory documentation is submitted.

The by-laws of the Commission were discussed. It was agreed that as a rule a member of the Organizing Committee should not serve for more than three terms unless he becomes Vice-President or President. The President is expected to serve one term after his office term has expired.

The Commission acknowledged the work done by Larsson-Leander and his staff together with the regional contributors on compiling and disseminating the Bibliography on Eclipsing Variables. Members consider this to be a valuable publication; Larsson-Leander reported that in fact the circulation is wider, 200 copies being distributed. He promised to continue the semiannual publication of the Bibliography for the next three years. The Commission then voted to ask the Executive Committee to grant a subsidy of $200 to defray partly the costs. A vote of thanks was also expressed to Drs Detre and Szeidel of the Budapest Observatory for the excellent services rendered to the Commission members by the frequent and rapid publication of the *Information Bulletin on Variable Stars.*

A. H. Batten then raised the question of need for the 7th Catalog of Orbital Elements of Spectroscopic Binary Systems, since the 6th Catalog will soon not be available. The Commission feels that the 7th Catalog will soon be needed and strongly recommends its preparation.

A Symposium on Be Stars and Shell Stars will be organized by Commissions 42, 29 and 36, probably in September 1975. *M. Plavec* will represent Commission 42 on the Organizing Committee. Co-sponsoring a joint Colloquium on Variable Stars in Budapest in September 1975 was also discussed, but further discussion on the program appears necessary.

Y. Kondo presented a brief report of the sub-committee on extraterrestrial observations. Members of this sub-committee are expected to work as local disseminators of information. *K. Gyldenkerne* discussed the problems of coordinated programs; these will be continued but no new stars have been added this time. *A. H. Batten* announced that the working group on Spectroscopic Binary Stars of Commissions 30 and 42 has now in fact been transferred to Commission 42.

The Commission discussed in detail the priorities for research in the field of the Commission. The Organizing Committee joined by M. V. Ovenden then prepared the following document, which was approved by the Commission.

Priorities for Research

Within the broad spectrum of the physical sciences, astrophysics is important as it studies matter under a variety of physical conditions not accessible in the laboratory. In the study of close binary stars, we approach most closely to laboratory experimentation since we examine the effects of one star upon another under a great variety of conditions.

When the stars are sufficiently close together to form an eclipsing pair, but sufficiently detached not to affect their structure significantly, such a non-interacting binary star provides the best available data on stellar masses, radii, luminosities, and other characteristics. When, on the other hand, the stars do interact strongly, their evolution will be completely different from the evolution of a single star. Since a large fraction of stars actually belong to binary or multiple systems, the study of the group of interacting binary systems is essential for our understanding of stars as a whole.

The non-interacting and interacting binary systems thus represent two different but closely related approaches to further research on close binary systems, each of which must be treated separately. The following projects and problems are considered to be of top priority in the near future. (The order in which they are listed does not imply relative importance.)

A. NON-INTERACTING CLOSE BINARY SYSTEMS

(1) Obtaining reliable masses of representative types of stars. While every mass determination, if reliable, is of importance, priority should be given to investigating early-type stars and stars evolved away from the main sequence. Infrared spectroscopy is important in this case.

(2) More emphasis should in general be placed on photometric and spectroscopic observations of eclipsing and spectroscopic binary systems of moderate and longer periods, since they are important for evolutionary studies and they provide data on evolved stars.

(3) A new and very promising method of obtaining dimensions of stars and binary systems is the Narrabri Intensity Interferometer. The successful studies of Alpha Virginis and Gamma Velorum convince us that a larger instrument of the same type would be of great value to the progress in studying not only single stars but binary systems as well.

(4) Since binary stars form a significant portion of the whole stellar population, and since their evolution strongly depends on masses and dimensions of the systems, it is essential to rediscuss the frequency distribution of binary stars with luminosity and spectral type. A new classification scheme is needed, in particular for eclipsing binaries.

B. INTERACTING CLOSE BINARY SYSTEMS

(1) Concentrated effort should be made to detect compact and collapsed objects in binary systems. Systems suspected of containing a black hole, such as Cyg X-1, Beta Lyrae, Epsilon Aurigae and others, should be studied by all available methods. Since the classical spectral observations in the

visible range have not given sufficient information, of particular importance become the observations from artificial satellites (far UV and X-rays), observations in the infrared, and radio observations.

(2) The existence of compact binary sources of X-rays emphasizes once more the necessity of investigating mass transfer in close binary systems, in particular the interaction of the gas streams with the mass-accreting component. Change of structure and evolution of this star as well as formation of disks and extended envelopes around it may explain a wealth of other phenomena, for example, the shell stars and many peculiar objects like novae, U Gem stars, symbiotic objects, etc.

(3) The studies mentioned sub B(1) and B(2) represent applications, of particular current importance, of general investigations on evolution of close binaries accompanied by mass exchange and mass loss. It has become clear that the results of such investigations apply to a wide variety of systems in which a great and unexpected range of physical phenomena occur. Hence it is necessary that investigations of the evolution of interacting systems be applied to an even wider range of conditions than before.

(4) Associated with the preceding problems are further theoretical topics such as hydrodynamical treatment of gas flow, time scales for synchronization of rotation of binary star components, physics of and radiation transfer in extended atmospheres and circumstellar disks. Resulting models should be compared with new spectrsocopic observations.

(5) The observed radio emission from several binary systems still represents a complete puzzle. More observations on radio frequencies as well as complex observational and theoretical studies of radio binaries are necessary.

(6) For the elucidation of many aspects of the physics of close binary systems, it is important that observations of various kinds be repeated on a number of selected systems over a considerable interval of time in order that the time-varying phenomena may be studied.

Scientific Meetings

Meeting of August 29, 1973, on *Current Trends in Binary Star Studies.* Eight short reports were presented as follows: K. O. Wright on spectral studies of VV Cephei; R. E. Wilson on his program computing photometric elements from light curves; J. B. Hutchings on the problem of reconciling mass ratios obtained photometrically and spectroscopically; D. M. Popper surveyed his work on deriving masses of detached systems, Algol systems, and binaries with H and K line emission; M. G. Fracastoro suggested that a search be made for new eclipsing binaries in the equatorial belt, within the magnitudes 7^m to 9^m. Y. Kondo reported on his observations of the He I line at $\lambda 10830$, and on his and McCluskey's analysis of LY Aur. F. B. Wood reported on the observations of γ^2 Velorum, which show fairly rapid variability in He II. No definite eclipses were detected. M. Plavec announced results of theoretical computations of mass loss from red giants with deep convective envelopes.

Joint meeting with Commissions 27, 40, and 44 on *X-Ray Sources, Radio Sources and Black Holes in Binary Systems.* It was held on August 23 and is reported by Commission 27. Contributions by members of Commission 42 included a new model of β Lyrae (R. E. Wilson) and a discussion of mass determination in X-ray binaries (Y. Kondo).

COMMISSION 44: ASTRONOMICAL OBSERVATIONS FROM OUTSIDE THE TERRESTRIAL ATMOSPHERE (OBSERVATIONS ASTRONOMIQUES AU-DEHORS DE L'ATMOSPHÈRE TERRESTRE)

Report of Meetings, 21, 22 and 28 August 1973

PRESIDENT: V. Prokof'ev.
SECRETARY: H. Lamers.

Administrative Session

The President gave a short review of recent results and a preview of future developments in space astronomy. He mentioned especially the planetary and interplanetary investigations, the solar observations by means of the OSO stations, Interkosmos satellites and Skylab, and the stellar observations by means of the OAO-2, TD-1A and OAO-3 satellites. The loss of the three Astronauts in USSR Dobrovolsky, Volkov and Patzaev was honoured by standing up and silence.

Colloquium 14 'UV and X-Ray Spectroscopy in Astrophysical and Laboratory Plasmas' in Utrecht in 1971, and Symposium 54 'Problems of Calibration of Absolute Magnitudes and Temperatures of Stars' in Geneva in 1972 were sponsored by Commission 44.

The following proposed Symposia and Colloquia will be sponsored by Commission 44: 'The Solar X- and Gamma-Radiation' in Buenos Aires to be held during the next COSPAR session in Brazil in 1974 (Commissions 10, 12 and 44), 'UV and X-Ray Spectroscopy of Astrophysical and Laboratory Plasmas' in Harvard Observatory on 9–11 September 1974 (Commission 14 and 44) and 'UV Spectra and Ground-Based Observations' in Utrecht, 1975 spring.

The Report of Commission 44 was prepared by the President with the addition of information obtained from A. Code, A. Dalgarno, A. Boggess and R. Bonnet. The information from G. A. Doschek and R. W. Noyes was received too late and therefore could not be added to the Report.

Before the election of the new President, Vice-President and Organizing Committee the chairman reported on some statistics on the membership of the Commission. There are two approximately equal groups of members working in the regions of solar and stellar space Astronomy. Therefore it will be worthwhile to nominate as President and Vice-President persons from these two groups respectively. A proposal was made to introduce in the Organizing Committee a specialist in the region of IR Astronomy, in particular P. J. Lena (Paris).

There was a discussion on the position of Commission 44 with respect to other Commissions of the IAU. Space research is well developed now and its work and results are included in various other Commissions dealing with specific topics (N. Roman). L. Gratton proposed that the technical matters should be dealt with in the Commission for instrumentation; however, Commission 9 is already too large. N. Roman proposed that Commission 9 should be divided in different groups. Commissions 44 and 9 should come together and redefine their tasks. A. Underhill thinks there is a reason to continue the work of Commission 44 in coordinating all space research and in co-sponsoring the discussions with other Commissions; it is a place where the space researchers come together.

A discussion about scientific priorities (L. Goldberg, R. Lüst, A. Code, N. Roman, J. Jefferies, A. Underhill, A. Gabriel) led to the following conclusions:

Commission 44 shares its scientific problems with other IAU Commissions, as it covers such aspects as the Sun, the Solar System and Galactic and Extragalactic research. On one hand there are scientific problems connected with the nature of Solar events, which need X- and UV-spectra of the Sun, and systematic observations of the emission of active regions and flares on the Sun. This study is connected with observations of the events in interplanetary space.

As regards stellar Astronomy from space, the spectral resolution in UV obtained by OAO-3 (Copernicus) (0.1–0.05 Å) is probably sufficient. It is necessary, however, to increase the sensitivity in order to record, both photometrically and spectrographically, UV and IR spectra of stars to limiting magnitudes fainter than 15 with sufficient spectral resolution. Further problems are: optical UV and IR observations and the identification of new stellar objects such as Pulsars, Quasars, and X-Ray Sources, and observations of UV and IR spectra of late-type stars in search for stellar chromospheres. Another high priority objective is diffraction-limited imaging with a powerful telescope, to be able to observe individual stars in galaxies.

The technical means for investigating all these problems are: Aircraft, Balloons, Rockets, Orbital Stations, Shuttle-Satellites or Free Flyers, and Interplanetary Probes.

Scientific Sessions, 22 and 28 August 1973

THE SOLAR X-RAY AND UV RADIATION

1. G. A. Newkirk: Observations of the Radiation and Polarization of the White Solar Corona (4000–7000 Å) at a distance of 1.6–6 solar radii, made from Skylab with a resolution of 8″ (Preliminary information).
2. V. K. Prokof'ev: X-Ray Observations of the Sun by means of the 'Interkosmos' Satellites, the X-Ray spectroheliograms in the region 8–12 Å and high resolution (4×10^{-4} Å) spectra in the region 1.8–9.2 Å.
3. R. C. Catura: Large Scale Structure of Solar Active Regions in X-Ray Emission Lines (Rocket Observations on 18 January 1973).
4. J. H. Parkinson: New Observations of the Solar X-Ray Spectrum in the region 9–22 Å with the Crystal Spectrometer.
5. Sh. Yousef: X-Ray Spectroheliograms and the Fine Structure of the Solar Coronal Active Regions.
6. A. H. Gabriel: Rocket Solar Astronomy (Review).
7. R. M. Bonnet: Balloon IR and UV Solar Astronomy (Review).
8. L. Goldberg on 28 August spoke about the preliminary results of the observations of the Sun from the orbital station Skylab.

STELLAR ASTRONOMY FROM SPACE

1. H. Lamers: A Short Review of the Symposium 'UV Stellar Radiation' during the COSPAR Meeting in Konstanz in May 1973.
2. A. B. Severny, E. I. Terez and A. M. Zvereva: Preliminary Results Obtained with an Astrophotometer installed on Lunokhod-2 (Read by V. Prokof'ev).
3. A. Monfils, K. Nandy and H. Lamers: ESRO TD-IA Satellite—Spectrophotometric and Spectroscopic Observations in the UV Region by the Sky Scanning Satellite.
4. E. B. Jenkins: A Review of the Technique and Results of the Observations of Interstellar Absorption by the OAO-3 Satellite (Copernicus).
5. R. L. F. Boyd: X-Ray Telescopes on the OAO-3 Satellite (Copernicus). Boyd spoke about the preliminary results obtained with these telescopes during a joint session of Commissions 27, 40, 42, 44 and 48 on 23 August.
6. M. Golay: Balloon Stellar Astronomy, near UV and IR (Review).
7. A. Code: The Current State of the UV Stellar Astronomy from Space and the Main Technical Problems.
8. N. Roman: The Perspectives of the Development of the UV Space Astronomy, the Space Shuttle Program and 3 m Large Astronomical Telescope (LAS).

COMMISSION 45: SPECTRAL CLASSIFICATIONS AND MULTI-BAND COLOUR INDICES (CLASSIFICATIONS SPECTRALES ET INDICES DE COULEUR À PLUSIEURS BANDES)

Report of Meetings

PRESIDENT: B. E. Westerlund.
SECRETARY: B. Hauck.

Administrative Session, 22 August 1973

The President opened the Administrative Session and started the discussion about the usefulness to publish Commission Reports. It was concluded that these Reports are useful, in particular to the members of other Commissions, students, etc. They permit a general view of the work accomplished in the fields of astronomy and astrophysics to be obtained. In order to allow a quick consultation, but without the obligation of publishing a large list of references, the quoted publications should be followed by the number in Astronomy and Astrophysics Abstracts.

The rules concerning the organization of the Commission will be discussed by the members of the new Organizing Committee and will be communicated to all members of the Commission one year before the next General Assembly.

The following by-law proposed by the President concerning the membership in the Organizing Committee, was accepted: 'A Member of the Organizing Committee serves two consecutive terms and then retires unless he becomes President or Vice-President. For the sake of continuity the Vice-President should succeed the retiring President who then serves on the Organizing Committee one more term'.

The new President and Vice-President as well as the Organizing Committee were elected with unanimity and the fourteen new members of the Commission were approved. K. Nandy was elected as new member of the Group on Spectroscopic and Photometric Data.

The text on priorities and expected developments in the fields of spectral classification and multi-band colour photometry was the subject of very long discussion. After some modification of the form, the text was accepted. Its essential points are:

(i) Improvement of criteria for spectral classification. It is noted that MK classification, high dispersion spectroscopy and photoelectric photometry complement each other in this refinement work; the slit spectrograms will continue to serve for monitoring photometric and objective prism surveys. Additional criteria are certainly to come from the research in the ultraviolet and infrared spectral ranges. It appears also likely that the stellar rotation will gain in importance as a specific classification parameter. Standard rotational-velocity stars will then define a system in the same way as the standard stars for spectral classification.

(ii) Calibration of the observed criteria in terms of the basic parameters mass, age and chemical composition.

Setting up a list of HR standard stars so that the various parameters of existing classification systems may be interpreted in terms of the basic astrophysical parameters.

(iii) Reclassification of HD stars into a homogeneous MK system.

(iv) Distribution of spectroscopic and photometric data, data centers.

New Organizing Committee:

President: C. O. R. Jaschek.

Vice-President: E. K. Kharadze.

Members: A. P. Cowley, B. Hauck, W. Iwanowska, P. C. Keenan, E. Mendoza V., U. Sinnerstad, V. L. Straizys, B. E. Westerlund.

Scientific Sessions, 28 August 1973

A. THE USE OF MULTI-BAND SYSTEMS FOR CLASSIFICATION OF FAINT STARS

A. G. D. Philip: Four-Colour Photometry of Early Type Stars.

R. F. Wing: Classification of Faint Red Stars by Narrow-Band Photometry.

V. Straizys (read by V. B. Nikonov): The Vilnius Photometric System for Three-Dimensional Classification of Stars and Determination of Interstellar Reddening.

B. EXAMINATION OF CLASSIFICATION CRITERIA

N. R. Walborn: The OB Stars.

C. B. Stephenson: Visual Classification Criteria for Later-Type Stars at Medium to Low Dispersion.

A. P. Cowley and N. Houk: The Michigan Spectral Catalogue, Part I and II.

C. DATA CENTERS

B. Hauck: Concept of a Data Center.

Joint Meeting of Commissions 37 and 45, 29 August 1973

STELLAR CONTENT OF ASSOCIATIONS AND CLUSTERS

R. Schild: The Peculiar B Stars in Orion.

R. F. Garrison: Peculiar B Stars in Scorpius, Orion and the Alpha Persei Cluster.

M. Breger: Delta Scuti Stars.

J. Dachs: On the Stellar Content of Open Clusters of the Pleiades Type.

R. Schild: The Nature of the Infrared Excesses in the Per Be Stars.

B. Hauck: Photometrical Ap Stars in Open Star Clusters.

K. C. Freeman: Some Globular Cluster Observations.

A. G. D. Philip: Bright Horizontal Branch Stars in the $\log g$, $\log T_{\rm eff}$ diagram.

COMMISSION 46: TEACHING OF ASTRONOMY (ENSEIGNEMENT DE L'ASTRONOMIE)

Report of Meetings

PRESIDENT: E. A. Müller.
SECRETARY: D. G. Wentzel.

Session 1, 22 August 1973

Fifteen Commission members or their representatives were present. The Commission Report was approved; the National Reports were extended by J.-C. Pecker who reported the introduction of astronomy as a regular part of the physics teaching in French schools from age 13 on.

The nomination of D. McNally as Commission President, the appointment of two additional Commission members (S. K. Trehan, India, and Th. Schmidt, German Federal Republic) and of two consulting members (B. Hajebi, Iran, and R. H. Wilkinson, Australia) were approved. The discussion on whether the Commission should propose to the IAU Executive Committee two vice-presidents or one involved (i) the difficulties of successful actions when correspondence is involved with a single vice-president in a country that is not used to rapid postal communications, (ii) the difficulty of the vice-president representing the president at a meeting when the vice-president cannot leave his country readily, and (iii) the difficulties inherent in choosing specific duties and establishing the succession to the presidency in case of two vice-presidents. The majority of voting members favored proposing two vice-presidents to the Executive Committee.

The Commission members took note of a meeting between Australian Science Teachers and astronomers, organized through Commission 46 on August 20, 1973, which proved stimulating to all participants. Discussion centered on how the resolution that emerged from the meeting with the teachers could be made effective in Australia and at the same time would encourage teaching efforts in other countries. The following two resolutions were passed with no dissenting votes:

Resolution 1: 'Considering the effort made by Australian teachers to improve the teaching of Astronomy, the IAU Commission No. 46 on Teaching of Astronomy earnestly recommends to the XVth General Assembly of the IAU that the Australian National Committee on Astronomy be requested to seek funds from the Australian Government to further the teaching of Astronomy in Australian Primary and Secondary Schools, through continuing the development of Curriculum Materials in Astronomy and the extension of in-service training in Astronomy for Australian Teachers.'

Resolution 2: 'The IAU Commission 46 (Teaching of Astronomy) earnestly recommends that the appropriate authorities in the various adhering countries be requested to actively support the efforts of teachers endeavouring to introduce curricular developments in astronomy.'

The Commission then discussed the financial support needed (i) to organize one International School for Young Astronomers per year, and (ii) to develop the project of the Visiting Professors. The following motion requesting financial support from the IAU Executive Committee was passed:

Recommendation: 'In order to support some of its educational projects Commission No. 46 requests from the IAU the following financial aid (in order of priority):

1. For the International Schools for Young Astronomers and/or for Astronomy Courses for Science Teachers in developing countries the sum of $20000.— over three years.
2. For the Visiting Lecturers' Project the sum of $6000.— over three years.'

Both these projects of the Commission are organized for the benefit of astronomers in remote institutions and in developing countries. Discussion of the motion suggested that item 2 should emphasize the educational benefits that might accrue to the country visited by a visiting lecturer, at whatever level of teaching this might be accomplished best.

Teleki presented the following proposals submitted to the IAU by Yugoslavia:

(1) 'The IAU, valueing the advancement of astronomical knowledge of young people, asks the Executive Committee and Commission 46 to investigate the possibility of organizing young astronomers' competitions, in different regions of the world, under the patronage and the organisation of the Union.'

The level of the proposed competitions would include secondary schools. Successful astronomy competitions exist already in Poland and Yugoslavia. President Müller and others stressed the difficulty of regional competitions given the great diversity of astronomy teaching in adjacent countries, and the relative large funding required to organize and run such competitions. It was pointed out that science fairs (which include all sciences) are already organized in many countries. The Commission came to the conclusion that this proposal was not feasible. Instead, the Commission proposed the following recommendation:

'Commission 46 recommends to its national delegates that they (or their national adhering organisations) support local and national competitions by making available information on how successful competitions are run, by helping science fairs find astronomers to judge astronomical entries, etc.'

(2) The second proposal by Yugoslavia refered to 'the establishment of a centre for exchange and loan of temporarily unused astronomical instruments and accessories.'

President Müller pointed out that a Working Group on Exchange of Equipment was formed at the XIVth General Assembly in Brighton (1970) consisting of representatives of Commissions 9 and 46. The chairman of the Working Group is N. N. Mikhelson, G. Abell and D. McNally were the Commission 46 representatives. D. McNally reported that in early 1972 the Working Group sent out a circular letter to ascertain those observatories which had equipment out of regular use that could be loaned to other observatories (preferably those of developing countries). No positive feedback was received to this circular. President Müller stressed the great difficulties involved in providing for packing and customs and other formalities if such equipment actually became available. The Commission came to the conclusion that also the second proposal by Teleki was not feasible. Instead, 'the Commission recommends that institutions which may have surplus equipment, which they are willing to donate or give on loan, prepare such lists and send these to the IAU Secretariat for publication in the IAU Information Bulletin. It should be stressed, however, that the receiving institutions must be in charge of all the custom, packing and shipping formalities.'

Sessions 2 and 3, 25 August 1973

The rules and guidelines proposed for the Commission were approved, subject to minor corrections suggested by Commission members. Copies of these rules and guidelines will be mailed to all Commission members.

President Müller summarized the various forms of cooperation that she has persued with the ICSU Committee on Science Teaching, with the UNESCO Division of Science Teaching, and with COSTED (Committee on Science and Technology in Developing Countries). She briefly reported on the ICSU/UNESCO Conference on "Education of Teachers for Integrated Science" held at the University of Maryland (USA) from 3 to 13 April 1973. The emphasis of this Conference was the education of teachers (both initial and in-service) to teach science in both integrated and coordinated ways. The Proceedings of the Conference will be published by UNESCO in the New Trend series. Among the future plans of the ICSU Committee on Science Teaching is a meeting of the full Committee dealing with the integration, separation or coordination of science teaching at university level. A subsequent discussion on the integration of astronomy into other science courses or curricula showed mainly that the needs and opportunities in the various countries differ tremendously. Those without astronomy courses look toward integration, those with courses do not. Yet in some countries more astronomy can be taught when integrated with physics than as a separate subject; this is the case for example in Bulgarian schools, as reported in a letter by N. S. Nikolov.

The collaboration with UNESCO concerns mainly the organization of both the schools for young astronomers and the elementary astronomy courses for school science teachers in developing countries. In future years, the elementary astronomy courses (such as the Kenya courses of August 1972) might be sponsored by UNESCO's Division of Pre-University Science and Technology. For the international schools for young astronomers financial support might be sought through UNESCO's Division of Scientific Research and Higher Education. President Müller stressed, however, that institutions interested in having Commission 46 organize either a school for young astronomers or elementary astronomy courses for school science teachers should make the first steps themselves by contacting both the Commission President and the UNESCO Field Service in their own country.

In a letter R. J. Chambers, who taught the 1972 elementary astronomy courses for Kenya science teachers, pointed out the lack of texts and materials keyed to various geographical regions, in particular to equatorial regions. He stressed the following remarks:

'The importance of observer-centered phenomena cannot be overemphasized. Aspects of the sky, rotation and revolution of the earth, weather and timekeeping, navigation, solar energy, and uses of artificial satellites for communications, meteorology, and geological surveys are vitally important to people in developing countries. A very large proportion of these people live in equatorial latitudes. If they are to participate in the application of this technology, and of the benefits which astronomy and space science potentially can bring to them, they must have adequate grounding in the fundamental concepts involved. To people in developing nations, the most relevant portion of the science for their lives is what generally comes under the heading of 'practical astronomy', to which might be added 'classical space science'. Yet these topics are the ones that are most susceptible to variation in presentation, depending on the geographical locale of the intended students.' Further on R. J. Chambers stressed that 'simple texts in astronomy are needed, including teachers' editions with lesson plans for all grade levels. Materials for secondary schools should have highest priority, because it is from this level that talented students are identified and selected for higher studies. Teacher's editions will serve to bridge the gap at teachers' colleges until college level materials can be prepared. The needs for primary schools are less important only in terms of scheduling. Most of the people in developing countries still attend only primary school. They won't be taught until their teachers are prepared, however, hence the suggested sequence of secondary, college, and finally primary level publications. Finally, the existence of television facilities in most countries opens the possibility of pre-packaged courses on astronomy which can short-cut the time necessary to train teachers.'

The training of astronomers in developing countries was treated by B. Hidayat (Indonesia) who emphasized that arrangements between small and advanced institutions should pay attention to the capability of the student upon return to his own institution, and that such exchanges need to be augmented by regional cooperation and through extended visits by professors so that they could profitably interact with the local students. S. Ferraz Mello pointed out that 'the main task for training scientists in developing countries is to provide a high-level scientific manpower; the two specific tasks concerning astronomers are (i) to form the appropriate scientists for the already existing and for the planned institutes for astronomical research, and (ii) to train the teaching staff for the universities in which astronomy is being taught at undergraduate and graduate level.' He then summarized how Brazil built up a core of astronomers starting in 1961 (i) by sending young people abroad for graduate training, (ii) by having visiting professors teaching in Brazil, and (iii) by starting graduate courses (since 1967) when the first graduated Brazilians returned from abroad. He emphasized that the young people sent abroad for graduate training in astronomy must have had a good training in physics and mathematics at their home institution. Furthermore it is important that the institutes hosting young people from developing countries integrate them in the research staff and not only have them follow some graduate courses. In addition S. Ferraz Mello stressed the need for Visiting Professors (both for shorter and longer stays) in developing countries.

Concerning the Young Astronomers' Schools President Müller expressed the deep appreciation of the Commission to the Secretary of the schools, J. Kleczek, and to the directors of the schools, Z. Kopal (Manchester), M. Rigutti (Arcetri), R. V. Karandikar (Hyderabad), J. Sahade (Cordoba), B. Hidajat (Lembang),—for their great efforts in organizing and running the schools. Our last School for Young Astronomers was held from July 23 to August 18, 1973, at the Bosscha Observatory in Java (Indonesia). The participants came from Indonesia (5), India (1), Japan (1), and South Africa (1). D. G. Wentzel was the principal lecturer who stayed for the entire duration of the school. The collaboration of five other foreign lecturers could be secured for up to one week each, due to the fact that they were on their way to the IAU General Assembly in Sydney. In a letter J. Kleczek mentioned that 'the fundamental aspect of all our schools was the enthusiastic endeavour of all the lecturers and assistants, and this was the impressive feature which made the schools so successful and useful for many young astronomers from many different countries.' If financial support is available, the schools for Young Astronomers are to be continued. It should be stressed that they consist not only of lecture courses but include practical and observing work and discussions, so as to maximize the benefits to the individual students who tend to come from very diverse backgrounds.

Various regional schools and advanced courses were mentioned, such as the Saas-Fee courses of the Swiss Society of Astronomy and Astrophysics. They are considered to be important and very useful, and they should be given more publicity. It was suggested that they should be announced in the *IAU Information Bulletin.*

D. McNally pointed out the increasing need to ensure that postgraduate students realize the structure of scientific life (organization, financial support, etc.).

The rest of the session was devoted to a discussion on teaching aids in Astronomy. The list of Astronomy Educational Material and its 1973 Addendum proved to be extremely useful. A second Addendum will be prepared for 1976 and should emphasize quality material. A careful selection of the available material must be made by each individual national representative after consultation with his colleagues, students and interested astronomy teachers in his own country. The members of the Organizing Committee in charge of preparing the 1976 Addendum are as follows: T. L. Swihart for part A (English), E. V. Kononovich for part B (Slavic languages), and H. E. Jørgensen for part C (other languages).

C. Zwaan summarized the series of Dutch television lessons in Astronomy. It was also mentioned that Astronomy lessons on television may become available through the British Open University.

Refering to the Contratype Project, President Müller congratulated Miss M. Gerbaldi, E. V. Kononovich, and D. G. Wentzel for the amount of work they have put into this project, and the French astronomers in particular are to be congratulated for the donations which they have made in this direction. It was, however, pointed out that in order to make the photographs directly useful, they must be accompanied by a detailed description of what they represent and what they are used for. In view of the difficulty of having materials returned if any general dissemination were begun, it was decided to reproduce the materials where they are, but only for individuals actively engaged in curriculum development and who would make the materials more generally useful. Later various centers may be established for wider distribution.

Further discussions concerned laboratory exercises, the self-paced Astronomy studies such as reported by L. Houziaux, and the better dissemination of astronomy educational opportunities via regional journals and journals of teachers' associations.

As for the Book Project it was suggested that, similar to the exchange of instructional equipment, announcements of available materials for donation or for loan should be made through the *IAU Information Bulletin.* The possible influence of UNESCO or IAU at customs of various countries was discussed.

Session 4, 28 August 1973

Reported by D. McNally.

A series of educational films were shown illustrating the types available for popular audiences, for higher school and for early university audiences. A single example of a research film was also screened. The choice of films reflected the effort put into the production of educational films in the Western Hemisphere and the ease of access. The suggestions of various Commission members were acknowledged, though not all of them could be adopted. The Commission is most grateful to the commercial firms who made the films freely available. The films shown were the following: Exploring the Milky Way; The Birth and Death of a Star; A New Look at the Sun; The Solar Atmosphere; Fields of Space; Storms on the Sun; The Star System; Ursae Majoris; The Motion of Stars; Algol; Force, Mass and Motion; Hubble's Variable Nebula. Details concerning these films may be obtained by writing to D. McNally.

Special Session with Australian Science Teachers, 20 August 1973

CHAIRMAN: D. McNally.
SECRETARY: M. B. Wilkinson.

The day preceding the opening of the XVth General Assembly of the IAU in Sydney, Commission 46 had a full day session together with Australian science teachers of primary and secondary school level. This meeting was arranged because Commission 46 considered it to be important that the Commission members should make contact with teachers and those concerned with curricular development in schools.

Approximately 70 teachers from N.S.W., 3 from Queensland and 1 from Victoria were present.

The papers read at the meeting were as follows:

D. McNally: Astronomy and the Physical Sciences.
D. G. Wentzel: Astronomy in Higher Education.
A. K. Carey: The Context of Australian Science Education.
R. H. Wilkinson: Astronomy in Australian Education.
J. Alexander: Astronomy in Elementary Education.
T. H. Dickson: Astronomy in Secondary Education in N.S.W.

The meeting permitted a valuable contact between the teachers and professional astronomers. Exchange of ideas on teaching aids and materials took place which was of value to the teachers. Of value to the astronomers was the reminder that they should be aware (i) of their actions in promoting the teaching of astronomy, and (ii) of the conceptional limitations of the children.

An interesting part of the programme was the reading of some reports of projects by children in the age group 8 to 10. These reports clearly demonstrated some of the conceptual problems involved at this level. Again the attention of the Commission was brought to the excellent astronomical teaching aids provided by the Junior School Science Project.

However, the greatest single point that was brought out was the need for in-service courses on astronomy for interested teachers. Most teachers meet astronomy when responding to the interests of the pupils and it was felt that teachers could be more responsive if they were aware of fundamental astronomical knowledge. It is important that Commission 46 itself should consider in the future the problems posed in providing such in-service training courses. While recognizing that such courses can be instituted in some countries, assistance to others through the project for teacher training and the visiting professor project should be actively pursued.

Finally, a resolution was framed, discussed and approved concerning the development of curriculum materials in Astronomy and the extension of in-service training in Astronomy for Australian science teachers. This resolution was discussed and passed in the first session of the Commission on August 22.

A more detailed account of the Proceedings of the meeting with Australian science teachers will be distributed to all Commission members and it is hoped that it will be published in the Journal of the Australian Science Teachers Association.

COMMISSION 47: COSMOLOGY (COSMOLOGIE)

Report of Meeting, 27 August 1973

ACTING PRESIDENT: M. J. Rees.
SECRETARY: M. S. Longair.

In the absence of Ya. B. Zel'dovich, M. J. Rees was nominated acting President for the duration of the XVth General Assembly. Only a single meeting of the Commission was held in view of the Symposia in Canberra and Cracow. At the meeting, a new President, Vice-President and Organizing Committee were elected. Following the business meeting the following papers were read.

G. de Vaucouleurs: Current cosmological research at the University of Texas.
R. D. Ekers: The 1972 Westerbork measurements of the gravitational deflection of radio waves by the Sun.
R. G. Conway and D. Stannard: The redshift-magnitude relation for quasars.
W. Davidson: The early stages of the big bang.
E. B. Jenkins: The galactic abundance of deuterium from observations with the Copernicus satellite.

COMMISSION 48: HIGH ENERGY ASTROPHYSICS (ASTROPHYSIQUE DE GRANDE ÉNERGIE)

Report of Meetings

ACTING PRESIDENT: M. J. Rees.

A business meeting of Commission 48 was held on Saturday, 25 August. After considering the proposed list of officers and new members of the Commission, there was a short discussion of the Executive Committee's request that the Commission should formulate a list of 'scientific priorities' within its field of interest. It was felt that Commission 48 covered such a broad range of topics that this task was neither feasible nor desirable. Consideration was also given to the optimal form of the Commission's contribution to the next edition of 'Reports on Astronomy'. It was agreed that the Commission should produce only a very short and general Report, supplemented by a selected list of review articles, conference proceedings, etc., wherein up-to-date reviews of topics in high energy astrophysics could be found. After the conclusion of the business, the following scientific papers were presented:

L. E. Peterson: Gamma-Ray Bursts of Cosmic Origin.

J. V. Jelley (read by M. J. Rees): Searches for Microwave Pulses from Supernovae.

E. E. Salpeter: Comets around Neutron Stars, and Their Possible Observational Consequences.

A further scientific session of the Commission was held on 29 August, at which the following papers were read:

M. P. Savedoff: New Methods for Detecting Rapid Optical Variables.

F. D. Drake: Pulsar Observations with High Time and Spectral Resolution.

L. E. Peterson: Gamma-Ray Background Measurements from Apollo 15 and 16.

R. C. Catura: Soft X-Ray Observations of M 87.

D. G. Wentzel: Instabilities in the Propagation of Low Frequency Waves.

F. Meyer: Gravitational Wave Experiments.

Commission 48 also participated, with other Commissions, in a session on close binary systems and X-ray sources.

PART 4

ASTRONOMER'S HANDBOOK

ASTRONOMER'S HANDBOOK

The Astronomer's Handbook (*IAU Transactions* **XIIC**, edited by J.-C. Pecker, *IAU Transactions* **XIIIB**, edited by L. Perek, and *IAU Transactions* **XIVB**, edited by C. de Jager and A. Jappel) contains documents of permanent importance. Only the parts changed after the XVth General Assembly are supplemented here.

I. SHORT HISTORY OF THE INTERNATIONAL ASTRONOMICAL UNION

Table 2 of *IAU Trans.* **XIIC** should be supplemented as follows:

XVth General Assembly, 1973, Sydney, N.S.W., Australia	
Members of the Union, after the General Assembly	3188
Adhering Countries	47
Participants in the General Assembly (Members 698)	840
Number of Commissions	40
Volumes of Transactions A	pp. VIII+760
B	pp. IX+334
Highlights 1973	pp. X+574
President: L. Goldberg	1973–1976
General Secretary: G. Contopoulos	1973–1976
Assistant General Secretary: Edith A. Müller	1973–1976
Vice-Presidents: J. G. Bolton	1973–1979
Ch. Fehrenbach	1973–1979
W. Iwanowska	1973–1979

II. ADMINISTRATION AND FINANCES OF THE UNION

A. ADMINISTRATION

The address of the Secretariat is:

International Astronomical Union
University
Thessaloniki
Greece

The Secretariat is headed by the General Secretary, G. Contopoulos. The staff of the office consists of the Executive Secretary, A. Jappel, the Administrative Assistant, Mrs J. Daňková, the Secretary to the office, Miss D. Mori and a typist.

All problems related to Symposia and Colloquia are dealt with by the Assistant General Secretary

Prof. Edith A. Müller
Observatoire de Genève
CH-1290 Sauverny
Suisse

Professor Müller also maintains the contact with the publisher, the D. Reidel Publishing Company, Dordrecht, Holland.

B. FINANCES

(a) The XVth General Assembly resolved to increase the unit of contribution to 1125 gold francs, that is $438.92 or 1329.93 Swiss francs (as on 28 June 1973), for the years 1974, 1975 and 1976.

III. THE INTERNATIONAL ASTRONOMICAL UNION AND THE OTHER INTERNATIONAL SCIENTIFIC ORGANIZATIONS

A. THE INTERNATIONAL COUNCIL OF SCIENTIFIC UNIONS

The terms of office of other international scientific bodies differ from those of the IAU. Information in this respect should be looked up in the Yearbook of the ICSU for the current year. Only IAU representatives are mentioned in the following paragraphs.

The IAU is represented by the General Secretary, G. Contopoulos, on the General Committee of the ICSU.

B. IAU REPRESENTATIVES ON SCIENTIFIC AND SPECIAL COMMITTEES AND COMMISSIONS OF ICSU

1. *Committee on Space Research (COSPAR):* C. de Jager
2. *ICSU Abstracting Board (IAB):* J. B. Sykes
3. *Federation of Astronomical and Geophysical Services (FAGS):* G. A. Wilkins and G. Contopoulos

   IAU representatives in FAGS services:
   Bureau International de l'Heure (BIH): H. Enslin and H. M. Smith
   International Polar Motion Service (IPMS): E. P. Fedorov and B. Guinot
   International Ursigrams and World Day Service (IUWDS): F. W. Jäger
   Scientific Ballooning and Radiations Monitoring Organizations (SBARMO): P. Simon and L. D. de Feiter

C. IAU REPRESENTATIVES ON INTER-UNION COMMITTEES AND IN COMMISSIONS OF ICSU

1. *Inter-Union Committee on Frequency Allocations for Radion Astronomy and Space Sciences (IUCAF):* F. G. Smith and R. Wielebinski
2. *Special Committee on Solar Terrestrial Physics (SCOSTEP):* Z. Švestka
3. *Inter-Union Commission for Science Teaching (IUCST):* D. MacNally
4. *Committee for Data on Science and Technology (CODATA):* G. A. Wilkins
5. *Inter-Union Commission on Spectroscopy (IUCS):* B. Edlén, J. G. Philips and M. J. Seaton
6. *Inter-Union Commission on the Moon (IUCM):* A. Dolfus and E. Anders
7. *Committee on Science and Technology in Developing Countries (COSTED):* D. MacNally

D. IAU REPRESENTATION IN OTHER ORGANIZATIONS

1. *La Fondation Internationale du Pic-du-Midi:* A. Lallemand
2. *Le Comité Consultative pour la Définition de la Seconde (CCDS) du Bureau International des Poids et Mesures:* Wm. Markowitz
3. *Le Comité Consultative pour la Définition du Mètre (CCDM) du Bureau International des Poids et Mesures:* A. H. Cook

4. *International Radio Consultative Committee (CCIR):* F. G. Smith and H. M. Smith
5. *European Physical Society:* Edith A. Müller

IV. SERVICES AND FUNCTIONS OF THE IAU

The text under D., pp. 32 through 38 of *IAU Trans.* XIIC should be replaced by the following text:

D. THE CENTRAL BUREAU FOR ASTRONOMICAL TELEGRAMS

Smithsonian Astrophysical Observatory, Cambridge, Massachusetts 02138, U.S.A.

Director of the Bureau: Dr B. G. Marsden
Associate Directors: Dr O. J. Gingerich, Dr Z. Sekanina, Dr R. B. Southworth

For more than a century, astronomers have maintained a telegraph and cable network for rapid communication of such celestial news as the discovery of a comet or a nova. When the IAU was established, the Central Bureau for Astronomical Telegrams was placed under its jurisdiction (through Commission 6), initially at the Brussels Observatory, and then for many years at the Copenhagen Observatory. Since 1965 the Bureau has been at the Smithsonian Astrophysical Observatory.

Telegrams or cables can be received any time of day or night at the Smithsonian Observatory. The cable address is SATELLITES, NEW YORK (and it is recommended that messages be specified for transmission VIA WUI), the telex number is 92-1428, and the Western Union address for astronomers in North America is RAPID SATELLITE, CAMBMASS. As far as is practicable, messages should be sent according to the code given in the Appendix. Persons telegraphing or cabling information to the Central Bureau are advised also to send the same (and appropriate additional) information by airmail letter at the same time. In many instances, it is quite sufficient to communicate information to the Central Bureau *only* by airmail letter.

Information is telegraphed or cabled out by the Central Bureau according to the following arrangements. In North America, telegrams are sent collect via Western Union to any astronomers agreeing to pay for the messages. Europe (except for the Netherlands and the U.S.S.R.), North Africa and the Middle East are served via the Meudon Regional Warning Center of the IUWDS; observatories there (except in Czechoslovakia, Sweden, Norway and Finland, whither the 'astrograms' are relayed via additional Regional Warning Centers) are expected to place a deposit with the Central Bureau. The Netherlands, the U.S.S.R., and also Australia and New Zealand receive the astrograms from SOLTERWARN and thence Regional Warning Centers. The rest of the world (Japan, South America, South Africa) is served by certain observatories that relay the astrograms directly from the Central Bureau.

It is not necessary for observatories to receive all the astrograms. Subscribers should specify which of the standard categories they desire:

A. (ALPHA) Comet* discoveries, brighter than magnitude 12
B. (BETA) Comet* discoveries, fainter than magnitude 12
C. (CHARLY) Comet* observations and brief finding ephemerides (primarily for observers)
D. (DELTA) Precise positions, orbits and ephemerides (primarily for computers)
E. (ECHO) Novae and supernovae, brighter than magnitude 12
F. (FRANCE) Novae and supernovae, fainter than magnitude 12
G. (GAMMA) All others, including planetary phenomena

In addition to the astrograms, the Central Bureau also publishes postcard *Circulars*. Issued at irregular intervals some 70 to 100 times per year, the IAU *Circulars* contain confirmation of the

*and unusual minor planets.

telegrams, as well as information sent to the Central Bureau by mail only: this information includes observed positions and predicted ephemerides for comets and unusual minor planets; further observations of novae, supernovae and unusual variable stars; predictions of occultations by planets; and other astronomical information that requires rather rapid dissemination. Beyond a certain distance from Cambridge the *Circulars* are sent by airmail. The 700 or so subscribers are billed for the *Circulars* in advance, and the accounts are diminished by 20¢ for each *Cicrular* issued (15¢ for those subscribers in North America who do not require invoices and will keep track of their own accounts).

APPENDIX I. ASTRONOMICAL TELEGRAM CODE

First groups for all telegrams

DISCOVERER OBJECT OBSERVER AAAAB

1. The name of the discoverer(s) and/or other designation, such as year and letter for a comet, constellation for a nova, galaxy for a supernova (N=NGC, I=IC, M=Messier), etc.
2. The type of object (COMET, OBJECT, NOVA, SUPERNOVA, variable star = VSTAR, etc.).
3. The name of the observer(s) and/or computer(s).
4. AAAA = equinox (assumed to be the mean equinox of the beginning of the year specified).
   B = 1 for an approximate position;
   = 2 for an accurate position;
   = 3 for orbital elements of an object moving around the Sun;
   = 4 for an ephemeris.

Note: when an ephemeris follows the elements, group 4 is replaced by the word EPHEMERIS, and the epoch of the ephemeris is assumed to be the same as that of the elements; see example 3.

Middle groups for a position

Approximate: CDDEE FFFGH IIJJJ LMMNN PQRRS TUUUU VWWXX
Accurate: CDDEE FFFGH IIJJK KKKLM MNNPP PQRRS TUUUU VWWXX

5. Date of observation:
   C = final digit of year;
   DD = month (01 = January, ..., 12 = December);
   EE = day (U.T.).
6. FFFGH = time of observation in decimals of a day (U.T.).
   This group may be omitted entirely in the case of observations of stationary objects.
7. *Approximate position:*
   Right ascension $II^{h}JJ^{m}_{.}J$
   Declination LMM°NN′
   P = 0;

   Accurate position
   Right ascension $II^{h}JJ^{m}KK^{s}_{.}KK$
   Declination LMM°NN′PP″.P
   L = sign: 2 = positive, 1 = negative.
   Q = 1 for 'total' magnitude (m_1) for a comet;
   = 2 for 'nuclear' magnitude (m_2) for a comet;
   = 3 for visual (m_v) } magnitude for objects of stellar appearance
   = 4 for photographic (m_{pg}) } (including minor planets);
   = 5 for photovisual (m_{pv}) }
   RR = magnitude; if the magnitude is negative, add 100;
   S = appearance, according to the table below; or, *if the object is not a comet*, S = tenth of a magnitude.

	Nothing reported about tail	Tail < 1°	Tail > 1°
Stellar appearance	0		
Nothing reported about appearance of object	1	2	3
Object diffuse, without central condensation	4	5	6
Object diffuse, with central condensation	7	8	9

8. For *comets and minor planets*, the daily motion:
   In right ascension T UU$^{m}_{\cdot}$UU
   In declination V WW°XX′
   For *supernovae*, the offset from the nucleus of the parent galaxy:
   In right ascension T UUUU″
   In declination V WWXX″
   T, V = sign: 2 = positive (east or north), 1 = negative (west or south).
   Group 8 may be omitted if unknown or irrelevant.

Middle groups for orbital elements

CDDEE FFFGH IIIII JJJJJ KKKKK TTTTT UUUUU

5. Date of perihelion passage:
   C = final digit of year;
   DD = month (01 = January, ..., 12 = December);
   EE = day.
6. FFF = time of perihelion pasage in decimals of a day (E.T.).
   G = time interval in days (rounded to the nearest integer) between the first and last positions used for the computation; 0 = 10 days or more.
   H = the number and quality of the observations on which the computation is based, or by which the elements have been checked, according to the following table:

	Maximum residuals > 5″	1″–5″	< 1″
Fewer than three accurate positions*	1	2	3
Three accurate positions	4	2	3
More than three accurate positions	7	5	6

*Or other factors decreasing the reliability of the orbit, such as approximate, doubtful or unsatisfactorily distributed positions

7. III°II = ω;
   JJJ°JJ = Ω;
   KKK°KK = i.
8. T.TTTT = q (in A.U.);
   U.UUUU = e.
   The group for e may be omitted if the orbit is parabolic ($e = 1$).

Middle groups for ephemerides

CDDEE (IIJJJ LMMNN 9TTTT 8UUUU) cddee

5. Date of the first position in the ephemeris:

C = final digit of year;
DD = month (01 = January, ..., 12 = December);
EE = day.

6. It is understood that ephemerides are given for 0^h E.T.
7. Right ascension $II^hJJ^m.J$
   Declination L MM°NN′
   L = sign: 2 = positive, 1 = negative.
8. T.TTT = geocentric distance in A.U., preceded by a 9;
   U.UUU = heliocentric distance in A.U., preceded by an 8.
   Groups 7 and 8 are repeated as often as is necessary to complete the ephemeris. Positions should be given at uniform intervals (10, 5, 2 or 1 days). Group 8 need not be given for every line in the ephemeris; see examples 3 and 4.
   cddee is the date of the last position in the ephemeris, in the same form as before.

Final groups for all telegrams

YYYYY ZZZZZ REMARKS COMMUNICATOR

9. YYYYY = the last five figures of the sum of all the groups of digits including and following group 4 (containing the equinox).
   ZZZZZ = the last five figures of the sum of the groups giving the right ascension, declination and magnitude only (or ω, Ω and i in the case of orbital elements). In practice, this 'second checksum' includes only the items under group 7 (or involving the letters I through S).

Note: any digit that is unknown, not significant, or otherwise to be withheld should be replaced by a slash (/); this is to count as zero in the summations. A group of five slashes (/////) should be avoided. If two or more observations are being telegraphed in the same message, checksums should follow each one. A checksum of all numerical information is useful even if the telegram is not sent in code.

10. Any additional remarks in qualification of the observation or the computation. In the case of an accurate position the location of the observing station could be specified.

11. The name of the communicator(s).

Examples

1. CLARK COMET CLARK 19501 30610 66/// 20540 13130 01135 2015/ 10002 81068 34805 GILMORE

Gilmore reports that Clark has discovered and observed a comet as follows:

1973 U.T.	α_{1950}	δ_{1950}	m_1
June 10.66	$20^h54^m.0$	$-31°30'$	13

Object diffuse without central condensation, tail $< 1°$. Daily motion: $\Delta\alpha = +1^m.5$, $\Delta\delta = -2'$.

2. BALLY CLAYTON 1968D COMET ROEMER SCHREUR 19502 80827 20246 18513 33623 22222 82157 77090 56515 19502 80827 20872 18513 16823 22225 7//// 48762 25761 CATALINA LPL

The Lunar and Planetary Laboratory communicates the following observations of comet Bally-Clayton (1968d), obtained by Roemer and Schreur at the Catalina station:

1968 U.T.	α_{1950}	δ_{1950}	m_2
Aug. 27.20246	$16^h51^m33^s.36$	$+32°22'22''.8$	15
27.20872	18 51 31.68	+32 22 25.7	

On the first plate the object was diffuse, with central condensation, nothing reported about a tail.

3. 1972F COMET CANDY 19503 20327 72656 25771 15959 12369 09275 75860 54099 EPHEMERIS 20403 00158 14433 91171 80934 00558 14741 01503 15007 90961 80972 03000 15042 20418 49301 64442 CANDY

Candy reports that he has calculated parabolic elements and an ephemeris for comet 1972f, as shown below. The elements are based on 3 accurate observations covering an arc of 5 days, and the residuals are all less than 1″.

$$T = 1972 \text{ Mar. } 27.726 \text{ E.T.} \quad \left.\begin{array}{l}\omega = 257^{\circ}\!.71 \\ \Omega = 159.59 \\ i = 123.69\end{array}\right\} 1950.0 \qquad q = 0.9275 \text{ A.U.}$$

1972 E.T.	α_{1950}	δ_{1950}	Δ	r
Apr. 3.0	$0^h15^m\!.8$	$-44°33'$	1.171	0.934
8.0	0 55.8	-47 41		
18.0	1 50.3	-50 07	0.961	0.972
18.0	3 00.0	-50 42		

4. KOHOUTEK OBJECT AKSNES 19504 11125 00412 11411 90325 81185 00362 11543 00316 11709 00272 11832 00231 11950 00192 12103 90344 81114 00157 12213 11207 69507 84703 APOLLO TYPE ASTEROID MAGNITUDE SEVENTEEN SEKANINA

Sekanina communicates the following ephemeris by Aksnes for the object discovered by Kohoutek. The object is an Apollo-type asteroid.

1971 E.T.	α_{1950}	δ_{1950}	Δ	r	Mag.
Nov. 25.0	$0^h41^m\!.2$	$-14°11'$	0.325	1.185	17
27.0	0 36.2	-15 43			
19.0	0 31.6	-17 09			
Dec. 1.0	0 27.2	-18 32			
3.0	0 23.1	-19 50			
5.0	0 19.2	-21 03	0.344	1.114	
7.0	0 15.7	-22 13			

5. HONDA SERPENS NOVA HONDA 19001 00215 8//// 18257 20238 03053 40764 41548 BRIGHTNESS INCREASING HIROSE

Hirose reports that Honda has discovered and observed a nova in Serpens, as shown below. The brightness is increasing.

1970 U.T.	α_{1900}	δ_{1900}	m_v
Feb. 15.8	$18^h25^m\!.7$	$+2°38'$	5.3

6. N3811 SUPERNOVA ROSINO 19501 09209 11386 24758 0412/ 20005 20003 89982 40264 ASIAGO

The Asiago Astrophysical Observatory communicates the observation by Rosino of a supernova in NGC 3811 ($\alpha = 11^h38^m\!.6$, $\delta = +47°58'$, equinox 1950.0), 5″ east and 3″ north of the nucleus. On 1969 Feb. 9 U.T. the photographic magnitude was 12.

V. PUBLICATIONS OF THE INTERNATIONAL ASTRONOMICAL UNION

Reference is made to the Report of the Executive Committee, pp. 27–29 and to *Information Bulletin* Nos. 28, 29 and 30.

VI. SYMPOSIA OF THE INTERNATIONAL ASTRONOMICAL UNION

Reference is made to the Report of the Executive Committee, pp. 6–7 and to *Information Bulletin* Nos. 28, 29 and 30.

EXECUTIVE COMMITTEE
1973–76

PRESIDENT

Professor L. Goldberg, Director, Kitt Peak National Observatory, P.O. Box 26732, Tucson, Arizona 85726, U.S.A.

VICE-PRESIDENTS

Professor B. J. Bok, Steward Observatory, University of Arizona, Tucson, Arizona 85721, U.S.A.
Mr J. G. Bolton, CSIRO, Division of Radiophysics, P.O. Box 76, Epping, N.S.W., Australia
Professor Ch. Fehrenbach, Directeur de l'Observatoire de Haute Provence, Saint Michel l'Observatoire, 04300 Forcalquier, France
Professor W. Iwanowska, Astronomical Observatory, N. Copernicus University, Sienkiewicza 30, 87–100 Toruń, Poland
Sir Bernard Lovell (Prof. A. C. B.), Nuffield Radio Astronomy Laboratories, Jodrell Bank, Macclesfield, Cheshire SK11 9DN, U.K.
Professor E. R. Mustel, Astronomical Council, USSR Academy of Sciences, Vavilov Str. 34, Moscow V-312, U.S.S.R.

GENERAL SECRETARY

Professor G. Contopoulos, Astronomical Department, University of Thessaloniki, Thessaloniki, Greece

ASSISTANT GENERAL SECRETARY

Professor Edith A. Müller, Observatoire de Genève, CH-1290 Sauverny, Switzerland

BY-LAWS

The introductory part in the English version of Article 24 reads as follows:
'24. Each Adhering Country pays annually to the Union a number of units of contribution according to its category as follows:'

WORKING RULES

The Executive Committee complemented the Working Rules as follows:
A second sentence has been added to article 1 reading as follows:
'The General Secretary shall publish a Style Book with rules covering the preparation of publications of the Union'
The following two articles have been added to the Working Rules:

V. SCIENTIFIC MEETINGS

32. The General Secretary shall publish rules for scientific meetings organized or sponsored by the Union.

VI. EXTERNAL CONTACTS

33. No dealings with third parties, attributable to the Union, shall be undertaken by any Member of the Union except on the authority of the General Secretary.

MEMBERSHIP OF COMMISSIONS AND ALPHABETICAL LIST OF MEMBERS

Reference is made to Appendix II.

LIST OF COUNTRIES ADHERING TO THE UNION

The following is a list of the 47 countries that adhered to the Union in September 1973, giving also the year of admission, the approximate number of Members, and the Adhering Organizations.

Country	*Year*	*Members*	*Adhering Organizations*
Arab Republic of Egypt	1925	11	National Committee of Astronomy, Helwan near Cairo
Argentina	1927	27	Comité Nacional de Astronomía, La Plata
Australia	1939	65	Australian National Committee for Astronomy, Sydney
Austria	1955	19	Österreichische Akademie der Wissenschaften, Wien
Belgium	1920	61	Académie Royale de Belgique, Bruxelles, Koninklijke Academie van België, Brussel
Brazil	1961	14	Comissão Brasileira de Astronomia (CBA), Conselho Nacional de Pesquisas, Rio de Janeiro
Bulgaria	1957	10	Bulgarian Academy of Sciences, Sofia
Canada	1920	120	National Committee of Astronomy for Canada, Vancouver
Chile	1947	7	Universidad de Chile, Santiago de Chile
Colombia	1967	1	Academia Colombiana de Ciencias Exactas, Físicas y Naturales, Bogotá
Cuba	1970	0	Instituto de Astronomía de la Academia de Ciencias de Cuba, La Habana
Czechoslovakia	1922	40	National Committee of Astronomy of Czechoslovakia, Czechoslovak Academy of Sciences, Praha
Denmark	1922	20	Danish National Committee for Astronomy, det Kongelike Danske Videskabernes Selskab, København
Finland	1948	13	Finnish Academy of Sciences and Letters, Helsinki
France	1920	226	Comité National Français d'Astronomie, Académie des Sciences, Paris
F.R.G.	1951	154	Rat Westdeutscher Sternwarten zu München
G.D.R.	1951	35	Nationalkomitee für Astronomie der DDR, Akademie der Wissenschaften der DDR, Berlin
Greece	1920	25	Greek National Committee for Astronomy, Academy of Athens, Athens
Hungary	1947	14	Hungarian Academy of Sciences, Budapest
India	1964	57	Indian National Committee of Astronomy, Indian National Science Academy, New Delhi
Iran	1969	1	Tehran University, Tehran
Ireland	1947	9	National Committee of Astronomy, Royal Irish Academy, Dublin
Israel	1954	21	The Israel Academy of Sciences and Humanities, Jerusalem
Italy	1921	111	Consiglio Nazionale delle Ricerche, Roma
Japan	1920	141	National Committee of Astronomy of Japan, Science Council of Japan, Tokyo
Korea (DPR)	1961	3	Academy of Sciences, Phyongyang
Korea (Republic)	1973	4	Korean Astronomical Society, Seoul
Mexico	1921	21	Universidad Nacional de Mexico, Mexico D.F.

Country	*Year*	*Members*	*Adhering Organizations*
Netherlands	1922	87	Koninklijke Nederlandse Academie van Wetenschappen, Amsterdam
New Zealand	1964	9	Royal Society of New Zealand, Wellington
Norway	1922	15	Det Norske Videnskaps-Akademii, Oslo
Poland	1922	50	Polska Akademia Nauk, Warszawa
Portugal	1924	15	Secção Portuguesa das Uniões Internacionais Astronómica, Geodésica e Geofísica, Lisboa
Roumania	1928	18	Romanian National Committee of Astronomy, Bucarest
South Africa	1938	13	Council for Scientific and Industrial Research, Pretoria
Spain	1922	18	Comisión Nacional de Astronomía, Madrid
Sweden	1925	51	Kungl. Vetenskapsakademien, Stockholm
Switzerland	1923	26	Schweizerische Naturforschende Gesellschaft, Genf
Taiwan	1959	10	Academia Sinica, Taipei
Turkey	1961	22	Türk Astronomi Dernegi, Istanbul
United Kingdom	1920	277	The Royal Society, London
Uruguay	1970	0	Comité Nacional de Astronomía, Montevideo
U.S.A.	1920	920	The National Academy of Sciences, Washington
U.S.S.R.	1935	394	Academy of Sciences of the U.S.S.R., Moscow
Vatican City State	1932	7	Pontificia Accademia delle Scienze, Città del Vaticano
Venezuela	1953	3	Academia de Ciencias Física, Matemáticas y Naturales, Caracas
Yugoslavia	1935	20	Savez društava matematičara, fizičara i astronoma Jugoslavije, Zagreb

PART 5

REPORT OF THE EXTRAORDINARY GENERAL ASSEMBLY, POLAND 1973

OPENING CEREMONY

4 September 1973

The Extraordinary General Assembly of the International Astronomical Union was held in commemoration of the quincentenary of the birth of Nicolaus Copernicus. The Executive Committee of the IAU was represented by its President, Professor L. Goldberg, its General Secretary, Professor G. Contopoulos and one Vice-President, Professor W. Iwanowska. The Opening Ceremony took place in the Congress Hall of the Palace of Culture and Science in Warsaw, on Tuesday, 4 September 1973, at 10 a.m.

The Program included a Word of Welcome on the part of the Chairman of the Local Organizing Committee, Professor J. Smak, an Address by the Chairman of the Council of Ministers of the Polish People's Republic, Mr P. Jaroszewicz, an Address by the President of the Polish Academy of Sciences, Professor W. Trzebiatowski, an Address by the Chairman of the Warsaw City Council, Mr J. Majewski, an Address by the Chairman of the National Committee for the IAU, Professor W. Iwanowska, and the Address and official opening of the Extraordinary General Assembly by the President of the IAU, Professor L. Goldberg.

After a musical interlude a special lecture was given by Professor O. Gingerich on the 'Astronomy and Cosmology of Copernicus'. This is included in the *Highlights of Astronomy 1973.*

Between 5 and 12 September five IAU Symposia and one joint IHUPS-IAU Symposium took place in Warsaw, Cracow and Toruń, as follows:

Symposium No. 62 = Copernicus Symposium I 'The Stability of the Solar System and of Small Stellar Systems', Warsaw, 5–8 September 1973.

Symposium No. 63 = Copernicus Symposium II 'Confrontation of Cosmological Theories with Observational Data', Cracow, 10–12 September 1973.

Symposium No. 64 = Copernicus Symposium III 'Gravitational Radiation and Gravitational Collapse', Warsaw, 5–8 September 1973.

Symposium No. 65 = Copernicus Symposium IV 'Exploration of the Planetary System', Toruń, 5–8 September 1973.

Symposium No. 66 = Copernicus Symposium V 'Late Stages of Stellar Evolution', Warsaw, 10–12 September 1973.

IUHPS/IAU Symposium = Copernicus Symposium VI 'The Astronomy of Copernicus and its Background', Toruń, 7–8 September 1973.

The Proceedings of the five IAU *Symposia* will be published as separate Volumes by the IAU, while the Proceedings of the IUHPS-IAU Symposium will be published in 'Studia Copernicana' by the Polish Academy of Sciences.

Three Invited Discourses were given on the topics:

G. B. Field: 'The Physics of the Interstellar Matter' (Warsaw).

V. A. Ambartsumian: 'Galactic Nuclei' (Cracow).

F. J. Low: 'Infrared Astronomy' (Toruń).

These will be also included in the *Highlights of Astronomy 1973.*

The program of the various Symposia was enlivened with several excursions and cultural events. We are all most grateful to our Polish hosts for organizing this very successful Extraordinary General Assembly.

G. Contopoulos
General Secretary

WORD OF WELCOME BY THE CHAIRMAN OF THE LOCAL ORGANIZING COMMITTEE. PROFESSOR J. SMAK

'Obywatelu Prezesie Rady Ministrów, Monsieur le Président de l'Union Astronomique Internationale, Distinguished Guests, Dear Members of the International Astronomical Union, Ladies and Gentlemen,

W imieniu Lokalnego Komitetu Organizacyjnego Nadzwyczajnego Kongresu Międzynarodowej Unii Astronomicznej mam zaszczyt powitać wszystkich naszych dostojnych i drogich gości przybyłych na ten Kongres, a w szczególności na dzisiejsze Uroczyste Otwarcie i Sesję Plenarną, związaną z obchodami 500-letniej Rocznicy Urodzin Mikołaja Kopernika.

On behalf of the Local Organizing Committee for the Extraordinary General Assembly of the International Astronomical Union I have the privilege to welcome all our distinguished and dear guests which came to participate in the Assembly, and particularly in this General Commemorative Session connected with the celebrations of the Quincentenary of Nicolaus Copernicus.

I have the honour now to request the Chairman of the Council of Ministers of the Polish People's Republic, Mr Piotr Jaroszewicz, to deliver his address.

Obywatelu Premierze, proszę o zabranie głosu.'

ADDRESS BY THE CHAIRMAN OF THE COUNCIL OF MINISTERS OF THE POLISH PEOPLE'S REPUBLIC, MR P. JAROSZEWICZ

'Mr President of the International Astronomical Union, Distinguished Members of the Extraordinary Session of the Union's General Assembly,

Among the many international meetings of scientists, the importance of which in the world of today can hardly be overestimated, the session we have just inaugurated undoubtedly assumes special significance. This is so, on account of the high rank astronomy enjoys as a branch of science, the high esteem for your Union in the international scientific community, and by no means incidental time and place of this meeting which crowns the international observances of Mikolaj Kopernik's 500th birthday anniversary on his native soil.

Astronomy is one of the most prominent and one of the oldest branches of science, yet it is abundant in new great discoveries of surprisingly large scale, constantly affecting man's cognitive abilities, world outlook and civilization. Just as in remote times astronomic knowledge had laid the ground for the development of navigation, due to which new seas and lands were discovered, our mother planet Earth became better known and its place in the universe was defined, today astronomy paves man's way in the cosmic space, fulfills his ages-long aspirations to fly in the air, steals away from Nature its greatest secrets.

Time calendar, in common use by all mankind today, owes its existence to astronomy. We use the calendar not only to measure time. We also use it to mark down the happenings which make up the history of humanity, which shape the awareness of successive generations and are the traces of their common heritage: development and progress.

On one page in this calendar the birthday of a genius was also marked down, a genius whose discovery opened a new chapter in the history of astronomy, a mile-stone in the history of human cognition, thus laying the foundations for a major breakthrough in human awareness and intellectual culture.

More than three hundred years had to elapse for that basic truth to become fully recognized, to become an element of education and popular awareness. The lapse of time, however, did not diminish the enormity, creative power and validity of Copernican traditions in present-day world; quite the opposite, it enhanced them. These traditions are alive today due to the scientific discoveries and the continually valid patterns of Kopernik as a man, as an explorer looking for truth everywhere, as a Pole and citizen who combined his service of the country with service of humanity.

The strength and significance of Copernican traditions are particularly notable in Poland.

The memory and loyalty to the legacy of Polish science and culture played an essential role in the history of our nation. They allowed us to preserve our national identity and the sense of community in the years of bondage, filled us with strength and consolation in the long struggle for national and social liberation, and following World War II, when there was a shortage of bread and roof over the head, they made us undertake the effort of reconstruction of science and cultural monuments parallel to the reconstruction of production plants in our towns and villages.

The worldwide observances of the 500th birthday anniversary of Mikolaj Kopernik, climaxed by the current debates of the extraordinary session of the general assembly of the International Astronomical Union, are a convincing confirmation of the significance and universal character of Copernican traditions.

For us Poles, who play host to this world meeting, the fact that such a distinguished group of scientists from the world over has come to our country is a source of great satisfaction. In greeting the members of this extraordinary session I wish to express, on behalf of the Government of the Polish People's Republic, our gratitude for the heart you have put in making this Copernican anniversary an even greater event and for promoting Copernican traditions in your homelands.

I also wish to convey sincere greetings extended to you by the First Secretary of the Central Committee of the Polish United Workers' Party, Edward Gierek, and by the President of the Council of State, Prof. Henryk Jablonski.

We wish the observances of the Copernican Year and your debates produced fertile and lasting effects, we wish they could help carry out the noble-minded mission of science in the contemporary world as a creative force in the service of mankind, the mission to which Kopernik was faithful and which today, in an era of international detente, should become a moral imperative determining the outlooks and pursuits of scientists in the whole world.

These outlooks and sense of responsibility are the major values on which largely depend the development of international collaboration, the initiation and realization of joint research programmes which meet half-way the needs and hopes of contemporary man, the success of science as a tool of progress in economy, social life and culture.

The efforts of our nation, our initiative in the international arena, our socio-economic, scientific and cultural policies are all aimed in that direction.

Ever since People's Poland came to exist, we have regarded science as one of the key instruments to be used in construction of our socialist economy and culture.

In the lifespan of one generation we have created modern science, we have developed quite a network of academic schools and research centres, we have brought up nearly 700 thousand specialists with higher education and a sizable scientific cadre. Polish science has successfully performed its function at the stage of reconstruction; today it helps create and partakes in the realization of an extensive programme of continual transformation of Poland and of us all. At this stage of development, which will multiply our country's economic and technical potentials, which will enrich its cultural life and increase the wealth of its people, we attach great importance to the cultivation of the best traditions of scientific progress and of the social service of science, to their correlation with the needs and tasks of today and tomorrow.

The fact that Kopernik's genius developed on the ground of Polish culture, in the reputed atmosphere of tolerance and freedom of creative pursuits, has charged us with an obligation to create favourable conditions for continued springy development of Polish Science, conditions in which it plays an increasing role in shaping socialist Poland, in multiplying its contribution to the sum-total of cognitive values and civilization of mankind.

That was the aim of the II Congress of Polish Science, held in Warsaw in June of this year, which mapped out the directions of its development today and in the future.

In a few years to come, it is our intention to considerably increase financial resources for astronomic research in this country. Soon, the Astronomical Copernican Centre will be open in Warsaw, and in Toruń, Kopernik's birthtown, a modern Centre of Radioastronomy is being developed.

We shall undertake more active and broader participation in international astronomical programmes, among others in the area of cosmic physics. We hope that the next decades will see

Poland's growing share in the immense task of exploring the Cosmos, in which our Great Compatriot was a pioneer.

We are glad this session of world astronomers will give them an opportunity to acquaint themselves with the achievements of our science, with the gains of post-war Poland, with the presence and inspiring strength of Copernican traditions in our country.

I wish all members of the extraordinary session of the General Assembly of the International Astronomical Union that your debates—thanks to the presentation of the latest research achievements, the exchange of opinions and creative inspirations—make a valuable contribution to the progress of cognitive thought and to the pool of scientific knowledge, to strengthening peace and cooperation among nations, to enriching Copernican traditions which unite us all and make us build human science in a human world.'

ADDRESS BY THE PRESIDENT OF THE POLISH ACADEMY OF SCIENCES, PROF. W. TRZEBIATOWSKI

'Mr Prime Minister, Mr President of the International Astronomical Union, Distinguished Guests,

The Extraordinary General Assembly of the International Astronomical Union which we are opening today is an occurrence of exceptional importance. In its international aspect, it marks the culminating point of this year's Copernican observances. For the participation in this session of astronomers from all parts of the world we accept, among other things, as a tribute paid to our Great Compatriot, the founder of modern astronomy, the bedrock of the present-day outlook on the world.

The session is at the same time a scientific gathering with an extensive working programme. At the symposia starting tomorrow, all the latest astronomic problems will be discussed. We are glad the symposia will be held in Poland's three academic centres: Warsaw, Toruń and Cracow, and will have active participation of Polish astronomers and physicists. That will provide a particularly important and valuable opportunity to strengthen the existing and establish new personal ties between Polish and foreign scholars.

Characteristically, two of the mentioned symposia, being a part of this Extraordinary General Assembly, will be devoted to subjects related to physics and astrophysics, while the third one will be dealing with exploratory problems of cosmic physics. That programme demonstrates the important role contemporary astronomy plays in the family of natural sciences, its growing correlation with physics and ever stronger feedback with all those branches of science and technology involved in the exploration and peaceful use of outer space. Aware of this, the Polish Academy of Sciences accords astronomy a prominent place in its long-range plans for the development of scientific agencies and the research conducted thereby. We wish that Polish astronomy make a contribution to world science that would be proportional to our ambitions, that it continue the valuable Copernican traditions.

On behalf of the Polish Academy of Sciences, and on my own behalf, I would like to most sincerely welcome all members of the Extraordinary General Assembly of the International Astronomical Union. I wish you a successful debate, inspiring scientific impressions, and a pleasant stay in this country.'

ADDRESS BY THE CHAIRMAN OF THE WARSAW CITY COUNCIL, MR. MAJEWSKI

'Distinguished Guests,

On behalf of Warsaw inhabitants and authorities, I have the honour to welcome in the capital of People's Poland members of the Extraordinary General Assembly of the International Astronomical Union. I wish to express my great satisfaction that we can play host to so many outstanding representatives of such a magnificent and universal domain of science as astronomy is and that your presence in our town is a sign of appreciation for the contribution to the history of human thought made by a great Pole, Mikolaj Kopernik.

Warsaw, which is your host town, is a city with a magnificent and at the same time dramatic history. Four days ago we celebrated the thirty-fourth anniversary of the outbreak of World War II. First bombs in that war fell on our city. But it was only the beginning of a tragedy. Over five years later, the final balance-sheet of destruction was horrifying: out of every ten buildings of the pre-war Warsaw nine had been destroyed, burned or blown up. The city centre, objects of culture and science, relics of the past, industry had all been completely destroyed. Over 800 thousand Warsaw inhabitants died under Nazi terror.

I mention this to emphasize the determination to survive, so characteristic of Warsaw inhabitants. Today, Warsaw reconstructed with the effort of the whole nation, is again a properly functioning, developing city. I hope that despite the extensive programme of your Assembly you find time to see that for your own. At the same time, however, I would like to turn your attention to Warsaw's achievements in the domain well-familiar to you, in science.

The traditions of our city in this domain are by no means insignificant. I think, for example, the traditions of the Warsaw mathematical school are not unknown to you. In the post-war period efforts at reconstruction of residential quarters and industrial establishments, relics of the past and monuments, were accompanied by a consistent strive to revive and subsequently develop the scientific base: higher schools, laboratories, and research centres. Today, nearly 30% of the country's scientific potential is concentrated in Warsaw and the achievements of our scientists, at least in proportion to the ratio of this concentration, are notable in more than one domain.

Astronomical traditions of Warsaw, which date back to over a hundred and fifty years ago, to the date of establishment of the first observatory, are being continued in a modern way. Construction work will begin shortly on the Copernican Centre, the first element of a big concentration of intellectual and research potential, known as the Warsaw Scientific Complex. I am very pleased to say that the Copernican Centre which is to serve the exchange of astronomical thought regardless of frontiers and social systems, will be constructed under international cooperation, the same noble-minded and peaceful cooperation which has brought you here to Warsaw.

I wish you, Distinguished Guests, fruitful debates and a pleasant stay in our city.'

ADDRESS BY THE CHAIRMAN OF THE NATIONAL COMMITTEE FOR THE IAU, PROFESSOR W. IWANOWSKA

'Mister Chairman of the Council of Ministers, Mister President of the International Astronomical Union, Dear Guests,

The reason why our National Committee for Astronomy was so anxious to have a Meeting of the International Astronomical Union in our country this year, was the 500th Anniversary of the birth of Nicolaus Copernicus. One could think that celebrating anniversaries of great men of science is a nice custom, but—in the present very busy life—it is something like a waste of time. One might also think that studying the contributions of great men to the development of science belongs to the domain of the history of science.

All this being true, we think—however—that it is useful for scientists, once per generation, to have a look back on the way already passed, in order to get a perspective for the future plans. The span of time between the Copernican "Revolutions" and the present day revolution in science and technology offers such an opportunity; I should say—an excellent opportunity, because both ends of this span are exceptional and unforeseen. The work of Copernicus was a starting point of a continuous, exponential development of modern science and modern technology, whereas the present day acceleration of their growth begins to raise problems to the mankind, how to manage and organize research work and its applications.

In both these moments astronomy played and plays a leading part. In fact, astronomy is a leading science among natural sciences, since it stores in the depths of the Universe the greatest resources of the Unknown. It was not really an accident that the development of modern science started in astronomy. To find the fundamental laws of mechanics, the law of universal gravitation, a laboratory at least as big as the planetary system was necessary and this laboratory should first be brought into order by Copernicus, before Johannes Kepler and the great Isaac Newton could read any kinematical or dynamical laws from it. Since then, the cosmic laboratory works and grows pushing its limits beyond stars, galaxies, and quasars, bringing a flow of new information about expected and unexpected phenomena. At the present time radio-astronomy and space research discover new properties of matter and energy in pulsars, quasars, active nuclei of galaxies, X-ray sources, relict radiation, multiatomic molecules in the interstellar medium. Using the most modern techniques of optics, electronics, computing machinery, the modern astronomy stimulates their development by increasing its demands for uttermost precision and perfection and yields to physics and technology—in return—new data on cosmic phenomena.

After a detailed analysis of the progress in all branches of astronomy over the last three years, made in Sydney at the ordinary General Assembly of the International Astronomical Union, within its more than forty commissions, you—astronomers from all the world—came to this country to discuss specifically the present day achievements in these branches of astronomy which took their origin directly from the work of Copernicus: as modern celestial mechanics, exploration of the planetary system, mysteries of gravity in collapsing cosmic masses or expanding envelopes of aging stars, and modern cosmology. These are the topics of the five symposia of this Extraordinary General Assembly. To see the span of progress in astronomy from Copernicus to the present-day astronomy, one more symposium—the sixth—is devoted to the beginnings of this progress, to the background, the birth, and the non-easy reception of the Copernican ideas. This is the plan of our recollection which could be labelled: 'From Copernicus to the present-day astronomy'. Where could we better remember and visualize the Copernican beginnings of our science and its deeply dramatic story, if not visiting these medieval houses, castles, colleges, cathedrals, and towers along the Vistula river, still existing and witnessing the life of the strange man called Nicolaus Copernicus.

There were several reasons why the Polish Academy of Sciences decided to extend their invitation to all astronomers from all countries to come here and to spend a week on fruitful meditations on the topic: 'From Copernicus to Modern Science'.

On behalf of the National Committee for Astronomy of the Polish Academy of Sciences I welcome you in this country and wish you both: scientific use and pleasure.'

ADDRESS BY THE PRESIDENT OF THE IAU, PROFESSOR L. GOLDBERG

'The International Astronomical Union is greatly honored to have been invited by the Polish Academy of Sciences to hold an Extraordinary General Assembly as a tribute to Nicolaus Copernicus during the celebration of the quincentenary of his birth. Here in the country where he lived and worked, we have come to pay our respects to the man who initiated research in the physical sciences while hoping at the same time to find new directions for further development of astronomy.

The five symposia to be held in Warsaw, Toruń and Cracow during the next ten days are among the most important organized by the IAU in recent years and may well have a decisive influence on the course of cosmological research over the next few years. On behalf of the Executive Committee, I would like to thank the Academy for their invitation, the symposia organizing committees for generating a brilliant scientific program and the local organizing committee for its masterful organization of the very intricate arrangements required to keep 1000 participants supplied with scientific cultural and material nourishments while moving about from one end to the other of this great and beautiful country.

One of the objects of the International Astronomical Union is to develop astronomy through international cooperation. In this we are following the example of Nicolaus Copernicus who, as Prof. Gingerich has pointed out, had no hesitation in acquiring from abroad the knowledge he needed to build his system and in sharing this knowledge with others of different nationality and religious and political beliefs. Although the years he spent in Italy were primarily for the purpose of studying canon law and medicine, he also rounded out his education in mathematics and astronomy and breathed deeply of the atmosphere of the Renaissance. There is considerable evidence that he became aware of the heliocentric ideas of Aristarchus during this time when the wisdom of the ancient classical period was being re-discovered. The point I wish to make, however, is that Copernicus was a master of foreign languages and that he could therefore initiate his research in astronomy by making a thorough search of the existing literature. Finally, in what was undoubtedly one of the earliest ventures in international scientific publication, the great *De revolutionibus* was published in Nuremberg through the initiative of the German mathematician Rheticus to whom Copernicus had given his manuscript. So did science progress through international cooperation in the time of Copernicus as it does today.

In assembling in Poland, the IAU also wishes to recognize the important contributions the astronomers of this country have made to the work of the Union. Poland was one of the first countries to satisfy the requirements for admission to the IAU and took part in the first General Assembly in 1922 under the leadership of Tadeusz Banachiewicz. At present the Polish astronomers are taking an active part in the work of the Executive Committee and that of numerous Commissions and are doing important research in a great many different fields, such as spectroscopy, stellar astronomy, radioastronomy, the interstellar medium, stellar evolution, solar physics and solar terrestrial relations. Their activities represent the extension and continuance of a tradition that began even before the birth of Copernicus. In fact, it is not at all surprising that the geocentric universe should have been overturned by a Pole in 16th Century Europe. Poland in the 16th century was one of the most advanced countries in Europe and was already famous for its love and learning and for its Jagiellonian University in Cracow which had been founded by Casimir II the Great in 1364. When Copernicus entered this University in 1491, it was already among the oldest and finest in Europe and the only one that could boast not one but two professors of astronomy.

And so, in assembling here during this jubilee year of 1973, the IAU recognizes its debt, both to Copernicus for setting mankind on the path of cosmic exploration and to our Polish colleagues for preserving and nourishing one of the world's oldest astronomical traditions. I now declare the Extraordinary General Assembly of 1973 officially open.'

APPENDIX I

LUNAR AND MARTIAN NOMENCLATURE

LUNAR AND MARTIAN NOMENCLATURE

A large number of names of craters and other features on the Moon and Mars, were accepted by the IAU General Assemblies X (Moscow, 1958), XI (Berkeley, 1961), XII (Hamburg, 1964), XIV (Brighton, 1970), and XV (Sydney, 1973).

The names were suggested by the appropriate IAU Commissions (16 and 17). In particular the Lunar names accepted at the XIVth and XVth General Assemblies were recommended by the 'Working Group on Lunar Nomenclature' under the Chairmanship of Dr D. H. Menzel. The Martian names were suggested by the 'Working Group on Martian Nomenclature' under the Chairmanship of Dr G. de Vaucouleurs. At the XVth General Assembly a new 'Working Group on Planetary System Nomenclature' was formed (Chairman: Dr P. M. Millman) comprising various Task Groups, one for each particular subject.

For further references see: *IAU Trans.* **X**, 259–263, 1960; **XIB**, 236–238, 1962; **XIIB**, 203–204, 1966; **XIIIB**, 99–105, 1968; **XIVB**, 63, 129, 139, 1971; *Space Sci. Rev.* **12**, 136–186, 1971.

Because at the recent General Assemblies some small changes, or corrections, were made, the complete list of Lunar and Martian Topographic Features is published here.

Table 1
Lunar Craters

Abbe	58S, 174E
Abbot	6N, 55E
Abel	34S, 85E
Abul Wáfa	2N, 117E
Adams	32S, 69E
Aitken	17S, 173E
Al-Biruni	18N, 93E
Alden	24S, 111E
Aldrin	1.4N, 22.1E
Alekhin	68S, 131W
Alter	19N, 108W
Amici	10S, 172W
Amundsen	85S, 85E
Anders	42S, 144W
Anderson	16N, 171E
Antoniadi	69S, 173W
Appleton	37N, 158E
Armstrong	1.5N, 25E
Arrhenius	55S, 91W
Artamonov	26N, 104E
Artem'ev	10N, 145W
Aston	33N, 88W
Avicenna	39N, 97W
Avogadro	64N, 165E
Baade	44S, 82W
Babakin	21S, 123E
Babcock	4N, 94E
Backlund	16S, 103E
Balboa	19N, 83W
Baldet	54S, 151W
Balmer	20S, 70E
Banachiewicz	5N, 80E
Banting	26N, 16E
Barbier	24S, 158E
Barnard	30S, 86E
Barringer	29S, 151W
Bartels	24N, 90W
Becquerel	41N, 129E
Bečvář	2S, 125E
Beijerinck	13S, 152E
Bel'kovich	61N, 90E
Bell	22N, 96W
Bellingsgauzen	61S, 164W
Belopol'skij	18S, 128W
Belyaev	23N, 143E
Bergstrand	19S, 176E
Berkner	25N, 105W
Berlage	64S, 164W
Bhabha	56S, 165W
Birkeland	30S, 174E
Birkhoff	59N, 148W
Bjerknes	38S, 113E
Blazhko	31N, 148W
Bobone	26N, 132W
Bohr	13N, 86W
Boltzmann	75S, 90W

Bolyai	34S, 125E
Boole	64N, 87W
Borman	39S, 149W
Bose	54S, 170W
Boss	46N, 88E
Bowen	18N, 9E
Boyle	54S, 178E
Brackett	18N, 24E
Bragg	42N, 103W
Brashear	74S, 172W
Bredikhin	17N, 158W
Brianchon	75N, 87W
Bridgman	44N, 137E
Brouwer	36S, 125W
Brunner	10S, 91E
Buffon	41S, 134W
Buisson	1S, 113E
Bunsen	42N, 86W
Butlerov	12N, 110W
Buys-Ballot	21N, 175E
Byrd	85N, 10E
Cabannes	61S, 171W
Cajal	13N, 31E
Cajori	48S, 168E
Cameron	6N, 46E
Campbell	45N, 152E
Cannizzaro	55N, 100W
Cannon	20N, 81E
Cantor	38N, 118E
Carmichael	20N, 40E
Carnot	52N, 144W
Carver	43S, 127E
Cassegrain	52S, 113E
Catalán	46S, 87W
Chaffee	39S, 155W
Chamberlin	59S, 96E
Champollion	37N, 175E
Chandler	44N, 171E
Chang Heng	19N, 112E
Chant	41S, 110W
Chaplygin	6S, 150E
Chapman	50N, 101W
Chappell	55N, 179W
Charlier	36N, 132W
Chaucer	3N, 140W
Chauvenet	12S, 137E
Chebyshev	34S, 133W
Chernyshev	47N, 174E
Chrétien	47S, 163E
Clark	38S, 119E
Clerke	22N, 29E
Coblentz	38S, 126E
Cockcroft	30N, 164W
Collins	1.3N, 23.7E
Compton	56N, 105E
Comrie	23N, 113W
Comstock	21N, 122W
Congreve	0, 168W
Cooper	53N, 176E
Coriolis	0, 172E
Coulomb	54N, 115W
Cremona	67N, 90W
Crocco	47S, 150E
Crommelin	68S, 148W
Crookes	11S, 165W
Curie	23S, 92E
Curtis	16N, 56E
Cyrano	20S, 157E
Daedalus	6S, 180
D'Alembert	52N, 164E
Dalton	17N, 84W
Daly	5N, 57E
Danjon	11S, 123E
Dante	25N, 180
Das	27S, 138W
Daubrée	16N, 15E
Da Vinci	10N, 44E
Davisson	38S, 175W
Dawson	67S, 134W
Debye	50N, 177W
De Forest	77S, 162W
Dellinger	7S, 140E
Delporte	16S, 121E
Denning	16S, 143E
De Roy	55S, 99W
Desargues	70N, 74W
De Sitter	80N, 38E
Deutsch	24N, 110E
De Vries	20S, 177W
Dewar	3S, 166E
Dirichlet	10N, 151W
Dobrovol'skij	13S, 129E
Donner	31S, 98E
Doppler	13S, 160W
Douglass	35N, 122W
Dreyer	10N, 97E
Drude	39S, 91W
Dryden	33S, 157W
Drygalski	80S, 80W
Dubyago	5N, 70E
Dufay	5N, 170E
Dugan	65N, 103E
Dunér	45N, 179E
Dyson	61N, 121W
Dziewulski	21N, 99E
Eckert	18N, 58E
Eddington	22N, 72W
Edison	25N, 99E
Ehngelgardt	5N, 159W
Ehrlich	41N, 172W
Einstein	17N, 88W
Eijkman	63S, 142W
Einthoven	5S, 110E
Ellerman	26S, 121W
Ellison	55N, 108W
Elvey	9N, 101W

Emden	63N, 176W
Eötvös	36S, 134E
Erro	6N, 98E
Esnault-Pelterie	47N, 142W
Espin	28N, 109E
Evans	10S, 134W
Evdokimov	35N, 153W
Evershed	36N, 160W
Fabry	43N, 100E
Fechner	59S, 125E
Fen'yi	45S, 105W
Feoktistov	31N, 140E
Fermi	20S, 122E
Fersman	18N, 126W
Fesenkov	23S, 135E
Firsov	4N, 112E
Fitzgerald	27N, 172W
Fizeau	58s, 133W
Fleming	15N, 109E
Focas	34S, 94W
Foster	23N, 142W
Fowler	43N, 145W
Fox	0, 98E
Franck	23N, 36E
Freud	26N, 52W
Freundlich	25N, 171E
Fridman	13S, 127W
Froelich	80N, 110W
Frost	37N, 119W
Gadomski	36N, 147W
Gagarin	20S, 149E
Galen	22N, 5E
Galois	15S, 153W
Gamow	65N, 144E
Ganskij	10S, 97E
Ganswindt	80S, 110E
Garavito	48S, 157E
Gavrilov	17N, 131E
Geiger	14S, 158E
Gerasimovich	23S, 124W
Gernsback	36S, 99E
Gibbs	18S, 84E
Gilbert	3S, 76E
Gill	64S, 75E
Ginzel	14N, 97E
Giordano Bruno	36N, 103E
Glazenap	2S, 138E
Goddard	15N, 89E
Golitsyn	25S, 105W
Golovin	40N, 161E
Grachev	3S, 108W
Graff	43S, 88W
Green	4N, 133E
Gregory	2N, 127E
Grigg	13N, 130W
Grissom	48S, 149W
Grotrian	66S, 128E
Gullstrand	45N, 129W
Gum	40S, 89E
Guthnick	48S, 94W
Guyot	11N, 117E
Hadley	25N, 3E
Hagen	48S, 135E
Haldane	2S, 84E
Hale	74S, 90E
Hamilton	43S, 84E
Harriot	33N, 114E
Hartmann	3N, 135E
Hartwig	6S, 80W
Harvey	19N, 147W
Hatanaka	29N, 122W
Hayford	13N, 176W
Hayn	65N, 83E
Healy	32N, 111W
Heaviside	11S, 167E
Hedin	3N, 77W
Helberg	22N, 102W
Henderson	5N, 152E
Hendrix	48S, 161W
Henyey	13N, 152W
Hermite	86N, 88W
Hertz	13N, 104E
Hertzsprung	0, 129W
Hess	54S, 174E
Heymans	75N, 144W
Hilbert	18S, 108E
Hill	21N, 41E
Hippocrates	71N, 146W
Hirayama	6S, 93E
Hoffmeister	15N, 137E
Hogg	34N, 122E
Hohmann	18S, 94W
Holetschek	28S, 151E
Hornsby	24N, 12E
Houtermans	9S, 87E
Houzeau	18S, 124W
Hubble	22N, 87E
Humason	31N, 57W
Hutton	37N, 169E
Huxley	20N, 5W
Ibn Yunus	14N, 91E
Icarus	6S, 173W
Idel'son	82S, 114E
Ingalls	26N, 153W
Innes	28N, 119E
Ioffe	15S, 129W
Izsak	23S, 117E
Jackson	22N, 163W
Jansky	9N, 89E
Jeans	56S, 91E
Jenner	42S, 96E
Joliot	26N, 94E
Joule	27N, 144W
Joy	25N, 7E
Jules Verne	35S, 147E
Kamerlingh Onnes	15N, 116W

Kapteyn	11S, 71E
Karpinskij	73N, 166E
Katchalsky	6N, 116E
Kearons	12S, 113W
Keeler	10S, 162E
Kekulé	16N, 138W
Khvol'son	14S, 112E
Kibal'chich	2N, 147W
Kidinnu	36N, 123E
Kiess	6S, 84E
Kimura	57S, 118E
King	5N, 120E
Kirkwood	69N, 157W
Klejmenov	33S, 141W
Klute	37N, 142W
Knox-Shaw	5N, 80E
Koch	43S, 150E
Kohlschütter	15N, 154E
Kolhörster	10N, 114W
Komarov	25N, 153E
Kondratyuk	15S, 115E
Konstantinov	20N, 159E
Kopff	17S, 90W
Korolev	5S, 157W
Kostinskij	14N, 118E
Kovalevskaya	31N, 129W
Koval'skij	22S, 101E
Kramers	53N, 128W
Krasnov	30S, 80W
Krasovskij	4N, 176W
Kreiken	9S, 84E
Krylov	35N, 167W
Kugler	53S, 104E
Kulik	42N, 155W
Kuo Shou Shing	8N, 134W
Kurchatov	38N, 142E
Lacchini	41N, 107W
Lamarck	23S, 70W
Lamb	43S, 101E
Lamé	14S, 64E
Lampland	31S, 131E
Landau	42N, 119W
Lane	9S, 132E
Langemak	10S, 119E
Langevin	44N, 162E
Langley	52N, 87W
Langmuir	36S, 129W
Larmor	32N, 180
Laue	28N, 97W
Lauritsen	27S, 96E
Lawrence	8N, 43E
Leavitt	46S, 140W
Lebedev	48S, 108E
Lebedinskij	8N, 165W
Leeuwenhoek	30S, 179W
Leibnitz	38S, 178E
Lemaitre	62S, 150W
Lents	3N, 102W
Leonov	19N, 148E
Leucippus	29N, 116W
Leuschner	1N, 109W
Levi-Civita	24S, 143E
Lewis	19S, 114W
Ley	43N, 154E
Lindblad	70N, 99W
Litke	17S, 123E
Lobachevskij	9N, 113E
Lodygin	18S, 147W
Lomonosov	27N, 98E
Lorentz	34N, 100W
Love	6S, 129E
Lovelace	82N, 107W
Lovell	37S, 143W
Lowell	13S, 103W
Lucian	15N, 37E
Lucretius	9S, 121W
Ludwig	7S, 97E
Lundmark	39S, 152E
Lyapunov	27N, 89E
Lyman	65S, 162E
Lyot	51N, 84E
Mach	18N, 149W
Maksutov	41S, 169W
Malyj	22N, 105E
Mandel'shtam	6N, 162E
Marci	22N, 169W
Marconi	9S, 145E
Mariotte	29S, 140W
Markov	53N, 63W
Maunder	14S, 94W
Maxwell	30N, 98E
McKellar	16S, 171W
McLaughlin	47N, 93W
McMath	15N, 167W
McNally	22N, 127W
Mechnikov	11S, 149W
Mees	14N, 96W
Meggers	24N, 123E
Meitner	11S, 113E
Mendel	49S, 110W
Mendeleev	6N, 141E
Merrill	75N, 116W
Meshcherskij	12N, 125E
Mezentsev	72N, 129W
Michelson	6N, 121W
Milankovič	77N, 170E
Millikan	47N, 121E
Mills	9N, 156E
Milne	31S, 113E
Mineur	25N, 162W
Minkowski	56S, 145W
Minnaert	67S, 179E
Mitra	18N, 155W
Möbius	16N, 101E
Mohorovičić	19S, 165W
Moiseev	9N, 103E

Montgolfier	47N, 160W
Moore	37N, 178W
Morozov	5N, 127E
Morse	22N, 175W
Moseley	21N, 90W
Moulton	61S, 97E
Nagaoka	20N, 154E
Nansen	81N, 93E
Nassau	25S, 177E
Nernst	36N, 94W
Neujmin	27S, 125E
Nicholson	26S, 85W
Nielsen	32N, 52W
Niepce	72N, 120W
Nijland	33N, 134E
Nikolaev	35N, 151E
Nishina	45S, 171W
Nobel	15N, 101W
Nöther	66N, 114W
Numerov	71S, 161W
Nunn	4N, 91E
Nušl	32N, 167E
Obruchev	39S, 162E
O'Day	31S, 157E
Ohm	18N, 114W
Olcott	20N, 117E
Omar Khayyam	58N, 102W
Oppenheimer	35S, 166W
Oresme	43S, 169E
Orlov	26S, 175W
Ostwald	11N, 122E
Paneth	63N, 95W
Pannekoek	4S, 140E
Papaleksi	10N, 164E
Paracelsus	23S, 163E
Paraskevopoulos	50N, 150W
Parenago	26N, 109W
Parkhurst	34S, 103E
Parsons	37N, 171W
Pascal	74N, 70W
Paschen	14S, 141W
Pasteur	12S, 105E
Patsaev	17S, 133E
Pauli	45S, 136E
Pavlov	29S, 142E
Pawsey	44N, 145E
Peary	89N, 30E
Pease	13N, 106W
Peek	3N, 87E
Perel'man	24S, 106E
Perepelkin	10S, 128E
Perkin	47N, 176W
Perrine	42N, 129W
Petrie	45N, 108E
Petropavlovskij	37N, 115W
Petrov	61S, 88E
Pettit	28S, 86W
Petzval	63S, 113W
Pirquet	20S, 140E
Pizzetti	35S, 119E
Planck	58S, 135E
Plaskett	82N, 175E
Plummer	25S, 155W
Pogson	42S, 111E
Poincaré	57S, 161E
Poinsot	79N, 145W
Polzunov	26N, 115E
Poncelet	76N, 54W
Popov	17N, 99E
Porter	56S, 10W
Poynting	17N, 133W
Prager	4S, 131E
Prandtl	60S, 141E
Priestley	57S, 108E
Purkyně	1S, 95E
Quételet	43N, 135W
Racah	14S, 180
Raimond	14N, 159W
Ramsay	40S, 145E
Rayet	45N, 114E
Rayleigh	29N, 90E
Razumov	39N, 114W
Ricco	75N, 177E
Riedel	49S, 140W
Riemann	40N, 88E
Rittenhouse	74S, 107E
Ritz	15S, 92E
Roberts	71N, 175W
Robertson	22N, 105W
Roche	42S, 135E
Röntgen	33N, 92W
Rowland	57N, 163W
Rozhdestvenskij	86N, 155W
Rumford	29S, 170W
Runge	2S, 87E
Russell	27N, 75W
Rydberg	47S, 96W
Rynin	47N, 104W
Saenger	4N, 102E
Safařik	10N, 177E
Saha	2S, 103E
St. John	10N, 150E
Sanford	32N, 139W
Sarabhai	25N, 21E
Sarton	49N, 121W
Scaliger	27S, 109E
Schaeberle	26S, 117E
Schjellerup	69N, 157E
Schlesinger	47N, 138W
Schliemann	2S, 155E
Schlüter	6S, 83W
Schneller	42N, 164W
Schönfeld	45N, 98W
Schorr	19S, 90E
Schrödinger	75S, 133E
Schuster	4N, 147E

Schwarzschild	71N, 120E
Scott	82S, 45E
Seares	74N, 145E
Sechenov	7S, 143W
Segers	47N, 128E
Seidel	33S, 152E
Seyfert	29N, 114E
Shajn	33N, 172E
Shaler	33S, 85W
Shapley	10N, 57E
Sharonov	13N, 173E
Shatalov	24N, 140E
Shi Shen	76N, 104E
Shternberg	19N, 117W
Siedentopf	22N, 135E
Sierpinski	27S, 155E
Sisakyan	41N, 109E
Sklodowska	18S, 96E
Slipher	50N, 160E
Smoluchowski	60N, 97W
Sniadecki	22S, 169W
Sommerfeld	65N, 161W
Spencer Jones	13N, 166E
Spiru Haret	59S, 176W
Spurr	26N, 3W
Stark	25S, 134E
Stebbins	65N, 143W
Stefan	46N, 109W
Stein	7N, 179E
Steklov	37S, 105W
Steno	33N, 162E
Stetson	40S, 119W
Stokes	52N, 88W
Stoletov	45N, 155W
Stoney	56S, 156W
Störmer	57N, 145E
Stratton	6S, 165E
Strömgren	22S, 133W
Struve	23N, 76W
Subbotin	29S, 135E
Sumner	37N, 109E
Sundman	11N, 91W
Swann	52N, 112E
Sylvester	83N, 80W
Szilard	34N, 106E
Tachinni	5N, 86E
Tebbutt	9N, 53E
Teisserenc	32N, 136W
Ten Bruggencate	9S, 134E
Tereshkova	28N, 144E
Tesla	38N, 125E
Theophrastos	17N, 39E
Thiel	40N, 134W
Thiessen	75N, 169W
Thomson	33S, 166E
Tikhomirov	25N, 162E
Tikhov	62N, 172E
Tiling	52S, 132W
Timiryazev	5S, 147W
Titius	27S, 101E
Titov	29N, 150E
Trumpler	28N, 168E
Tsander	5N, 149W
Tseraskij	49S, 141E
Tsinger	57N, 176E
Tsiolkovskij	20S, 129E
Tsu Chung-Chi	17N, 144E
Tyndall	35S, 117E
Väisälä	26N, 48W
Valier	7N, 174E
Van de Graaff	27S, 172E
Van den Bergh	31N, 159W
Van der Waals	44S, 119E
Van Gent	16N, 160E
Van Maanen	36N, 127E
Van Rhijn	53N, 145E
Van 't Hoff	62N, 132W
Van Wijk	63S, 119E
Vashakidze	44N, 93E
Vavilov	1S, 139W
Vening Meinesz	0, 163E
Ventris	5S, 158E
Vernadskij	23N, 130E
Very	26N, 25E
Vesalius	3S, 115E
Vestine	34N, 94E
Vetchinkin	10N, 131E
Vil'ev	6S, 144E
Volkov	13S, 131E
Volta	54N, 85W
Volterra	57N, 131E
Von der Pahlen	25S, 133W
Von Kármán	45S, 176E
Von Neumann	40N, 153E
Von Zeipel	42N, 142W
Voskresenskij	28N, 88W
Walker	26S, 162W
Wan-Hoo	11S, 139W
Waterman	26S, 128E
Watson	63S, 124W
Watts	9N, 46E
Weber	50N, 124W
Wegener	45N, 114W
Wells	41N, 122E
Wexler	69S, 90E
Weyl	16N, 120W
White	45S, 160W
Widmannstätten	6S, 85E
Wiechert	84S, 165E
Wiener	41N, 146E
Wilsing	22S, 155W
Winkler	42N, 179W
Winlock	35N, 106W
Woltjer	45N, 160W
Wood	44N, 121W
Wright	32S, 86W

Wyld	1S, 98E	Zelinskij	29S, 167E
Yablochkov	61N, 127E	Zernike	18N, 168E
Yamamoto	59N, 161E	Zhiritskij	25S, 120E
Yangel'	17N, 5E	Zhukovskij	8N, 167W
Zanstra	3N, 124E	Zinner	27N, 59W
Zeeman	75S, 135W	Zsigmondy	59N, 105W

Table 2
List of Reference Points on the Moon

No.	Name	ξ	η	ζ
1	W. Bond B	+0.055	+0.906	+0.420
2	Rhaeticus A	.091	.030	.995
3	Egede A	.113	.782	.611
4	Sulpicius Gallus	.191	.336	.924
5	Cayley	.260	.069	.963
6	Bessel	.286	.370	.883
7	Luther	.342	.547	.864
8	Dawes	.424	.296	.856
9	Hercules G	.435	.723	.537
10	Maury	.510	.604	.614
11	Sina	.518	.153	.842
12	Berzelius F	.605	.542	.583
13	Tralles B	.688	.458	.564
14	Taruntius A	.758	.126	.640
15	Picard	.789	.251	.557
16	Bode A	−0.020	+0.156	+0.989
17	Archimedes A	.098	.470	.877
18	Pico B	.182	.724	.665
19	La Condamine Q	.246	.795	.555
20	Gambart A	.321	.017	.947
21	Carlini	.339	.555	.760
22	Draper C	.350	.293	.887
23	Sharp A	.456	.738	.495
24	Diophantus	.499	.463	.732
25	Brayley	.561	.356	.746
26	Kepler A	.584	.124	.802
27	Marius A	.701	.218	.679
28	Herodotus A	.734	.366	.572
29	Reiner A	.778	.089	.621
30	Galilaei A	.872	.203	.445
31	Mösting A	−0.090	−0.056	+0.994
32	Orontius D	.083	.634	.767
33	Alpetragius B	.115	.261	.958
34	Gauricus D	.162	.576	.822
35	Clavius J	.164	.847	.501
36	Guericke C	.196	.200	.959
37	Scheiner A	.232	.868	.439
38	Hesiodus B	.268	.456	.856
39	Fra Mauro A	.355	.095	.930
40	Epimenides A	.365	.685	.629

No.	Name	ξ	η	ζ
41	Darney C	.425	.244	.872
42	Mee F	.435	.686	.584
43	Herigonius	.543	.231	.806
44	Drebbel D	.598	.615	.514
45	Wichmann	.610	.131	.781
46	Flamsteed A	.674	.137	.725
47	Billy D	.722	.255	.644
48	Vico C	.829	.352	.436
49	Damoiseau E	.847	.091	.524
50	Licetus H	+0.038	−0.718	+0.695
51	Airy A	.128	.293	.949
52	Jacobi B	.140	.813	.562
53	Hipparchus C	.142	.129	.981
54	Pontanus F	.178	.467	.866
55	Nicolai A	.296	.675	.676
56	Alfraganus C	.309	.106	.950
57	Pons B	.311	.481	.820
58	Janssen K	.466	.720	.512
59	Rosse	.545	.307	.780
60	Rheita P	.553	.615	.564
61	Isidorus D	.559	.074	.825
62	Reichenbach K	.598	.483	.645
63	Gaudibert J	.620	.193	.761
64	Biot	.718	.385	.580
65	Messier A.	.730	.035	.682
66	Langrenus C	.862	.098	.497
67	Bruce	+0.007	+0.020	+0.999
68	Hadley A	.104	.423	.901
69	Cassini M	.049	.660	.750
70	W. Bond C	.059	.910	.408
71	Aristillus A	.066	.553	.829
72	Hyginus D	.074	.198	.977
73	Hyginus B	.088	.132	.987
74	Egede B	.098	.771	.627
75	Sulpicius Gallus G	.104	.338	.935
76	Manilius D	.119	.229	.967
77	Aratus D	+0.136	+0.412	+0.899
78	Godin D	.144	.017	.989
79	Sulpicius Gallus A	.144	.375	.916
80	Linné B	.211	.507	.833
81	Menelaus A	.221	.293	.929
82	Silberschlag A	.227	.121	.965
83	Eudoxus A	.240	.717	.654
84	Bessel E	.251	.336	.906
85	Linné E	.253	.447	.856
86	Aristoteles M	.272	.802	.529
87	Taquet	.315	.286	.905
88	Bessel A	.325	.418	.848
89	Deseilligny	.328	.360	.872
90	Tacquet C	.350	.233	.906
91	Arago B	.355	.060	.932

No.	Name	ξ	η	ζ
92	Baily B	.362	.776	.513
93	Arago D	.378	.120	.915
94	Le Monnier C	.411	.380	.827
95	Posidonius A	.419	.525	.739
96	Hercules C	.425	.678	.600
97	Janssen G	.433	.162	.886
98	Janssen D	.458	.271	.845
99	Maskelyne K	.494	.056	.866
100	Römer L	.523	.395	.755
101	Maskelyne H	.532	.085	.842
102	Maraldi A	.556	.342	.757
103	Maraldi B	.580	.248	.775
104	Secchi B	.661	.064	.746
105	Picard X	.856	.227	.464
106	Marco Polo B	−0.031	+0.295	+0.955
107	Archimedes D	.040	.532	.844
108	Piazzi Smyth B	.045	.649	.758
109	Bode G	.061	.110	.992
110	Pico C	.078	.733	.675
111	Spitzberg C	.128	.542	.828
112	Pico D	.142	.687	.710
113	Eratosthenes B	.143	.321	.935
114	Schröter D	.165	.079	.984
115	Plato E	.180	.762	.621
116	Timocharis B	.186	.467	.864
117	Schröter M	−0.200	+0.121	+0.971
118	Gambart C	.208	.034	.978
119	Le Verrier E	.215	.673	.705
120	Stadius B	.230	.205	.951
121	Carlini D	.231	.544	.806
122	Timocharis A	.240	.419	.874
123	Helicon B	.286	.615	.734
124	Copernic H	.311	.120	.942
125	Laplace A	.326	.690	.642
126	La Hire A	.349	.477	.807
127	Pytheas A	.346	.349	.869
128	Reinhold F	.364	.059	.929
129	Heis D	.439	.524	.727
130	Tobias Mayer D	.440	.212	.872
131	Hortensius C	.446	.103	.888
132	Diophantus B	.469	.485	.736
133	Mairan E	.477	.612	.628
134	Euler E	.508	.417	.751
135	Brayley D	.509	.342	.788
136	Landsberg A	.516	.003	.856
137	Angström	.575	.498	.647
138	Encke E	.644	.006	.763
139	Bessarion C	.650	.276	.707
140	Aristarchus C	.650	.469	.596
141	Schiaparelli A	.813	.390	.432
142	Alphonsus H	−0.008	−0.269	+0.962

No.	Name	ξ	η	ζ
143	Herschel C	.055	.087	.995
144	Thebit E	.074	.392	.916
145	Hell E	.088	.567	.819
146	Regiomontanus E	.095	.473	.875
147	Maginus Y	.098	.786	.610
148	Palisa P	.126	.168	.978
149	Mösting B	.128	.047	.989
150	Lassell E	.168	.312	.934
151	Lippershey T	.174	.426	.886
152	Gauricus N	.186	.536	.823
153	Tycho H	.192	.710	.677
154	Turner F	.243	.028	.969
155	Longomontanus R	.269	.792	.547
156	Cichus K	.274	.596	.754
157	Gould A	.276	.329	.902
158	Opelt K	.285	.235	.929
159	Kies D	.288	.421	.859
160	Guericke A	.291	.193	.936
161	Cichus H	.321	.541	.777
162	Darney J	−0.353	−0.248	+0.902
163	Bonpland E	.379	.169	.910
164	Marth	.419	.517	.746
165	Darney E	.418	.215	.882
166	Euclides D	.428	.163	.888
167	Ramsden A	.434	.551	.712
168	Landsberg B	.470	.043	.882
169	Hippalus A	.495	.402	.770
170	Vittelo E	.510	.487	.707
171	Hérigonius E	.565	.238	.787
172	Gassendi Y	.580	.356	.731
173	Gassendi L	.623	.348	.698
174	Gassendi F	.682	.259	.683
175	Flamsteed D	.704	.055	.706
176	De Gasparis B	.707	.454	.542
177	Parrot L	+0.015	−0.310	+0.051
178	Zach J	.044	.842	.537
179	Seeliger	.053	.039	.998
180	Albategnius C	.064	.179	.982
181	Aliacensis G	.069	.550	.833
182	Blanchinus K	.080	.419	.904
183	Pickering B	.129	.037	.991
184	Airy P	.140	.275	.951
185	Gemma Frisius K	.151	.608	.779
186	Ritchey C	.157	.190	.969
187	Mutus K	.195	.846	.495
188	-	.196	.701	.683
189	Geber E	.210	.351	.912
190	Dollond D	.215	.143	.966
191	Abulfeda R	.219	.221	.950
192	Buch B	+0.230	−0.614	+0.755
193	Hommel M	.232	.864	.445

No.	Name	ξ	η	ζ
194	Barocius E C	.255	.745	.615
195	Zagut F	.260	.505	.824
196	Theon Senior A	.266	.003	.964
197	Hommel H A	+0.313	−0.788	+0.529
198	Fermat E	.320	.341	.885
199	Zagyt S	.322	.551	.772
200	Tacitus C	.329	.236	.916
201	Alfraganus D	.343	.070	.936
202	Polybius D	.418	.452	.788
203	Cyrillus G	.432	.270	.860
204	Polybius L	.438	.374	.817
205	Watt A	.462	.769	.440
206	Neander Y	.509	.568	.647
207	Rheita P	.553	.615	.562
208	Bohnenberger G	.616	.295	.730
209	Lubbock G	.632	.064	.772
210	Messier D	.722	.062	.687
211	Holden V	.838	.317	.442

The craters 1–66 have a diameter larger than 8 km, while the rest are smaller.

Table 3
Large Craters on Mars

Adams	31N, 197	Comas Sola	20S, 158
Agassiz	70S, 89	Copernicus	50S, 169
Airy	5S, 0	Crommelin	5N, 10
Antoniadi	22N, 299	Cruls	43S, 197
Arago	10N, 330	Curie	29N, 5
Arrhenius	40S, 237	Daly	66S, 22
Bakhuysen	23S, 344	Dana	73S, 32
Baldet	23N, 295	Darwin	57S, 20
Barabashov	47N, 69	Da Vinci	2N, 39
Barnard	61S, 298	Dawes	9S, 322
Becquerel	22N, 8	Denning	18S, 326
Beer	15S, 8	Douglass	52S, 70
Bianchini	64S, 97	Du Martheray	6N, 266
Bjerknes	43S, 189	Du Toit	72S, 46
Boeddicker	15S, 197	Eddie	12N, 218
Bond	33S, 36	Ejriksson	19S, 174
Bouguer	19S, 333	Escalante	0, 245
Brashear	54S, 120	Eudoxus	44S, 147
Briault	10S, 270	Fesenkov	22N, 87
Burroughs	72S, 243	Flammarion	26N, 312
Burton	14S, 156	Flaugergues	17S, 341
Campbell	54S, 195	Focas	34N, 347
Cassini	24N, 328	Fontana	64S, 73
Cerulli	32N, 338	Fournier	4S, 287
Chamberlin	66S, 124	Gale	6S, 222
Charlier	69S, 169	Galileo	6N, 27
Clark	56S, 134	Galle	51S, 31
Coblentz	55S, 91	Gilbert	68S, 274
Columbus	29S, 166	Gill	16N, 354

Gledhill	53S, 273	Mie	48N, 220
Graff	21S, 206	Milankovič	55N, 147
Green	52S, 8	Millochau	21S, 275
Hadley	19S, 203	Mitchel	68S, 284
Haldane	53S, 231	Molesworth	28S, 211
Hale	36S, 36	Moreux	42N, 315
Halley	49S, 59	Müller	26S, 232
Hartwig	39S, 16	Nansen	50S, 141
Heaviside	71S, 95	Newcomb	24S, 358
Helmholtz	46S, 21	Newton	40S, 158
Henry	11N, 336	Nicholson	0, 166
Herschel	14S, 230	Niesten	28S, 302
Hipparchus	44S, 151	Oudemans	10S, 92
Holden	26S, 34	Pasteur	19N, 335
Holmes	75S, 292	Perepelkin	52N, 65
Hooke	45S, 44	Peridier	26N, 276
Huggins	49S, 204	Pettit	12N, 174
Hussey	54S, 127	Phillips	67S, 45
Hutton	72S, 255	Pickering	34S, 133
Huxley	63S, 259	Playfair	78S, 125
Huygens	14S, 304	Porter	50S, 114
Janssen	3N, 322	Priestley	54S, 228
Jarry-Desloges	9S, 276	Proctor	48S, 330
Jeans	70S, 206	Ptolemaeus	46S, 158
Joly	75S, 42	Quenisset	34N, 319
Jones	19S, 20	Rabe	44S, 325
Kaiser	46S, 340	Radau	17N, 5
Keeler	61S, 152	Rayleigh	76S, 240
Kepler	47S, 219	Redi	61S, 267
Knobel	6S, 226	Renaudot	42N, 297
Korolev	73N, 196	Reuyl	10S, 193
Kunowsky	57N, 9	Reynolds	74S, 160
Lambert	20S, 335	Richardson	73S, 181
Lamont	59S, 114	Ritchey	29S, 51
Lampland	36S, 79	Ross	58S, 108
Lassell	21S, 63	Rossby	48S, 192
Lau	74S, 107	Rudaux	38N, 309
Le Verrier	38S, 343	Russell	55S, 348
Liais	75S, 253	Rutherford	19N, 11
Li Fan	47S, 153	Schaeberle	24S, 310
Liu Hsin	53S, 172	Schiaparelli	3S, 343
Lockyer	28N, 199	Schmidt	72S, 79
Lohse	43S, 16	Schroeter	2S, 304
Lomonosov	65N, 8	Secchi	58S, 258
Lowell	52S, 81	Sharonov	27N, 59
Lyell	70S, 15	Sklodowska	34N, 3
Lyot	50N, 331	Slipher	48S, 84
Madler	11S, 357	Smith	66S, 103
Magelhaens	32S, 174	South	77S, 339
Maggini	28N, 350	Spallanzani	58S, 273
Main	77S, 310	Steno	68S, 115
Maraldi	62S, 32	Stokes	56N, 189
Mariner	35S, 164	Stoney	69S, 134
Marth	13N, 3	Suess	67S, 179
Martz	34S, 217	Teisserenc de Bort	1N, 315
Maunder	50S, 358	Terby	28S, 286
McLaughlin	22N, 22	Tikhov	51S, 254
Mendel	59S, 199	Trouvelot	16N, 13

Trumpler	62S, 151
Tycho Brahe	50S, 214
Tyndall	40N, 190
Very	50S, 177
Vinogradsky	56S, 217
Vogel	37S, 13
Von Karman	64S, 59
Wallace	53S, 249
Wegener	65S, 4
Weinbaum	66S, 245
Wells	60S, 238
Williams	18S, 164
Wirtz	49S, 26
Wislicenus	18S, 349
Wright	59S, 151

Table 4
Topography on Mars

Catena	λ	φ
Coprates	66° to 56°	−15°
Ganges	71° to 67°	−02° to −03°
Tithonia	98° to 80°	−06°
Chasma		
Australis	270°	−80° to −88°
Borealis	65° to 30°	+85°
Candor	78° to 73°	−04° to −06°
Capri	52° to 32°	−14° to −03°
Coprates	68° to 54°	−11° to −14°
Eos	51° to 32°	−16° to −17°
Ganges	52° to 48°	−08°
Hebes	81° to 73°	+01° to −01°
Ius	98° to 80°	−07°
Juventae	61°	−04°
Melas	78° to 70°	−08° to −12°
Ophir	77° to 64°	−03° to −09°
Tithonia	90° to 80°	−04°
Dorsum		
Argyre	70°	−61° to −65°
Fossa		
Alba	117° to 109°	+38° to +49°
Ceraunìae	107°	+25°
Claritas	108° to 105°	−19° to −32°
Elysium	225° to 219°	+28° to +26°
Hephaestus	240° to 233°	+22° to +18°
Mareotis	85° to 69°	+41° to +48°
Medusae	162°	−08°
Memnonia	158° to 140°	−22° to −15°
Nili	284° to 279°	+20° to +26°
Sirenum	163° to 138°	−36° to −27°
Tantalus	105° to 99°	+34° to +47°
Tempe	80° to 62°	+35° to +46°
Thaumasia	100° to 80°	−36° to −40°
Labyrinthus		
Noctis	110° to 92°	−05° to −08°
Mensa		
Deuteronilus	346° to 340°	+42° to +45°
Nilosyrtis	290°	+32°
Protonilus	315°	+38°

Mons		
Arsia	121°	−09°
Ascraeus	104°	+12°
Elysium	213°	+25°
Olympus	133°	+18°
Pavonis	113°	+01°
Montes		
Charitum	50° to 32°	−57°
Hellesponti	315°	−45° to −48°
Nereidum	57° to 43°	−48° to −38°
Phlegra	195°	+31° to +46°
Tharsis	125° to 101°	−12° to +16°
Planitia		
Acidalia	30°	+48°
Arcadia	155°	+48°
Amazonis	160°	+13°
Argyre	43°	−49°
Chryse	45°	+17°
Elysium	210°	+15°
Hellas	290°	−45°
Isidis	270°	+15°
Syrtis	290°	+15°
Utopia	235°	+35°
Patera		
Alba	110°	+40°
Amphitrites	299°	−59°
Apollinaris	186°	−08°
Biblis	124°	+02°
Hadriaca	267°	−31°
Orcus	181°	+14°
Pavonis	121°	+03°
Tyrrhena	253°	−22°
Uranius	93°	+26°
Planum		
Aurorae	52° to 48°	−10° to −11°
Hesperia	258° to 242°	+10° to −35°
Lunae	70° to 60°	+05° to +20°
Ophir	61° to 55°	−09° to −12°
Solis	98° to 88°	−20° to −30°
Syria	105° to 100°	−10° to −18°
Sinai	90° to 70°	−10° to −20°
Tholus		
Albor	210°	+19°
Australis	323°	−57°
Ceraunius	97°	+24°
Hecates	210°	+32°
Hippalus	89°	+76°
Iaxartes	15°	+72°
Jovis	117°	+18°
Kison	358°	+73°
Ortygia	8°	+70°
Tharsis	91°	+14°
Uranius	98°	+26°

Vallis		
Al-Qahira	202° to 194°	−23° to −15°
Ares	23° to 14°	+02° to +10°
Auqakuh	298°	+28°
Huo Hsing	295° to 292°	+32° to +28°
Ma'adim	183°	−27° to −20°
Mangala	151°	−10° to −4°
Nirgal	44° to 36°	−32° to −27°
Kasei	70° to 56°	+21
Shalbatana	45°	+01° to +15°
Simud	40° to 37°	+00° to +14°
Tiu	32°	+10° to +18°
Valles Marineris	95° to 45°	−05° to −15°
Vastitas		
Borealis	continuous	+55° to +67°

Table 5
Crater Names on Phobos and Deimos

Phobos	*Deimos*
1 – Roche	A – Swift
7 – Wendell	B – Voltaire
8 – Todd	
10 – Sharpless	
14 – D'Arrest	
18 – Stickney	
34 – Hall	

APPENDIX II

MEMBERSHIP OF COMMISSIONS

ALPHABETICAL LIST OF MEMBERS

COMPOSITION DES COMMISSIONS
MEMBERSHIP OF COMMISSIONS

4. Commision des Ephémérides. (Ephemerides)

Président: R. L. Duncombe.

Vice-Président: V. K. Abalakin.

Comité d'Organisation: J. Kovalevsky, T. Lederle, B. Morando, J. D. Mulholland, A. M. Sinzi, G. A. Wilkins.

Membres: Aoki S., Bec A., Chebotarev G. A., Clemence G. M., de Greiff G. A., Deprit A., Fricke W., Fursenko M. A., Gondolatsch F., Haupt R. F., Herrick S., Hertz H. G., Klepczynski W. J., Lahiri N. C., Lieske J. H., Morrison L. V., Oesterwinter C., O'Handley D. A., Rodriguez M., Sadler D. H., Seidelmann P. K., Shapiro I. I., Taylor D. J., van Flandern T. C., Walter H. G., Williams J. G.

5. Commission de Documentation (Documentation)

Président: J.-C. Pecker.

Vice-Président: W. D. Heintz.

Comité d'Organisation: F. Henn, J. Kleczek, A. Maxwell, T. S. Shcherbina-Samojlova, J. B. Sykes, G. A. Wilkins.

Membres: Aoki S., Beer A., Bourgeois P., Bouska J., Dewhirst D. W., Fricke W., Hirose H., Lantos P., Martynov D. Ya., McNally D., Meadows A. J., Mein P., Ogorodnikov K. F., Radlova L. N., Velghe A. G., Weidemann V., Wempe J., Witkowski J.

6. Commission des Télégrammes Astronomiques. (Astronomical Telegrams)

(Comité du Comité Exécutif)

Président: P. Simon.

Vice-Président: E. Roemer.

Sans Comité d'Organisation.

Membres: Candy M. P., Cunningham L. E., Hers J., Kozai Y., Marsden B. G. (Secrétaire de la Commission), Martynov D. Ya., Mrkos A.

7. Commission de la Mécanique Céleste. (Celestial Mechanics)

Président: P. J. Message.

Vice-Président: V. Szebehely.

Comité d'Organisation: E. P. Aksenov, V. A. Brumberg, M. S. Davis, A. Deprit, G. N. Duboshin, G. E. O. Giacaglia, G. Hori, Y. Kozai, B. Morando, L. Sehnal.

Membres: Antonacopoulos G., Aoki S., Batrakov Yu., Bec (Miss) A., Belorizky D., Böhme S., Bozis G., Broucke R., Candy M. P., Cesco R. P., Chapront J., Chebotarev G. A., Clemence G. M., Contopoulos G., Cook A. H., Counselman C. C., Cox J. F., Cunningham L. E., Danby J. M. A., Demin V. G., Duncombe R. L., Everhart E., Fabre H., Ferraz-Mello S., Galibina I. V., Garfinkel B., Gaska S., Goldreich P., Goudas C. I., Brebenikov E. A., Hadjidemetriou J. D., Hagihara Y., Hamid S. E., Hénon M., Herget P., Herrick S., Hertz H., Jefferys W. H., Jeffreys H., Jupp A., Katsis D. N., Kaula W. M., Kovalevsky J., Kustaanheimo P. E., Lundquist C. A., Magnaradze (Miss) N. C., Marsden B., Meffroy J.,

Merman G. A., Møller O., Musen P., Mulholland J. D., Nahon F., Nacozy P. E., O'Handley D., Orlov A. A., Peale S. J., Petrovskaya M. S., Popović B., Rabe E., Roy A. E., Ryabov J. A., Sadler D. H., Sagnier J. L., Schubart J., Sconzo P., Sharaf S. G., Shapiro I. I., Shteins K. A., Sibahara R., Siegel C. L., Slavenas P. V., Standish E. M., Stiefel E., Tapley B. D., Tatevyan S. K., Thiry Y., Vinti J. P., Williams C. A., Yarov-Yarovoy M. S., Zagar F.

8. Commission de l'Astronomie de Position. (Positional Astronomy)

PRÉSIDENT: G. van Herk.

VICE-PRÉSIDENT: R. H. Tucker.

COMITÉ D'ORGANISATION: C. A. Anguita, W. Fricke, B. L. Klock, P. Lacroute, A. A. Nemiro, J. L. Schombert, K. N. Tavastsherna, H. W. Wood, H. Yasuda.

MEMBRES: Adams A. N., Argyracos J. G., Atkinson R. d'E., Bacchus P., Bagildinski B. K., Billaud G., Blackwell K. C., Bohrman A., Bykov M. F., Chernega N. A., Dambara T., Débarbat (Miss) S., Dejaiffe R., De Vegt Ch., Dieckvoss W., Duncombe R. L., Eichhorn-von Wurmb H. K., Gliese W., Gordon Ya. E., Guinot B., Haas J., Hansson N., Harris B. J., Heintz W. D., Hoffleit (Miss) E. D., Høg E., Hughes V. A., Isobe S., Jackson P., Konin V. V., Korol A. K., Kosin G. S., Larink J., Latypov A. A., Laustsen S., Lévy J. R., López J. A., Maître V., Marcus (Miss) E., Melchior P. J., Mitić L. A., Murray C. A., Nefed'eva A. I., Nikoloff I., Okuda T., Olsen H. J. F., Osório J. J. S. P., Petrov G. M., Pilowski K., Podobed V. V., Polozhentsev D. D., Postoiev A., Proverbio E., Quijano L., Reiz A., Requième Y., Rousseau J. M., Rybka P., Sandig H.-U., Scheepmaker A. C., Schmeidler F., Schuler W., Scott F. P., Slaucitajs S., Stoy R. H., Symms L. S. T., Ševerlić B. M., Teleki G., Thomas D. V., Tsubokawa I., Tuzi K., Verbaandert J., von der Heide J. E. B., Walter H., Woolsey E. G., Yamazaki A., Zimmermann G. K., Zverev M. S.

9. Commission des Instruments Astronomiques. (Astronomical Instruments)

PRÉSIDENT: A. B. Meinel.

VICE-PRÉSIDENT: J. Ring.

COMITÉ D'ORGANISATION: K. L. Hallam, I. M. Kopylov, W. C. Livingston, N. N. Mikhel'son, V. B. Nikonov, E. H. Richardson, M. F. Walker, G. Wlérick.

MEMBRES: Aly M. K. M., Andrews D. H., Babcock H. W., Bakos G. A., Baranne A., Baum W. A., Beck N. G., Behr A., Bhattacharyya J. C., Bijaoui A., Blitzstein W., Boyce P. B., Brahde R., Brealey G. A., Brosterhus E. B. F., Brück H., Brückner G., Carpenter G. J., Charvin P., Chincarini G., Clarke D., Connes P., Couder A., Cousins A. W. J., Crawford D. L., Deharveng M., De Jonge J. K., Delbouille L., Delsemme A. H., Dobronravin P. P., Dollfus A., Duchesne M., Dunham T. Jr., Dunkelman L., Evans J. V., Fellgett P. B., Fischer R. R., Fitch W. S., Ford W. K. Jr, Giovanelli R. G., Gottlieb K., Gratton L., Gunn J. E., Hall J. S., Hilliard R., Hoag A. A., Høg E., Honneycutt R. K., Hooghoudt B. G., Ingrao H. C., Ioannisiani B. K., Jayarajan A. P., Jensch A., Karandikar R. V., Kissel K. E., Köhler H., Lallemand A., Lasker B. E., Linfoot E. H., Linnik V. P., Luud L. S., McGee J. D., McMullan D., Melnikov O. A., Merman (Mrs) N. V., Minnet H. C., Mohler O. C., Monin G. A., Moroz V. I., Odgers G. I., Petford A. D., Pierce A. K., Pope J. D., Prokof'ev V. K., Purgathofer A. Th., Rakos K. D., Reay N. K., Robinson L. B., Roddier F., Rösch J., Saïssac J., Schmahl G., Schöneich W., Schroeder D. J., Seddon H., Shcheglov P. V., Sisson G. M., Smyth M. J., Strand K. Aa., Sukharev L. A., Taylor D. J., Texereau J., Treanor P. J. S. J., Tull R. C., Ulrich B., Valníček B., Wallenquist Å. A. E., Walraven Th., Wayman P. A., Wellgate G. B., Wilcock W. L., Wynne Ch. G., Zacharov.

10. Commission de l'Activité Solaire. (Solar Activity)

PRÉSIDENT: K. O. Kiepenheuer.

VICE-PRÉSIDENT: G. A. Newkirk.

COMITÉ D'ORGANISATION: L. W. Acton, V. Bumba, L. D. de Feiter, J. T. Jefferies, V. A. Krat, F. Moriyama, P. Simon.

MEMBRES: Altschuler M. D., Aly M. K. M., Anderson K. A., Athay R. G., Balli E., Banin V. G., Beckers J. M., Bell B., Bhatnagar A., Billings D. E., Bonov A. D., Bray R. J., Bruckner G., Bruzek A., Carlquist P., Cimino M., Coutrez R., Covington A. E., de Groot T., de Jager C., Deubner F. L., Dezsö L., Dizer M., Dodson-Prince H., Dollfus A., Dubov E. E., Dunn R. B., Elste G. H., Falciani R., Fokker A. D., Fortini T., Friedman H., Fritzova-Svestkova L., Gaizauskas V., Giovanelli R. G., Gleissberg W., Gnevyshev M. N., Gnevysheva R. S., Godoli G., Gopasjuk S. I., Gurtovenko E. A., Hagen J. P., Hanasz J., Hansen R. T., Hedeman E. R., Hiei E., Hirayama T., Howard R. F., Hyder C., Iospa B. A., Jager F. W., Jakimiec J., Jensen E., Kai K., Kjeldseth M. O., Kleczek J., Kopecky M., Krivsky L., Kruger A., Kuklin G. V., Kundu M. R., Kunzel H., Leroy J. L., Letfus V., Lincoln J. V., Livshits M. A., Loughhead R. E., McKenna Lawlor S., Macris C. J., Makita M., Malitson H. H., Maltby P., Malville J. M., Mandel'shtam S. L., Martres M. J., Mattig W., Maxwell A., Menzel D. H., Mergentaler J., Michard R., Mogilevskij E. I., Mohler O. C., Moiseev I. G., Moreton G. E., Mustel E. R., Nagasawa S., Nakagawa Y., Neupert W. M., Nishi K., Obridko V. N., Ohman Y., Orrall F. Q., Parkinson W. H., Pick M., Piddington J. H., Pneuman G. W., Popovici C., Rayrole J., Razmadze T. S., Reeves E. M., Rigutti M., Rompolt B., Rösch J., Topolová-Euzicková B., Saito K., Sakurai K., Sarabhai V. A., Sawyer C., Schluter A., Schmidt H. U., Schroeter E. H., Servajean R., Severny A. B., Shapley A. H., Sheeley N. R. Jr., Sitnik G. F., Slonim E. M., Smerd S. F., Smith E. v. P., Smith Henry J., Smith S., Stenflo J. O., Stepanov W. E., Steshenko N. V., Stix M., Sturrock P. A., Suemoto Z., Švestka Z., Takakura T., Tanaka H., Tandberg-Hanssen E., Teske R. G., Tifrea E., Tlamicha A., Trellis M., Tuominen J., Valnicek B., Van Allen J. A., Uchida Y., Van't Veer F., Vojtech L., Vitinskij Ju. I., Waldmeier M., Wilcox J. M., Wiehr E., Wild J. P., Wilson P. R., Xanthakis J., Yakovkin N. A., Yoshimura H., Zirin H., Zwaan C.

12. Commission de la Radiation et de la Structure de l'Atmosphère Solaire. (Radiation and Structure of the Solar Atmosphere)

PRÉSIDENT: R. G. Giovanelli.

VICE-PRÉSIDENT: M. K. V. Bappu.

COMITÉ D'ORGANISATION: R. G. Athay, J. M. Beckers, P. Delache, C. Jordan, G. M. Nikolsky, E. H. Schröter.

MEMBRES: Acton L., Adam (Miss) M. G., Allen C. W., Altschuler M., Billings D. E., Blackwell D. E., Blamont J. E., Böhm K.-H., Bohm-Vitense E., Bonnet R., Bray R. J., Bruzek A., Cox A. N., Delbouille L., de Jager C., Dubov E. E., Dumont S., Dunn R. B., Edlén B., Edmonds F. N., Elliott I., Elste G., Evans J. W., Falciani R., Friedman H., Gabriel A., Gingerich O. J., Gokdogan (Mrs) N., Goldberg L., Gopasyuk S. I., Gnevyshev M. N., Grevasse N., Haupt R. F., Hiei E., Hotinli M., House L., Houtgast J., Jager F. W., Jefferies J. T., Kanno M., Kato S., Kawaguchi I., Kiepenheuer K. O., Kononovich E V., Kopecký M., Krat V. A., Kuklin G. V., Laborde G., Labs D., Leighton R. B., Leroy J.-L., Linsky J., Locke J. L., Lopez Arroyo M., Loughhead R. E., Lüst R., Mathias O. F. W., Matsushima S., Mattig W., Mein P., Mergentaler J., Michard R., Migeotte M. V., Moore-Sitterly Ch., Mouradian Z., Mugglestone D., Muller (Miss) E. A., Namba O., Neven L., Newkirk G. A., Nicolet M., Nishi K., Orrall F. Q., Oster L., Pande C., Pajdušákova L., Parkinson W. H., Pasachoff J. M., Pecker J.-C., Pecker-Wimel (Mrs) C., Peyturaux R., Pierce A. K., Reeves E. M., Righini G., Rigutti M., Roddier F., Roland G., Seaton M. J., Severny A. B., Sivaraman K. R., Sitnik G. F., Sobolev V. M., Suemoto Z., Švestka Z., Swensson J. W., Tandberg-Hanssen E., Teplitskaya R. B., Thomas R. N., Tousey R., Uchida Y., Unsold A., Waldmeier M., Warwick J. W., Wilson P. R., Wlérick G., Wyller A., Zirin H., Zirker J. B., Zwaan C.

14. Commission des Données Spectroscopiques Fondamentales. (Fundamental Spectroscopic Data)

PRÉSIDENT: R. H. Garstang.

VICE-PRÉSIDENT: W. Lochte-Holtgreven.

COMITÉ D'ORGANISATION: K. M. Baird, A. H. Cook, S. L. Mandel'shtam, Ch. Moore-Sitterly, J. G. Phillips, R. Tousey.

MEMBRES: Allen C. W., Andrew K. L., Baird K. M., Barrow R., Bates D. R., Bely O., Blaha M., Branscomb L. M., Burgess A., Carroll P. K., Carver J. H., Corliss C. H., Dalgarno A., Dobronravin P. P., Douglas A. E., Dressler K., Dufay M., Edlén B., Engelhardt E., Essen L., Felenbok P., Flower D., Gabriel A. H., Garton W. R. S., Glagolewskij Ju. V., Grant I. P., Green L. C., Herman R., Herzberg G., Hesser J. E., Hindmarsh W. R., House L. L., Huber M., Humphreys C. J., Jacquinot P., Joly F., Jordan H. L., Junkes J., Kessler K. G., Khokhlova V. L., King R. B., Lawrence G., Layzer D., Littlefield T. A., Mandel'shtam S. L., Martin W. C., Mel'nikov O. A., Migeotte M. V., Milford S. N., Mohler O. C., Monfils A., Naqvi A. M., Nevin T. E., Nicholls R. W., Nussbaumer H., Obi S., Oetken L., Omont A., Peach G., Petrini D., Pfenning H., Prokof'ev V. K., Querci F., Richter J., Sahal S., Salpeter E. W., Schadee A., Seaton M. J., Shore B. W., Smit J. A., Smith G., Somerville W. B., Sørensen N. G., Swings J. P., Swings P., Takayanagi K., Tech J. L., Terrien J., Traving G., Trefftz E., van Bueren H. G., van Regemorter H., van Rensbergen W., Varsavsky C. M., Wares G. W., Weniger S., Wiese W. L., Wilson R., Zirin H.

15. Commission pour l'Étude Physique des Comètes, des Petites Planètes et des Météorites. (Physical Study of Comets, Minor Planets and Meteorites)

PRÉSIDENT: A. H. Delsemme.

VICE-PRÉSIDENT: N. B. Richter.

COMITÉ D'ORGANISATION: E. Anders, C. Arpigny, L. Biermann, O. V. Dobrovolsky, B. D. Donn, T. Gehrels, J. Rahe, E. Roemer, V. Vanýsek, F. L. Whipple.

MEMBRES: Bertaud Ch., Beyer M., Blamont J. E., Bobrovnikoff N. T., Bouška J., Brandt J. C., Brown P. Lancaster, Candy M. P., Ceplecha Z., Cherednichenko V. L., Demenko (Mrs) A. A., Dossin F., Douglas A. E., Greenberg J. M., Harwit M. O., Haser L., Herzberg G., Huebner W. F., Hunaerts J. J., Konopleva (Mrs) V. P., Levin B., Liller W. F., Lüst R., Lyttleton R. A., Malaise D. J., Maran S. P., Marsden B. G., Martel-Chossat M. T., McCord C. D., Miller F. D., Mrkos A., Ness N. F., O'Dell C. R., O'Keefe J. A., Öpik E. J., Poloskov S. M., Proisy P., Rémy-Battiau L., Riives V., Sekanina Z., Sivaraman K. R., Swings P., Vsekhsvyatsky S. K., Waterfield R. L., Wetherill G., Woszczyk A., Wurm K.

16. Commission pour l'Étude Physique des Planètes et des Satellites. (Physical Studies of Planets and Satellites)

PRÉSIDENT: C. H. Mayer.

VICE-PRÉSIDENT: T. C. Owen.

COMITÉ D'ORGANISATION: J. Connes, I. K. Koval, S. Miyamoto, G. H. Pettengill, S. K. Runcorn, B. A. Smith.

MEMBRES: Arthur D. W. G., Ashbrook J., Barreto L. M., Baum W. A., Belton M. J. S., Berge G. L., Bobrov M. S., Boyce P. B., Broadfoot A. L., Brunk W. E., Bullen K. E., Camichel H., Chamberlain J. W., Collinson E. H., Colombo G., Cook A. F., Counselman C. C., deMarcus W. C., de Mottoni G., de Vaucouleurs G., Dickel J. R., Dollfus A., Drake F. D., Eshleman V. R., Fink U., Fox W. E., Gehrels T., Giclas H. L., Goldstein R. M., Goody R. M., Greyber H. D., Guerin P., Guest J. E., Hagfors T., Hall J. S., Halliday I., Haupt J. F., Herzberg G., Hovenier J. W., Hubbard W. B., Hunt G. E., Hunten D. M., Ingrac H. C., Irvine W. M., Jeffreys H., Jelley J., Kaplan L., Karandikar R. V., Kellogg W. W., Kondo Y., Kopal Z., Kumar S. S., Kuzmin A. D., Leblanc Y., Levin B. J., Liddell U., Link F., Lipskij Y. N., Martynov D. Ya., McCord T. B., McElroy M. B.,

Menzel D. H., Middlehurst B., Millman P. M., Moroz V. E., Morrison D. D., Narayana J. V., O'Leary B. T., Öpik E. J., Ottelet I., Parkinson T., Rogers A. E. E., Rösch J., Safronov V., Sagan C., Saissac J., Salisbury J. W., Shapiro I. I., Shimizu T., Shoemaker E., Sinton W., Smith H. J., Smoluchowski R., Strong J. D., Tejfel' V. G., Tombaugh C., Trafton L., Tyler G. L., Van Allen J. A., Veverka J., Wallace L. V., Whitaker E. A., Wildey R. L., Wildt R., Williams I. P., Wilson A. G., Young A. T.

17. Commission de la Lune. (The Moon)

Président: S. K. Runcorn.

Vice-Présidents: E. Anders, K. P. Florensky.

Comité d'Organisation: A. Dollfus, V. P. Dzhapiashvili, K. Koziel, J. D. Mulholland, J. A. O'Keefe, C. P. Sonett.

Membres: Alley C. O., Arthur D. W. G., Baldwin R. B., Bastin J., Beckmann J. E., Bender P., Cameron (Mrs) W., Counselman C. C., Elston W. E., Evans D. J., Ezersky V. I., Fielder G., Gavrilov I. Y., Gold T., Goldreich P., Goudas C. L., Green J., Guest J., Gurstein A. A., Habibullin S. T., Hagfors T., Jeffreys H., Kaula W. M., Kopal Z., Kuzmin A. D., Leikin G. A., Levin B., Link F., Lipsky V. N., Martynov Ja., Menzel D. H., Mikhailov A. A., Millman P., Mills A. A., Miyamoto S., Morrison U. V., Moutsoulas M., Nefed'ev A. A., Pettengill G., Potter K. I., Rösch J., Sagan C., Shoemaker E., Strom R. O., Troïtsky V. S., Urey H. C., van Flandern T. C., Whipple F. L., Weimer T., Wetherill G. W., Whitaker E. A., Williams J. G., Wilson U.

19. Commission de la Rotation de la Terre. (Rotation of the Earth)

Président: C. Sugawa.

Vice-Président: R. O. Vicente.

Comité d'Organisation: H. J. Abraham, P. L. Bender, M. H. Cohen, K. Lambeck, V. Milovanović, P. Paquet, H. M. Smith, Ya. S. Yatskiv, S. Yumi (ex officio).

Membres: Atkinson R., Barlier F., Billaud G., Bonanomi J., Buchar E., Cecchini G., Davies J. G., Débarbat S. V., Dramba C., Elsmore B., Enslin H., Fedorov E. P., Fichera E., Fleckenstein J. O., Guinot B., Hall R. G., Hers J., Hosoyama K., Hurukawa K., Iijima S., Jeffreys H., Kakuta Ch., Kalmykov A. M., Knowles S. B., Kostina (Mrs) L. D., Kulikov K. A., Lederle T., Lefebvre M., Levallois J. J., Locke J. L., Markowitz Wm., Meinig M., Melchior P. J., Mikhailov A. A., Miyadi M., Nicolini T., O'Hora N. P. J., Okuda T., Opalski W., Orlin H., Orlov B. A., Orte D. A., Oterma L., Parijski N. N., Pavlov N. N., Popov N. A., Postoiev A., Proverbio E., Randic L., Rice D. A., Rochester M. G., Sakharov V. I., Sekiguchi N., Sevarlic B. M., Shapiro I. I., Smith F. G., Stoyko A., Stoyko N., Takagi S., Teleki G., Thomas D. V., Torao M., Tsao Mo., Wako Y., Williams J. G., Winkler G., Witkowski J.

20. Commission des Positions et des Mouvements des Petites Planètes, des Comètes et des Satellites. (Positions and Motions of Minor Planets, Comets and Satellites)

Président: L. Kresák.

Vice-Président: B. G. Marsden.

Comité d'Organisation: F. K. Edmondson, P. Herget, E. I. Kazimirchak-Polonskaya, B. Morando, W. H. Robertson, E. Roemer, J. Schubart.

Membres: Ahmed I. I., Aksnes K., Arend S. J. V., Batrakov Yu. V., Bielicki M., Bruwer J. A., Candy M. P., Chebotarev G. A., Cristescu C., Cunningham L. E., Debehogne H., Evdokimov Yu. V., Everhart E., Ferraz-Mello S., Galibina I. V., Gehrels T., Giclas H. L., Gutiérrez-Moreno A., Harrington R. S., Herrick S., Hertz H. G., Hirose H., Itsigsohn M., Klemola A. R., Klepczynski W. J., Kohoutek L., Kovalevsky J., Lindblad B. A., McCrosky R. E., Michkovitch V. V., Milet B., Mintz Blanco B., Missana N., Orlov A. A., Oterma L., Popović B., Porter J. G., Protitch M. B., Rabe E., Rasmusen H.

Q., Sagnier J. L., Schmitt A., Schrutka-Rechtenstamm G., Sekanina Z., Shtejns K. A., Sitarski G., Strobel W., Sultanov G. F., Tomita Ko., Torroja J. M., van Houten C. J., van Houten-Groeneveld I., Vsekhsvyatskij S. K., Whipple F. L., Wild P., Wilkins G. A., Williams J. G., Wood H. W., Yakhontova N. S., Zadunaisky P. E.

21. Commission de la Luminescence du Ciel. (The Light of the Night Sky)

PRÉSIDENT: J. L. Weinberg.

VICE-PRÉSIDENT: R. Dumont.

COMITÉ D'ORGANISATION: A. S. Asaad, H. Elsässer, R. H. Giese, H. Tanabe, Yu. L. Trutse, G. Weill, R. D. Wolstencroft.

MEMBRES: Anderson K., Bates D. R., Blackwell D. E., Chamberlain J. W., Cook A. F., Divari N. B., Dufay M., Fesenkova E. V., Galperin G. I., Haug U., Henry R. C., Huruhata M., Hunten D. M., James J. F., Jarrett A. H., Kaplan J., Karandikar R. V., Karyagina Z. V., Kastler A., Knaflich (Mrs) H. B., Koutchmy S., Krasovskij I., Kulkarni P. V., Leinert Ch., Megrelishvilli T. G., Neuzil L., Nicolet M., Pearse R. W. B., Roach F. E., Robley R., Roosen R. G., Sanchez-Martinez F., Saxena P. P., Schmidt Th., Soberman R. K., Steiger W. R., van Allen J. A., Wallace L. V., Weniger S.

22. Commission on Meteors and Meteorites

NEW PRESIDENT: B. A. Lindblad.

NEW VICE-PRESIDENT: I. Halliday.

NEW ORGANIZING COMMITTEE: A. F. Cook, W. G. Elford, T. R. Kaiser, L. Kresak, B. J. Levin, R. E. McCrosky, F. F. Verniani.

MEMBERS: Abbott W. N., Anders E., Astapovich I. S., Babadzhanov P. B., Ceplecha Z., Davies J. G., Davis J., Ellyett C. D., Fedinskij V. B., Fireman E. L., Guth V., Hajduk A., Hawkins G. S., Hemenway C. L., Hey J. S., Hirose H., Hodge P. W., Hoppe J. A., Jacchia L., Jaynel A. A., Kashscheev B. L., Katasev L. A., Keay C. S. L., Kizilirmak A., Kostylev K. V., Kramer K. N., Kresakova M., Krinov E. L., Lebedinets V. N., Lovell B., McDonnell J. A. M., McIntosh B. A., Millman P. M., Nazarova T. N., O'Keefe J. A., Olivier C. P., Öpik E. J., Plavcova Z., Rajchl J., Roosen R. G., Russell J. A., Sekanina Z., Shao C. Y., Simek M., Simonenko A. N., Soberman R. K., Southworth R. B., Stohl J., Wetherill G. W., Whipple F. L., Wood J. A.

24. Commission d'Astrométrie Photographique. (Photographic Astrometry)

PRÉSIDENT: P. Lacroute.

VICE-PRÉSIDENT: C. A. Murray.

COMITÉ D'ORGANISATION: A. Blaauw, A. N. Deutsch, W. Dieckvoss, W. Gliese, K. Aa. Strand, W. F. van Altena, H. W. Wood.

MEMBRES: Abhyankar K. D., Bouigue R., Brosche P., Churms J., Clube S. V. M., Crézé M., Delhaye J., de Vegt Ch., Dommanget J., Eichhorn H. K. v. W., Fatchikhin N. W., Fracastoro M. G., Franz O. G., Fredrick L. W., Gallouët L., Giclas H. L., Goyal A. N., Haas J., Harrington R. S., Harris B. J., Heintz W. D., Herget P., Hoffleit (Miss) E. D., Klemola A. R., Lapushka K. K., Lavdovskij V. V., Lippincott (Miss) S. L., Lozinskij A. M., Luyten W. J., Mennessier M. O., Meurers J., Morgan W. W., O'Connel D. J. K., Oja T., Potter H. I., Prochazka F. V., Quijano L., Roemer E., Sanders W. L., Schilt J., Smart W. M., Stoy R. H., Thomas D. V., Upgren A., van de Kamp P., Vasilevskis S., Wagman N. E., Wesselink A. J.

25. Commission de Photométrie et de Polarimétrie Stellaires. (Stellar Photometry and Polarimetry)

PRÉSIDENT: M. Golay.

VICE-PRÉSIDENT: A. Hiltner.

COMITÉ D'ORGANISATION: A. W. J. Cousins, D. L. Crawford, J. A. Graham, A. R. Hyland, M. F. McCarthy, K. Osawa, V. Straižys, T. Gehrels.

MEMBRES: Argue A. N., Bahng J. D. R., Becker W., Behr A., Bigay J. H., Bok B. J., Borgman J., Breger M., Brück H. A., Chis G., Chugajnov P. F., Dachs J., Eelsalu H. T., Feinstein A., Fernie J. D., Fitch W. S., Gallouët L., Gutiérrez-Moreno A., Haffner H., Hardie R. H., Hauck B., Hill P. W., Holmberg E. B., Irwin J. B., Jerzykiewicz M., Johnson H. L., Joshi S. C., Kron J. E., Landolt A. U., Lasker B. M., Lockwood G. W., Masani A., Mayer P., Mendoza V. E. E., Mianes P., Mitchell R. I., Moreno H., Morris S. C., Muller A. B., Mumford G. S., Nikonov V. B., Notni P., Perry C. L., Philip A. G. D., Roslund C., Rufener F. G., Rybka E., Sarma M. B. K., Schmidt H., Schöneich W., Smyth M. J., Steinlin U. W., Stock J., Strohmeier W., Todoran I., Ulrich B., Ureche V., Velghe A. G., Wallenquist A. A. E., Walraven T., Wesselink A. J., Willstrop R. V., Young A. T.

26. Commission des Étoiles Doubles. (Double Stars)

PRÉSIDENT: Miss S. L. Lippincott.

VICE-PRÉSIDENT: P. Muller.

COMITÉ D'ORGANISATION: A. H. Batten, P. Couteau, A. N. Deutsch, J. Dommanget, O. G. Franz, K. Aa. Strand.

MEMBRES: Arend S., Baize P., Cester B., Da Silva A. V., Djurković P. M., Finsen W. S., Fracastoro M. G., Freitas-Mourão R. R. de, Güntzel-Lingner U., Hadjidemetriou J. D., Harrington R. S., Heintz W. D., Hidajat B., Holden F., Hopmann J., Jeffers H. M., Johnson M. C., Jonckheere R., Kiselev A. A., Kulikovsky P. G., Kumar S. S., Luyten W. J., Nather R. E., Scarfe C. D., The P. S., van de Kamp P., Walker R. L., Wierzbinski St., Wieth-Knudsen N. P., Worley C. E., Zagar F.

27. Commission des Étoiles Variables. (Variable Stars)

PRÉSIDENT: M. W. Feast.

VICE-PRÉSIDENT: W. Wenzel.

COMITÉ D'ORGANISATION: N. Baker, J. D. Fernie, H. Gursky, B. V. Kukarkin, L. Rosino, J. Smak.

MEMBRES: Aizenman M. L., Alania I. F., Andrews A. D., Arkhipova V. P., Ashbrook J., Bakos G. A., Balazs-Detre J., Bateson F. M., Bath G. T., Bertaud Ch., Bessell M. S., Bhatnagar P. L., Boyarchuk A. A., Breger M., Butler C. J., Chavira E., Cherepashchuk A. M., Chis G., Chugainov P. F., Cox A. N., de Kock R. P., Dickens R. J., Edwards P. J., Efremov Yu. I., Eskioglu A. N., Fitch W. S., Friedjung M., Frolov M. S., Gaposhkin S., Gascoigne S. C. B., Geyer E. H., Gershberg R. E., Godoli G., Gorbatsky V. G., Harwood M., Hansen C. J., Haro G., Heidmann N., Heiser A., Hersig G. H., Herr R. B., Hesser J. E., Heyden F. J., Hoffleit E. D., Houk N., Huffer C. M., Huruhata M., Hutchings J. B., Iben I., Jarzebowski T., Jones A. F., Kippenhahn R., Kholopov P. N., Kopylov I. M., Kordylewski K., Kraft R. P., Krzeminski W., Kuhi L. V., Kumsishvili J. I., Kunkel W. E., Kwee K. K., Landolt A. U., Ledoux P., Lesh J. Rountree, Leung K. C., Lockwood G. W., Lortet-Zuckermann M. C., Luud L. S., Maffei P., Mannino G., Masani A., Mavridis L. N., Mayall M. W., McNamara D. H., Miller W. J., Mirzoyan L. V., Morguleff N., Mumford G. S., Murdin P. G., Nicolov N. S., O'Connell D. J. K., Odgers G. J., Oosterhoff P. T., Opolski A., Osawa K., Oskanyan V., Payne-Gaposchkin C. H., Peltjer L. C., Piotrowski S., Plaut L., Popova M., Rakos K. D., Renson P., Richter G., Rodgers A. W., Romano G., Ruslev S. V., Sarma M. B. K., Sawyer-Hogg H. B., Shobbrook R. R., Sinvhal S. D., Smeyers P., Smith, Halan J., Stepien K., Strohmeier W., Swope H. H., Szeidl B., Takeati M., Tammam G. A., Tempesti P., Tremko J., Tsesevich V. P., Usher P. D., van Agt S., Vandekerkhove E., van Genderen A. M., van Hoof A., Wachmann A. A., Walker M. F., Wallerstein G. N., Walraven T., Warner B., Wehlau A., Wesselink A. J., Wright F. W., Zwicky F.

28. Commission des Galaxies. (Galaxies)

PRÉSIDENT: E. B. Holmberg.

VICE-PRÉSIDENT: B. E. Markarian.

COMITÉ D'ORGANISATION: F. Bertola, J. Bolton, G. R. Burbidge, R. D. Davies, K. C. Freeman, M. S. Roberts, J. L. Sérsic, S. van den Bergh, B. A. Vorontsov-Velyaminov, B. E. Westerlund.

MEMBRES: Abell G. O., Ables H., Alfvén H., Ambartsumian V. A., Andrillat (Mrs) Y., Arakelyan M. A., Arp H. C., Bahcall J. N., Barbieri C., Baum W. A., Bigay J. H., Bok B. J., Bondi H., Bonnor W. B., Bottinelli L., Braccesi A., Chincarini G., Contopoulos G., Courtès G., Dessy J. Landi, de Vaucouleurs G., Dibaj E. A., Einasto J. E., Elvius (Mrs) A. M., Esipov V. F., Evans D. S., Fehrenbach C., Field G. B., Ford W. K., Gascoigne S. C. B., Gouguenheim L., Haro G., Heckmann O., Heeschen D. S., Heidmann J., Hodge P. W., Hoyle F., Johnson H. M., Kalinkov M., Kellerman K. I., Kiang T., King I. R., Kopylov I. M., Kundu M. R., Kustaanheimo P. E., Larson R. B., Layzer D., Lequeux J., Lin C. C., Lindblad P. O., Low F. J., Lynden-Bell D., Lynds C. R., McClure R. D., McCrea W. H., MacKay D. C., McVittie G. C., Mayall N. U., Mills B. Y., Minkowski R. L., Morgan W. W., Narlikar J. V., Neugebauer G., Neyman J., Noonan T. W., Oke J. B., Oleak H., Omer G. C., Oort J. H., Osterbrock D. E., Pachner J., Pacholczyk A. G., Page T. L., Peimbert M., Pishmish de Recillas (Mrs) P., Poveda A., Prendergast K. H., Pronik V. I., Reaves G., Reddish V. C., Richter N., Rindler W., Robinson I., Rood H. J., Rosino L., Rubin V. C., Rudnicki K., Sargent W. L. W., Saslaw W. C., Schmidt M., Schücking E., Schürer M., Sciama D. W., Scott E. L., Setti B., Shakeshaft J. R., Shane C. D., Shen B. S. P., Spinrad H., Stein W., Stibbs D. W. N., Takarada K., Takase B., Tammann G. A., Thackeray A. D., Tifft W. G., Toomre A., Tovmassian H., Treder H. J., Ulrich (Mrs) M.-H., Vandekerkhove E., van de Kruit P. C., van Woerden H., Visvanathan N., Whitford A. E., Wild P., Williams R. E., Wills D., Wilson A. G., Zel'dovich Ya. B., Zwicky F.

29. Commission des Spectres Stellaires. (Stellar Spectra)

PRÉSIDENT: A. A. Boyarchuk.

VICE-PRÉSIDENT: M. Hack.

COMITÉ D'ORGANISATION: M. K. V. Bappu, W. K. Bonsack, M. W. Feast, Y. Fujita, C. O. R. Jaschek, D. C. Morton, J. P. Swings, K. O. Wright.

MEMBRES: Abhyankar K. D., Abt H. A., Aller L. H., Andrillat H., Andrillat (Mrs) Y., Asaad A. S., Aslanov I. A., Berger J., Bertaud Ch., Bessel M. S., Bidelman W. P., Bloch (Miss) M., Boesgaard (Mrs) A. M., Boggess A. III, Bouigue R., Brück H. A., Buscombe W., Cayrel R., Cayrel-de-Strobel (Mrs) G., Chalonge D., Climenhaga J. L., Code A. D., Conti P. S., Cowley (Mrs) A. P., Divin (Miss) L., Dobronravin P. P., Dolidze M. V., Dunham Jr T., Edmonds Jr F. N., Evans D. S., Faraggiana (Miss) R., Friedjung M., Fringant (Miss) A. M., Garrison R. F., Gershberg R. E., Glagolevskij Ju. V., Gollnow H., Gorbatskij V. G., Gratton L., Gray D. F., Greenstein J. L., Griffin R. E. M., Griffin R. F., Groth H. G., Guthrie B. N. G., Haro G., Heintze J. R. W., Henize K. G., Herbig G. H., Herman (Mrs) R., Houziaux L., Huang Su-Shu., Hyland A. R., Jaschek (Mrs) M., Johnson H. R., Jugaku J., Keenan P. C., Khokhlova V. L., Kienle H., King R. B., Kodaira K., Kogure T., Kopylov I. M., Kowatchew B. J., Kraft R. P., Labs D., Larsson-Leander G., Luud L. S., McNamara D. H., Mannino G., Mel'nikov O. A., Milligan J. E., Morguleff N., Mustel E. R., Myerscough (Mrs) V. P., Nicholls R. W., Nikitin A. A., Nishimura S., Oetken L., Oke J. B., Osawa K., Pagel B. E. J., Parsons S. B., Pasinetti L. E., Payne-Gaposchkin (Mrs) C. H., Peat D. W., Pedoussaut A., Preston G. W. III., Przybylski A., Querci F., Ringuelet A. E., Rodgers A. W., Rublev S. V., Sahade J., Sanwal N. B., Schild R. E., Sinnerstad U., Slettebak A., Smak J., Smoliński J., Stawikowski A., Stecher T. P., Svolopoulos S. N., Swensson J. W., Swings P., Taffara S., Thackeray A. D., Thompson

G. I., Tsuji T., Underhill (Miss) A. B., Utsumi K., Vandekerkhove E., Van't Veer (Mrs) C., Voigt H. H., Wallerstein G. N., Waterworth M. D., Wehlau W. H., Wellmann P., Wempe J., Weniger S., Williams P. M., Wilson O. C., Wilson R., Wolff S. C., Wood H. J., Wyller A. A., Yamashita Y.

30. Commission des Vitesses Radiales. (Radial Velocities)

PRÉSIDENT: R. F. Griffin.

VICE-PRÉSIDENT: A. H. Batten.

COMITÉ D'ORGANISATION: R. Bouigue, M. Duflot, H. Gollnow, J. F. Heard, L. Oetken, A. G. D. Philip, V. C. Rubin.

MEMBRES: Abt H. A., Barbier-Brossat (Mrs) M., Beardsley W. R., Bertiau F. C., Blaauw A., Boulon J., Crampton D., De Jonge J. K., de Vaucouleurs G., Edmondson F. K., Eelsalu H. T., Fehrenbach Ch., Fletcher J. M., Garrison R. F., Georgelin Y., Gratton L., Harding G. A., Heintze J. R. W., Hube D. P., Kharadze E. K., Kraft R. P., Martin (Miss) M., Pedoussaut A., Perry C. K., Preston G. W. III, Sahade J., Serkowski K., Thackeray A. D., Willstrop R. V.

31. Commission de L'Heure. (Time)

PRÉSIDENT: H. Enslin.

VICE-PRÉSIDENT: A. Orte.

COMITÉ D'ORGANISATION: D. J. Belocerkovskij, T. Gökmen, B. Guinot, G. Hemmleb, S. Iijima, Wm. Markowitz, P. Melchior, J. D. Mulholland, H. M. Smith, G. M. R. Winkler.

MEMBRES: O'Handley D. A., Abraham H. J., Bakulin P. I., Bender P. L., Billaud G., Blinov N., Bonanomi J., Brkić Z. M., Caprioli G., Dingle H., Drâmbă G., Gama L. I., Gougenheim A., Hall R. G., Hers J., Klepczynski W., Lacombe C. G., Lederle T., Madwar M. R., Matsunami N., Mikhailov A. A., Miyadi M., Opalski W., Pâquet P., Parcelier P., Pavlov N. N., Penny (Miss) C. J. A., Pensado J., Postoiev A., Proverbio E., Randić L., Sadler D. H., Sandig H. U., Schuler W., Sinzi A. M., Stoyko (Mme) A., Stoyko N., Sceglov V. P., Sirjaev A. V., Šternberk B., Takagi S., Thomson M. M., Torao M., Tsao M., Verbaandert J., Yumi S., Zagar F.

33. Commission de la Structure et de la Dynamique du Système Galactique. (The Structure and Dynamics of the Galactic System)

PRÉSIDENT: L. Perek.

VICE-PRÉSIDENT: F. J. Kerr.

COMITÉ D'ORGANISATION: T. A. Agekjan, J. Einasto, T. Elvius, K. C. Freeman, S. W. McCuskey, P. G. Mezger, K. Ogorodnikov, B. E. Westerlund.

MEMBRES: Aarseth S., Ambartsumian V. A., Andrle P., Aoki S., Baldwin J. E., Barbanis B., Becker W., Beer A., Bigay J. H., Blaauw A., Blanco V. M., Bok B. J., Boulon J., Burke B. F., Burton W. B., Clube S. V. M., Contopoulos G., Courtès G., Crawford D. L., Crézé M., Cuperman S., Delhaye J., Dickel (Mrs) H. R., Dickel J. R., Dieter (Mrs) N. H., Drilling J. S., Dzigvashvili R. M., Edmondson F. K., Elsässer H., Elvius (Mrs) A. M., Fehrenbach Ch., Fenkart R. P., FitzGerald M. P., Fricke W., Fujimoto M., Genkin I. L., Georgelin Y., Goldreich P., Gyldenkerne K., Haffner H., Haug U., Hayli A., Heckmann O., Hénon M., Hohl F., Hori G., Hunter C., Idlis G. M., Innanen K. A., Irwin J. B., Iwaniszewska-Lubienska (Mrs) C., Iwanowska (Miss) W., Jaschek C. O. R., Johnson H. M., Kaburaki M., Kalnajs A. J., Kato S., Kharadze E. K., Kholopov P. N., King I. R., Kinman T. D., Klare G., Kukarkin B. V., Kulsrud R., Kurth R., Kuz'min G. G., Larson R. B., Lecar M., Lin C. C., Lindblad P. O., Lodén (Mrs) K., Lodén L. O., Lunel (Mrs) M., Luyten W. J., Lynden-Bell D., Lynga G., McCarthy S. J., MacConnell D. J., MacRae D. A., Maksumov M. N., Malmquist K. G., Mark J. W. K., Marochnik L., Martinet L., Mavridis L. N., Mayor M., Menessier (Mme) M. O., Miller R. H., Mirzoyan L. V., Miyamoto M., Mohr

J. M., Monnett G., Morgan W. W., Münch G., Murray C. A., Nahon F., Neckel Th., Niimi H., Oja T., Ollongren A., Oort J. H., Ostriker J. P., Ovenden M. W., Palmer P. E., Pavlovskaya (Mrs) E. D., Perry C. L., Pesch P., Philip A. G. D., Pilowski K., Plaut L., Priester W., Ramberg J. M., Riegel K. W., Roberts W. W., Roman (Miss) N. G., Rubin (Mrs) V. C., Rudnicki K., Rybicki G. B., Sanduleak N., Schilt J., Schmidt H., Schmidt K. H., Schmidt M., Schmidt-Kaler Th., Seggewiss W., Shane W. W., Sharov A. S., Sher D., Shimuzu T., Shu F., Simonson S. C., Slettebak A., Spiegel E. A., Steinlin U. W., Stephenson C. B., Stibbs D. W. N., Sturch C., Svolopoulos S. N., Szebehely V., Tammann G. A., The P. S., Toomre A., Upgren A. R., Vanderlinden H. L., Vandervoort P. O., van Hoof A., van Woerden H., Varsavsky C. M., Velghe A. G., Verschuur G. L., Vetešník M., Wayman P. A., Weaver H. F., Westerhout G., White R. E., Wielen R., Woltjer L., Woolley R., Yuan C.

34. Commission de la Matière Insterstellaire et des Nébuleuses Planétaires. (Interstellar Matter and Planetary Nebulae)

PRÉSIDENT: H. van Woerden.

VICE-PRÉSIDENT: G. B. Field.

COMITÉ D'ORGANISATION: J. E. Baldwin, F. D. Kahn, G. Khromov, B. T. Lynds, T. K. Menon, D. C. Morton, M. Peimbert, B. J. Robinson.

MEMBRES: Akabane K., Aller L. H., Altenhoff W., Andrillat H., Andrillat Y., Arhipova V. P., Arny T. T., Axford W. I., Barrett A. H., Behr A., Bergeron J., Berkhuijsen E. M., Biermann L., Bless R. C., Boggess A., Bok B. J., Borgman J., Braes L. L. E., Brück M. T., Burgess A., Burke B. F., Burtou W. B., Buscombe W., Bystrova N. V., Capriotti E. R., Carruthers R. G., Cesarsky C., Cesarsky D., Chopinet M., Churchwell W., Chvojkova E., Clark B.G., Code A. D., Coffin-Souffrin S., Courtès G., Cox D. P., Coyne G. V., Crézé M., Cruvellier P., Cudaback D., Cugnon P., Czyzak S. J., Dalgarno A., Davies R. D., de Jong T., Dibaj E. A., Dickel H. R., Dieter N. H., Divan L., Donn B. D., Dorschner J., Downes D., Dunham T., Dupree A., Dyson J. E., Elvius A. M., Encrenaz P., Faulkner D., Fejes I., Felten J. E., Flower D., Friedemann C., Gardner F. F., Gaustad J. E., Gehrels T., Georgelin Y., Goldberg L., Goldsmith D. W., Goldstein S. J., Goldsworthy F. A., Goldwire H. C., Gosachinsky I. V., Goss W. M., Greenberg J. M., Grewing M., Grubissich C., Grzedzielski S., Guélin M., Gurzadjan G. A., Habing H. J., Hall J. S., Hardebeck E. G., Havtasi J. S., Heiles C. E., Helfer H. L., Henize K. G., Herbig G. H., Herzberg G., Hidajat B., Higgs L. A., Hiltner W. A., Hjellming R. M., Hobbs L. M. Höglund B., Hulsbosch A. N. M., Hummer D. G., Hutchings J. B., Jenkins E. B., Johnson H. M., Joly F., Kaftan-Kassim M. A., Kaler J. B., Kamijo F., Kaplan S. A., Kazes I., Kerr F. J., Kharadze E. K., Kipper A. J., Ko H. C., Kogure T., Kohoutek L., Kostyakova E. B., Krishna Swamy K. S., Kundu M. R., Lambrecht H., Lasker B. M., Lequeux J., Liller W., Lin C. C., Lodén L. O., Lortet-Zuckermann M. C., Louise R., Low F. J., McCray R. A., McCrea W. H., McGee R. X., McNally D., Martel-Chossat M. T., Mathews W. G., Mathewson D. S., Mathis J. S., Meaburn J., Mezger P. G., Mills B. Y., Miller J. S., Minin I. N., Minkowski R. L., Münch G., Nandy K., Neugebauer G., O'Dell R. C., Okuda H., Oort J. H., Osaki T., Osterbrock D. E., Ozernoj L. M., Palmer P., Parker E. N., Pecker J. C., Penston M. V., Penzias A., Perinotto M., Petrosian V., Pikel'ner S. B., Pottasch S. R., Price R. M., Pronik V. I., Radhakrishnan V., Raimond E., Razmadze N. A., Riegel W., Roberts M. S., Roberts W. W., Rogers A. E. E., Rohlfs K., Rose W. K., Rosino L., Rozkovskij D. A., Rubin R. H., Rubin V. C., Rublev S. V., Safronov V. S., Salpeter E. E., Sancisi R., Sato Y., Savage B. D., Savedoff M. P., Schalen C., Schatzman E., Scheuer P., Schmidt K. H., Schmidt Th., Schmidt-Kaler Th., Schwarz U. J., Seaton M. J., Seddon H., Serkowski K., Shah G. A., Shakeshaft J. R., Shane W. W., Shao C. Y., Shapiro M. M., Sharpless S. L., Shu F., Shuter W. W., Sklovskij I. S., Smith M. G., Sobolev V. V., Soboleva N. S., Sofia S., Solomon P. M., Somerville W. B., Spitzer L.,

Stecher T. P., Strömgren B., Sullivan W. T., Swing J. P., Syrovatskij S. I., Takakubo K., Tamura S., Terzian Y., Thackeray A. D., Thaddeus P., Thompson A. R., Tolbert C. R., Townes C. H., Tovmasyan G. M., Treanor P. J., Turner K. C., Ulrich M. H., Unno W., van de Hulst H. C., van der Laan H., van Horn H. M., Vanysek V., Verschuur G. L., Viswanathan N., von Hoerner S., Vorontsov-Vel'yaminov B. A., Walker G. A. H., Weaver H. F., Webster B. L., Weiler K. W., Wendker H. J., Wentzel D. G., Wesselius P. R., Westerhout G., Weymann R. J., Whiteoak J. B., Wickramasinghe N. C., Williams D., Williams D. A., Williams I. P., Williams R. E., Wilson R. W., Witt A. N., Wolstencroft R. D., Woltjer L., Woolf N. J., Wrixon G. T.

35. Commission de la Constitution des Étoiles. (Stellar Constitution)

PRÉSIDENT: L. Mestel.

VICE-PRÉSIDENT: B. Paczyński.

COMITÉ D'ORGANISATION: A. N. Cox, P. R. Demarque, P. Giannone, C. Hayashi, I. Iben, R. Kippenhahn, A. G. Massevitch, G. Ruben, E. Schatzman, R. Tayler.

MEMBRES: Aizenmann M., Anand S. P. S., Arnett W. D., Appenzeller I., Baglin A., Bhatnagar P. L., Biermann L., Bodenheimer P., Böhm K. H., Bondi H., Boury A., Brownlee R. R., Burbidge G. R., Cameron A. G. W., Carson T. R., Castellani V., Castor J. I., Chandrasekhar S., Chevalier C., Chin H. Y., Christy R. F., Cohen J. M., Cowling T. G., Cox J. P., Deinzer W., Dingens P. S. A., Dluzhnevskaya O. B., Dziembowski W. A., Eggleton P. P., Eminzade T. A., Epstein I., Ezer D., Faulkner D. J., Faulkner J., Ferraro V. C. A., Fowler W. A., Forbes J. E., Frantzman Yu. L., Fricke K. J., Gabriel M., Gaughlan G., Gjellstad C., Hamada I., Hitotuyanagi Z., Hoshi R., Hoyle F., Imshennik V. S., James R. A., King D. S., Kothari D. S., Krook M., Kumar S. S., Kushwaha R., Larson R. B., Latour J., Lauterborn D., Lebovitz N., Ledoux P., McCrea W. H., Marx G., Masani A., Meyer-Hofmeister E., Monaghan J. J., Morris S. C., Moss D. L., Nadyozhin D. K., Nakano T., Nishida M., Ohyama N., Okamoto I., Osaki Y., Popova M. D., Profir'jev V. V., Poveda A., Reeves H., Reiz A., Rood R., Rosseland S., Rouse C. A., Roxburg I., Sakashita S., Sakurai T., Salpeter E. E., Savedoff M., Schramm D. N., Schwarzschild M. P., Sears R. S., Sengbusch K. V., Smeyers P., Smith R. C., Souffrin P., Spiegel E., Stibbs D. W. N., Strittmatter P. A., Strömgren B., Suda K., Sugimoto D., Sweet P. A., Sweigart A. V., Temesvary S., Thomas H. C., Toomre J., Truran J. W., Tuominen I. V., Tuominen J. V., Tutukov A. V., Uchida J., Ulrich R., Uus U., Van der Borght R., Van Horn H. M., Vardya M. S., Vauclair G., Vila S. C., Vilhu O., Weigert A., Zahn J. P., Zhevakin S. A., Ziolkowski J.

36. Commission de la Théorie des Atmosphères Stellaires. (The Theory of Stellar Atmospheres)

PRÉSIDENT: R. Cayrel.

VICE-PRÉSIDENT: D. Mihalas.

COMITÉ D'ORGANISATION: A. Avrett, K. Hunger, J. Hutchings, V. L. Khoklova, B. E. J. Pagel, F. Praderie, R. N. Thomas, M. Vardya.

MEMBRES: Abhyankar K. D., Aller L. H., Arpigny C., Athay R. G., Auer L. H., Auman J. R., Avrett E. H., Baschek B., Biermann L., Bless R. C., Böhm K. H., Böhm-Vitense E., Busbridge I W., Carson T. R., Cayrel-de-Strobel C., Conti P. S., Cuny (Miss) Y., de Feiter L. D., de Jager C., Delache Ph., Dumont (Miss) S., Edmonds F. N., Elste G. H. E., Faraggiana (Mrs) R., Finn G. D., Fischel D., Frisch U., Gebbie K. B., Gingerich O., Gökdogan N., Gordon Ch., Grant I. P., Greenstein J. L., Grevesse N., Groth H. G., Gussmann E. A., Hack (Mrs) M. H., Hardrop J., Hearn A. G., Hitotuyanagi Z., Hotinli M., House L. L., Houtgast J., Houziaux L., Huang S. S., Hummer D. G., Ivanov V. V., Jefferies J. T., Johnson H. R., Kalkofen W., Kandel R. S., Kopylov I. M., Krishna Swamy K. S., Kumar S. S., Kushwaha R. S., Matsushima S., Menzel D. H., Mirzoyan L. V., Miyamoto S., Mugglestone D., Müller (Miss) E. A., Münch G., Mustel E. R., Mutschlecner

J. P., Myerscough V. P., Naiai K., Neff J. S., Neven L., Orrall F. Q., Pecker J.-C., Phillips, J. G., Pottasch S. R., Rucinski S., Rudkjobing M., Rubicki G. B., Saito S., Sapar A., Schatzman E. L., Schmalberger D. C., Seaton M. J., Skumanish A., Sitnik G. F., Sobolev V. V., Sobouti Y., Spiegel E. A., Spite M., Stibbs D. W. N., Strom S. E., Strömgren B., Swihart T. L., Traving G., Ueno S., Uesugi A., Underhill (Miss) A. B., Unno W., Unsöld A., van Regemorter H., van 't Veer (Mrs) C., van 't Veer F., Weidemann V., Wellmann P., Wilson P. R., Wright K. O., Wyller A. A., Zirker J. B., Zwaan C.

37. Commission des Amas Stellaires et des Associations. (Star Clusters and Associations)

PRÉSIDENT: I. R. King.

VICE-PRÉSIDENT: S. van den Bergh.

COMITÉ D'ORGANISATION: Mrs K. A. Barkhatova, S. C. B. Gascoigne, G. Larsson-Leander, C. A. Murray, J. Ruprecht.

MEMBRES: Aarseth S., Agekyan T. A., Alksnis A. K., Artiukhina (Mrs) N. M., Balázs B., Becker W., Bijaoui A., Blaauw A., Bouvier P. B., Breger M., Burnichon M.-L., Cannon R. D., Cuffey J., Eggen O. J., Einasto J., Feast M. W., Feinstein A., Golay M., Haffner H., Hénon M., Hesser J. E., Johnson H. L., Kadla (Mrs) Z., Kholopov P. N., Kron G. E., Laval (Miss) A., Lavdovskij V. V., Lindoff U., Lodén L. O., Lohmann W., Lynden-Bell D., Lyngå G., Menon T. K., Meurers J., Morgan W. W., Pişmiş de Recillas (Mrs) P., Poveda A., Reddish V. C., Rosino L., Rountree Lesh (Mrs) J., Sawyer Hogg (Mrs) H. B., Serkowski K., Sharov A. S., Simoda M., Swope (Miss) H. H., Terzan A., Thé Pik Sin, Upgren A., van Altena W. F., Vanderlinden H. L., von Hoerner S., Walker G. A. H., Walker M. F., Wallenquist Å. A. E., Weaver H. F., White R., Wielen R., Woolley R.

38. Commission pour l'Échange des Astronomes. (Exchange of Astronomers)

PRÉSIDENT: P. M. Routly.

VICE-PRÉSIDENT: D. A. MacRae.

COMITÉ D'ORGANISATION: J. Delhaye, G. S. Khromov, A. Reiz, J. Sahade, F. G. Smith, F. B. Wood.

MEMBRES: Abetti G., Bappu M. K. V., Bok B. J., Keller G., Kienle H., Kourganoff V., Martynov D. Ya., Mikhajlov A. A., Miyamoto S., Mohr J. M., Page T. L., Righini G., Rosseland S., Stoy R. H., Swings P., Wild J. P., Witkowski J., Wyatt Jr. S. P.

40. Commission de la Radioastronomie. (Radio Astronomy)

PRÉSIDENT: Yu. N. Parijskij.

VICE-PRÉSIDENT: H. van der Laan.

COMITÉ D'ORGANISATION: E. J. Blum, D. S. Heeschen, A. D. Kuzmin, D. J. McLean, T. K. Menon, P. G. Mezger, H. P. Palmer, B. J. Robinson, B. M. Zuckermann.

MEMBRES: Akabane K., Alexander J. K., Allen R. J., Altenhoff W. J., Andrew B. H., Argyle P. E., Baars J. W. M., Baart E. E., Baldwin J. E., Barrett A. H., Berge G. L., Biraud F., Bohme A., Boischot A., Bolton J. G., Booth R. S., Bracewell R. N., Braude S. Ya., Broten N. W., Brouw W. N., Brown R. H., Burbidge G. R., Burke B. F., Caroubalos C., Carr T. D., Ceccarelli M., Christiansen W. N., Clark G. B., Cohen M. H., Conway R. G., Cooper B. F. C., Costain C. H., Coutrez R. A. J., Covington A. E., Croom D. L., Cudaback D. D., Daintree E. G., Davies J. G., Davies R. D., Davis M. M., de Groot T., Denisse J. F., Dennison P. A., Dent W. A., Dickel J. R., Dickel H. R., Dieter N. H., Douglas J. N., Drake F. D., Dravskikh A. F., Dulk G. A., Dyson F. J., Ekers R. D., Elgaroy O., Ellis G. R. A., Elsmore B., Elwert G., Epstein E. E., Erickson W. C., Eriksen G., Eshleman V. R., Feix G., Felten J. E., Field G. B., Findlay J. W., Fleischer R., Fokker A. D., Fomalont E. B., Foster P. R., Friedman H., Galt J., Gardner F. F., Gent H., Ginzburg V. L., Gold T., Goldstein S. J., Goldwire H. C., Gonze R., Gordon M. A., Gorgolewski S., Gosachinsky I. V., Goss W. M., Gower J. F. R., Grewing M., Guelin M.,

Gulkis S., Hachenberg O., Haddock F. T., Hagen J. P., Hanasz J., Hartz T. R., Hazard C., Heidmann J., Heiles C., Hewish A., Hey J. S., Higgs L. A., Hill E. R., Hjellming R. M., Hoang Binh D., Hobbs R. W., Hogbom J. A., Hoglund B., Hogg D. E., Hooghoudt B. G., Howard W. E., Hughes V. A., Ihsanova V. N., Jauncey D. L., Jennision R. C., Johnston K., Joly F., Joshi M. N., Kaftan-Kassim M., Kahn F. D., Kai K., Kakinuma T., Kardashev N. S., Kaufmann P., Kawabata Kin-Aki, Kazes I., Kellermann K. I., Kenderdine S., Kerr F. J., Kislyakov A. G., Ko Hsien-Ching, Komesaroff M., Kotelnicov V. A., Kraus J. D., Krishnan T., van der Kruit P., Kundu M. R., Laffineur M., Lang K. R., Lantos P., Large M. I., Leblanc Y., Lequeux J., Le Squeren A. M., Lilley A. E., Lindblad P. O., Little A. G., Little L. Th., Locke J. L., Longair M. S., Lovell Sir Bernard, Lyne A. G., McGee R. X., McKenna-Lawlor S., McVittie G. C., McDonald G. H., MacRae D. A., Maran S. P., Matveyenko L. I., Maxwell A., Mayer C. H., Meeks M. L., Miley G. K., Mills B. Y., Minkowski R. L., Moffet A. T., Moriyama F., Morimoto M., Muller C. A., Nicolson G. D., Oort J. H., Osterbrock D. E., Pacholozyk A. G., Palmer P., Parker E. A., Pasachoff J. M., Pauliny-Toth I. I. K., Penzias A. A., Pettengill G. H., Pick-Gutmann M., Ponsonby J. E. B., Pooley G. G., Price R. M., Priester W., Rabben H. H., Radhakrishnan V., Raimond E., Ramaty R., Gopala Rao U. V., Razin V. A., Reber G., Ribes J. C., Riegel K. W., Righini G., Roberts J. A., Roberts M. S., Roeder R. C., Roger R. S., Rogers A. E. E., Rogstad D. H., Rohlfs K., Rydbeck O. E. H., Ryle (Sir) Martin, Ryzhkov N.F., Salomonovitch A. E., Salpeter E. E., Sanamjan V. A., Sato F., Scalise E. Jr., Scheuer P. A. G., Schmidt M., Schwarz U. J., Scott P. F., Seaquist E. R., Seielstad G. E., Shakeshaft J. R., Sheridan K. V., Shimmins A. J., Shklovskij J. S., Sholomitsky G. B., Shuter W. L. A., Simon P., Simonson S. C., Slee O. B., Slysh V. I., Smerd S. F., Smith A. G., Smith F. G., Smith H. J., Soboleva N. S., Sorochenko R. L., Stanley G. J., Steinberg J. L., Stewart P., Stone R. G., Swarup G., Swenson G. E. Jr., Tabara H., Takakura T., Tanaka H., Tanaka R., Terzian Y., Thompson A. R., Tlamicha A., Tolbert C. R., Tovmassian G. M., Townes C. H., Troickij V. S., Tsuchiya A., Tsuruta S., Turlo Z., Turner B. E., Turner K. C., Turtle A. J., Ulrich B. T., Demoulin Ulrich M. H., van de Hulst H. C., van Woerden H., Varsavsky C. M., Veron P., Verschuur G. L., Wade C. M., Walsh D., Warwick J. W., Weaver H. F., Weliachew L., Wendker H. J., Westerhout G., Westfold K. C., Whiteoak J. B., Wickramasinghe N. C., Wielebinski R., Wild J. P., Williams D., Wills D., Wilson R. W., Winnberg A., Woltjer L., Zeleznjakov V. V.

41. Commission de l'Histoire de l'Astronomie. (History of Astronomy)

PRÉSIDENT: O. Gingerich.

VICE-PRÉSIDENT: J. Dobrzycki.

COMITÉ D'ORGANISATION: W. Hartner, M. A. Hoskin, Z. Horsky, O. Pedersen (representative appointed by the International Union for History and Philosophy of Science), P. V. Slavenas.

MEMBRES: Abetti G., Ansari S. M. R., Argyrakos J. G., Ashbrook J., Baehr U., Beer A., Berendzen R., Brunet J. P., Chenakal V. L., Cimino M., Collinder P., Dadic Z., Deeming T. J., Dewhirst D., Douglas (Miss) A. V., Eddy J., Erpylev N. P., Evans D. S., Ferrari d'Occhieppo K., Filliozat J., Fleckenstein J. O., Forbes J. E., Freiesleben M. C., Freitas-Mourao R. R. De, Hawkins G. S., Hayli A. F., Hirose H., Kennedy J. E., Kiang T., King H. C., Kotsakis D., Kulikovsky P. G., Levy J. R., Link F., Mckenna-Lawlor (Mrs) S. M. P., Meadows A. J., Michel H., Michkovitch V. V., Nakayama S., Norlind W., Norlund N. E., Omer G. C., Pelseneer J., Petri W., Pogo A., Ronan C., Rosen E., Rybka E., Samaha A. H. M., Scheglov V. P., Shukla K. S., Sticker B., Wattenberg D., Whitaker E. A., Whitrow G. J., Wright (Miss) H., Yabuuti K., Zagar F.

42. Commission des Étoiles Doubles Serrées. (Close Binary Stars)

PRÉSIDENT: T. Herczeg.

VICE-PRÉSIDENT: G. Larsson-Leander.

COMITÉ D'ORGANISATION: A. H. Batten, M. G. Fracastoro, K. Gyldenkerne, M. Kitamura, M. Plavec, D. M. Popper, J. Smak, E. P. J. van den Heuvel, B. Warner, F. B. Wood.

MEMBRES: Abhyankar K. D., Bappu M. K. V., Beer A., Binnendijk L., Blitzstein W., Broglia P., Catalano S., Cester B., Chen K.-Y., Cherepashchuk A. M., Chis G., Cillié G. G., Cousins A. W. J., Cowley (Mrs) A. P., Cristaldi S., Dadaev A. N., de Kort J., Doughty N. A., Ebbighausen E. G., Ferrari d'Occhieppo K., Fracassini M, Fredrick L. W., Frantsman Yu. L., Fresa A., Gaposchkin S., Giannone P., Grygar J., Hall D. S., Heard J. F., Hill G., Hjellming R. M., Horák T., Huang S. S., Huffer C. M., Hutchings J., Irwin J. B., Jurkevich I., Jaschek C., Kawabata S., Kizilirmak A., Koch R. H., Kondo Y., Knipe G. F. G., Kopal Z., Kraft R. P., Krat V. A., Kristenson H., Kříž S., Kruszewski A., Kwee K. K., Landolt A. U., Lavrov M. I., Linnell A. P., Magalashvili N. L., Mammano A., Martynov D. Ya., Mauder H., Mayer P., McCluskey G. E., Merrill J. E., Mumford G. S., Nelson B., O'Connell D. J. K., Ovenden M. V., Paczynski B., Piotrowski S., Popovici C., Purgathofer A., Rakos K. D., Refsdal S., Reuning E. G., Rucinski S., Sahade J., Schmidt H., Serkowski K., Shapley (Mrs) M. B., Shulberg A. R., Sinvhal S. D., Sobieski S., Strohmeier W., Szafraniec (Miss) R., Todoran I., Tremko J., Tsessevich V. P., Ureche V., Van Hoof A., Vetesnik M., Walter K., Wehlau W. H., Weigert A., Wellmann P., Wesselink A. J., Whelan J. A. J., Widorn T., Wilson R. E., Wood D. B., Wright K. O.

44. Commission des Observations Astronomiques au-dehors de L'Atmosphère Terrestre. (Astronomical Observations from outside the Terrestrial Atmosphere)

PRÉSIDENT: A. D. Code.

VICE-PRÉSIDENT: R. M. Bonnet.

COMITÉ D'ORGANISATION: J. E. Blamont, C. de Jager, A. H. Gabriel, R. Giacconi, P. Léna, K. A. Pounds, V. K. Prokof'ev, R. Tousey, R. Wilson.

MEMBRES: Acton L. W., Alexander J. K., Bless R. C., Boksenberg A., Boggess A. III, Boneff N., Bogd R. L. F., Brandt J. C., Bruekner G. E., Bumba V., Burton W. M., Butler H. E., Corver J. H., Chamberlain J. W., Chubb T. A., Clark T. A., Courtès G., Culhane J. L., Danielson R. E., Davies L., Davis R. J., Dollfus A., Fazio G. G., Fichtel C. E., Fisher D. C., Fredga K., Friedman H., Gehrels T., Glaser H., Goed T., Goldberg L., Gratton L., Greisen K. J., Gursky H., Haddock F., Hollam K. L., Hartz T. R., Hajakawa S., Hearn A. C., Helmker H., Henize K. G., Henoux J. C., Herzberg G., Hinteregger H. E., Houziaux L., Jechlee C., Jenkins E. B., Jordan C., Kondo J., Kraushaar J. E., Kupperian J. E., Kurt V. G., Lindblad B. A., Lovell B., Lüst R., McCracken K. G., McWhirter R. W. P., Manara A., Mandel'shtam S. L., Massevich A. G., Matsvoka M., Mehltretter J. P., Migeotte M. V., Mikhajlov A. A., Milford S. N., Monfils A., Morton D. C., Müller (Miss) E. A., Ness N. F., Neupert W. M., Newkirk G. A., Nicolet M., Nikolsky G. M., Oda M., Papagiannis M., Parkinson W. H., Parkinson J. H., Rao U. R., Reeves E. M., Rense W. A., Righini G., Rogerson J. B., Roman N. G., Sato H., Savage B., Severnyi A. B., Smith Henry J., Sofia S., Speer R., Stecher T. P., Steinberg J. L., Stone R. G., Takakura T., Underhill A. B., Valniček B., van de Hulst H. C., van Duinen R. J., Walsh D., Weinberg J. L., Yamashita K., Zhongolovich I. D., Zwicky F.

45. Commission des Classifications Spectrales et Indices de Couleur à plusieurs Bandes. (Spectral Classifications and Multi-band Colour Indices)

PRÉSIDENT: C. O. R. Jaschek.

VICE-PRÉSIDENT: E. K. Kharadze.

COMITÉ D'ORGANISATION: Mrs A. P. Cowley, B. Hauck, Miss W. Iwanowska, P. C. Keenan, E. Mendoza V., U. Sinnerstad, V. L. Straižys, B. E. Westerlund.

MEMBRES: Ardeberg A. L., Bahng J. D. R., Bappu M. K. V., Barbier-Brossat (Mrs) M., Bartaya (Mrs) R. A., Beer A., Bernacca G., Bidelman W. P., Blanco V. M., Boyce P. B., Buscombe W., Cester B., Cherepashchuk A. M., Crawford D. L., Dessy J. L., Divan

(Miss) L., Duflot (Mrs) M., Elvius T., Evans D. S., Feast M. W., Fehrenbach Ch., Fracassini M., Garrison R. F., Geyer E. H., Glagolevskij Ju. V., Golay M., Gyldenkerne K., Hack (Mrs) M. H., Hallam K. L., Henize K. G., Herman R., Hiltner W. A., Hoag A. A., Houk N., Jaschek (Mrs) M., Johnson H. L., Kron G. E., Lodén (Mrs) K., Low F. J., Lynga G., McCarthy M. F., McClure R. H., MacConnell D. J., McCuskey S. W., McNamara D. H., Mavridis L. N., Morgan W. W., Nandy K., Nikonov V. B., Notni P., Oja T., Osawa K., Perry C. L., Philip A. G. D., Preston G. W. III., Reddish V. C., Roman (Miss) N. G., Rountree-Lesh (Mrs) J., Rudkjobing M., Sanduleak N., Sanwal N. B., Schild R. E., Schmidt-Kaler Th., Seitter (Miss) W. C., Sharpless S. L., Sinvhal S. D., Slettebak A., Steinlin U. W., Stephenson C. B., Strömgren B., Treanor P. J., Upgren A. R., van den Bergh S., Walborn N. R., Walker G. A. H., Warner B., West R. M., Williams J. A., Wilson O. C., Wing R., Yoss K. M.

46. Commission pour l'Enseignement de l'Astronomie. (The Teaching of Astronomy)

PRÉSIDENT: D. McNally.

VICE-PRÉSIDENTS: E. V. Kononovich, D. G. Wentzel.

COMITÉ D'ORGANISATION: L. Houziaux, H. E. Jørgenson, J. Kleczek, E. A. Müller, M. Rigutti, T. L. Swihart.

MEMBRES: Abell G. D., Abhyankar K. D., Andrillat H. L., Asaad A. S., Atanasijevic I., Buscombe W., Climenhaga J. L., Dinulescu N., Dominko F., Doughty N. A., Elgarøy Ø., Ferraz-Mello S., Gouguenheim L., Grushinskij N. P., Hack M., Hauck B., Haupt H. F., Hazer S., Hidajat B., Ivanov V. V., Iwaniszewska C., Jarret A. H., Kourganoff V., Lambrecht H., Marik M., Marsh J. C. D., Mavridis L., Meadows A. J., Moreno H., Nicolov N. S., Osborn W., Osorio J., Ovenden M. W., Owaki N., Riihimaa J., Ringuelet-Kaswalder A., Rodgers A. W., Sandquist A., Scheffler H., Schmidt Th., Schmitter E., Steinitz R., Torroja J. M., Trehan S. K., Zwaan C.

47. Commission de la Cosmologie. (Cosmology)

PRÉSIDENT: M. S. Longair.

VICE-PRÉSIDENT: I. D. Novikov.

COMITÉ D'ORGANISATION: G. de Vaucouleurs, L. Gratton, W. H. McCrea, G. C. McVittie, C. W. Misner, H. Nariai, Y. Ne'eman, D. W. Sciama, Ya. B. Zel'dovich.

MEMBRES: Abell G. O., Aizu K., Alfvén H., Ambartsumian V. A., Andrillat H., Baldwin J. E., Bardeen J. M., Bertola F., Bondi H., Bonnor W. B., Burbridge G. R., Cohen J. M., Davidson W., Dautcourt G., Dicke R. H., Ellis G. F. R., Felten J. E., Field G. B., Fujimoto M., Godart O., Gold T., Goldsmith D. W., Harrison E. R., Hawking S. W., Hayakawa S., Hayashi C., Heckmann O., Heidmann J., Hoyle F., Icke V., Jauncey D. L., Kasper U., Kato S., Kawabata K., Kellermann K. I., Kovetz A., Kozlovsky B. Z., Kustaanheimo P. E., Layser D., Narlikar J. V., Neyman J., Nishida M., Noerdlinger P. D., O'Connell R. F., Oort J. H., Ozernoi L. M., Pachner J., Peebles P. J. E., Persides S., Petrosian V., Rees M. J., Rindler W., Robinson I., Roeder R. C., Ryle M., Sapar A. A., Sato H., Schatzman E., Scheuer P. A. G., Schmidt M., Schücking E., Scott E. L., Seielstad G. A., Sersic J. L., Setti G., Shariv G., Simon R. L. E., Tauber G. E., Tayler R. J., Thorne K. S., Tifft W. G., Treder H. J., Wheeler J. A., Whitrow G. J., Wilson A. G., Zelmanov A. L.

48. Commission de l'Astrophysique de Grande Énergie. (High Energy Astrophysics)

PRÉSIDENT: M. J. Rees.

VICE-PRÉSIDENT: I. S. Shklovski.

COMITÉ D'ORGANISATION: A. G. W. Cameron, H. Friedman, V. L. Ginzburg, T. Gold, K. I. Greisen, P. Morrison, F. Pacini, D. W. Sciama, M. M. Shapiro, F. G. Smith.

MEMBRES: Adams D. J., Alfvén H., Alvarez L., Arons J., Asseo (Mrs) E., Axford W. I.,

Bergeron (Miss) J., Biswas S., Bonometto S., Boyd R. L. F., Burbidge G. R., Cavaliere A., Chandrasekhar S., Chitre S. N., Chubb T. A., Clark G., Cohen J. M., Collin-Souffrin S., Dautcourt G., Davidson W., Davis L. Jr., De Felice F., de Graaf T., De Witt B. S., De Young D. S., Dicke R. H., Disney M., Drake F. D., Duthie J. G., Edwards P. J., Fazio G., Fichtel C., Felten J. E., Fenton K. B., Field G. B., Fisher P. C., Fowler W. A., Garmire G. P., Giacconi R., Goldsmith D. W., Gratton L., Greyber H., Grewing M., Gunn J. E., Gursky H., Harwit M. O., Hayakawa S., Henriksen R. N., Henry R. C., Hoang Binh Dy, Hoyle F., Ipser J., Israel W., Ito K., Jackson J. C., Jelley J. V., Jokipii J. R., Kafka P., Kahn F. D., Kondo M., Kozlovsky B.-Z., Kreisel E., Kristiansson K. G. B., Kulsrud R. M., Kurt V. G., Longair M. S., Lust R., Lynden-Bell D., McCray R. A., Matsuoka M., Mestel L., Meyer F., Michel F. C., Miyamoto S., Ne'eman Y., O'Connell R. F., Oda M., Okoye S. E., Ostriker J. P., Pacholczyk A. G., Parker E. N., Parkinson J. H., Peterson L. E., Petrosian V., Piddington J. H., Pinkau K., Porter N. A., Pounds K. A., Reeves H., Rossi B., Salpeter E. E., Sartori L., Saslaw W. C., Savedoff M. P., Scargle J. D., Schatten R. H., Schatzman E., Scheuer P. A. G., Seielstad G., Setti G., Shaviv G., Sofia S., Spitzer L., Steigman G., Sturrock P. A., Sunyaev R. A., Syrovatskij S. I., Tayler R. J., Thorne K. S., Truran J., Tsuruta (Miss) S., Vidal N., Wentzel D. G., Westfold K. C., Wheeler J. A., Woltjer L., Zel'dovich Ya. B.

49. Commission du Plasma Interplanétaire et de l'Héliosphère. (The Interplanetary Plasma and the Heliosphere)

Président: W. I. Axford.
Vice-Président: A. Hewish.
Comité d'Organisation: to be appointed.

50. Commission de la Protection des Sites d'Observatoires Existant et Potentiels. (Protection of Existing and Potential Observatory Sites)

Président: M. F. Walker.
Vice-Président: R. Cayrel.
Comité d'Organisation: to be appointed.

LISTE ALPHABETIQUE DES MEMBRES

ALPHABETICAL LIST OF MEMBRES

Les noms figurent à la première lettre du premier mot; par exemple, le nom Van Houten se trouve à la lettre V.

Les noms des nouveaux membres élus par la XVe Assemblée Générale sont annotés d'un astérisque.

Les noms des pays sont donnés sous leur forme anglaise.

Les noms des astronomes utilisant l'alphabet cyrillique sont donnés selon les règles de translitération publiées dans les *Transactions* **XIVA**, 13.

Les Membres de l'Union sont priés de bien vouloir informer le Secrétaire Exécutif de l'UAI, Département d'Astronomie, Université, Thessaloniki, Grèce, de toutes les erreurs qu'ils pourraient relever dans la présente liste.

Names are given under the first letter of the first word: for example, the name Van Houten will be found under V.

The names of new members, elected at the XVth General Assembly, are prefixed by an asterisk.

The names of the countries are given in the English form.

Names of astronomers using the Cyrillic alphabet are transliterated according to the rules published in *Transactions* **XIVA**, 13.

Please inform the Executive Secretary of the IAU, Astronomy Department, University, Thessaloniki, Greece, of any errors you may find in this list.

AARSETH (Dr S.), Institute of Theoretical Astronomy, Madingley Road, Cambridge, U.K.
ABALAKIN (Dr V.K.), Institute of Theoretical Astronomy, USSR Academy of Sciences, Leningrad, U.S.S.R.
ABBOTT (Prof. Dr W.N.), 42 Leophoros A. Michalacopoulou, Ilisia, Athens 612, Greece
ABDALA (Dr J.), Edificio 'Pio XII' primer piso No. 1, San Francisquito a Puente Ayacucho, Caracas, Venezuela
* ABELE (Dr M.K.), Latvian State University, bulv. Rainisa, Riga, U.S.S.R.
ABETTI (Prof. G.), Instituto Naz. Ottica-Arcetri, Largo Enrico Fermi 7, 50125 Firenze, Italy
ABELL (Dr G.O.), Department of Astronomy, University of California, Los Angeles 24, California 90024, U.S.A.
ABHYANKAR (Dr K.D.), Professor of Astronomy, Astronomy Department, Osmania University, Hyderabad 7, A.P., India
* ABLES (Dr H.D.), U.S. Naval Observatory, Flagstaff Station, Flagstaff, Arizona 86001, U.S.A.
ABRAHAM (H.J.), Mount Stromlo Observatory, Canberra, A.C.T., Australia
* ABRAHAM (Dr Z.), Instituto de Fisica, UFRGS, Depto. Astronomía, R. Luiz Englert S/N, 9000 Porto-Alegre, RGS, Brazil
ABRAMI (Prof. A.), Osservatorio Astronomico, via G.B. Tiepolo No. 11, Trieste (409), Italy
ABREU (Dr J.C. de Brito e), Observatorio Astronomico de Lisboa, Tapada, Lisbon, Portugal
ACTON (Dr L.W.), Orgn. 52–14, Bldg. 202, Lockheed Palo Alto Research Laboratory, 3251 Hanover Street, Palo Alto, California 94304, U.S.A.
ABT (Dr H.A.), Kitt Peak National Observatory, Box 26732, Tucson, Arizona 85726, U.S.A.
ADAM (Miss Dr M.G.), University Observatory, South Parks Road, Oxford, U.K.
ADAMOPOULOS (Dr G.), Rue Lamias No. 8, Athens (6), Greece
ADAMS (A.N.), U.S. Naval Observatory, Washington 25, D.C., U.S.A.

* ADAMS (Dr D.J.), Astronomy Department, University of Leicester, University Road, Leicester, U.K.
* ADAMS (Dr Th.), Yerkes Observatory, Williams Bay, Wisconsin 53191, U.S.A.
ADEL (Prof. A.), Atmospheric Research Observatory, Arizona State College, Flagstaff, Arizona, U.S.A.
ADOLFSSON (Dr T.), Lund Observatory, S-222 24 Lund, Sweden
AFANAS'EVA (Dr P.M.), Main Astronomical Observatory, USSR Academy of Sciences, Pulkovo, U.S.S.R.
AGEKJAN (Prof. Dr T.A.), University of Leningrad, Leningrad, U.S.S.R.
* AGRINIER (Dr B.), C.E.N. Saclay – S.E.P., B. P. No. 2, 91 Gif-sur-Yvette, France
AHMED (I.I.), Helwan Observatory, Helwan near Cairo, Arab Rep. of Egypt
AHNERT (Dr P.), Zentralinstitut für Astrophysik, Sternwarte Sonneberg, DDR 64 Sonneberg, G.D.R.
* AIZENMAN (Dr M.L.), Département de Physique, Université de Montréal, C.P. 6128, Montréal 101, Canada
AIZU (Prof. Dr K.), Physics Department, Rikkyo University, Nishi-Ikebukuro 3-chome, Toshima-ku, Tokyo, 171 Japan
AKABANE (Dr K.), Tokyo Astronomical Observatory, Mitaka, Tokyo, Japan
AKSENOV (Dr E.P.), Sternberg Astronomical Institute, Moscow, U.S.S.R.
* AKSNES (Dr K.), Smithsonian Astrophysical Observatory, 60 Garden Street, Cambridge, Massachusetts 02138, U.S.A.
AKYOL (Dr M.Ü.), Department of Astronomy, P.K. 21, Bornova, Izmir, Turkey
ALANIA (Dr I.F.), Abastumani Astrophysical Observatory, Abastumani, Georgian SSR, U.S.S.R.
* ALBANO (Dr J.), Observatorio Astronómico, La Plata, Argentina
* ALBRECHT (Dr R.), University-Observatory, Türkenschanzstr. 17, A-1180 Vienna, Austria
ALEXANDER (J.B.), Royal Greenwich Observatory, Herstmonceux Castle, Hailsham, Sussex, U.K.
ALEXANDER (J.K.), Code 693, Laboratory for Extraterrestrial Physics, NASA-Goddard Space Flight Center, Greenbelt, Maryland 20771, U.S.A.
ALFVÉN (Prof. H.), Royal Institute of Technology, S-100 44 Stockholm 70, Sweden
ALKSNIS (Dr A.K.), Astrophysical Laboratory, Latvian Academy of Sciences, Riga, U.S.S.R.
ALLADIN (S.M.), Department of Astronomy, Osmania University, Hyderabad, India
ALLEN (Prof. C.W.), University of London Observatory, Mill Hill Park, London, NW7 2QS, U.K.
ALLEN (Dr R.J.), Kapteyn Astronomical Laboratory, Postbus 800, Groningen, The Netherlands
* ALLER (Dr H.D.), Department of Astronomy, University of Michigan, Ann Arbor, Michigan 48104, U.S.A.
ALLER (Prof. L.H.), Astronomy Department, University of California, Los Angeles, California 90024, U.S.A.
* ALLER (Dr M.F.), Department of Astronomy, The University of Michigan, Ann Arbor, Michigan 48104, U.S.A.
ALMÁR (Dr I.), Institute of Geodesy (Földmérési Intézet), Budapest 1051, Guszev u 19, Hungary
ALTENHOFF (Dr W.), Max-Planck-Institut für Radioastronomie, Argelanderstr. 3, 5300 Bonn, F.R.G.
* ALTROCK (Dr R.), Sacramento Peak Observatory, Air Force Cambridge Research Labs, Sunspot, New Mexico 88349, U.S.A.
ALTSCHULER (Dr M.), High Altitude Observatory, Boulder, Colorado 80302, U.S.A.
ALVAREZ (Dr L.), Lawrence Radiation Laboratory, University of California, Berkeley, California 94720, U.S.A.
ALY (Prof. M.K.), Director, Rabitatul Aalum Al-Islami, Islamic Observatory, Mecca Almukarrama, Saudi Arabia
AMBARTSUMIAN (Prof. Dr V.A.), President of the Academy of Sciences of Armenian SSR, Erevan, Armenia, U.S.S.R.
ANAND (Prof. S.P.S.), Department of Astronomy, University of Toronto, Toronto 5, Ontario, Canada

ANDERS (Prof. E.), Enrico Fermi Institute, University of Chicago, 5640 S. Ellis Avenue, Chicago, Illinois 60637, U.S.A.

ANDERSON (Dr B.), Nuffield Radio Astronomy Laboratories, Jodrell Bank, Macclesfield, Cheshire, U.K.

* ANDERSON (Dr Chr.M.), University of Wisconsin-Madison, Madison, Wisconsin 53706, U.S.A.

ANDERSON (Dr K.), Space Science Laboratory, University of California, Berkeley, California 94728, U.S.A.

ANDREW (Dr B.H.), Astrophysics Branch, National Research Council, Ottawa 7, Canada

ANDREW (Prof. K.L.), Purdue University, Department of Physics, Lafayette, Indiana 47907, U.S.A.

ANDREWS (A.D.), Armagh Observatory, Armagh, U.K.

ANDREWS (D.H.), Dominion Astrophysical Observatory, R.R.7, Victoria, B.C., Canada

ANDREWS (Dr P.J.), Radcliffe Observatory, P.O. Box 373, Pretoria, South Africa

* ANDRIENKO (Dr D.A.), Astronomical Department, Kiev 252053, Observatornaja 3, U.S.S.R.

* ANDRIESSE (Dr Ir C.D.), Kapteyn Institute, P.O. Box 800, Groningen, The Netherlands

ANDRILLAT (Prof. Dr H.), Laboratoire d'Astronomie de la Faculté des Sciences, de Montpellier, Hérault, France

ANDRILLAT (Mrs Dr Y.), Laboratoire d'Astronomie de la Faculté des Sciences, de Montpellier, Hérault, France

ANDRLE (Dr P.), Astronomical Institute, Czechoslovak Academy of Sciences, Budečská 6, Prague 2, Czechoslovakia

* ANGEL (Dr R.), Columbia University, New York, N.Y., U.S.A.

* ANGIONE (Dr R.), California State University, San Diego, California 92115, U.S.A.

ANGUITA (Dr C.A.), Universidad de Chile, Observatorio Astronómico Nacional, Casilla 36-D, Santiago, Chile

* ANSARI (Dr S.M.R.), Department of Physics, Aligarh Muslim University, Aligarh, U.P., India

ANTONACOPOULOS (Dr G.), Laboratory of Astronomy, University of Athens, 57, Solonos Str., Athens, Greece

* ANZER (Dr U.), Max-Planck-Institut für Physik und Astrophysik, 8000 München, Föhringer Ring 6, F.R.G.

AOKI (Prof. Dr S.), Tokyo Astronomical Observatory, University of Tokyo, Mitaka, Tokyo, Japan

AOUDOUZE (Dr J.), Institut d'Astrophysique, 98bis, boulevard Arago, 75014 Paris, France

APPENZELLER (Dr I.), Universitätssternwarte, Geismarlandstr. 11, 3400 Göttingen, F.R.G.

ARAKELIAN (Dr M.A.), Byurakan Astrophysical Observatory, Byurakan, Armenia, U.S.S.R.

ARDEBERG (A.L.), Institute of Astronomy, S-222 24 Lund, Sweden

AREND (Prof. Dr S.J.V.), Astronome Honoraire, Observatoire Royal de Belgique, avenue Circulaire, 3, B-1180 Bruxelles, Belgium

ARGUE (A.N.), The Observatories, Madingley Road, Cambridge, U.K.

ARGYLE (Dr P.E.), Dominion Radio Astrophysical Observatory, Box 248, Penticton, B.C., Canada

ARGYRACOS (Prof. Dr J.G.), 193 Patission Street, Athens (816), Greece

ARIAS-DE GREIFF (Prof. J.), Observatorio Astronómico Nacional, Apartado Nacional 2584, Bogóta 1, D.E., Colombia

ARKHIPOVA (Dr V.P.), Sternberg Astronomical Institute, Moscow, U.S.S.R.

ARNETT (Dr W.D.), Department of Space Science, Rice University, Houston, Texas 77001, U.S.A.

ARNOULD (M.L.L.), chaussée de Tubize, 146, B-1430 Wauthier-Braine, Belgium

ARNQUIST (Dr W.N.), 8127 Delgany Avenue, Playa Del Rey , California 90291, U.S.A.

ARNY (Dr T.T.), Department of Physics and Astronomy, University of Massachusetts, Amherst, Mass. 01002, U.S.A.

* ARONS (Dr J.), Berkeley Astronomy Department, University of California, Berkeley, California 94720, U.S.A.

ARP (Dr H.C.), Mount Wilson and Palomar Observatories, 813 Santa Barbara Street, Pasadena, California 91106, U.S.A.

ARROYO (Dr M.L.), Observatorio Astronómico, Alfonso XII, 3, Madrid 7, Spain

ARPIGNY (Prof. Dr C.), Institut d'Astrophysique, Université de Liège, avenue de Cointe, 5, B-4200 Cointe-Ougrée, Belgium

ARTHUR (D.W.G.), Lunar and Planetary Laboratory, University of Arizona Tucson, Arizona, U.S.A.

ARTIUKHINA (Mrs Dr N.M.), Sternberg Astronomical Institute, Moscow, U.S.S.R.

ASAAD (Dr A.S.), Helwan Observatory, Helwan near Cairo, Arab Rep. of Egypt

ASHBROOK (Dr J.), 16 Summer Street, Weston 93, Massachusetts, U.S.A.

ASLANOV (Dr I.A.), Shemaha Astrophysical Observatory, Azerbajdzhan Academy of Sciences, Shemaha, U.S.S.R.

* ASSEO (Dr E.), Observatoire de Paris, Section d'Astrophysique de Meudon, 92190 Meudon, France

ASTAPOVICH (Dr I.S.), 15/17a Nikolsko-Botanicheskaya, log. 45, Kiev-33, U.S.S.R.

ATANASIJEVIC (Dr I.), Sterrenkundig Instituut, Toernooiveld, 6805 Nijmegen, The Netherlands

ATHAY (Dr R.G.), High Altitude Observatory of the University of Colorado, Boulder, P.O. Box 1470, Colorado 80302, U.S.A.

ATKINSON (R.d'E.), Astronomy Department, Indiana University, Swain Hall West, Bloomington, Indiana 47401, U.S.A.

AUER (Dr L.H.), Department of Astronomy, Watson Astronomy Center, Box 2023 Yale Station, Yale University, New Haven, Connecticut 06520, U.S.A.

AUMAN (Dr J.R.), Department of Geophysics, University of British Columbia, Vancouver 8, B.C., Canada

* AULUCK (Prof. F.C.), Department of Physics & Astrophysics, University of Delhi, Delhi-110007, India

* AVERY (Dr L.W.), National Research Council of Canada, Astrophysics Branch, Ottawa, Ontario, Canada

AVIGNON (Miss Dr Y.), Observatoire de Paris, Section d'Astrophysique, 92190 Meudon, France

AVRETT (Dr E.H.), Smithsonian Astrophysical Observatory, 60 Garden Street, Cambridge, Massachusetts 02138, U.S.A.

AXFORD (Dr W.I.), Director, Max-Planck Institute for Aeronomy, 3411 Lindau/Harz, F.R.G.

AYDIN (Dr C.), Faculty of Science, Department of Astronomy, Ankara, Turkey

BAARS (Dr J.W.M.), Netherlands Foundation for Radio Astronomy, Radiosterrewacht, Dwingeloo 7514, The Netherlands

BAART (Dr E.E.), Rhodes University, P.O. Box 94, Grahamstown, South Africa

BABADZHANOV (Dr P.B.), Director, Astrophysical Institute, Dushanbe, U.S.S.R.

BABCOCK (Dr H.W.), Mount Wilson and Palomar Observatories, 813 Santa Barbara Street, Pasadena, California 91106, U.S.A.

BACCHUS (Prof. P.), 1, impasse de l'Observatoire, 59-Lille (Nord), France

BAEHR (Dr U.), Astronomisches Rechen-Institut, Quickestr. 44a, 69 Heidelberg, F.R.G.

BAERENTZEN (Dr J.), Ole Roemer Observatory, DK-8000 Aarhus C, Denmark

BAGILDINSKIJ (Dr B.K.), Main Astronomical Observatory, USSR Academy of Sciences, Pulkovo, U.S.S.R.

BAGLIN (Miss Dr A.), Institut d'Astrophysique, 98bis, boulevard Arago, 75014 Paris, France

BAHCALL (Dr J.N.), Institute for Advanced Study, Princeton, N.J. 08540, U.S.A.

BAHNER (Dr K.), Landessternwarte, Königstuhl-Heidelberg, F.R.G.

BAHNG (Prof. J.D.R.), Dearborn Observatory, Northwestern University, Evanston, Illinois 60201, U.S.A.

BAIRD (Dr K. M.), Division of Applied Physics, National Research Council, Sussex Drive, Ottawa, Ontario, Canada

BAIZE (Dr P.), 6, rue Daubigny, 75-Paris 17e, France
* BAJAJA (Dr E.), Instituto Argentino de Radioastronomía, Casilla No. 5, Villa Elisa, Prov. Bs.As., Argentina
BAKER (Dr J.G.), 60 Garden Street, Cambridge, Massachusetts 03138, U.S.A.
BAKER (Dr N.), Astronomy Department, Columbia University, New York, N.Y. 10027, U.S.A.
BAKOS (Prof. G.A.), Department of Physics, University of Waterloo, Waterloo, Ontario, Canada
BAKULIN (Dr P.I.), Sternberg Astronomical Institute, Moscow, U.S.S.R.
BALÁZS (Dr B.), Konkoly Observatory, P.O. 114, Box 67, Budapest XII, Hungary
BALÁZS-DETRE (Mrs Dr J.), Konkoly Observatory, Szabadsághegy, P.O. 114, Box 67, Budapest XII, Hungary
BALDWIN (Dr J.E.), Cavendish Laboratory, Free School Lane, Cambridge, U.K.
BALDWIN (Dr R.), Oliver Machinery Co, 445 Sixth Street, NW Grand Rapids, Michigan 49502, U.S.A.
BALKLAVS (Dr A.E.), Astrophysical Laboratory, Latvian Academy of Sciences, Riga, U.S.S.R.
* BALL (Dr J.A.), Harvard College Observatory, Cambridge, Massachusetts 02138, U.S.A.
BALLARIO (Prof. M.C.), Osservatorio Astrofisico di Arcetri, Via S. Leonardo 75, Firenze, Italy
BALLI (Prof. Dr E.), University Observatory, Beyazit, Istanbul, Turkey
BANIN (Dr V.G.), Post Office Meget, Sibizmiran, Irkutsk, U.S.S.R.
* BANOS (Dr C.), Astronomical Institute, National Observatory of Athens, Athens (306), Greece
BANOS (Dr G.), Astronomical Institute, National Observatory of Athens, Athens (306), Greece
BAPPU (Dr M.K.V.), Director, Indian Institute of Astrophysics, Kodaikanal 3, India
BARANNE (A.), Observatoire de Marseille, 2 place Le Verrier, 13004 Marseille, France
BARBANIS (Dr B.), University of Patras, Department of Astronomy, Patras, Greece
* BARBARO (Dr G.), Osservatorio Astronomico, Vicolo dell 'Osservatorio, Padova, Italy
BARBER (D.R.), 'Craigwell', Coulsdon Road, Sidmouth, Devon, U.K.
BARBIER-BROSSAT (Mrs M.), Chargée de Recherche, Observatoire de Marseille, 2, place Le Verrier, 13004 Marseille, France
BARBIERI (Dr C.), Osservatorio Astronomico di Padova, Padova, Italy
BARBON (Dr R.), Osservatorio Astrofisico Asiago (Vincenza), Italy
BARDEEN (Dr J.M.), Department of Astronomy, University of Washington, Seattle, Washington 98105, U.S.A.
* BARKAT (Dr Z.), Cornell University, CRSR, Israel
* BARKER (Dr E.), McDonald Observatory, Fort Davis, Texas 79734, U.S.A.
BARKHATOVA (Mrs Dr K.A.), State University, Sverdlovsk, U.S.S.R.
* BARLETTI (Ing. R.), Osservatorio Astrofisico di Arcetri, Largo Enrico Fermi, 5, Firenze, Italy
BARLIER (F.), Observatoire de Paris, Section d'Astrophysique, 92190 Meudon, France
BARNEY (Miss Dr I.), Yale University Observatory, Box 2023 Yale Station, New Haven, Connecticut, U.S.A.
BARNOTHY (Prof. Dr J.M.), 833 Lincoln Street, Evanston, Illinois 60201, U.S.A.
BARRETO (Prof. L.M.), Observatorio Nacional, Rua General Bruce, 586, São Cristóvão, Rio de Janeiro, Brazil
BARRETT (Prof. A.H.), Research Laboratory of Electronics, Massachusetts Institute of Technology, Cambridge, Mass. 02139, U.S.A.
* BARROW (Dr R.), Physical Chemistry Laboratory, Oxford University, South Parks Road, Oxford OX1 30Z, U.K.
* BARRY (Dr D.C.), Dept. of Astronomy, University of Southern California, Los Angeles, California 90007, U.S.A.
BARTAYA (Mrs Dr R.A.), Astrophysical Observatory, Abastumani, Georgia, U.S.S.R.
* BARTH (Dr Ch.), Scientific Director LASP, University of Colorado, Boulder, Colorado 80302, U.S.A.
BARTOLINI (Dr C.), Osservatorio Astronomico Universitario, via Zamboni 33, Bologna, Italy
* BASART (Dr J.P.), Electrical Engineering Dept., Iowa State University, Ames, Iowa 50010, U.S.A.

BASCHEK (Prof. Dr B.), Lehrstuhl für Theoretische Astrophysik, 69 Heidelberg, Berliner Strasse 19, F.R.G.

* BASH (Dr F.), University of Texas at Austin, Austin, Texas 78712, U.S.A.

BASTIN (Dr J.A.), Physics Department, Queen Mary College (London University), Mile End Road, London E.1, U.K.

* BASU (Dr D.), Universidade de São Paulo, Instituto Astronômico e Geofisico, Av. Miguel Stefano, s/n – Parque de Estado, Caixa Postal no 30.627, São Paulo, Brasil

BATES (Prof. D.R.), Department of Applied Mathematics, Queen's University, Belfast, U.K.

BATESON (F.M.), 18 Pooles Road, Greerton, Tauranga, New Zealand

* BATH (Dr G.T.), Department of Astrophysics, South Parks Road, Oxford OX1 3RQ, U.K.

BATTEN (Dr A.H.), Dominion Astrophysical Observatory, R.R.7., Victoria, B.C., Canada

BATRAKOV (Dr J.W.), Institute of Theoretical Astronomy, Leningrad, U.S.S.R.

* BATTISTINI (Dr P.), Osservatorio Astronomico, Universitario, Via Zamboni, 33, Bologna, Italy

BAUER (C.A.), Pennsylvania State University, Department of Physics, State College, Pennsylvania, U.S.A.

BAUM (Dr W.A.), Planetary Research Center, Lowell Observatory, Flagstaff, Arizona 86002, U.S.A.

BAUSTIAN (W.W.), Kitt Peak National Observatory, 950 North Cherry Avenue, P.O. Box 26732, Tucson, Arizona 85726,U.S.A.

* BAUTZ (Dr L.P.), Dearborn Observatory, Northwestern University, Evanston, Illinois 60201, U.S.A.

BEALS (Dr C.S.), Manotick, Ontario, Canada

BEARDSLEY (W.R.), Allegheny Observatory, Riverview Park, Pittsburgh 14, Pennsylvania, U.S.A.

* BEAUDET (Dr G.), 1770 Edgewood, St-Bruno, P.Q., Canada

BEAVERS (Dr W.), Iowa State University, Ames, Iowa, U.S.A.

BEC (Miss A.), Bureau des Longitudes, 3, rue Mazarine, Paris 8e, France

BECK (H.-G.), Carl Zeiss Observatory, Jena, G.D.R.

BECKER (Prof. Dr F.), Klingsorstrasse 3/96, 8 München 61, F.R.G.

BECKER (Dr R.A.), Aerospace Corporation, P.O. Box 95085, Los Angeles, California, U.S.A.

BECKER (Prof. W.), Astronomisches Institut der Universität Basel, Venusstrasse 7, CH-4102 Binningen, Switzerland

BECKERS (Dr J.M.), Sacramento Peak Observatory, Sunspot, New Mexico 833349, U.S.A.

* BECKMAN (Dr J.E.), Department of Physics, Queen Mary College, London E1 4NS, U.K.

BEER (Dr A.), The Observatories, Madingley Road, Cambridge, U.K.

BEGGS (D.W.), The Observatories, Madingley Road, Cambridge, U.K.

BEHR (Prof. Dr A.), Director, Hamburg Observatory, 205 Hamburg 80, Gojenbergsweg 112, F.R.G.

* BELKOVICH (Dr O.I.), Engelgardt Astronomical Observatory, Observatory Station, Kazan 422526, U.S.S.R.

BELL (Miss Dr B.), Harvard College Observatory, Cambridge, Massachusetts 02138, U.S.A.

BELL (Miss Dr N.), Institut d'Astrophysique, 98bis, boulevard Arago, 75014 Paris, France

* BELL (M.B.), Astrophysics Branch, Radio and Electrical Engineering Division, National Research Council of Canada, Ottawa, Canada

BELL (Dr R.A.), Astronomy Program, University of Maryland, College Park, Maryland 20742, U.S.A.

BELOTSERKOVSKIJ (Dr D.J.), Institute of Physico-Technical Measurements, Moscow, U.S.S.R.

BELTON (Dr M.S.), Kitt Peak National Observatory, 950 North Cherry Avenue, P.O. Box 26732, Tucson, Arizona 85726, U.S.A.

BELY (O.), Observatoire de Nice, Le Mont-Gros, 06300 Nice, France

* BELY-DUBAU (Dr F.), Observatoire de Nice, Le Mont-Gros, 06300 Nice, France

BENDER (Dr P.), Joint Institute for Laboratory Physics, University of Colorado, Boulder, Colorado 80302, U.S.A.

BENEVIDES-SOARES (Dr P.), Instituto Astronómico e Geofísico, Caixa Postal 30 627, São Paulo OO-SP, Brazil

* BERENDZEN (Dr R.), Dept. of Astronomy, Boston University, Boston, Massachusetts, U.S.A.

BERGE (Dr G.L.), Owens Valley Radio Observatory, California Institute of Technology, Pasadena, California 91109, U.S.A.

BERGER (Dr J.), Institut d'Astrophysique, 98bis, boulevard Arago, 75014 Paris, France

* BERGERON (Dr J.), Observatoire de Paris, Section d'Astrophysique de Meudon, 92190 Meudon, France

* BERKHUIJSEN (Dr E.M.), MPI für Radioastronomie, 3 Argelanderstrasse, 53 Bonn, F.R.G.

* BERNACCA (Dr P.), Osservatorio Astrofisico, Asiago (Vicenza), Italy

BERTAUD (Dr Ch.), Observatoire de Paris, Section d'Astrophysique, 92190 Meudon, France

* BERTHOMIEU (Dr G.), Observatoire de Nice, Le Mont-Gros, 06300 Nice, France

BERTIAU (Dr F.C.), Specola Vaticana, Castel Gandolfo, Vatican City State

BERTOLA (Dr F.), Osservatorio Astrofisico di Asiago (Vicenza), Italy

* BESSELL (Dr M.S.), Mount Stromlo and Siding Spring Observatories, Private Bag, Woden P.O., A.C.T. 2606, Australia

BEYER (Dr M.), Justus-Brinckmann-Strasse 101, Hamburg-Bergedorf, F.R.G.

* BHATIA (Dr P.K.), Department of Mathematics, Faculty of Engineering, University of Jodhpur, Jodhpur (Rajasthan), India

* BHATIA (Dr V.B.), Department of Physics & Astrophysics, University of Delhi, Delhi-110007, India

BHATNAGAR (Dr A.), Vedhshala (Astronomical Observatory), Naranpura, Ahmedabad-13, India

BHATNAGAR (Dr P.L.), Department of Mathematics, Himachal Pradesh University, Simla, H.P., India

BHATTACHARYYA (J.C.), Astrophysical Observatory, Kodaikanal 3, India

BHONSLE (R.V.), Physical Research Laboratory, Ahmedabad, India

BIDELMAN (Dr W.P.), Warner and Swasey Observatory, 1975 Taylor Road, East Cleveland, Ohio 44112, U.S.A.

BIELICKI (Dr M.), Astronomical Observatory of Warsaw University, Al. Ujazdowskie 4, Warsaw, Poland

BIERMANN (Prof. Dr L.), Max-Planck-Institut für Physik und Astrophysik, Institut für Astrophysik, Föhringer Ring 6, München 23, F.R.G.

BIGAY (Dr J.H.), Directeur de l'Observatoire de Lyon, 69 Saint-Genis-Laval, France

* BIJAOUI (Dr A.), Observatoire de Nice, Le Mont-Gros, 06300 Nice, France

BILLAUD (G.), Observatoire de Paris, 61, avenue de l'Observatoire, Paris 14e, France

BILLINGS (Dr D.E.), High Altitude Observatory of the University of Colorado, Boulder, Colorado, U.S.A.

BINGHAM (Dr R.G.), Royal Greenwich Observatory, Herstmonceux Castle, Hailsham Sussex, U.K.

BINNENDIJK (Prof. Dr L.), Flower and Cook Observatory, University of Pennsylvania, Philadelphia 4, Pennsylvania, U.S.A.

* BIRAUD (Dr F.), Observatoire de Paris, Section d'Astrophysique de Meudon, 92190 Meudon, France

BISNOVATYI-KOGAN (Dr G.S.), Institute of Applied Mathematics, USSR Academy of Sciences, Moscow, U.S.S.R.

* BISWAS (Prof. S.), Tata Institute of Fundamental Research, Homi Bhabha Road, Bombay 5, India

BLAAUW (Prof. Dr A.), Director General, European Southern Observatory, 131 Bergedorfer Strasse, 205 Hamburg 80, F.R.G.

BLACKETT (Prof. P.M.S.), Imperial College of Science and Technology, South Kensington, London SW7, U.K.

BLACKWELL (Prof. D.E.), Dept. of Astrophysics, University Observatory, South Parks Road, Oxford, U.K.

BLAHA (Dr M.), Department of Physics and Astronomy, University of Maryland, College Park, Maryland 20742, U.S.A.

BLAMONT (Prof. J.E.), Service d'Aéronomie, Réduit de Verrières, 91-Verrières-le-Buisson, France

* BLANCO (Dr C.), Osservatorio Astrofisico, Città Universitaria, Viale A. Doria, I-95125 Catania, Italy

BLANCO (Dr V.M.), Observatorio Inter-Americano de Cerro Tololo, Casilla 63-D, La Serena, Chile

* BLEEKER (Dr Ir J.A.M.), Cosmic Ray Working Group, Kamerlingh Onnes Laboratorium, Nieuwsteeg 18, Leiden, The Netherlands

BLESS (Dr R.C.), Washburn Observatory, University of Wisconsin, 475 North Charter Street, Madison, Wisconsin 53706, U.S.A.

BLINOV (Dr N.S.), Sternberg Astronomical Institute, Moscow, U.S.S.R.

BLITZSTEIN (Dr W.), Flower and Cook Observatory, University of Pennsylvania, Philadelphia 4, Pennsylvania, U.S.A.

BLOCH (Miss Dr M.), Observatoire de Lyon, 69230-Saint-Genis-Laval (Rhône), France

BLUM (Dr E.J.), Observatoire de Paris, Section d'Astrophysique, 92190 Meudon, France

BOBROV (Dr M.S.), Astronomical Council, USSR Academy of Sciences, Vavilov Str. 34, Moscow V-312, U.S.S.R.

BOBROVNIKOFF (Dr N.T.), 1623 Visalia Avenue, Berkeley, California 94707, U.S.A.

BODENHEIMER (Dr P.), Lick Observatory, University of California, Santa Cruz, California 95060 U.S.A.

BOESGAARD (Mrs Dr A.M.), Institute of Astronomy, University of Hawaii, 2525 Correa Road, Honolulu, Hawaii 96822, U.S.A.

BOGGESS (Dr A.III), Goddard Space Flight Center, Washington 25, D.C., U.S.A.

BOGORODSKIJ (Dr A.F.), Astronomical Observatory of the Kiev University, Kiev, U.S.S.R.

* BOHLIN (Dr R.C.), NASA, Goddard Space Flight Center, Greenbelt, Maryland 20771, U.S.A.

BÖHM (Prof. Dr K.-H.), Astronomy Department, University of Washington, Seattle, Washington 98105, U.S.A.

BÖHM-VITENSE (Mrs Prof. Dr E.), Astronomy Department, University of Washington, Seattle, Washington, 98105, U.S.A.

* BÖHME (Mrs Dr A.), Zentralinstitut für solar-terrestrische Physik, Bereich C 1, Telegrafenberg, DDR 15, Potsdam, G.D.R.

BÖHME (Dr S.), Astronomisches Rechen-Institut, Mönchhofstrasse 12-14, 69 Heidelberg, F.R.G.

BOHRMANN (Prof. Dr A.), Angelhofweg 31, 6901 Wilhelmsfeld, F.R.G.

BOISCHOT (Dr A.), Observatoire de Paris, Section d'Astrophysique, 92190 Meudon, France

BOK (Prof. B.J.), Steward Observatory, University of Arizona, Tucson, Arizona 85721, U.S.A.

BOK (Mrs Dr P.F.), Steward Observatory, University of Arizona, Tucson, Arizona 85721, U.S.A.

* BOKSENBERG (Dr A.), University College London, Gower Street, London WC1, U.K.

BOLDT (Dr E.A.), NASA, Goddard Space Flight Center, Greenbelt, Maryland 20771, U.S.A.

BOLTON (J.G.), CSIRO, Division of Radiophysics, P.O. Box 76, Epping, N.S.W. 2121, Australia

BONANOMI (Dr J.), Directeur de l'Observatoire de Neuchâtel, Neuchâtel, Switzerland

* BOND (Dr H.E.), Louisiana State University, Baton Rouge, Louisiana 70803, U.S.A.

* BONDARENKO (Mrs Dr L.N.), Sternberg Astronomical Institute, Universitetskij Prospekt 13, Moscow 117234, U.S.S.R.

BONDI (Dr H.), King's College, Strand, London WC2, U.K.

BONEFF (Prof. Dr N.), Bulgarian Academy of Sciences, Section of Astronomy, '7 November' 1, Sofia, Bulgaria

BONEV (B.), Observatoire Populaire de Stara-Zagora, Bulgaria

BONNEAU (Prof. M.), Faculté des Sciences, place Victor Hugo, 13 Marseille, France

BONNET (Dr R.), L.P.S.P., 91-Verrières-le-Buisson, France

BONNOR (Prof. W.B.), Queen Elisabeth College, Campden Hill Road, London W8, U.K.

* BONOMETTO (Dr S.), Osservatorio Astronomico, Vicolo dell' Osservatorio, Padova, Italy

Bonov (A.D.), Université de Sofia, Sofia, Bulgaria

Bonsack (Prof. W.K.), Institute for Astronomy, 2840 Koluwalu Street, University of Hawaii, Honolulu, Hawaii 96822, U.S.A.

* Booth (Dr R.S.), Nuffield Radio Astronomy Laboratory, Jodrell Bank, Macclesfield, Cheshire, U.K.

Borgmann (Dr J.), 'Kapteyn' Sterrewacht, Mensingenweg 20, Roden (Drente), The Netherlands

* Börner (Dr G.), Max-Planck-Institut für Physik und Astrophysik, 8000 München, Föhringer Ring 6, F.R.G.

Bosman-Crespin (Mrs D.), place d'Italie, 4, B-4000 Liège, Belgium

Botelheiro (Dr A.), Observatorio Astronomico, Lisboa, Portugal

Botez (Mrs Dr E.), Pedagogical Institute, Suceava, Roumania

* Bottinelli (Dr L.), Observatoire de Paris, Section d'Astrophysique de Meudon, 92190 Meudon, France

Bouigue (Prof. R.), Directeur de l'Observatoire de Toulouse, 1, avenue Camille-Flammarion, 31-Toulouse (Haute-Garonne), France

Boulon (Dr J.), Observatoire de Paris, 61, avenue de l'Observatoire, 75014 Paris, France

Bourgeois (Prof. Dr P.E.), Directeur Honoraire, Observatoire Royal de Belgique, rue Paul Hankar, 31, B-1180 Bruxelles, Belgium

Boury (Dr A.J.J.L.), Institut d'Astrophysique, Université de Liège, rue Chinrue, 9, B-4870 Theux, Belgium

Bouška (Dr J.), Department of Astronomy and Astrophysics, Charles University, Švédská 8, Prague 5-Smíchov, Czechoslovakia

Bouvier (Prof. P.B.), Observatoire de Genève, CH-1290 Sauverny, Switzerland

Bowen (Dr E.G.), Councellor (Scientific), Embassy of Australia, 1601 Massachusetts Avenue N.W., Washington, D.C. 20036, U.S.A.

Bowyer (Dr C.S.), Astronomy Department, University of California, Berkeley, California 94720, U.S.A.

Boyce (Dr P.B.), P.O. Box 1269, Flagstaff, Arizona 86001, U.S.A.

Boyarchuk (Dr A.A.), Crimean Astrophysical Observatory, USSR Academy of Sciences, P/O Nauchny, Crimea, U.S.S.R.

Boyd (Prof. R.L.F.), Department of Physics, University College London, Mullard Space Science Laboratory, Holmbury St Mary, Dorking, Surrey, U.K.

Bozis (Dr G.), Department of Theoretical Mechanics, University of Thessaloniki, Thessaloniki, Greece

* Bozkurt (Dr S.), Ege University Observatory, Bornova, Izmir, Turkey

Braccesi (A.), Laboratorio di Radioastronomia, Istituto di Fisica, Bologna, Italy

Bracewell (Prof. R.N.), Radio Astronomy Institute, Stanford University, Stanford, California 94305, U.S.A.

Braes (Dr L.L.E.), Sterrewacht, Leiden, The Netherlands

Brahde (R.), Institute of Theoretical Astrophysics, University of Oslo, Blindern, Oslo 3, Norway

Brandt (Dr J.C.), Code 680, Laboratory for Solar Physics, Goddard Space Flight Center, Greenbelt, Maryland 20771, U.S.A.

Brandt (Dr V.E.), Sternberg Astronomical Institute, Moscow 117234, U.S.S.R.

Branscomb (Dr L.M.), National Bureau of Standards, Washington 25, D.C., U.S.A.

Branson (Dr N.J.B.A.), The Old Schools, Cambridge University, Cambridge, U.K.

Braude (Prof. Dr S.Ya.), Institute of Radio Physics and Electronics, Academy of Sciences of the Ukrainian SSR, Khar'kov-85, U.S.S.R.

* Brault (Dr J.W.), Kitt Peak National Observatory, 950 North Cherry Ave., Tucson, P.O. Box 26732, Arizona 85726, U.S.A.

Bray (Dr R.J.), CSIRO, Division of Physics, University Grounds, Chippendale, N.S.W. 2008, Australia

Brealey (G.A.), Dominion Astrophysical Observatory, R.R.7, Victoria, B.C., Canada

* Brecher (Prof. K.), Dept. of Physics, Massachusetts Institute of Technology, Cambridge, Mass. 02139, U.S.A.
* Breger (Dr M.), Dept. of Astronomy, University of Texas, Austin, Texas 78712, U.S.A.

Brejdo (Mrs Dr I.I.), Main Astronomical Observatory, USSR Academy of Sciences, Pulkovo, U.S.S.R.

* Bridle (Dr A.H.), Astronomy Group, Department of Physics, Queen's University of Kingston, Ontario, Canada

Brini (Prof. D.), Laboratorio T.E.S.R.E., Via Castagnoli 1, Bologna, Italy

* Brinkman (Dr A.C.), Laboratorium voor Ruimteonderzoek, Beneluxlaan 21, Utrecht, The Netherlands

Brkić (Z.M.), Astronomska Observatorija, Veliki Vracar, Beograd, Yugoslavia

Broadfoot (Dr L.), Kitt Peak National Observatory, 950 North Cherry Ave, Tucson, Arizona 85726, U.S.A.

Broglia (Dr P.), Osservatorio Astronomico, Merate (Como), Italy

Brosche (Dr P.), Sternwarte der Universität, D-53 Bonn, Auf dem Hügel 71, F.R.G.

Brosterhus (Dr E.B.F.), Dominion Astrophysical Observatory, R.R.7, Victoria, B.C., Canada

Broten (N.W.), Astrophysics Branch, National Research Council, Ottawa, Ontario, Canada

Broucke (Dr R.), 5603 Stardust Road, La Cañada, California 91011, U.S.A.

Brouw (W.N.), Leiden Observatory, Leiden, The Netherlands

Brown (Dr Harrison), Division of Geological Sciences, California Institute of Technology, Pasadena 4, California, U.S.A.

Brown (P. Lancaster), Karlsvik, 30 Enghams Wood Road, Beaconsfield, Bucks., U.K.

Brown (Prof. R. Hanbury), Chatterton Astronomy Department, School of Physics, University of Sydney, Sydney, N.S.W. 2006, Australia

* Brown (Dr R.L.), National Radio Astronomy Observatory, Charlottesville, Virginia 22901, U.S.A.

Brownlee (Dr R.R.), P.O. Box 1663, Los Alamos, New Mexico, U.S.A.

Brück (Mrs Dr M.T.), Royal Observatory, Blackford Hill, Edinburgh 9, U.K.

Brück (Prof. H.A.), Astronomer Royal for Scotland, Royal Observatory, Blackford Hill, Edinburgh 9, U.K.

Brückner (Dr G.), U.S. Naval Research Laboratory, Plasma Physics Section, Code 7700, Washington, D.C. 20390, U.S.A.

Brumberg (Dr V.A.), Institute of Theoretical Astronomy, USSR Academy of Sciences, Leningrad, U.S.S.R.

* Brunet (Dr J.-P.), Observatoire de Marseille, 2, place le Verrier, 13004 Marseille, France

Brunk (W.), Lunar and Planetary Programs, NASA Headquarters, Washington, D.C. 20546, U.S.A.

Bruwer (J.A.), Republic Observatory, Johannesburg, South Africa

Bruzek (Dr A.), Fraunhofer Institut, Schöneckstrasse 6, Freiburg i. Br., F.R.G.

Buchar (Prof. Dr E.), Astronomical Institute, Technical University, Karlovo nám. 13, Prague 2, Czechoslovakia

* Buhl (Dr D.), National Radio Astronomy Observatory, Edgemont Road, Charlottesville, Virginia 22901, U.S.A.

Bullard (Sir Edward C.), University of Cambridge, Department of Geodesy and Geophysics, Madingley Rise, Madingley Road, Cambridge, U.K.

Bullen (Prof. K.E.), Department of Applied Mathematics, University of Sydney, Sydney, N.S.W., Australia

Bumba (Dr V.), Astronomical Institute, Czechoslovak Academy of Sciences, Observatory Ondřejov, Czechoslovakia

Burbidge (Mrs Dr E.M.), University of California, La Jolla, California 92038, U.S.A.

Burbidge (Dr G.R.), University of California, La Jolla, California 92038, U.S.A.

Burch (Dr C.R.), H.H. Royal Fort Wills Physics Laboratory, Bristol 8, U.K.

* BURGER (Dr Ir J.J.), Werkgroep Kosmische Straling, Kamerlingh Onnes Laboratorium, Nieuwsteeg 18, Leiden, The Netherlands

BURGESS (Dr A.), University of Cambridge, Department of Applied Mathematics and Theoretical Physics, Silver Street, Cambridge, U.K.

BURGESS (Dr D.D.), Physics Department, Imperial College of Science and Technology, London SW7, U.K.

BURKE (Dr B.F.), Room 26-459, Massachusetts Institute of Technology, Cambridge, Mass. 02139, U.S.A.

BURKE (Prof. J.A.), Department of Physics, University of Victoria, Victoria, B.C., Canada

* BURKHEAD (Dr M.S.), Department of Astronomy, Swain Hall West 319, Indiana University, Bloomington, Indiana 47401, U.S.A.

* BURNICHON (Dr M.L.), Institut d'Astrophysique, 98bis, boulevard Arago, F75014 Paris, France

* Burns (Dr J.A.), Thurston Hall, Cornell University, Ithaca, New York 14850, U.S.A.

* BURTON (Dr W.B.), National Radio Astronomy Observatory, Edgemont Road, Charlottesville, Virginia 22901, U.S.A.

* BURTON (B.Sc. W.M.), Astrophysics Research Unit, Culham Laboratory, Abingdon, Berkshire, U.K.

BUSBRIDGE (Miss Dr I.W.), 'Haremere', Westerham Road, Keston, Kent BR2 6HH, U.K.

BUSCOMBE (Dr W.), Astronomy Department, Northwestern University, Evanston, Illinois 60201, U.S.A.

* BUTLER (Dr C.J.), Armagh Observatory, Armagh, Northern Ireland

BUTLER (Dr H.E.), Royal Observatory, Blackford Hill, Edinburgh 9, U.K.

BYKOV (Dr M.F.), Astronomical Institute of the Usbek SSR Academy of Sciences, Tashkent, U.S.S.R.

BYSTROV (Dr N.F.), Main Astronomical Observatory, USSR Academy of Sciences, Pulkovo, U.S.S.R.

BYSTROVA (Mrs Dr N.V.), Special Astrophysical Observatory, USSR Academy of Sciences, Leningrad Branch of the SAO, Pulkovo 196140, U.S.S.R.

* CABRITA (E.M.L.), Tapada da Ajuda, Lisboa 3, Portugal

CACCIANI (Dr A.), Osservatorio Astronomico di Roma, Rome, Italy

* CAHN (Dr J.), University of Illinois, Urbana, Illinois 61801, U.S.A.

CALAMAI (Prof. G.), Osservatorio Astrofisico di Arcetri, Firenze, Italy

* CALOI (Dr V.), Laboratorio di Astrofisica, Casella Postale 67, Frascati (Roma), Italy

CAMERON (Prof. A.G.W.), Center for Astrophysics, Harvard College Observatory, 60 Garden Str., Cambridge, Mass, 02138, U.S.A.

CAMERON (Mrs W.), NASA, Goddard Space Flight Center, Greenbelt, Maryland 20771, U.S.A.

CAMICHEL (Dr H.), Observatoire de Toulouse, 31-Toulouse (Hte-Garonne), France

CAMM (Dr G.L.), Department of Mathematics, The University, Manchester 13, U.K.

* CAMPBELL (B.Sc. J.W.), Science Research Council, Royal Observatory Edinburgh, Edinburgh EH9 3HJ, U.K.

CANAVAGGIA (Miss Dr R.), Observatoire de Paris, 61, avenue de l'Observatoire, 75014 Paris, France

CANDY (M.P.), Perth Observatory, Bickley, Western Australia 6076

* CANFIELD (Dr R.C.), Sacramento Peak Observatory, Air Force Cambridge Research Labs., Sunspot, New Mexico 88349, U.S.A.

* CANNON (Dr R.D.), UK 48-inch Schmidt Telescope Project, Royal Observatory, Edinburgh EH9 3HJ, U.K.

* CANTU' (Dr A.M.), Osservatorio Astrofisico di Arcetri, Largo Enrico Fermi, 5, Firenze, Italy

* CAPACCIOLI (Dr M.), Osservatorio Astrofisico, Asiago (Vicenza), Italy

CAPRIOLI (Dr G.), Osservatorio Astronomico di Monte Mario, Via Trionfale 204, Roma, Italy

Capriotti (Dr E.R.), Astronomy Department, The Ohio State University, 174 West 18th Avenue, Columbus, Ohio 43210, U.S.A.

Cardús (J.O.), Astronome de l'Observatorio del Ebro, Roquetas, Tarragona, Spain

* Carleton (Dr N.), Smithsonian Institution, Astrophysical Observatory, Cambridge, Massachusetts 02138, U.S.A.

Carlqvist (P.A.), Royal Institute of Technology, Division of Plasma Physics, Stockholm 70, Sweden

Caroubalos (Dr C.), Observatoire de Meudon, 92190 Meudon, France

* Carpenter (C.Eng. G.J.), Royal Observatory, Edinburgh EH9 3HJ, U.K.

Carpenter (Dr Lloyd H.), NASA, Goddard Space Flight Center, Greenbelt, Maryland 20771, U.S.A.

Carr (Dr T.D.), Department of Physics, University of Florida, Gainesville, Florida, U.S.A.

* Carranza (Dr G.J.), Laprida 880, Córdoba, Argentina

Carrasco (Dr R.), Observatorio Astronómico Nacional de Madrid, Alfonso XII, 3, Madrid (7), Spain

Carroll (Sir John A.), 14, Belvedere Grove, Wimbledon, London, SW 19, U.K.

Carroll (Prof. P.K.), Department of Physics, University College, Belfield, Dublin 4, Ireland

Carson (Dr T.R.), University Observatory, Buchanan Gardens, St. Andrews, Fife, Scotland, U.K.

Carver (Prof. J.H.), Department of Physics, University of Adelaide, Adelaide, Australia

* Casanovas Corderroure (J.), Instituto Universitario de Astrofísica, Universidad de La Laguna, Tenerife, Spain

* Cassinelli (Dr J.), University of Wisconsin, Washburn Observatory, Madison, Wisconsin 53706, U.S.A.

Castellani (Dr V.), Laboratorio di Astrofisica, Casella Postale 67, Frascati (Roma), Italy

* Castelli (J.P.), AFCRL (LIR/Stop 30), L.G. Hanscom Field, Bedford, MA 01730, U.S.A.

* Castor (Dr J.I.), Joint Institute for Laboratory Astrophysics, University of Colorado, Boulder, Colorado 80302, U.S.A.

* Catalano (Dr F.A.), Osservatorio Astrofisico, Città Universitaria, Viale A. Doria, I-95125 Catania, Italy

Catalano (Dr S.), Osservatorio Astrofisico, Città Universitaria, Viale A. Doria, I-95125 Catania, Sicilia, Italy

Caughlan (Mrs Dr G.), Department of Physics, Montana State University, Bozeman, Montana 59715, U.S.A.

* Cavaliere (Dr A.), Laboratorio del C.N.E.N., Frascati (Roma), Italy

Cayrel (Dr R.), Observatoire de Paris, Section d'Astrophysique, 92190 Meudon, France

Cayrel-de-Strobel (Mrs Dr G.), Institut d'Astrophysique, 98bis, boulevard Arago, 75014 Paris, France

* Cazzola (Dr P.), Osservatorio Astronomico, Vicolo dell'Osservatorio, Padova, Italy

Ceccarelli (Prof. M.), Instituto di Fisica generale, Università di Bologna, Italy

Cecchini (Prof. G.), Via Roma, 56011 Calci (Pisa), Italy

Ceplecha (Dr Z.), Astronomical Institute, Czechoslovak Academy of Sciences, Observatory Ondřejov, Czechoslovakia

Cesco (Dr C.U.), Observatorio Félix Aguilar, Facultad de Ingeniería, Universidad Nacional de Cuyo, San Juan, Argentina

Cesco (Dr R.P.), Observatorio Astronómico, La Plata, Argentina

Cester (Prof. Dr B.), Osservatorio Astronomico, Via G.B. Tiepolo, 15, Trieste (409), Italy

* Chaffee (Dr F.), Smithsonian Astrophysical Observatory, Cambridge, Massachusetts 02138, U.S.A.

Chalonge (Dr D.), Institut d'Astrophysique, 98bis, boulevard Arago, 75014 Paris, France

Chamberlain (Prof. J.W.), Department of Space Physics and Astronomy, Rice University, Houston, Texas 77001, U.S.A.

* CHAMBLISS (Dr C.), Kutztown State College, Kutztown, Pennsylvania 19530, U.S.A.
CHANDRA (S.), General Electric Co., Room M 1325, P.O. 8555, Philadelphia, Pennsylvania 19101, U.S.A.
CHANDRASEKHAR (Prof. S.), Laboratory for Astrophysics & Space Research, 933 East 56th Street, Chicago, Illinois 60637, U.S.A.
* CHANG (Dr Shao-Chang), Geophysics Institute of National Central University, Chung-li, Taiwan,
CHAPMAN (Dr R.D.), Solar Physics Group, Goddard Space Flight Center, Greenbelt, Maryland 20771, U.S.A.
CHAPRONT (Dr J.), Bureau des Longitudes, 3, rue Mazarine, Paris 8e, France
CHARVIN (P.), Observatoire de Meudon, 5, place Janssen, 92190 Meudon, France
* CHAU (Prof. W.Y.), Department of Physics, Queen's University of Kingston, Ontario, Canada
CHAVIRA (E.), Instituto Nacional de Astrofísica Optica y Electrónica, Tonantzintla, Pue., Apartados Postales Nos. 216 y 51, Puebla, Pue. Mexico
CHEBOTAREV (Prof. G.A.), Institute of Theoretical Astronomy, Leningrad B-164, U.S.S.R.
CHEKIRDA (Dr A.T.), Kharkov Observatory, Kharkov, U.S.S.R.
CHEN (Dr Kwan-Yu), Astronomy Department, Rosemary Hill Observatory, University of Florida, Gainesville, Florida 32601, U.S.A.
CHENAKAL (Dr V.L.), Director of Lomonosov Museum, Universitetskaja nab.3, Leningrad 199164, U.S.S.R.
CHEREDNICHENKO (Dr V.I.), Kiev Polytechnical Institute, Kiev 56, U.S.S.R.
CHEREPASHCHUK (Dr A.M.), Sternberg Astronomical Institute, Moscow 117234, U.S.S.R.
CHERNEGA (Dr N.A.), Astronomical Observatory, Kiev University, Kiev, U.S.S.R.
CHERTOPRUD (Dr V.E.), Astronomical Council, USSR Academy of Sciences, Vavilov Str. 34, Moscow V-312, U.S.S.R.
* CHEVALIER (Dr C.), Observatoire de Paris, Section d'Astrophysique de Meudon, 92190 Meudon, France
CHINCARINI (G.L.), Department of Astronomy, P.M.A. Building/Room 16.336, The University of Texas, Austin, Texas 78712, U.S.A.
* CHIOSI (Dr C.), Osservatorio Astronomico, Vicolo dell'Osservatorio, Padova, Italy
CHIS (Prof. G.), Observatoire Astronomique, Rue de la République 109, Cluj, Roumania
CHITRE (Dr S.N.), Tata Institute of Fundamental Research, Homi Bhabha Road, Colaba, Bombay-5, India
CHIU (Dr H.Y.), Institute for Space Studies, 475 Riverside Drive, New York, N.Y. 100 N.Y. 10027, U.S.A.
* CHIUDERI (Dr C.), Osservatorio Astrofisico di Arcetri, Largo Enrico Fermi, 5, Firenze, Italy
CHIUDERI (Dr F.D.), Osservatorio Astrofisico di Arcetri, Largo Enrico Fermi, 5, Firenze, Italy
* CHOI (Bon Chol), Pyongyang Astronomical Observatory, Academy of Sciences, Pyongyang, Democratic People's Republic of Korea
CHOPINET (Miss M.), Observatoire de l'Université de Bordeaux, 33-Floirac, France
* CHOU (Dr Kyong Chol), Department of Astronomy and Meteorology, Yonsei University, Seoul, Republic of Korea
CHRISTIANSEN (Prof. W.N.), School of Electrical Engineering, The University of Sydney, Sydney, N.S.W. 2006, Australia
CHRISTY (Prof. R.F.), Physics Department, California Institute of Technology, Pasadena, California, U.S.A.
CHUBB (Dr T.A.), 5023 N 38th Street, Arlington 7, Virginia, U.S.A.
CHUDOVICHEVA (Mrs Dr N.A.), Engelhardt Observatory, Kazan, U.S.S.R.
CHUGAJNOV (Dr P.F.), Crimean Astrophysical Observatory, USSR Academy of Sciences, Crimea, U.S.S.R.
* CHURCHWELL (Dr E.B.), D-53 Bonn 1, Auf dem Hügel 69, F.R.G.
CHURMS (J.), Royal Observatory, Observatory, Cape, South Africa
CHUVAEV (Dr K.K.), Crimean Astrophysical Observatory, USSR Academy of Sciences, P/O

Nauchny, Crimea, U.S.S.R.

CHVOJKOVA (Dr E.), Astronomical Institute, Czechoslovak Academy of Sciences, Budečská 6, Prague 2, Czechoslovakia

* CIATTI (Dr F.), Osservatorio Astrofisico, Asiago (Vicenza), Italy

CID PALACIOS (Dr R.), Facultad de Ciencias, Ciudad Universitaria, Zaragoza, Spain

CILLIÉ (Prof. G.G.), Department of Mathematics, University of Stellenbosch, Cape Province, South Africa

CIMINO (Prof. Dr M.), Direttore, Osservatorio Astronomico su Monte Mario, Rome, Italy

* CLARK (Dr A.), Dept. of Mechanical and Aerospace Sciences, University of Rochester, Rochester, New York 14627, U.S.A.

CLARK (Dr B.G.), National Radio Astronomy Observatory, P.O. Box 2, West Green Bank, West Virginia 24944, U.S.A.

CLARK (Dr G.), Physics Department, Massachusetts Institute of Technology, Cambridge, Mass. 02139, U.S.A.

* CLARK (Dr Th.A.), Laboratory for Extraterrestrial Physics, University of Maryland, College Park, Maryland 20742, U.S.A.

* CLARKE (Dr D.), Department of Astronomy, The University, Glasgow G12 8QW, U.K.

* CLARKE (Dr T.R., Jr.), Royal Ontario Museum, Toronto, Ontario, Canada

CLAYTON (Dr D.D.), Space Science Department, Rice University, Houston, Texas, U.S.A.

* CLEGG (Dr P.E.), Physics Department, Queen Mary College, Mile End Road, London E1 4NS, U.K.

CLEMENCE (Dr G.M.), Yale University Observatory, Box 2023, Yale Station, New Haven, Connecticut, U.S.A.

CLEMENT (Prof. Dr M.J.), David Dunlap Observatory, Richmond Hill, Ontario, Canada

CLIMENHAGA (Dr J.L.), Department of Physics, University of Victoria, Victoria, B.C., Canada

CLUBE (Dr S.V.M.), Royal Greenwich Observatory, Herstmonceux Castle, Hailsham, Sussex, U.K.

COCKE (Dr J.), Steward Observatory, University of Arizona, Tucson, Arizona 85721, U.S.A.

CODE (A.D.), Washburn Observatory, University of Wisconsin, 475 North Charter Street, Madison, Wisconsin 53706, U.S.A.

COELHO BALSA (Dr M.C.), Observatório Astronómico de Universidade de Coimbra, Coimbra, Portugal

* COFFEEN (Dr D.), Lunar & Planetary Laboratory, University of Arizona, Tucson, Arizona 85721, U.S.A.

COHEN (Dr J.M.), Institute for Advanced Study, School of Natural Sciences, Princeton, New Jersey 08540, U.S.A.

COHEN (Prof. M.H.), Department of Astronomy, California Institute of Technology, Pasadena, California 91109, U.S.A.

COLBURN (Dr D.S.), Ames Research Center, Moffet Field, California 94035, U.S.A.

COLGATE (Dr S.), New Mexico Institute of Mining and Technology, Socorro, New Mexico, U.S.A.

COLLINDER (Dr P.), Odensgata 5 B, S-752 22 Uppsala, Sweden

COLLINS II (Dr G.W.), Department of Astronomy, Ohio State University, Columbus, Ohio 43210, U.S.A.

COLLINSON (E.H.), Fox's Corner, Snape, Saxmundham Suffolk IP17 1SD, U.K.

* COLOMB (Dr F.R.), Instituto Argentino de Radioastronomía, Casilla No. 5, Villa Elisa, Prov. Bs. As., Argentina

COLOMBO (Prof. Dr G.), Instituto Meccanica Applicata Universita di Padova, via F. Marzolo 9, Padova, Italy

CONDON (Dr E.U.), Joint Institute for Laboratory Astrophysics, University of Colorado, Boulder, Colorado 80310, U.S.A.

CONNES (Mrs J.), Service de Calcul Numérique, Observatoire de Meudon, 5, place Janssen, 92190 Meudon, France

CONNES (P.), Laboratoire du C.N.R.S., Bellevue, France
CONTI (Dr P.S.), JILA, University of Colorado, Boulder, Colorado 80302, U.S.A.
CONTOPOULOS (Prof. Dr G.), Astronomical Department, University of Thessaloniki, Thessaloniki, Greece
CONWAY (Dr R.G.), Nuffield Radio Astronomy Laboratories, Jodrell Bank, Macclesfield, Cheshire, U.K.
COOK (Dr A.F.), Smithsonian Astrophysical Observatory, Cambridge, Massachusetts 02138, U.S.A.
COOK (Dr A.H.), The Cavendish Laboratory, Free School Lane, Cambridge CB2 3RQ, U.K.
COOKE (Dr B.A.), X-Ray Astronomy Group, Physics Department, Leicester University, Leicester, U.K.
COOPER (B.F.C.), CSIRO, Division of Radiophysics, P.O. Box 76, Epping, N.S.W. 2121, Australia
CORLISS (Dr C.H.), National Bureau of Standards, Washington, D.C. 20234, U.S.A.
* CORNEJO (A.), Instituto Nacional de Astrofísica, Optica y Electrónica, Tonantzintla, Puebla, Mexico
COSTAIN (Dr C.H.), Dominion Radio Astrophysical Observatory, P.O. Box 248, Penticton, B.C., Canada
COUDER (Dr A.), Observatoire de Paris, 61, avenue de l'Observatoire, 75014 Paris, France
COUDERC (Dr P.), Observatoire de Paris, 61, avenue de l'Observatoire, 75014 Paris, France
* COUNSELMAN III (Dr Ch.), Dept. of Earth & Planetary Sciences, Massachusetts Institute of Technology, Cambridge, Mass. 02139, U.S.A.
COURTÈS (Dr G.), Observatoire de Marseille, 2, place Le Verrier, 13004 Marseille, France
COUSINS (Dr A.W.J.), Royal Observatory, Observatory, C.P., South Africa
COUTEAU (Dr P.C.), Observatoire de Nice, Le Mont-Gros, 06300, France
COUTREZ (Prof. Dr R.A.J.), Directeur, Institut d'Astronomie et d'Astrophysique, Université Libre de Bruxelles, avenue F.D. Roosevelt, 50, B-1050 Bruxelles, Belgium
* COUTTS (Miss Dr C.M.), Department of Astronomy, University of Toronto, Toronto, Ontario M5S 1A7, Canada
COVINGTON (A.E.), Astrophysics Branch, National Research Council, Ottawa, Ontario, Canada
COWLEY (Mrs Dr A.P.), Astronomy Department University of Michigan, Ann Arbor, Michigan 48104, U.S.A.
COWLEY (Dr Ch.), Astronomy Department, University of Michigan, Ann Arbor, Michigan 48104, U.S.A.
COWLING (Prof. T.G.), Department of Mathematics, The University, Leeds 2, U.K.
COX (Dr A.N.), Box 1663, Los Alamos, New Mexico, U.S.A.
* COX (Dr D.P.), Department of Physics, University of Wisconsin, Madison, Wisconsin 53706, U.S.A.
COX (Dr J.P.), Joint Institute for Laboratory Astrophysics, 1511 University Avenue, Boulder, Colorado 80302, U.S.A.
COYNE (Dr G.V., S.J.), Specola Vaticana, Castel Gandolfo, 00040 Rome, Italy
CRAMPTON (Dr D.), Dominion Astrophysical Observatory, R.R.7, Victoria, B.C., Canada
CRAWFORD (Dr D.L.), Kitt Peak National Observatory, 950 North Cherry Avenue, P.O. Box 26732, Tucson, Arizona 85726, U.S.A.
* CREZE (Dr M.), Observatoire de Paris, 61, avenue de l'Observatoire, 75014 Paris, France
CRISTALDI (Prof. S.), Viale A. Doria, Città Universitaria, 95123 Catania, Italy
CRISTESCU (Dr C.), Observatoire Astronomique, str. Cutitul de Argint, 5, Bucarest 28, Roumania
CROOM (Dr D.L.), Radio and Space Research Station, Ditton Park, Slough, U.K.
* CRUIKSHANK (Dr D.), University of Hawaii, Institute of Astronomy, Honolulu, Hawaii 96872, U.S.A.

CRUVELLIER (Dr P.), Observatoire de Marseille, place Le Verrier, 13004 Marseille, France

* CRUZ-GONZALEZ (C.), Instituto de Astronomía, UNAM, Apartado Postal 70–264, México, D.F., Mexico

CSADA (Dr I.K.), Konkoly Observatory, Konkoly Thege U. 13–17, Budapest XII, Hungary

CUDABACK (Dr D.D.), Radio Astronomy Laboratory, 633 Campbell Hall, University of California, Berkeley, California 94720, U.S.A.

CUFFEY (J.), New Mexico State University, Dept. of Earth Sciences and Astronomy, University Park, New Mexico 88001, U.S.A.

* CUGNON (Dr P.), Observatoire Royal de Belgique, av. Circulaire 3, B-1180 Bruxelles, Belgium

CULHANE (Dr J.L.), Mullard Space Science Laboratory, Physics Department, University College London, Gower Street, London WC1, U.K.

CUNNINGHAM (Dr L.E.), Department of Astronomy, University of California, Berkeley 4, California 94720, U.S.A.

CUNY (Miss Dr Y.), Observatoire de Meudon, 92190 Meudon, France

CUPERMAN (Dr S.), Department of Physics and Astronomy, Tel-Aviv University, Ramat-Aviv, Israel

CUREA (Prof. Dr I.), Recteur de l'Université de Timisoara, Timisoara, Roumania

CZYZAK (Prof. S.J.), Astronomy Department, The Ohio State University, Columbus, Onio 43210, U.S.A.

DACHS (Dr J.), Astronomisches Institut der Ruhr-Universität, Postfach 2148, 4630 Bochum-Querenburg, F.R.G.

DADAEV (Dr A.N.), Pulkovo Observatory, Leningrad, U.S.S.R.

DADIĆ (Dr Ž.), Institut za povijest nauka Jazu, Demetrova 18, Zagreb 3, Yugoslavia

DAENE (Dr H.), Heinrich-Hertz-Institut für Solar-Terrestrische Physik, Berlin-Adlershof, Rudower Chaussee, G.D.R.

DAINTREE (Dr E.J.), Nuffield Radio Astronomy Laboratories, Jodrell Bank, Macclesfield, Cheshire, U.K.

DALGARNO (Prof. A.), Smithsonian Astrophysical Observatory, 60 Garden Street, Cambridge, Massachusetts 02138, U.S.A.

DALLAPORTA (Prof. N.), Istituto di Fisica teorica, Università di Padova, Italy

* DALTABUIT (Dr E.), Instituto de Astronomía, UNAM, Apartado Postal 70–264, México D.F., Mexico

DAMBARA (Dr T.), Geographical Survey Institute, 7–1000 Kamimeguro, Meguro-ku, Tokyo, Japan

DANBY (Dr J.M.A.), Yale University Observatory, Box 2023, Yale Station, New Haven, Connecticut, U.S.A.

DANIELSON (Dr R.E.), Princeton University Observatory, Peyton Hall, Princeton, N.J. 08540, U.S.A.

DANZIGER (Dr I.J.), Harvard College Observatory, 60 Garden Street, Massachusetts 02138, U.S.A.

* DA ROCHA VIEIRA (Dr E.), Instituto de Física, Universidade Federal do Rio Grande do Sul, 90000 Porto-Alegre, RS, Brazil

DA SILVA (Dr A.V.C.S.), Observatorio Astronomico da Universidade Santa Clara, Coimbra, Portugal

DA SILVA (Prof. Dr I.F.), Faculdade de Ciencias, Lisbon, Portugal

DAUBE-KURZEMNIECE (Mrs Dr I.A.), Astrophysical Laboratory of the Academy of Sciences of the Latvian SSR, Riga, U.S.S.R.

DAUTCOURT (Dr G.), Zentralinstitut für Astrophysik, Telegrafenberg, 15 Potsdam, G.D.R.

* DAVIDSON (Dr K.), Princeton University Observatory, Princeton, New Jersey 08540, U.S.A.

DAVIDSON (Prof. W.), Mathematics Department, University of Otago, P.O. Box 56, Dunedin, New Zealand

DAVIES (Dr J.G.), Nuffield Radio Astronomy Laboratories, Jodrell Bank, Macclesfield Cheshire, U.K.

DAVIES (Dr R.D.), Nuffield Radio Astronomy Laboratories, Jodrell Bank, Macclesfield Cheshire, U.K.

* DAVIS (Dr C.G.), University of California, Los Alamos Scientific Lab., Los Alamos, New Mexico 87544, U.S.A.

DAVIS (Dr J.), Department of Physics, University of Sydney, Sydney, N.S.W. 2006, Australia

DAVIS JR (Prof. L.), Physics Department, California Institute of Technology, Pasadena, California 91109, U.S.A.

DAVIS (Dr M.M.), National Science Foundation, 1800 G Street, Washington, D.C. 20050, U.S.A.

DAVIS (Prof. M.S.), The University of North Carolina, Department of Physics, Chapel Hill, N.C. 27514, U.S.A.

DAVIS (Dr R.J.), Smithsonian Astrophysical Observatory, 60 Garden Street, Cambridge, Massachusetts 02138, U.S.A.

DAWE (Dr A.J.), Department of Astronomy, Leicester University, Leicester, LE1 7RH, U.K.

DÉBARBAT (Miss S.), Observatoire de Paris, 61, av. de l'Observatoire, Paris-14e, France

DEBEHOGNE (Dr H.), Observatoire Royal de Belgique, avenue Circulaire, 3, B-1180 Bruxelles, Belgium

* DE BIASE (Dr G.A.), Osservatorio Astronomico, Viale del Parco Mellini, 84, Roma, Italy

DECAUX (B.), Bureau International de l'Heure, Observatoire de Paris, 61 avenue de l'Observatoire, Paris-14, France

DEEMING (Dr T.J.), Astronomy Department, 404 Physics Building, University of Texas, Austin, Texas 78712, U.S.A.

DE FEITER (Dr L.D.), The Astronomical Institute, Space Research Laboratory, Beneluxlaan 21, Utrecht, The Netherlands

* DE FELICE (Dr F.), Osservatorio Astronomico, Vicolo dell'Osservatorio, Padova, Italy

* DE FREITAS PACHECO (Dr J.A.), Observatorio de São Paulo – USP, Caixa Postal 30627, São Paulo, Brazil

DE GRAAF (Dr T.), Kapteyn Astronomical Laboratory, Postbus 800, Groningen, The Netherlands

DE GRAAFF (Dr W.), Laboratorium voor Ruimte-Onderzoek, Beneluxlaan 21, Utrecht, The Netherlands

DE GROOT (M.J.H.), European Southern Observatory, Bergedorfer Strasse 131, 205 Hamburg 80, F.R.G.

DE GROOT (Dr T.), Sterrewacht 'Sonnenborgh', Servaas Bolwerk 13, Utrecht, The Netherlands

* DEHARVENG (Dr M.), L.A.S. – Traverse du Siphon, Les Trois Lucs, 13012, Marseille, France

DEINZER (Dr W.), Universitäts-Sternwarte, 34 Göttingen, Geismarlandstrasse 11, F.R.G.

DE JAGER (Prof. Dr C.), Director, The Astronomical Institute, Space Research Laboratory, Beneluxlaan 21, Utrecht, The Netherlands

* DEJAIFFE (Dr R.J.), Observatoire Royal de Belgique, Av. Circulaire, 3, B-1180 Bruxelles, Belgium

* DE JONG (Dr T.), Harvard College Observatory, Cambridge, Mass. 02138, U.S.A.

DE JONGE (Dr J.K.), Department of Astronomy, University of Pittsburgh, Riverview Park, Pittsburgh, Pa. 15214, U.S.A.

DE KOCK (R.P.), Royal Observatory, Observatory, C.P., South Africa

DE KORT (S.J.) (Rev. Dr J.), Department of Astronomy, Faculty of Science, Catholic University, Toernooiveld, Nijmegen, 6805, The Netherlands

DELACHE (Dr Ph.), Observatoire de Nice, Le Mont-Gros, 06-Nice, France

DELANNOY (J.), Observatoire de Paris, Section d'Astrophysique, 92190-Meudon, France

DELBOUILLE (Prof. Dr L.), Institut d'Astrophysique, Université de Liège, avenue de Cointe, 5, B-4200 Cointe-Ougrée, Belgium

DELHAYE (Prof. J.), Observatoire de Paris, 61, avenue de l'Observatoire, Paris 14-e, France

DE LOORE (Dr C.W.H.), Astrofysisch Instituut, Vrije Universiteit Brussel, F. D. Rooseveltlaan, 50, B-1050 Brussel, Belgium

* Delplace (Dr A.-M.), Observatoire de Paris, Section d'Astrophysique de Meudon, 92190 Meudon, France

Delsemme (Dr A.H.), Department of Physics and Astronomy, University of Toledo, Toledo, Ohio 43606, U.S.A.

De Marcus (Prof. W.C.), Department of Physics, University of Kentucky, Lexington, Kentucky, U.S.A.

Demarque (Prof. P.R.), Yale University Observatory, Box 2023, Yale Station, New Haven, Connecticut 06520, U.S.A.

Demenko (Mrs Dr A.A.), Astronomical Observatory of the University, Ukrainian SSR, Kiev, U.S.S.R.

Demers (Prof. Dr S.), Institute of Astronomy, Laurentian University, Sudbury, Ontario, Canada

* Demin (Prof. Dr V.G.), Moscow University, Leninskie Gory, Moscow V-234, U.S.S.R.

De Mottoni (Dr G.), Via Fratelli Rosselli 15/23, Genova, Italy

Denisse (Dr J.F.), Observatoire de Paris, 61, avenue de l'Observatoire, Paris-14e, France

Dennison (Dr E.W.), Mount Wilson and Palomar Observatories, 813 Santa Barbara Street, Pasadena, California, U.S.A.

* Dennison (Dr P.A.), Department of Physics, The University of Adelaide, Adelaide, S.A. 5001, Australia

Dent (Dr W.A.), Physics and Astronomy Department, University of Massachusetts, Amherst, Massachusetts 01002, U.S.A.

De Sabbata (Prof. Dr V.), Istituto di Fisica, Università di Bologna, via Irnerio 46, Bologna, Italy

Deprit (Prof. A.), Department of Mathematics, University of South Florida, Tampa, Florida 33620, U.S.A.

* Desesquelles (Dr J.), Laboratoire de Spectrométrie Ionique et Moléculaire de l'Université de Lyon, 43, boulevard du 11 Novembre 1918, 69 Villeurbanne, France

Dessy (Dr J. Landi), Observatorio Nacional Argentino, Cordoba, Argentina

Detre (Dr L.), Director, Konkoly Observatory, Szabadsághegy, Budapest XII, Hungary

Deubner (Dr F.L.), Fraunhofer Institut, Schöneckstrasse 6, 7800 Freiburg, F.R.G.

Deutsch (Prof. Dr A.N.), Pulkovo Observatory, Leningrad-M 140, U.S.S.R.

* Deutschman (Dr W.A.), Smithsonian Institution, Astrophysical Observatory, Cambridge, Massachusetts 02138, U.S.A.

De Vaucouleurs (Prof. G.), Department of Astronomy, University of Texas, Austin 12, Texas 78412, U.S.A.

De Vegt (Dr Ch.), Sternwarte Bergedorf, Gojenbergsweg 112, 205 Hamburg 80, F.R.G.

* Devinney (E.J., Jr.), University of Southern Florida, Tampa, Florida 33612, U.S.A.

Dewhirst (Dr D.W.), The Observatories, Madingley Road, Cambridge, U.K.

* DeWitt (Dr B.S.), Department of Physics, University of Texas at Austin, Austin, Texas 78712, U.S.A.

DeWitt (Dr J.H.), Arthur J. Dyer Observatory, Vanderbilt University, Nashville 5, Tennessee, U.S.A.

* DeYoung (Dr D.S.), National Radio Astronomy Observatory, Edgemont Road, Charlottesville, Virginia 22901, U.S.A.

Dezsö (Prof. L.), Heliophysical Observatory, Hungarian Academy of Sciences, Debrecen 10, Hungary

Dibay (Dr E.A.), Sternberg Astronomical Institute, Moscow, U.S.S.R.

Dicke (Dr R.H.), P.O. Box 708, Jadwin Hall, Princeton University, Princeton, New Jersey 08540, U.S.A.

Dickel (Mrs Dr H.R.), University of Illinois Observatory, Urbana, Illinois 61801, U.S.A.

Dickel (Dr J.R.), University of Illinois Observatory, Urbana, Illinois 61801, U.S.A.

Dickens (R.J.), Royal Greenwich Observatory, Herstmonceux Castle, Hailsham, Sussex, U.K.

Dieckvoss (Prof. Dr W.), Hamburger Sternwarte, Hamburg-Bergedorf, F.R.G.

DIETER (Mrs Dr N.H.), Radio Astronomy Laboratory, University of California, Berkeley, California 94720, U.S.A.
DIMITROFF (Prof. G.Z.), R.F.D. No. 1, Hartland, Vermont 05048, U.S.A.
DIMOV (Dr N.A.), Moscow D-481, Poste restante, U.S.S.R.
DINESCU (Dr A.), Observatorul Astronomic, Strada Cutitul de Argint 5, Bucuresti 28, Roumania
DINGENS (Prof. Dr P.), Kortrijksesteenweg 523, B-9000 Gent, Belgium
DINGLE (Prof. H.), 104, Downs Court Road, Purley, Surrey, U.K.
DINULESCU (Prof. N.), Rue Corbeni, 30, Secteur 2, Bucarest, Roumania
DIRIKIS (Dr M.A.), Astronomical Observatory of the Latvian University, Latvian SSR, Riga, U.S.S.R.
* DISNEY (Dr M.), Mount Stromlo and Siding Spring Observatories, Private Bag, Woden P.O., A.C.T. 2606, Australia
DIVAN (Miss Dr L.), Institut d'Astrophysique, 98bis, boulevard Arago, 75014 Paris, France
DIVARI (Dr N.B.), Odessa Politechnical Institute, Odessa, U.S.S.R.
DIXON (Dr M.E.), Department of Astronomy, University of Edinburgh, Edinburgh, U.K.
* DIXON (Dr R.S.), The Ohio State University Radio Observatory, 2015 Neil Avenue, Columbus, Ohio 43210, U.S.A.
DIZER (Prof. Dr M.), Astronomical Observatory, Kandilli, Çengelköy, Istanbul, Turkey
DJURKOVIĆ (P.M.), Astronomical Observatory, Volgina 7, Beograd, Yugoslavia
DLUZHNEVSKAYA (Mrs Dr O.B.), Astronomical Council, USSR Academy of Sciences, Vavilov Str. 34, Moscow V-312, U.S.S.R.
DOAN (Dr N.H.), Observatoire de Lyon, 69, St. Genis Laval, France
DOAZAN (Mrs V.), Observatoire de Paris, 61, avenue de l'Observatoire, 75-Paris-14e, France
DOBRITSCHEV (V.M.), Department of Astronomy, '7th November' Street 1, Sofia, Bulgaria
DOBRONRAVIN (Dr P.P.), Crimean Astrophysical Observatory, USSR Academy of Sciences, p/o Nauchny, Crimea, U.S.S.R.
DOBROVOL'SKIJ (Dr O.V.), Institute of Astrophysics, Dushanbe, U.S.S.R.
DOBRZYCKI (Dr J.), Institute of the History of Science and Technics, Nowy Swiat 72, Warszawa, Poland
* D'ODORICO (Dr S.), Osservatorio Astrofisico, Asiago (Vicenza), Italy
DODSON-PRICE (Mrs Dr H.W.), McMath Hulbert Observatory, University of Michigan, Pontiac 4, Michigan, U.S.A.
DOĞAN (Dr N.), Faculty of Science, Department of Astronomy, Ankara, Turkey
DOHERTY (L.H.), Astrophysics Branch, National Research Council, Ottawa 7, Ontario, Canada
DOHERTY (Dr L.R.), Astronomy Department, Sterling Hall, University of Wisconsin, Madison, Wisconsin 53706, U.S.A.
* DOLAN (Dr J.F.), Warner Swasey Observatory, East Cleveland, Ohio 44112, U.S.A.
DOLGINOV (Prof. Dr A.Z.), Physical-Technical Institute, USSR Academy of Sciences, Leningrad, U.S.S.R.
DOLIDZE (Dr M.V.), Abastumani Observatory, Abastumani, Georgia, U.S.S.R.
DOLLFUS (Dr A.), Observatoire de Paris, Section d'Astrophysique, 92190 Meudon, France
DOMINKO (Prof. Dr F.), Saranovićeva 11, Ljubljana, Yugoslavia
DOMMANGET (Dr J.), Chef du Département d'Astrométrie et de Mécanique Céleste, Observatoire Royal de Belgique, avenue Circulaire, 3, B-1180 Bruxelles, Belgium
DONITCH (N.), address unknown
DONN (Dr B.D.), Code 613, Goddard Space Flight Center, Greenbelt, Maryland, U.S.A.
DOROSHKEVICH (Dr A.G.), Institute of Applied Mathematics, USSR Academy of Sciences, Moscow, U.S.S.R.
DORSCHNER (Dr J.), University Observatory, Schillergässchen 2, 69 Jena, G.D.R.
DOS REIS (Prof. M.), Director of the Observatory, Coimbra, Portugal
DOSSIN (Dr F.), European Southern Observatory, Am Bahnhof 21, 205 Hamburg 80, F.R.G.

DOUGHTY (Dr N.A.), Department of Physics, University of Canterbury, Private Bag, Christchurch, New Zealand

DOUGLAS (Dr A.E.), Division of Physics, National Research Council, Ottawa, Ontario, Canada

DOUGLAS (Miss Dr A.V.), 127 King St. W., Kingston, Ontario, Canada

DOUGLAS (Prof. J.N.), Department of Astronomy, The University of Texas, Austin, Texas, U.S.A.

DRAKE (Prof. F.D.), Department of Astronomy, Cornell University, Ithaca, New York 14850, U.S.A.

DRÂMBĂ (Prof. Dr C.), Observatoire de Bucarest, 5, rue Cutitul de Argint, Bucarest 28, Roumania

* DRAVSKIKH (Dr A.F.), Leningrad Branch of Special Astrophysical Observatory, Academy of Sciences, Pulkovo, Leningrad 196140, U.S.S.R.

DRESSLER (Dr K.), Labor. Phys. Chem. E.T.H., Zürich, Switzerland

DRILLING (Dr J.S.), Louisiana State University, Baton Rouge, Louisiana 70803, U.S.A.

* DRYER (Dr M.), Space Environment Laboratory, National Oceanic & Atmospheric Administration, Boulder, Colorado 80302, U.S.A.

DROFA (Dr V.K.), Astronomical Observatory, Kiev University, Kiev, U.S.S.R.

DUBINSKIJ (Dr B.A.), Scientific Council for 'Radio Astronomy', USSR Academy of Sciences, Moscow, U.S.S.R.

DUBOSHIN (Prof. Dr G.N.), Sternberg Astronomical Institute, Moscow B-234, U.S.S.R.

DUBOV (Dr E.E.), Crimean Astrophysical Observatory, USSR Academy of Sciences, p/o Nauchny, Crimea, U.S.S.R.

DUCHESNE (Dr M.), Observatoire de Paris, 61, avenue de l'Observatoire, Paris-14e, France

DUFAY (M.), Institut de Physique Générale de la Faculté des Sciences de Lyon, 18 quai Claude Bernard, Lyon, France

DUFLOT (Mrs Dr M.), Observatoire de Marseille, 2 Place Le Verrier, 13004 Marseille, France

DUFLOT (R.), Chargée de Recherches CNRS, Observatoire de Marseille, 2, place Le Verrier, 13004 Marseille, France

* DULEY (Prof. W.W.), York University, Toronto, Ontario, Canada

DULK (Dr G.A.), Department of Astro-Geophysics, University of Colorado, Boulder, Colorado 80302, U.S.A.

DUMONT (R.), Observatoire de Bordeaux, 33-Floirac, France

DUMONT (Miss Dr S.), Institut d'Astrophysique, 98bis, boulevard Arago, 75014 Paris, France

DUNCOMBE (Dr R.L.), Director, Nautical Almanac Office, U.S. Naval Observatory, Washington D.C. 20390, U.S.A.

DUNHAM JR (Dr T.), P.O. Box 135, Chocorua, New Hampshire 03817, U.S.A.

DUNKELMAN (Dr L.), Astrophysics Branch, Code 613, Goddard Space Flight Center, Greenbelt, Maryland 20771, U.S.A.

DUNN (Dr R.B.), Sacramento Peak Observatory, Sunspot, New Mexico, U.S.A.

* DUPREE (Dr A.), Harvard College Observatory, Cambridge, Massachusetts 02138, U.S.A.

DURNEY (Dr B.), High Altitude Observatory, Boulder, Colorado 80302, U.S.A.

* DURRANT (Dr Ch.J.), Fraunhofer Institut, Schöneckstrasse 6, D78 Freiburg-im-Breisgau, F.R.G.

* DUTHIE (Dr J.G.), The University of Rochester, Department of Physics and Astronomy, Rochester, New York 14627, U.S.A.

DVORYASHIN (Dr A.S.), Crimean Astrophysical Observatory, USSR Academy of Sciences, Crimea, U.S.S.R.

* DYCK (Dr M.), Kitt Peak National Observatory, Tucson, P.O. Box 26732, Arizona 85726, U.S.A.

DYER (Dr E.R.), National Academy of Sciences, 2101 Constitution Avenue, N.W., Washington D.C. 20418, U.S.A.

DYSON (Dr F.J.), Institute for Advanced Studies, Princeton, New Jeresy 08540, U.S.A.

DYSON (Dr J.E.), Department of Astronomy, The University, Manchester M13 9PL, U.K.

DZHAPIASHVILI (Dr V.P.), Vice-Director, Abastumani Astrophysical Observatory, Abastumani, Georgia, U.S.S.R.

* DZIEMBOWSKI (Dr W.), Polish Academy of Sciences, Institute of Astronomy, Al. Ujazdowskie 4, Warszawa, Poland

DZIGVASHVILI (Dr R.M.), Abastumani Astrophysical Observatory, Georgian Academy of Sciences, Abastumani, U.S.S.R.

EBBIGHAUSEN (Prof. E. G.), Department of Physics, University of Oregon, Eugene, Oregon, U.S.A.

EDDY (Dr J.), High Altitude Observatory, Boulder, Colorado 80302, U.S.A.

EDLÉN (Prof. B.), Institute of Physics, Sölvegatan 14, S-223 62 Lund, Sweden

EDMONDS JR (Dr F.N.), Department of Astronomy, The University of Texas, Austin, Texas 78712, U.S.A.

EDMONDSON (Dr F.K.), Swain Hall West 319A, Indiana University, Bloomington, Indiana 47401, U.S.A.

* EDWARDS (Prof. P.J.), Department of Physics, University of Otago, P.O. Box 56, Dunedin, New Zealand

EELSALU (Dr H.), W. Struve Astrophysical Observatory, Tartu, Tôravere, Estonia, U.S.S.R.

EFREMOV (Dr Ju.I.), Astronomical Council, USSR Academy of Sciences, Vavilov Str. 34, Moscow V-312, U.S.S.R.

* EFREMOV (Dr Yu.N.), Sternberg Astronomical Institute, Moscow V-234, 117234, U.S.S.R.

EGGLETON (Dr P.P.), Institute of Theoretical Astronomy, Madingley Road, Cambridge, U.K.

EICHHORN-VON WURMB (Prof. H.K.), Department of Astronomy, University of South Florida, Tampa, Florida 33549, U.S.A.

EINASTO (Dr J.), W. Struve Astrophysical Observatory, Tartu, Tôravere, Estonia, U.S.S.R.

* EKERS (Dr R.D.), Kapteyn Laboratorium, Postbus 800, Groningen, The Netherlands

ELFORD (Dr W.G.), Department of Physics, University of Adelaide, Adelaide, Australia

ELGARÖY (Dr Ö.), Institute of Theoretical Astrophysics, University of Oslo, Blindern, Oslo 3, Norway

ELLIOTT (Dr I.), Dunsink Observatory, Castleknock, Co. Dublin, Ireland

ELLIS (Dr G.F.R.), Department of Applied Mathematics and Theoretical Physics, Silver Street, Cambridge, U.K.

ELLIS (Prof. G.R.A.), University of Tasmania, P.O. Box 2520, Hobart, Tasmania, Australia

ELLYETT (Prof. C.D.), Department of Physics, University of Newcastle, Newcastle, Australia

ELSÄSSER (Prof. Dr H.), Max-Planck-Institut für Astronomie und Landessternwarte, 6900 Heidelberg-Königstuhl, F.R.G.

ELSMORE (B.), Mullard Radio Observatory, Cavendish Laboratory, Free School Lane, Cambridge, U.K.

ELSTE (Dr G.), Department of Astronomy, University of Michigan, Ann Arbor, Michigan 48104, U.S.A.

ELSTON (Prof. W.E.), University of New Mexico, Department of Geology, Albuquerque, New Mexico 87106, U.S.A.

ELVEY (Dr C.T.), 5359 E. Hawthorne Street, Tucson, Arizona 85711, U.S.A.

ELVIUS (Mrs Dr A.M.), Stockholm Observatory, S-133 00 Saltsjöbaden, Sweden

ELVIUS (Prof. T.), Director, Astronomical Observatory, S-222 24 Lund, Sweden

ELWERT (Prof. Dr G.), Astronomisches Institut der Universität, Gmelinstrasse 6, Tübingen, F.R.G.

EL'YASBERG (Prof. Dr P.E.), Institute of Cosmical Research, USSR Academy of Sciences, Moscow, U.S.S.R.

EMINZADE (Dr T.A.), Shemakha Astrophysical Observatory, Azerbajdzhan Academy of Sciences, Baku, U.S.S.R.

* ENCRENAZ (Dr P.), Observatoire de Meudon, 92190 Meudon, France

ENGELHAD (Prof. Dr E.J.G.), 33 Braunschweig, Sackring 34, F.R.G.

ENGIN (Dr S.), Faculty of Science, Department of Astronomy, Ankara, Turkey
ENGVOLD (O.), Institute of Theoretical Astrophysics, University of Oslo, P.O. Box 1029, Blindern, Oslo 3, Norway
* ENOME (Dr S.), Toyokawa Observatory, Toyokawa, 442 Japan
ENSLIN (Dr H.), Deutsches Hydrographisches Institut, 2 Hamburg 4, F.R.G.
EPSTEIN (Dr E.E.), Aerospace Corporation, P.O. Box 95085, Los Angeles, California 90045, U.S.A.
EPSTEIN (Dr I.), Rutherfurd Observatory, Columbia University, New York 27, N.Y., U.S.A.
ERICKSON (Dr W.C.), University of Maryland, Department of Physics and Astronomy, College Park, Maryland, U.S.A.
ERIKSEN (G.), Institute of Theoretical Astrophysics, University of Oslo, Blindern, Oslo 3, Norway
ERPYLEV (Dr N.P.), Astronomical Council, USSR Academy of Sciences, Vavilov Str. 34, Moscow V-312, U.S.S.R.
ERUSHEV (Dr N.N.), Crimean Astrophysical Observatory, USSR Academy of Sciences, Crimea, U.S.S.R.
ESHLEMAN (Prof. V.R.), Stanford University, Stanford, California, U.S.A.
ESIPOV (Dr V.P.), Sternberg Astronomical Institute, Moscow, U.S.S.R.
ESKIOĞLU (Dr A.N.), P.K. 7, Fener–Istanbul, Turkey
ESSEN (Dr L.), National Physical Laboratory, Teddington, Middlesex, U.K.
EVANS (Dr D.S.), Department of Astronomy, University of Texas, Austin, Texas 78712, U.S.A.
EVANS (Dr J.V.), MIT Lincoln Laboratory, Millstone/Haystack Observatory, Lexington, Massachusetts 02173, U.S.A.
EVANS (Dr J.W.), Director, Sacramento Peak Observatory, Sunspot, New Mexico, U.S.A.
EVDOKIMOV (Dr Yu.V.), Astronomical Observatory, Kazan University, Kazan, U.S.S.R.
EVERHART (Dr E.), Department of Physics, University of Denver, 2115 S. University Blvd, Denver, Colorado 80210, U.S.A.
* EVIATAR (Prof. A.), Tel Aviv University, Ramat Aviv, Israel
EWEN (Dr H.I.), 60 Beaver Road, Weston, Massachusetts, U.S.A.
EZER (Mrs Prof. Dr D.), Department of Physics, Middle East Technical University, Ankara, Turkey
EZERSKIJ (Dr V.I.), Astronomical Observatory, Kharkov University, Kharkov, U.S.S.R.

FABRE (Dr H.), Observatoire de Nice, Le Mont-Gros, 06-Nice, France
FALCIANI (Dr R.), Osservatorio Astronomico di Capodimonte, Via Moiariello, 16,80131 Naples, Italy
* FALLER (Dr J.), Joint Institute for Laboratory Astrophysics, Boulder, Colorado 80302, U.S.A.
* FÄLTHAMMAR (Prof. C.-G.), Department of Plasma Physics, Royal Institute of Technology, S-100 44 Stockholm, Sweden
FANTI (Dr R.), Physical Institute of the University of Bologna, via Irnerio 46, Bologna, Italy
FARAGGIANA (Miss Dr R.), Astronomical Observatory Trieste, via G.B. Tiepolo 11, Trieste, Italy
FATCHIKHIN (Dr N.V.), Pulkovo Observatory, Leningrad, U.S.S.R.
* FAULKNER (Dr D.J.), Mount Stromlo and Siding Spring Observatories, Private Bag, Woden P.O., A.C.T. 2606, Australia
FAULKNER (Dr J.), Lick Observatory, University of California, Santa Cruz, California 95060, U.S.A.
* FAY (Dr T.), Indiana University, Department of Astronomy, Bloomington, Indiana 47401, U.S.A.
FAZIO (Dr G.G.), Smithsonian Astrophysical Observatory, 60 Garden Street, Cambridge, Massachusetts 02138, U.S.A.
FEAST (Dr M.W.), Radcliffe Observatory, P.O. Box 373, Pretoria, South Africa

* FEAUTRIER (Dr N.), Observatoire de Paris, Section d'Astrophysique de Meudon, 92190 Meudon, France

FEDOROV (Prof. E.P.), The Main Astronomical Observatory of the Ukrainian Academy of Sciences, Kiev-127, Goloseevo, U.S.S.R.

FEDYNSKIJ (Prof. Dr V.V.), Astronomical Council, USSR Academy of Sciences, Vavilov Str. 34, Moscow V-312, U.S.S.R.

FEHRENBACH (Prof. Ch.), Directeur de l'Observatoire de Haute Provence, Saint Michel l'Observatoire, 04300 Forcalquier, France

FEINSTEIN (Dr A.), Observatorio Astronómico, La Plata, Argentina

FEIX (Dr G.), Astronomisches Institut der Ruhr-Universität, Postfach 2148, 4630 Bochum-Querenburg, F.R.G.

* FELDMAN (Dr P.A.), CRESS, Petrie Science Building, York University, 4700 Keele Street, Downsview (Toronto), M3J-1P3, Ontario, Canada

FELDMAN (Dr U.), Department of Physics and Astronomy, Tel-Aviv University, Ramat-Aviv, Israel

FELENBOK (P.), Observatoire de Paris, Section d'Astrophysique, 92190 Meudon, France

FELLGETT (Prof. P.B.), Department of Applied Physical Sciences, University of Reading, Whiteknights Park, Reading, Berks, U.K.

* FELLI (Dr M.), Osservatorio Astrofisico di Arcetri, Largo Enrico Fermi, 5, Firenze, Italy

FELTEN (Dr J.E.), Steward Observatory, University of Arizona, Tucson, Arizona 85721, U.S.A.

FENKART (Dr R.P.), Astronomisches Institut der Universität Basel, Venusstrasse 7, CH-4102 Binningen, Switzerland

* FENTON (Dr K.B.), Department of Physics, University of Tasmania, Hobart, Tasmania 7001, Australia

FERNIE (Dr J.D.), David Dunlap Observatory, Richmond Hill, Ontario, Canada

FERRARI D'OCCHIEPPO (Prof. Dr K.), Universitäts-Sternwarte, Türkenschanzstrasse 17, A-1180 Wien, Austria

FERRAZ-MELLO (Dr S.), Instituto Technológico de Aeronáutica, Observatório, São José dos Campos, São Paulo, Brazil

FESENKOVA (Mrs Prof. Dr E.V.), Astronomical Council, USSR Academy of Sciences, Vavilov Street 34, Moscow V-312, U.S.S.R.

FICHERA (Prof. E.), Instituto Universitario Navale, Astronomia Generale e Sferica, Via Amm. Acton - Napoli, Italy

FIELD (Dr G.B.), Harvard College Observatory, 60 Garden Str., Cambridge, Mass. 02138, U.S.A.

FICHTEL (Dr C.E.), NASA, Goddard Space Flight Center, Greenbelt, Maryland 20771, U.S.A.

FIELDER (Dr G.), Lunar and Planetary Unit, Department of Environmental Sciences, University of Lancaster, Bailrigg, Lancaster, U.K.

FILLIOZAT (Dr J.), Professeur au Collège de France, 11, place Marcelin Berthelot, 75-Paris-5e, France

FINDLAY (Dr J.W.), National Radio Astronomy Observatory, Edgemont Road, Charlottesville, Virginia, U.S.A.

FINK (Dr U.), Lunar and Planetary Laboratory, University of Arizona, Tucson, Arizona 85721, U.S.A.

FINN (Dr G.D.), Institute for Astronomy, 2525 Correa Road, Honolulu, Hawaii 96822, U.S.A.

FINSEN (Dr W.S.), P.O. Box 4204, Johannesburg, South Africa

FIREMAN (Dr E.L.), Smithsonian Astrophysical Observatory, 60 Garden Str., Cambridge, Massachusetts 02138, U.S.A.

* FIRMANI (Dr C.), Instituto de Astronomía, UNAM, Apartado Postal 70–264, México D.F., Mexico

* FIRNEIS (Dr M.G.), Institut für Theoretische Astronomie, Türkenschanzstrasse 17, 1180 Wien, Austria

FIROR (Dr J.W.), High Altitude Observatory, University of Colorado, Boulder, Colorado, U.S.A.

FISCHEL (Dr D.), Code 613, NASA, Goddard Space Flight Center, Greenbelt, Maryland 20771, U.S.A.

FISCHER (Prof. Dr P.L.), Universitäts-Sternwarte Wien, A-1180 Wien, Türkenschanzstrasse 17, Austria

FISCHER (P.C.), Lockheed Palo Alto Research Laboratory, 3251 Hanover Street, Palo Alto, California 94304, U.S.A.

FISCHER (Dr R.R.), Institute for Astronomy, University of Hawaii, 2525 Correa Road, Honolulu, Hawaii 96822, U.S.A.

FISHKOVA (Mrs Dr L.M.), Abastumani Astrophysical Observatory, Abastumani, Georgia, U.S.S.R.

FITCH (Dr W.S.), Steward Observatory, University of Arizona, Tucson, Arizona, U.S.A.

* FITZGERALD (Dr M.P.), University of Waterloo, Waterloo, Ontario, Canada

* FIX (Dr J.D.), The University of Iowa, Iowa City, Iowa 52240, U.S.A.

FLECKENSTEIN (Prof. Dr J.O.), Rebgasse 32, 4000 Basel, Switzerland

FLEISCHER (Dr R.), 1733 Church Street, N.W., Washington, D.C. 20036, U.S.A.

FLETCHER (J.M.), Dominion Astrophysical Observatory, R.R. 7, Victoria B.C., Canada

FLORENSKIJ (Dr K.P.), Institute for Space Research, USSR Academy of Sciences, Profsojuznaja Str. 88, Moscow W-485, U.S.S.R.

* FLORIDES (Dr P.S.), School of Mathematics, 39 Trinity College, Dublin 2, Ireland

* FLORSCH (Dr A.), Observatoire de Strasbourg, 11, rue de l'Université, 67000 Strasbourg, France

* FLOWER (Dr D.), Observatoire de Paris, Section d'Astrophysique de Meudon, 92190 Meudon, France

* FOGH OLSEN (H.J.), Copenhagen University Observatory, Brorfelde, DK-4340 Tølløse, Denmark

FOKKER (Dr A.D.), Sterrewacht 'Sonnenborgh', Servaas Bolwerk 13, Utrecht, The Netherlands

FOMALONT (Dr E.B.), National Radio Astronomy Observatory, Green Bank, W.Va. 24944, Box 2, U.S.A.

* FORBES (Dr E.G.), History Department, University of Edinburgh, William Robertson Building, 50, George Square, Edinburgh EH8 9JY, U.K.

FORBES (Dr J.E.), Washburn Observatory, University of Wisconsin, 475 North Charter Street, Madison, Wisconsin 53706, U.S.A.

FORD JR (Dr W.K.), Department of Terrestrial Magnetism, Carnegie Institution of Washington, Washington, D.C. 30015, U.S.A.

* FORTI (Dr G.), Osservatorio di Arcetri, Largo Enrico Fermi, 5, Firenze, Italy

FORTINI (Prof. T.), Osservatorio Astronomico, Rome, Italy

FOSTER (Mrs Dr P.R.), Department of Natural Philosophy, University of Aberdeen, Aberdeen AB9 2UE, U.K.

* FOUKAL (Dr P.), Harvard College Observatory, 60 Garden Street, Cambridge, Massachusetts 02138, U.S.A.

FOWLER (Prof. W.A.), California Institute of Technology, 1201 E. California Blvd, Pasadena, California, U.S.A.

FOX (W.E.), 40, Windsor Road, Newark, Nottinghamshire, U.K.

FRACASSINI (Prof. Dr M.), Osservatorio Astronomico di Milano, via Brera 28, Milano, Italy

FRACASTORO (Prof. M.G.), Direttore, Osservatorio Astronomico di Torino, Pino Torinese, Torino 10025, Italy

FRANKLIN (Dr F.A.), Smithsonian Institution, Astrophysical Observatory, 60 Garden Street, Cambridge, Massachusetts 02138, U.S.A.

FRANTSMAN (Dr Yu.L.), Radio Astrophysical Observatory, Latvian Academy of Sciences, Riga, Latvian SSR, U.S.S.R.

FRANZ (Dr O.G.), Lowell Observatory, Flagstaff, Arizona 86002, U.S.A.

* FRASER (Dr C.W.), University Observatory, Buchanan Gardens, St Andrews Fife KY16 9LZ, U.K.

FREDGA (Miss Dr K.), Institute of Plasma Physics, Royal Institute of Technology, S-100 44 Stockholm 70, Sweden

FREDRICK (Prof. L.W.), Leander McCormick Observatory, University of Virginia, Charlottesville, Virginia 22903, U.S.A.

FREEMAN (Dr K.C.), Mount Stromlo and Siding Spring Observatories, Research School of Physical Sciences, The Australian National University, Private Bag, Canberra 2600, Australia

FREIESLEBEN (Dr H.C.), Hydrographisches Institut der Universität, Alt-Oberweg 24, 2000 Hamburg 4, F.R.G.

FREITAS-MOURÃO (Dr R.R. de), Observatorio Nacional, Rua General Bruce 586, São Cristóvão, Rio de Janeiro, Brazil

FRESA (Prof. A.), Via Porto 23, Salerno, Italy

FRICKE (Dr K.J.), Universitätssternwarte, Geismarlandstrasse 11, 3400 Göttingen, F.R.G.

FRICKE (Prof. Dr W.), Direktor des Astronomischen Rechen-Instituts, Mönchhofstrasse 12–14, 69 Heidelberg 1, F.R.G.

FRIEDEMANN (Dr Ch.), Universitätssternwarte und Astrophysikalisches Institut der Friedrich-Schiller-Universität, Jena, G.D.R.

* FRIEDJUNG (Dr M.), Institut d'Astrophysique, 98bis, Boulevard Arago, 75014 Paris, France

FRIEDLANDER (Dr M.W.), Washington University, Department of Physics, St. Louis, Missouri 63130, U.S.A.

FRIEDMAN (Dr H.), Code 7100, U.S. Naval Research Laboratory, Washington D.C. 20390, U.S.A.

FRINGANT (Miss Dr A.M.), Institut d'Astrophysique, 98bis, Boulevard Arago, 75014 Paris, France

* FRISCH (Dr H.), Observatoire de Nice, Le Mont-Gros, 06300 Nice, France

FRISCH (Dr U.), Institut d'Astrophysique, 98bis, Boulevard Arago, 75014 Paris, France

FRITZOVÁ-ŠVESTKOVÁ (Mrs Dr L.), 7809 Buchholz, Kandelstrasse 19, F.R.G.

* FRYE (Dr G.M.), Case Institute of Technology, Cleveland, Ohio, U.S.A.

* FROESCHLE (Dr C.), Observatoire de Nice, Le Mont-Gros, 06300 Nice, France

FROLOV (Dr M.S.), Astronomical Council, USSR Academy of Sciences, Vavilov Street 34, Moscow V-312, U.S.S.R.

FUCHS (Prof. Dr J.), Universitäts-Sternwarte, Universitätsstrasse 4, Innsbruck, Austria

FUJIMOTO (Dr M.), Department of Physics, Faculty of Science, Nagoya University, Furocho, Chikusa-ku, Nagoya, Japan

FUJITA (Dr Y.), Department of Astronomy, Faculty of Science, University of Tokyo, Bunkyo-ku, Tokyo, 113 Japan

* FURENLID (Dr K.I.), Kitt Peak National Observatory, P.O. Box 26732, Tucson, Arizona 85726, U.S.A.

FURSENKO (Mrs Dr M.A.), Institute of Theoretical Astronomy, USSR Academy of Sciences, Leningrad, U.S.S.R.

GABRIEL (Dr A.H.), Astrophysics Research Unit, Culham Laboratory, Abingdon, Berkshire, U.K.

GABRIEL (Dr M.R.L.), Institut d'Astrophysique, Université de Liège, avenue de Cointe, 5, B-4200 Cointe-Ougrée, Belgium

* GAHM (Dr G.F.), Stockholm Observatory, S-133 00 Saltsjöbaden, Sweden

GAIDE (A.), Observatoire Cantonal de Genève, Genève, Sauverny, 1290, Suisse

GAIZAUSKAS (Dr V.), Astrophysics Branch, National Research Council, M-50, Ottawa 7, Ontario, Canada

GALIBINA (Mrs Dr I.V.), Institute of Theoretical Astronomy, USSR Academy of Sciences, Leningrad, U.S.S.R.

GALKIN (Dr L.S.), Crimean Astrophysical Observatory, USSR Academy of Sciences, Crimea, U.S.S.R.

GALLET (R.M.), National Bureau of Standards, Boulder, Colorado, U.S.A.

GALLOUËT (Dr L.), Observatoire de Paris, 61, avenue de l'Observatoire, 75-Paris 14e France
GALPERIN (Dr G.I.), Institute of Physics and of the Atmosphere, B. Gruzinskaya 10, Moscow, U.S.S.R.
GALT (Dr J.A.), Dominion Radio Astrophysical Observatory, Box 248, Penticton, B.C., Canada
* GAPOSCHKIN (Dr E.), Smithsonian Institution, Astrophysical Observatory, Cambridge, Massachusetts 02138, U.S.A.
GAPOSCHKIN (Dr S.), Harvard College Observatory, Cambridge, Massachusetts 02138, U.S.A.
GARDNER (Dr F.F.), CSIRO, Division of Radiophysics, P.O. Box 76, Epping, N.S.W. 2121, Australia
GARFINKEL (Dr B.), Department of Astronomy, Yale University, New Haven, Connecticut 06520, U.S.A.
* GARLICK (Dr G.F.J.), Department of Physics, University of Hull, Hull HU6 7RX, U.K.
GARMIRE (Dr G.P.), 328 Downs, California Institute of Technology, Pasadena, California 91109, U.S.A.
* GARNIER (Ing. R.), European Southern Observatory, Casilla 16317, Correo 9, Santiago de Chile, Chile
GARRISON (Prof. R.F.), Department of Astronomy, University of Toronto, Toronto 5, Ontario, Canada
GARSTANG (Dr R.H.), Joint Institute for Laboratory Astrophysics, University of Colorado, Boulder, Colorado 80304, U.S.A.
GARTON (Prof. W.R.S.), Department of Physics, Imperial College, London S.W. 7, U.K.
* GARZOLI (Mrs Dr S.L.), Marcelo T. de Alvear 2412, 4° H, Buenos Aires, Argentina
GASCOIGNE (Dr S.C.B.), Mount Stromlo Observatory, Canberra, A.C.T., Australia
GASKA (Dr S.), Astronomical Institute, N. Copernicus University, Sienkiewicza 30, Toruń, Poland
GAUSTAD (Dr J.E.), Astronomy Department, University of California, Berkeley, California 94720, U.S.A.
GAUTIER (Dr D.), Observatoire de Meudon, 92190 Meudon, France
GAVIOLA (Dr E.), Instituto Fisico Balseiro, Bariloche, Rio Negro, Argentina
GAVRILOV (Dr I.V.), Main Astronomical Observatory, Ukrainian Academy of Sciences, Kiev, U.S.S.R.
* GAY (Dr J.), Observatoire de Paris, Section d'Astrophysique de Meudon, 92190 Meudon, France
GEAKE (Dr J.E.), Faculty of Technology, University of Manchester, Manchester 1, U.K.
GEBBIE (Dr K.), Joint Institute for Laboratory Astrophysics, University of Colorado, Boulder, Colorado 80302, U.S.A.
GEHRELS (Dr T.), Space Science Building, University of Arizona, Tucson, Arizona 85721, U.S.A.
* GEISS (Prof. J.), Physikalisches Institut der Universität, Sidlerstrasse 5, 3012 Bern, Switzerland
GEL'FREJKH (Dr G.B.), Main Astronomical Observatory, USSR Academy of Sciences, Pulkovo, U.S.S.R.
* GENKIN (Dr I.L.), Astrophysical Institute of the Academy of Science of Kazakh SSR, Alma-Ata 480068, U.S.S.R.
GENT (H.), Royal Radar Establishment, St Andrews Road, Great Malvern, Worcestershire, U.K.
GENTILI DE GIUSEPPE (M.), Observatoire du Pic-du-Midi, 65 Bagnères-de-Bigorre, France
GEORGELIN (Dr Y.), Observatoire de Marseille, Place Le Verrier, 13004 Marseille, France
* GERLEI (O.), Heliophysical Observatory, Debrecen, Hungary
* GEROLA (Prof. H.), Instituto de Astronomía y Fisica del Espacio, Casilla de Correo 67, Suc. 28, Buenos Aires, Argentina
GERSHBERG (Dr R.E.), Crimean Astrophysical Observatory, USSR Academy of Sciences, Crimea, U.S.S.R.
GEYER (Dr E.H.), Observatorium Hoher List (Universität Bonn), 5568 Daun/Eifel, F.R.G.
GHABRUS (Dr R.A.), Helwan Observatory, Helwan near Cairo, Arab Rep. of Egypt
GIACAGLIA (Prof. G.E.O.), 227 Taylor Hall, The University of Texas at Austin, Austin, Texas 78712, U.S.A.

GIACCONI (Dr R.), Center for Astrophysics, Smithsonian Astrophysical Observatory / Harvard College Observatory, 60 Garden Str., Cambridge, Mass. 02138, U.S.A.
GIANNONE (Dr P.), Osservatorio Astronomico, Viale Parco Mellini 84, 00185 Roma, Italy
GIANNUZZI (Prof. M.A.), Corso Matteotti 190, Albano, Rome, Italy
GICLAS (H.L.), Lowell Observatory, Flagstaff, Arizona, U.S.A.
* GIERASCH (Dr P.), Cornell University, Space Sciences Building, Ithaca, New York 14850, U.S.A.
GIESE (Dr R.H.), Ruhr-Universität Bochum, Bereich Extraterrestrische Physik, NA Gebäude, 463 Bochum, F.R.G.
GINGERICH (Dr O.J.), Smithsonian Astrophysical Observatory, 60 Garden Street, Cambridge, Massachusetts 02138, U.S.A.
GINZBURG (Prof. V.L.), Corresponding Member of the USSR Academy of Sciences, Physical Institute, Leninsky Prospect 53, Moscow, U.S.S.R.
GIOVANELLI (Dr R.G.), CSIRO Division of Physics, University Grounds, Chippendale, N.S.W. 2008, Australia
GJELLESTAD (Miss G.), Universitetet i Bergen, Geofysisk Institutt, Avd. C, Bergen, Norway
GLAGOLEVSKIJ (Dr Ju V.), Special Astrophysical Observatory, USSR Academy of Sciences, St. Zelenchukskaya, Stavropolsky Kraj, U.S.S.R.
GLASER (Dr H.), Solar Physics Program, Physics and Astronomy Programs NASA, Office of Space Science and Applications, Washington D.C. 20546, U.S.A.
GLEDHILL (Dr J.A.), Department of Physics, Rhodes University, P.O. Box 94, Grahamstown, South Africa
GLEISSBERG (Prof. Dr W.), Senckenberganlage 23, 6000 Frankfurt, F.R.G.
GLIESE (Dr W.), Astronomisches Rechen-Institut, Mönchhofstrasse 12–14, 69 Heidelberg, F.R.G.
* GLUSHNEVA (Mrs Dr I.N.), Sternberg Astronomical Institute, Universitetskij prospekt 13, Moscow 117234, U.S.S.R.
GNEVYSHEV (Dr M.N.), Astronomical Observatory, Leningrad-M 140, U.S.S.R.
GNEVYSHEVA (Mrs Dr R.S.), Main Astronomical Observatory, USSR Academy of Sciences, Pulkovo, U.S.S.R.
* GÖBEL (Dr E.), Institut für Theoretische Astronomie, Türkenschanzstrasse 17, A-1180 Wien, Austria
GODART (Prof. Dr O.), Institut d'Astronomie et de Géophysique Georges Lemaitre, Sciences I., Avenue Baudouin 1er, B-1348 Louvain-la-Neuve, Belgium
GODOLI (Prof. Dr G.), Direttore, Osservatorio Astrofisico, Catania 95123, Italy
GÖKDOĞAN (Mrs Prof. Dr N.), Director, University Observatory, Beyazit, Istanbul, Turkey
GÖKMEN (Dr T.), Kandilli Observatory, Çengelköy, Istanbul, Turkey
GOLAY (Prof. Dr M.), Directeur de l'Observatoire de Genève, 1290 Sauverny (GE), Switzerland
GOLD (Prof. T.), Space Science Building, Cornell University, Ithaca, N.Y. 14850, U.S.A.
GOLDBERG (Prof. L.), Director, Kitt Peak National Observatory, P.O. Box 26732, Tucson, Arizona 85726, U.S.A.
GOLDREICH (Dr P.), California Institute of Technology, Pasadena, California 91109, U.S.A.
* GOLDSMITH (Dr D.W.), State University of New York, Department of Earth and Space Sciences, 2644 Cedar Street, Berkeley, Calif. 94708, U.S.A.
* GOLDSMITH (Dr S.), Department of Physics and Astronomy, Tel-Aviv University, Israel
GOLDSTEIN (Dr R.M.), Jet Propulsion Laboratory, California Institute of Technology, 4800 Oak Grove Drive, Pasadena, California 91103, U.S.A.
GOLDSTEIN (Dr S.J.), Leander McCormick Observatory, University of Virginia, Charlottesville, Virginia 22903, U.S.A.
GOLDSWORTHY (Prof. F.A.), Department of Mathematics, University of Leeds, Leeds 2, Yorkshire, U.K.
GOLDWIRE JR (Dr H.C.), Space Science Department, Rice University, Houston, Texas 77001, U.S.A.
GOLLNOW (Dr H.), Mount Stromlo Observatory, Canberra, A.C.T., Australia

GOMES (Dr A.M.), University of Brazil, Rua Ipiranga, 25–40 Andar, Rio de Janeiro, ZC-01, Brazil

GONDOLATSCH (Prof. Dr F.), Astronomisches Recheninstitut, Mönchhofstrasse 12–14, 69 Heidelberg, F.R.G.

GONZÁLEZ (Miss G.), Instituto Nacional de Astrofísica, Optica y Electrónica, Tonantzintla, Pue., Apartados Postales Nos. 216 y 51, Puebla, Pue., Mexico

GONZE (Ir. R.F.J.), Observatoire Royal de Belgique, avenue Circulaire, 3, B-1180 Bruxelles, Belgium

GOODY (Dr E.M.), Director, Blue Hill Observatory, Harvard University, Cambridge, Massachusetts 02138, U.S.A.

GOPALA RAO (U.V.), Astrophysical Observatory, Kodaikanal-3, India

GOPASYUK (Dr S.I.), Crimean Astrophysical Observatory, USSR Academy of Sciences, Crimea, U.S.S.R.

GORBATSKIJ (Dr V.G.), Leningrad University Observatory, Leningrad, U.S.S.R.

GORDON (Mrs Dr C.-Pecker-Wimel), IBM Thomas J. Watson Research Center, P.O. Box 218, Yorktown Heights, New York 10598, U.S.A.

* GORDON (Dr C.), Hampshire College, Amherst, Massachusetts 01002, U.S.A.

GORDON (Dr I.M.), Institute of Radiophysics and Electronics, Academy of Sciences of the Ukrainian SSR, Kharkov 85, U.S.S.R.

GORDON (Dr J.E.) Director, Nikolaiev Observatory, Nikolaiev, U.S.S.R.

* GORDON (Dr K.), Hampshire College, Amherst, Massachusetts 01002, U.S.A.

* GORDON (Dr M.A.), National Radio Astronomy Observatory, 950 North Cherry Avenue, Tucson, Arizona 85717, U.S.A.

* GORENSTEIN (Dr P.), American Science & Engineering, 955 Massachusetts Avenue, Cambridge, Massachusetts 02139, U.S.A.

GORGOLEWSKI (Dr S.), Astronomical Observatory, N. Copernicus University, Sienkiewicza 30, Poland

GORSHKOV (Prof. Dr P.M.), State University, Leningrad, U.S.S.R.

GORYNYA (Dr A.A.), Kiev University Observatory, Kiev, U.S.S.R.

GOSACHINSKIJ (Dr I.V.), Special Astrophysical Observatory, USSR Academy of Sciences, Leningrad Branch of the SAO, Pulkovo 196140, U.S.S.R.

* GOSLING (Dr J.), High Altitude Observatory, Boulder, Colorado, U.S.A.

GOSS (Dr W.M.), Max-Planck-Institut für Radioastronomie, Argelanderstrasse 3, 53 Bonn, F.R.G.

GOSSNER (Mrs Dr S.D.), 415 E 50th Street, New York, N.Y., U.S.A.

* GOTTLIEB (Dr C.A.), Harvard College Observatory, Cambridge, Massachusetts 02138, U.S.A.

GOTTLIEB (K.), Mount Stromlo Observatory, Canberra, A.C.T., Australia

GOUDAS (Dr C.L.), University of Patras, Patras, Greece

GOUGENHEIM (Prof. A.), Directeur du Service Hydrographique de la Marine, 30 Boulevard Flandrin, 75-Paris-16e, France

GOUGH (Dr D.O.), Institute of Theoretical Astronomy, Madingley Road, Cambridge, U.K.

* GOUGUENHEIM (Dr L.), Observatoire de Paris, Section d'Astrophysique de Meudon, 92190 Meudon, France

GOULD (Dr R.), Physics Department, University of California-San Diego, La Jolla, California, U.S.A.

GOWER (Mrs A.Ch.), University of British Columbia, Department of Physics, Vancouver 8, B.C., Canada

GOWER (Prof. Dr J.F.R.), Department of Physics, University of British Columbia, Vancouver 8, B.C., Canada

GOY (G.), Observatoire Cantonal de Genève, Genève, Sauverny 1290, Switzerland

GOYAL (A.N.), Department of Mathematics, University of Rajasthan, Jaipur, India

* GRABOSKE (Dr H.), Lawrence Livermore Laboratory, University of California, Livermore, California, U.S.A.

* GRABOWSKI (Dr B.), Opole, ul. Oleska 45 m 16, Poland
GRADSZTAJN (Dr E.), Institut de Physique Nucléaire d'Orsay, 91-Gif s/Yvette, France
GRAHAM (Dr J.A.), Cerro Tololo Inter-American Observatory, Casilla 63-D, La Serena, Chile
GRAHL (Dr B.H.), Universitäts-Sternwarte Bonn/Rhein, Poppelsdorfer Allee 49, 53 Bonn, F.R.G.
GRAINGER (Dr J.), University of Manchester, Institute of Science and Technology, Sackville Street, Manchester 1, U.K.
GRANT (Dr I.P.), Pembroke College, Oxford, U.K.
GRATTON (Prof. L.), Laboratorio di Astrofisica, Casella Postale 67, 00044 Frascati, Rome, Italy
GRAY (Dr D.F.), Department of Astronomy, The University of Western Ontario, London 72, Ontario, Canada
* GREBENIKOV (Prof. Dr E.A.), Sternberg Astronomical Institute, Universitetskij prospekt 13, Moscow 117234, U.S.S.R.
GREEN (Dr J.), 941 Via Nogales, Palos Verdes Estate, California 90274, U.S.A.
GREEN (Prof. L.C.), Strawbridge Observatory, Haverford College, Haverford, Pennsylvania 19041, U.S.A.
GREEN (Dr R.M.), Department of Astronomy, The University, Glasgow, W.2, U.K.
GREENBERG (Dr J.M.), Department of Astronomy and Space Science, ES 316, State University of New York at Albany, 1400 Washington Avenue, Albany, New York 12222, U.S.A.
* GREENSTEIN (Dr G.), Amherst College, Astronomy Department, Amherst, Massachusetts 01002, U.S.A.
GREENSTEIN (Prof. J.L.), Hale Observatories, California Institute of Technology, 1201 E. California Street, Pasadena, California 91109, U.S.A.
* GREGORY (Dr P.C.), Department of Physics, University of British Columbia, Vancouver 8, B.C., Canada
GREGUL (Mrs Dr A.J.), Astronomical Observatory of the University, Kiev, Ukrainian SSR, U.S.S.R.
GREISEN (Dr K.I.), Cornell University, Physics Department, Clark Hall, Ithaca, New York 14850, U.S.A.
GREVESSE (Dr N.), Institut d'Astrophysique, Université de Liège, avenue de Cointe, 5, B-4200 Cointe-Ougrée, Belgium
GREWING (Dr M.), Institut für Astrophysik und Extraterrestrische Forschung, Poppelsdorfer Allee 49, 5300 Bonn, F.R.G.
GREYBER (Dr H.D.), The American Association for the Advancement of Science, 1515 Massachusetts Avenue, N.W., Washington D.C. 20005, U.S.A.
GRIFFIN (Mrs Dr R.E.), The Observatories, Madingley Road, Cambridge, U.K.
GRIFFIN (Dr R.F.), The Observatories, Madingley Road, Cambridge, U.K.
* GRIFFITH (Prof. J.S.), Department of Mathematical Sciences, Lakehead University, Thunder Bay, P7B 5E1, Ontario, Canada
GRIGOREVSKIJ (Dr V.M.), Astronomical Observatory, Shevchenko Park, Odessa, GSP-714, 270609, U.S.S.R.
* GRISHCHUK (Dr L.P.), Sternberg Astronomical Institute, Universitetskij prospekt 13, Moscow 117234, U.S.S.R.
* GROSSMAN (Dr A.S.), Erwin W. Fick Observatory, Iowa State University, Ames, Iowa 50010, U.S.A.
* GROSSMAN-DOERTH (Dr U.), Fraunhofer Institut, 7800 Freiburg im Breisgau, F.R.G.
GROTH (Dr H.G.), Universitäts-Sternwarte München, Sternwartstrasse 23, 8 München, F.R.G.
GRUBISSICH (Prof. Dr C.), via Aosta 34/5, 35100 Padova, Italy
* GRUEFF (Dr G.), Laboratorio di Radioastronomia, Via Irnerio, 46, Bologna, Italy
GRUSHINSKIJ (Prof. Dr N.P.), Sternberg Astronomical Institute, Moscow, U.S.S.R.
GRYGAR (Dr J.), Astronomical Institute, Czechoslovak Academy of Sciences, Observatory Ondřejov, Czechoslovakia

GRZEDZIELSKI (Dr S.), Astronomical Observatory of Warsaw University, Al. Ujazdowskie 4, Warsaw, Poland

* GUBANOV (Dr V.S.), Pulkovo Observatory, Leningrad 196140, U.S.S.R.

* GÜDÜR (Dr N.), address unknown

* GUELIN (Dr M.), Observatoire de Paris, Section d'Astrophysique de Meudon, 92190 Meudon, France

GUÉRIN (Dr P.), Institut d'Astrophysique, 98bis, Boulevard Arago, 75014 Paris, France

GUEST (Dr J.E.), University of London Observatory, Mill Hill Park, London N.W.7, U.K.

GUINOT (Dr B.), Observatoire de Paris, 61 avenue de l'Observatoire, Paris 14e, France

GULKIS (Dr S.), Jet Propulsion Laboratory, 1836–365, 4800 Oak Grove Drive, Pasadena, California 91103, U.S.A.

* GÜLMEN (Dr O.), Ege University Observatory, Bornova, Izmir, Turkey

* GULYAEV (Dr A.P.), Sternberg Astronomical Institute, Universitetskij prospekt 13, Moscow 117234, U.S.S.R.

* GULYAEV (Dr R.A.), Institute of Terrestrial Magnetism, Ionosphere and Radio Wave Propagation, Akademischeskij gorodok, Moscow region 142092, U.S.S.R.

GUNN (Dr J.E.), Hale Observatory, California Institute of Technology, Pasadena, California 91109, U.S.A.

GÜNTZEL-LINGNER (Dr U.), Astronomisches Rechen-Institut, Mönchhofstrasse 12–14, 69 Heidelberg, F.R.G.

* GURM (Dr H.S.), Department of Physics, Panjabi University, Patiala (Panjab), India

* GURSHTEIN (Dr A.A.), Institute for Space Research, USSR Academy of Sciences, Profsojuznaya Str. 88, Moscow GSP-312, U.S.S.R.

GURSKY (Dr H.), Center for Astrophysics, Smithsonian Astrophysical Observatory / Harvard College Observatory, 60 Garden Str., Cambridge, Mass. 02138, U.S.A.

GURTOVENKO (Dr E.A.), Main Astronomical Observatory, Ukrainian Academy of Sciences, Kiev, U.S.S.R.

GURZADIAN (Dr G.A.), Byurakan Astrophysical Observatory, Armenian S.S.R., Erevan, Armenia, U.S.S.R.

* GUSEINOV (Dr O.K.), Schemakha Astrophysical Observatory, Schemakha, U.S.S.R.

GUSEJNOV (Dr R.E.), Shemaha Astrophysical Observatory, Academy of Sciences of the Azerbaidzhan SSR, Shemaha, U.S.S.R.

GUSSMANN (Dr E.-A.), Zentralinstitut für Astrophysik, DDR 15 Potsdam, Telegrafenberg, G.D.R.

* GUSTAFSSON (Dr B.), Astronomical Observatory, Box 515, S-751 20 Uppsala 1, Sweden

GUTH (Dr V.), Astronomical Institute, Czechoslovak Academy of Sciences, Observatory Ondřejov, Czechoslovakia

GUTHRIE (B.N.G.), The Royal Observatory, Edinburgh, 9, U.K.

GUTIÉRREZ-MORENO (Mrs Dr A.), Observatorio Astronómico de la Universidad de Chile, Casilla 36-D, Santiago de Chile, Chile

GYLDENKERNE (Mr K.), Universitets Observatoriet, Brorfelde, DK-4340 Tølløse, Denmark

HAAS (Prof. Dr J.), Lipschitz-Str. 1, 53 Bonn, F.R.G.

HABIBULLIN (Prof. Dr Sh.T.), Director, Astronomical Observatory of State University, Lenina 18, Kazan, U.S.S.R.

HABING (Dr H.J.), Sterrewacht, Leiden, The Netherlands

HACHENBERG (Prof. Dr O.), Direktor, Radioastronomisches Institut, Universität Bonn, Auf dem Hügel 71, 53 Bonn, F.R.G.

HACK (Mrs Prof. M.), Director, Astronomical Observatory, Via Tiepolo 11, 34131 Trieste I, Italy

* HACYAN (Dr S.), Instituto de Astronomía, UNAM, Apartado Postal 70–264, México D.F., Mexico

HADDOCK (Dr F.), The Observatory, University of Michigan, Ann Arbor, Michigan, U.S.A.

HADJIDEMETRIOU (Dr J.D.), Department of Theoretical Mechanics, University of Thessaloniki, Thessaloniki, Greece

HAERENDEL (Dr G.), Institut für extraterrestrische Physik des Max-Planck-Instituts für Physik und Astrophysik, 8046 Garching b. München, F.R.G.

HAFFNER (Prof. H.), Astronomisches Institut der Universität, 87 Würzburg, Büttnerstrasse 72, F.R.G.

HAGEN (Dr J.P.), Professor of Radio Astronomy, The Pennsylvania State University, University Park, Pennsylvania, U.S.A.

HAGFORS (Dr T.), M.I.T. Lincoln Laboratory, Millstone, Haystack, Lexington, Massachusetts 02173, U.S.A.

HAGIHARA (Prof. Y.), Department of Astronomy, Faculty of Science, University of Tokyo, Bunkyo-ku, Tokyo, Japan

HAJDUK (Dr A.), Astronomical Institute, Slovak Academy of Sciences, Dúbravská Cesta, Bratislava, Czechoslovakia

* HALL (Dr D.), Kitt Peak National Observatory, 950 North Cherry Avenue, Tucson, Arizona 85726, U.S.A.

* HALL (Dr D.S.), Associate Professor, Dyer Observatory, Vanderbilt University, Nashville, Tenessee 37235, U.S.A.

HALL (Dr J.S.), Director, Lowell Observatory, Flagstaff, Arizona, U.S.A.

HALL (Dr R.G.), U.S. Naval Observatory, Washington D.C. 20390, U.S.A.

HALLAM (Dr K.L.), Laboratory for Optical Astronomy, Code 273, Goddard Space Flight Center, Greenbelt, Maryland 20771, U.S.A.

HALLIDAY (Dr I.), Upper Atmosphere Research Section, Astrophysics Branch, National Research Council, Ottawa 7, Ontario, Canada

HAMADA (Prof. T.), Department of Physics, Ibaraki University, Mito 310, Japan

HÄMEEN-ANTTILA (Prof. Dr A.), Oulun Yliopiston Tähtitieteen Laitos, (The Astronomical Institute of the Oulu University), Oulu, Finland

HAMID (Dr S.El-Din), Department of Astronomy, Faculty of Science, University Fouad, Giza, Cairo, Arab Rep. of Egypt

* HAMMERSCHLAG (Dr Ir R.H.), Sterrewacht, Servaas Bolwerk 13, Utrecht, The Netherlands

HANASZ (Dr J.), Astronomical Observatory, Toruń-Piwnice, Poland

HANSEN (Dr C.J.), Department of Astrophysics and Physics, University of Colorado, Boulder, Colorado, U.S.A.

HANSEN (Dr R.), High Altitude Observatory, Boulder, Colorado 80302, U.S.A.

HANSSON (N.), Lund Observatory, S-222 24 Lund, Sweden

HARDEBECK (Mrs Dr E.G.), 110 Terrace Drive, Big Pine, California 93513, U.S.A.

HARDIE (Dr R.H.), Dyer Observatory, Vanderbilt University, Nashville, 5, Tennessee, U.S.A.

HARDING (G.A.), Royal Greenwich Observatory, Herstmonceux Castle, Hailsham, Sussex, U.K.

HARDORP (Prof. Dr J.), Department of Earth and Space Sciences, State University of New York, Stony Brook, New York 11790, U.S.A.

HARO (Dr G.), Instituto Nacional de Astrofísica Optica y Electrónica, Tonantzintla, Pue., Apartados Postales Nos. 216 y 51, Puebla, Pue., Mexico

* HARRINGTON (Dr J.P.), Dept. of Physics and Astronomy, University of Maryland, College Park, Maryland 20742, U.S.A.

* HARRINGTON (Dr R.S.), U.S. Naval Observatory, 950 North Cherry Avenue, Tucson, Arizona 85717, U.S.A.

HARRIS (B.J.), Perth Observatory, Bickley, 6076, Western Australia

HARRISON (Prof. E.R.), Department of Physics and Astronomy, University of Massachusetts, Amherst, Massachusetts 01002, U.S.A.

HARROWER (Prof. G.A.), Department of Physics, Queen's University, Kingston, Ontario, Canada

* HARTEN (Dr R.H.), Sterrewacht, Kaiserstraat 57, Leiden, The Netherlands

HARTNER (Dr W.), Institut für Geschichte der Naturwissenschaften, Johann Wolfgang Goethe Universität, Frankfurt am Main, F.R.G.

HARTWICK (Prof. Dr F.D.A.), Department of Physics, University of Victoria, Victoria, B.C., Canada

HARTZ (Dr T.R.), Communications Research Centre, Dept. of Communications, P.O. Box 490, Terminal A, Ottawa 2, Ontario, Canada

HARVEY (Dr Ch.), Observatoire de Meudon, 92190 Meudon, France

* HARVEY (Dr J.W.), Kitt Peak National Observatory, 950 North Cherry Avenue, Tucson, Arizona 85726, U.S.A.

HARWIT (Dr M.), Cornell University, Center for Radiophysics and Space Research, Clark Hall, Ithaca, New York 14850, U.S.A.

HARWOOD (Miss M.), Harvard College Observatory, Cambridge, Massachusetts 02138, U.S.A.

HASER (Dr L.), Max-Planck-Institut für Physik und Astrophysik, Institut für Extraterrestrische Physik, 8046 Garching bei München, F.R.G.

* HASSAN (Dr S.M.), Helwan Institute of Astronomy and Geophysics, Helwan, Cairo, Arab Rep. of Egypt

HATTORI (Dr A.), Kwasan Observatory, University of Kyoto, Yamashina, Kyoto, Japan

HAUCK (B.), Observatoire Cantonal de Genève, Genève, Sauverny 1290, Switzerland

HAUG (Dr U.), Hamburger Sternwarte, 205 Hamburg-Bergedorf, G.F.R.

HAUGE (Dr Ö.), Institute of Theoretical Astrophysics, University of Oslo, P.O. Box 1029, Blindern, Oslo 3, Norway

HAUPT (Prof. Dr H.), Director, Universitäts-Sternwarte, Universitätsplatz 5, A-8010 Graz, Austria

HAUPT (R.), Assistant Director of the American Ephemeris and Nautical Almanac, U.S. Naval Observatory, Washington, D.C., U.S.A.

HAUPT (R.F.), U.S. Naval Observatory, Washington, D.C., 20390, U.S.A.

* HAUPT (Dr W.), Astronomisches Institut der Ruhr-Universität Bochum, 4630 Bochum-Querenburg, Postfach 2148, F.R.G.

* HAVLEN (Dr R.J.), European Southern Observatory, Bergedorferstrasse 131, 205 Hamburg 80, F.R.G.

* HAVNES (Dr. O.), Institut of Theoretical Astrophysics, University of Oslo, P.O. Box 1029, Blindern, Oslo 3, Norway

HAWKING (Dr S.W.), Institute of Theoretical Astronomy, Madingley Road, Cambridge, U.K.

HAWKINS (Dr G.S.), 'Fairfields', Horstead, Norwich 66Y, U.K.

HAYAKAWA (Prof. Dr S.), Department of Physics, Faculty of Science, Nagoya University, Furocho, Chikusa-ku, Nagoya, Japan

HAYASHI (Prof. C.), Department of Nuclear Science, Faculty of Science, Kyoto University, Kyoto, Japan

HAYLI (Dr A.F.), Observatoire de Besançon, 43, Avenue de l'Observatoire, 25000 Besançon, France

* HAYMES (Dr R.C.), Rice University, Houston, Texas 77001, U.S.A.

HAZARD (Dr C.), Institute of Theoretical Astronomy, Madingley Road, Cambridge, U.K.

HAZER (Dr S.), Faculty of Science, Department of Astronomy, P.K. 21, Bornova, Izmir, Turkey

HAZLEHURST (Dr J.), The Astronomy Centre, University of Sussex, Falmer, Nr. Brighton, Sussex, U.K.

HEARD (Dr J.F.), David Dunlap Observatory, Richmond Hill, Ontario, Canada

HEARN (Dr A.G.), Lehrstuhl für Theoretische Astrophysik, Berliner Strasse 19, 69 Heidelberg 1, F.R.G.

HECKMANN (Prof. Dr O.), Schmiedesberg 2b, 2057 Reinbek, F.R.G.

HEDDLE (Dr D.W.O.), Head, Department of Physics, Royal Holloway College, University of London, Egham Hill, Egham, Surrey TW20 OEX, U. K.

HEDEMAN (Miss E.R.), McMath-Hulbert Observatory, Lake Angelus Road, Pontiac, Michigan, U.S.A.

HEESCHEN (Dr D.S.), National Radio Astronomy Observatory, Edgemont Road, Charlottesville, Virginia 22901, U.S.A.

HEIDMANN (Dr J.), Observatoire de Paris, Section d'Astrophysique, 92190 Meudon, France

* HEIDMANN (Dr N.), Institut d'Astrophysique, 98bis, Boulevard Arago, 75014 Paris, France

HEILES (Dr C.E.), Astronomy Department, University of California, Berkeley, California 94720, U.S.A.

HEINTZ (Prof. W.D.), Sproul Observatory, Swarthmore, Pa. 19081, U.S.A.

HEINTZE (Dr J.R.W.), Sterrewacht 'Sonnenborgh', Servaas Bolwerk 13, Utrecht, The Netherlands

HEISER (Dr A.), A.J. Dyer Observatory, Box 1803, Vanderbilt University, Nashville, Tenn. 37203, U.S.A.

HELFER (Dr H.L.), Department of Physics and Astronomy, University of Rochester, Rochester, N.Y. 14628, U.S.A.

* HELLER (Dr M.), ul. Powstanców Warszawy 13/94, 33–110 Tarnów, Poland

HELMKEN (Dr H.), Smithsonian Astrophysical Observatory, 60 Garden Street, Cambridge, Massachusetts 02138, U.S.A.

HEMENWAY (Dr C.L.), 100 Fuller Road, Albany, New York 12205, U.S.A.

HEMMLEB (Dipl. Ing. G.), Zentralinstitut Physik der Erde, DDR 15 Potsdam, Telegrafenberg, G.D.R.

HENIZE (Dr K.G.), Astronaut Office, NASA Manned Spacecraft Center, 2101 Webster-Seabrook Road, Houston, Texas 77058, U.S.A.

HENN (Miss F.), Astronomisches Rechen-Institut, Mönchhofstrasse 12–14, 69 Heidelberg, G.F.R.

HÉNON (M.), Observatoire de Nice, Le Mont-Gros, 06-Nice, France.

* HENOUX (Dr J.-C.), Observatoire de Paris, Section d'Astrophysique de Meudon, 92190 Meudon, France

* HENRARD (Dr J.), Facultés Universitaires de Namur, rue de Bruxelles, 61, 5000 Namur, Belgium

HENRIKSEN (Prof. Dr R.N.), Astronomy Group, Department of Physics, Queen's University, Kingston, Ontario, Canada

HENRY (Dr R.C.), E.O. Hulbert Center for Space Research, Department of Physics, The Johns Hopkins University, Baltimore, Maryland 21218, U.S.A.

HERBIG (Dr G.H.), Lick Observatory, University of California, Santa Cruz, California 95060, U.S.A.

HERCZEG (Prof. Dr T.), Department of Physics and Astronomy, University of Oklahoma, Norman, Oklahoma 73069, U.S.A.

HERGET (Dr P.), Cincinnati Observatory, Observatory Place, Cincinnati, Ohio 45208, U.S.A.

HERMAN (Mrs Dr R.), Observatoire de Paris, Section d'Astrophysique, 92190 Meudon, France

HERR (Dr R.B.), Physics Department, University of Delaware, Newark, Delaware 19711, U.S.A.

HERRICK (Dr S.), University of California, Los Angeles, California 90024, U.S.A.

HERS (J.), 48, Central Road, Linden Extension, Randburg, South Africa

HERTZ (Dr H.G.), 2301 E Street N.W., Apt. A 608, Washington, D.C. 20037, U.S.A.

HERZBERG (Dr G.), Distinguished Research Scientist, National Research Council, Ottawa, Ontario, Canada

HESSER (Dr J.E.), Cerro Tololo Inter-American Observatory, Casilla 63-D, La Serena, Chile

HEWISH (Dr A.), Cavendish Laboratory, Free School Lane, Cambridge, U.K.

* HEWITT (Dr A.V.), U.S. Naval Observatory, Flagstaff Station, Flagstaff, Arizona, U.S.A.

HEY (Dr J.S.), 4 Shortlands Close, Willingdon, Sussex, U.K.

HEYDEN S.J. (Dr F.J.), Director of Solar Division, Manila Observatory, P.O. Box 1231, Manila, Philippines D-404

* HEYVAERTS (Dr J.-F.), Observatoire de Paris, Section d'Astrophysique de Meudon, 92190 Meudon, France

HIBBS (A.R.), Jet Propulsion Laboratory, California Institute of Technology, Pasadena, California 91109, U.S.A.

HIDAJAT (Dr B.), Acting Director, Bosscha Observatory, Lembang, Java, Indonesia
HIEI (Dr E.), Tokyo Astronomical Observatory, University of Tokyo, Mitaka, Tokyo, Japan
HIGGS (Dr L.A.), Astrophysics Branch, National Research Council, Ottawa, Ontario, Canada
HILL (Dr G.), Dominion Astrophysical Observatory, R.R.7, Victoria, B.C., Canada
HILL (Dr P.W.), University Observatory, St Andrews, Fife, U.K.
HILLIARD (Dr R.), Steward Observatory, University of Arizona, Tucson, Arizona 85721, U.S.A.
* HILLS (Dr J.G.), The University of Michigan, Ann Arbor, Michigan 48104, U.S.A.
HILTNER (Dr W.A.), Yerkes Observatory, Williams Bay, Wisconsin, U.S.A.
HINDMARSH (Prof. W.R.), Physics Department, The University, Newcastle upon Tyne, 1, U.K.
HINTEREGGER (Dr H.E.), Hq. AFCRL (CRUU) Stop 30, L.G. Hanscom Field, Bedford, Massachusetts 01731, U.S.A.
HIRAYAMA (Dr T.), Tokyo Astronomical Observatory, University of Tokyo, Mitaka, Tokyo, Japan
HIROSE (Dr H.), Tokyo Astronomical Observatory, Mitaka, Tokyo, Japan
HIRST (Mr P.W.), 1, Clifford Crescent, Pekalmy Township, Bergvliet, Cape Province, South Africa
HITOTUYANAGI (Prof. Dr Z.), Department of Astronomy, Faculty of Science Tohoku University, Sendai, Japan
HJELMING (Dr R.M.), National Radio Astronomy Observatory, Edgemont Road, Charlottesville, Va. 22901, U.S.A.
HOAG (Dr A.A.), Kitt Peak National Observatory, P.O. Box 26732, Tucson, Arizona 85726, U.S.A.
* HOANG BINH (Dr D.), Observatoire de Paris, Section d'Astrophysique de Meudon, 92190 Meudon, France
HOBBS (Dr L.M.), Yerkes Observatory, Williams Bay, Wisconsin 53191, U.S.A.
HOBBS (Dr R.W.), NASA/Goddard Space Flight Center, Code 614, Solar Physics Branch, Laboratory for Space Sciences, Greenbelt, Maryland 20771, U.S.A.
HODGE (Prof. P.W.), Astronomy Department, University of Washington, Seattle, Washington 98105, U.S.A.
* HOEKSTRA (Dr R.), Laboratorium voor Ruimteonderzoek, Beneluxlaan 21, Utrecht, The Netherlands
HOFFLEIT (Miss Dr E.D.), Yale University Observatory, Box 2023, Yale Station, New Haven, Conn., U.S.A.
HØG (E.), Hamburger Sternwarte, 204 Hamburg 80, F.R.G.
HÖGBON (Dr J.A.), Stockholm Observatory, S-133 00 Saltsjöbaden, Sweden
HOGG (Dr D.E.), National Radio Astronomy Observatory, Edgemont Road, Charlottesville, Virginia 22901, U.S.A.
HÖGLUND (Dr B.), Onsala Space Observatory, 43034, Onsala, Sweden
HOHL (Dr F.), NASA Hampton, NASA Langley Research Center, Virginia, U.S.A.
HOLDEN (F.), 302 S. Market Street, San Jose, California 95113, U.S.A.
HOLMBERG (Prof. E.B.), Astronomical Observatory, Box 515, S-751 20 Uppsala 1, Sweden
HOLWEGER (Dr H.), Institut für Theoretische Physik, Oldhausenstr., 2300 Kiel, F.R.G.
* HONEYCUTT (Dr R.K.), Indiana University, Bloomington, Indiana 47401, U.S.A.
HOOGHOUDT (B.G.), University Observatory, Sterrewacht, Leiden, The Netherlands
HOPMANN (Prof. Dr J.), Meckenheimer Allee 153, D 53 Bonn, F.R.G.
HOPPE (Prof. Dr J.A.), Heinrich-Hertz Institut für Solar-Terrestrische Physik, Berlin-Adlershof, Rudower Chaussee, G.D.R.
HORÁK (Dr T.B.), Astronomical Institute, Czechoslovak Academy of Sciences, Budečská 6, Prague 2, Czechoslovakia
HORÁK (Dr Z.), Technical University, Technická 4, Prague 6, Czechoslovakia
HORI (Dr G.), Yale University Observatory, Box 2023, Yale Station, New Haven, Connecticut, U.S.A.

HOSHI (Dr R.), Department of Physics, Kyoto University, Kyoto, Japan
HOSKIN (Dr M.A.), Churchill College, Cambridge, U.K.
* HOSKING (Dr R.J.), Dept. of Applied Mathematics, University of Waikato, Hamilton, New Zealand
HOSOKAWA (Dr Y.), Yamagata University, Koshirakawa-cho, Yamagata-shi, Yamagata, Japan
* HOSOYAMA (Dr K.), International Latitude Observatory, Mizusawa, Iwate, Japan
HOTINLI (Dr M.), University Observatory, Beyazit, Istanbul, Turkey
* HOUCK (Dr J.), Cornell University, Space Sciences Building, Ithaca, New York 14850, U.S.A.
HOUCK (Dr T.E.), Washburn Observatory, University of Wisconsin, Madison, Wisconsin 53706, U.S.A.
* HOUK (Dr N.), Dept. of Astronomy, University of Michigan, Ann Arbor, Michigan 48104, U.S.A.
HOUSE (Dr L.L.), High Altitude Observatory, Boulder, Colorado 80302, U.S.A.
HOUTGAST (Dr J.), Astronomical Observatory 'Sonnenborgh', Zonnenburg 2, Utrecht, The Netherlands
HOUZIAUX (Prof. Dr L.N.), Faculté des Sciences, Département d'Astrophysique, Plaine de Nimy, B-7000 Mons, Belgium
HOVENIER (J.W.), Wiskundig Seminarium, Vrije Universiteit, De Boelelaan 1081, Amsterdam 11, The Netherlands
HOWARD (Dr R.F.), Mount Wilson and Palomar Observatories, 813 Santa Barbara Street, Pasadena, California 91106, U.S.A.
HOWARD (Dr W.E.III), National Radio Astronomy Observatory, Edgemont Road, Charlottesville, Virginia 22901, U.S.A.
HOYLE (Prof. F.), St. John's College, Cambridge, U.K.
HUANG (Dr Su-Shu), Dearborn Observatory, Northwestern University, Evanston, Illinois 60201, U.S.A.
* HUBBARD (Dr W.B.), Department of Planetary Sciences, University of Arizona, Tucson, Arizona 85721, U.S.A.
HUBE (Dr D.P.), Department of Physics, University of Alberta, Edmonton, 7, Alberta, Canada
HUBENET (Dr H.), Sterrewacht 'Sonnenborgh', Servaas Bolwerk 13, Utrecht, The Netherlands
* HUBER (Dr M.), Laboratorium für Atomphysik und Astrophysik, Eidg. Technische Hochschule, Hutterstrasse 34, 8006 Zürich, Switzerland
* HUBER (Dr M.C.E.), Harvard College Observatory, Cambridge, Mass 02138, U.S.A.
* HUEBNER (Dr W.F.), University of California, Los Alamos Scientific Laboratory, Los Alamos, New Mexico 87544, U.S.A.
HUFFER (Dr C.M.), 2942 Olive View Road, Alpine, California 92001, U.S.A.
HUGHES (Dr J.A.), U.S. Naval Observatory, Washington, D.C. 20390, U.S.A.
HUGHES (Dr V.A.), Department of Physics and Astronomy, Queen's University, Kingston, Ontario, Canada
HUGUENIN (Dr G.R.), Dept. of Physics and Astronomy, University of Massachusetts, Amherst, Mass. 01002, U.S.A.
* HULSBOSCH (Dr A.N.M.), Max-Planck-Institut für Radioastronomie, Argelanderstrasse 3, 53 Bonn, F.R.G.
HUMMER (Dr D.), Joint Institute for Laboratory Astrophysics, University of Colorado, Boulder, Colorado, U.S.A.
HUMPHREYS (Dr C.J.), Purdue University, Physics Department, Lafayette, Indiana 47907, U.S.A.
* HUMPHREYS (Dr R.), The University of Minnesota, Minneapolis, Minn. 55455, U.S.A.
HUNAERTS (Dr J.J.), Observatoire Royal de Belgique, avenue Circulaire, 3, B-1180 Bruxelles, Belgium
* HUNDHAUSEN (Dr A.), High Altitude Observatory, Boulder, Colorado, U.S.A.
HUNGER (Prof. Dr K.), Lehrstuhl für Astrophysik, Technische Universität Berlin, 1 Berlin 10, Ernst-Reuter-Platz 7, F.R.G.
* HUNT (Dr G.E.), Meteorological Office, London Road, Bracknell, Berkshire, U.K.
HUNTEN (Dr D.M.), Kitt Peak National Observatory, Box 26732, Tucson, Arizona 85726, U.S.A.

HUNTER (Dr A.), Royal Greenwich Observatory, Herstmonceux Castle, Hailsham, Sussex, U.K.
HUNTER (Dr Chr.), Department of Mathematics, The Florida State University, Tallahasse, Florida 32306, U.S.A.
HUNTER (Dr J.H.), Yale University Observatory, Box 2023, Yale Station, New Haven, Connecticut 06520, U.S.A.
HURUHATA (Prof. M.), Tokyo Astronomical Observatory, Mitaka, Tokyo, Japan
* HUTUKAWA (Dr K.), Tokyo Astronomical Observatory, Mitaka, Tokyo, Japan
HUTCHINGS (Dr J.B.), Dominion Astrophysical Observatory, R.R. 7, Victoria, B.C., Canada.
HYDER (Dr C.L.), NASA Solar Group, University of New Mexico, 800 Yale Boulevard, N.E., Albuquerque, New Mexico 87106, U.S.A.
HYNEK (Dr J.A.), Dearborn Observatory, Northwestern University, Evanston, Illinois 60201, U.S.A.
* HYUN (Prof. Jong June), Department of Astronomy and Meteorology, College of Arts and Sciences, Seoul National University, Seoul, Republic of Korea.

IANINI (Dr G.), Observatorio Astronómico, Cordoba, Argentina
* IBANOGLU (Dr C.) Ege University Observatory, Bornova, Izmir, Turkey
IBEN JR (Prof. I.), University of Illinois, Urbana-Champaign, Illinois, U.S.A.
* ICKE (Dr V.), Institute of Astronomy, Madingley Road, Cambridge CB3 OEZ, U.K.
IDLIS (Dr G.M.), Astrophysical Institute, Alma-Ata, U.S.S.R.
IIJIMA (Dr S.), Tokyo Astronomical Observatory, Mitaka, Tokyo, Japan
IKEDA (Dr T.), International Latitude Observatory, Mizusawa-shi, Iwate-Ken, Japan
IKHSANOV (Dr R.N.), Main Astronomical Observatory, USSR Academy of Sciences, Pulkovo, U.S.S.R.
IKHSANOVA (Mrs Dr V.N.), Main Astronomical Observatory, USSR Academy of Sciences, Pulkovo, U.S.S.R.
* ILL (Dr M.), Astronomical Observatory, Baja, Hungary
IMSHENNIK (Dr V.S.), Institute of Applied Mathematics, USSR Academy of Sciences, Moscow, U.S.S.R.
INGHAM (Dr M.F.), Department of Astrophysics, University Observatory, South Parks Road, Oxford, U.K.
INGRAO (H.C.), 58 Hundreds Road, Wellesley Hills, Massachusetts 02181, U.S.A.
INNANEN (Dr K.A.), Physics Department and Centre for Research in Experimental Space Science, York University, 4700 Keele St., Downsview, Toronto, Canada
* INOUE (Dr T.), Kyoto-Sangyo-University, Kamigamo, Kyoto, 603 Japan
IONNISIANI (Dr B.K.), Main Astronomical Observatory, Academy of Sciences, Pulkovo, U.S.S.R.
IOSHPA (Dr B.A.), Institute of Terrestrial Magnetism, Ionosphere, and Propagation of Radiowaves, USSR Academy of Sciences, Moscow, U.S.S.R.
* IPSER (Dr J.R.), Laboratory for Astrophysics and Space Research, University of Chicago, Chicago, Illinois 60637, U.S.A.
IRELAND (Dr J.G.), c/o 13, Gordon Road, Belvedere, Kent, U.K.
IRIARTE (B.), Instituto Nacional de Astrofísica Optica y Electrónica, Tonantzintla, Pue., Apartados Postales Nos. 216 y 51, Puebla, Pue., Mexico
IRVINE (Dr W.M.), Department of Physics and Astronomy, University of Massachusetts, Amherst, Massachusetts 01002, U.S.A.
IRWIN (Dr J.B.), Department of Earth and Planetary Science, Newark State College, Union, N.J. 07083, U.S.A.
ISHIDA (Dr K.), Tokyo Astronomical Observatory, The University of Tokyo, 2–21–1 Osawa, Mitaka, Tokyo, Japan
* ISHIZAWA (Dr T.), Department of Astronomy, University of Kyoto, Kyoto, Japan
* ISOBE (Dr S.), Tokyo Astronomical Observatory, Mitaka, Tokyo, Japan
* ISRAEL (Prof. W.), University of Alberta, Edmonton, Alberta, Canada

* ITO (Dr K.), Department of Physics, Rikkyo University, Nishi-Ikebukuro, Tokyo, Japan
ITZIGSOHN (M.), Observatorio Astronómico, La Plata, Argentina
IVANCHUK (Dr V.I.), Kiev University, Observatornaya 3, Kiev 53, Ukrainian SSR, U.S.S.R.
IVANOV (Dr V.V.), Astronomical Observatory, Leningrad University, Leningrad, U.S.S.R.
IVANOV-KHOLODNY (Dr G.S.), Institute of Applied Geophysics, Academy of Sciences of the USSR, Moscow, U.S.S.R.
IWANISZEWSKA-LUBIENSKA (Mrs Dr C.), Astronomical Observatory, N. Copernicus University, Sienkiewicza 30, Toruń, Poland
IWANOWSKA (Miss Prof. W.), Astronomical Observatory, N. Copernicus University, Sienkiewicza 30, 87–100 Toruń, Poland
IZVEKOV (Dr V.A.), Institute of Theoretical Astronomy, USSR Academy of Sciences, Leningrad, U.S.S.R.

* JAAKKOLA (T.S.), University of Helsinki, Observatory and Astrophysics Laboratory, Tähtitornimäki, SF-00130 Helsinki 13, Finland
JACCHIA (Dr L.), Smithsonian Astrophysical Observatory, 60 Garden Street, Cambridge, Mass. 02138, U.S.A.
JACKSON (C.), Yale-Columbia Station, Mount Stromlo Observatory, Canberra, A.C.T., Australia
* JACKSON (Dr J.C.), Department of Mathematics, University College London, Gower Street, London WC1E 6BT, U.K.
JACKSON (Dr P.), Universitäts-Sternwarte Wien, Türkenschanzstrasse 17, A 1180-Vienna, Austria
* JACKSON (W.M.), Astrochemistry Branch, Code 691, NASA-Goddard Space Flight Center, Greenbelt, Maryland 20771, U.S.A.
JACOBSEN (Prof. T.S.), University of Washington, Seattle 5, Washington 98115, U.S.A.
JACQUINOT (P.), Laboratoires de Bellevue, 1 Place Aristide-Briand, Bellevue, 92-Meudon, France
JÄGER (Prof. Dr F.W.), Heinrich-Hertz-Institut für Solar-Terrestrische Physik, DDR 15 Potsdam, Telegrafenberg, G.D.R.
JAKIMIEC (Dr J.), Wrocław University Astronomical Institute, Kopernika 11, Wrocław 9, Poland
* JAMES (J.F.), The Schuster Physical Laboratory, The University, Manchester M 13 9PL, U.K.
JAMES (Dr R.A.), Department of Astronomy, The University, Manchester 13, U.K.
* JAMESON (Dr R.F.), Department of Astronomy, The University, Leicester, U.K.
JÁNOSSY (Prof. L.), Director of the Central Physical Institute of Budapest, Budapest, Hungary
JÄRNEFELT (Prof. G.J.), Ohjaajantie 3 B 18, Helsinki 40, Finland
JARRETT (Prof. A.H.), Professor-Director of the Boyden Observatory, Boyden Observatory, P.O. Box 334, Bloemfontein, Republic of South Africa
JARZEBOWSKI (Dr T.), Astronomical Observatory, Kopernika 11, Wrocław, Poland
JASCHEK (Dr. C.O.R.), Observatorio Astronómico, La Plata, Argentina
JASCHEK (Mrs Dr M.), Observatorio Astronomico, La Plata, Argentina
JASTROW (Dr R.), NASA, Institute for Space Studies, 2880 Broadway, New York, N.Y. 10025, U.S.A.
JAUNCEY (Dr D.L.), Center for Radiophysics and Space Research, Cornell University, Ithaca, New York 14850, U.S.A.
JAVET (Prof. Dr P.), Chantemerle 19, 1010 Lausanne, Switzerland
JAYARAJAN (A.P.), Astrophysical Observatory, Kodaikanal-3, India
JEFFERIES (Dr J.T.), Institute for Astronomy, University of Hawaii, 2840 Kolowalu Street, Honolulu, Hawaii 96822, U.S.A.
JEFFERS (Dr H.M.), P.O. Box 5305, Carmel, California 93921, U.S.A.
* JEFFERS (Dr S.), Physics Department, York University, Downsview, Ontario M3J 1P3, Canada
JEFFREYS (Lady), Girton College, Cambridge, U.K.
JEFFREYS (Sir Harold), 160 Huntingdon Road, Cambridge, U.K.

JEFFREYS (Dr W.), University of Texas, Department of Astronomy, Austin, Texas 78712, U.S.A.
JELLEY (Dr J.V.), Nuclear Physics Division, Building 8, Atomic Energy Research Establishment, Harwell, Didcot, Berkshire, U.K.
JENKINS (Dr E.B.), Princeton University Observatory, Princeton, New Jersey 08540, U.S.A.
JENKINS (Miss L.F.), Yale University Observatory, Box 2023 Yale Station, New Haven, Connecticut, U.S.A.
JENNISON (Prof. R.C.), The Electronics Laboratory, University of Kent, Canterbury, Kent, U.K.
JENSCH (A.), Carl Zeiss Observatory, Jena, G.D.R.
JENSEN (Prof. E.), Director, Institute of Theoretical Astrophysics, University of Oslo, Blindern, Oslo 3, Norway
JERZYKIEWICZ (Dr M.), Astronomical Institute, University of Wrocław, Ul. Kopernika 11, Wrocław 9, Poland
JOHANSEN (Dr K.T.), University Observatory, Brorfelde, DK-4340, Tølløse, Denmark
* JOHNSON (Dr F.M.), Dept. of Physics, California State University, Fullerton, Calif. 92634, U.S.A.
JOHNSON (Dr H.L.), Laboratorio de Fisica Aplicada, Centro de Investigacion Cientifica y Educacion Superior Ensenada, Baja California, Mexico
JOHNSON (Dr H.M.), Dept. 52–14, Bldg. 202, Lockheed Missiles & Space Co., 3251 Hanover Street, Palo Alto, California 94304, U.S.A.
JOHNSON (Prof. H.R.), University of Indiana, Department of Astronomy, Swain Hall West 319, Bloomington, Indiana 47401, U.S.A.
JOHNSON (Dr M.C.), Department of Physics, University, Edgbaston, Birmingham 15, U.K.
* JOHNSTON (Dr K.), Naval Research Laboratory, Washington, D.C. 20390, U.S.A.
JOKIPII (Dr J.R.), 405–47 Downs Lab., California Institute of Technology, Pasadena, California 91109, U.S.A.
* JOLY (Dr F.), Laboratoire d'Astrophysique de l'Université de Bordeaux, 40, rue Lamartine, 33400 Talence, France
JONES (A.F.), 11 Hereford Street, Stoke, Nelson, New Zealand
* JONES (Dr B.F.), Royal Greenwich Observatory, Herstmonceux Castle, Hailsham, Sussex, U.K.
JONES (Dr D.H.P.), Royal Greenwich Observatory, Herstmonceux Castle, Hailsham, Sussex, U.K.
JORDAN (Dr C.), The Culham Laboratories, Abingdon, Berkshire, U.K.
JORDAN (Dr H.L.), Direktor am Institut für Plasmaphysik der Kernforschungsanlage Jülich G.M.B.H., Postfach 365, 517 Jülich 1, F.R.G.
* JORDAN (Dr S.D.), Goddard Space Flight Center, Greenbelt, Maryland 20771, U.S.A.
JØRGENSEN (Dr H.E.), University Observatory, Øster Voldgade 3, DK-1350, Copenhagen K, Denmark
* JOSEPH (Dr J.H.), Dept. of Environmental Sciences, Israel
JOSHI (M.N.), Tata Institute of Fundamental Research, Bombay, India
JOSHI (S.C.), Uttar Pradesh State Observatory, Manora Peak, Naini Tal, India
JUGAKU (Dr J.), Tokyo Astronomical Observatory, Mitaka, Tokyo, Japan
JUNG (Dr J.), Observatoire de Strasbourg, 11, rue de l'Université, 67000 Strasbourg, France
JUNKES S.J. (Dr J.), Specola Vaticana, Castel Gandolfo, Vatican City State
* JUPP (Dr A.H.), Department of Applied Mathematics, University of Liverpool, P.O. Box 147, Liverpool L69 3BX, U.K.
JURKEVICH (Dr I.), Naval Research Laboratory, Code 7850, Washington, D.C. 20390, U.S.A.

KABURAKI (Prof. M.), Department of Astronomy, Faculty of Science, University of Tokyo, Yayoicho, Bunkyo-ku, Tokyo, Japan
KADLA (Mrs Dr Z.), Pulkovo Observatory, Leningrad M-140, U.S.S.R.
* KAFKA (P.), Max-Planck-Institut für Physik und Astrophysik, 8000 München, Föhringer Ring 6, F.R.G.
KAFTAN-KASSIM (Dr M.A.), State University of New York, 1400 Washington Avenue, Albany, New York 12203, U.S.A.

KAHN (Dr F.D.), Department of Astronomy, The University, M13 9PL Manchester, U.K.
KAI (Dr K.), Tokyo Astronomical Observatory, University of Tokyo, Mitaka, Tokyo, Japan
KAIJDANOVSKIJ (Dr N.L.), Special Astrophysical Observatory, USSR Academy of Sciences, Leningrad Branch of the SAO, Pulkovo 196140, U.S.S.R.
KAISER (Prof. T.R.), Department of Physics, The University, Sheffield 10, U.K.
KAKINUMA (Dr T.), Research Institute of Atmospherics, Nagoya University, Toyokawa, Aichi, Japan
KAKUTA (Dr Ch.), International Latitude Observatory of Mizusawa, Mizusawa, Japan
KALANDADZE (Mrs Dr N.B.), Astrophysical Observatory, Abastumani, Georgia, U.S.S.R.
KALER (Dr J.B.), University of Illinois Observatory, Urbana, Illinois 31801, U.S.A.
* KALIKHEVICH (Dr N.S.), Nikolaev Astronomical Observatory, Nikolaev 327000, Ukraine, U.S.S.R.
KALINIAK (Dr A.A.), Main Astronomical Observatory, Academy of Sciences, Pulkovo, U.S.S.R.
KALINKOV (M.), Section d'Astronomie de l'Académie Bulgare des Sciences, Rue 7 Novembre No. 1, Sofia, Bulgaria
KALKOFEN (Dr W.), Smithsonian Astrophysical Observatory, 60 Garden Street, Cambridge, Massachusetts 02138, U.S.A.
KALLOGLIAN (Dr A.T.), Byurakan Astrophysical Observatory, Byurakan, Armenia, U.S.S.R.
KALMYKOV (Dr A.M.), Tashkent Astronomical Observatory, Uzbek Academy of Sciences, Tashkent, U.S.S.R.
KALNAJS (Dr A.J.), Mt. Stromlo and Siding Spring Observatories, School of Physical Sciences, The Australian National University, Box 4, P.O., Canberra, A.C.T., 2600, Australia
KAMIJO (Dr F.), Department of Astronomy, University of Tokyo, Yayoi, Bunkyo-ku, Tokyo, Japan
KAMINISHI (Prof. Dr K.), Department of Physics, Faculty of Science, Kumamoto University, Kumamoto, Japan
KANDEL (Dr R.S.), Department of Astronomy, Boston University, Boston, Massachusetts 02215, U.S.A.
* KANE (Dr S.R.), Space Sciences Laboratory, University of California, Berkeley, California 94720, U.S.A.
KANNO (Dr M.), Hida Observatory, University of Kyoto, Kamitakara, Gifu-ken, Japan
* KANYO (S.), Konkoly Observatory, 1121 Budapest-Szabadsághegy, Hungary
KAPLAN (Dr J.), Department of Physics, University of California at Los Angeles, Los Angeles, California, U.S.A.
KAPLAN (L.), Jet Propulsion Laboratory, 4800 Oak Grove Drive, Pasadena, California, U.S.A.
KAPLAN (Prof. Dr S.A.), Radiophysical Research Institute, Gorkij, U.S.S.R.
* KARACHENTSEV (Dr I.D.), Special Astrophysical Observatory, USSR Academy of Sciences, ul. Berezhnogo 167, st. Zelenchukskaja, Stravropolsky Kraj 357140, U.S.S.R.
KARANDIKAR (Dr R.V.), Director, Nizamiah Observatory, Begumpet, Hyderabad-16, India
KARDASHEV (Dr N.S.), Sternberg Astronomical Institute, Moscow, U.S.S.R.
KARLSSON (Dr B.A.R.), Institute of Astronomy, S-222 24 Lund, Sweden
KARPINSKIJ (Dr V.N.), Main Astronomical Observatory, USSR Academy of Sciences, Pulkovo, U.S.S.R.
KARYAGINA (Miss Dr Z.V.), Astrophysical Institute, Alma-Ata, U.S.S.R.
KASHSCHEEV (Prof. Dr B.L.), Kharkov Polytechnical Institute, Kharkov, U.S.S.R.
* KASPER (Dr U.), Zentralinstitut für Astrophysik, Telegrafenberg, DDR 15, Potsdam, G.D.R.
KASTLER (Prof. A.), Ecole Normale Supérieure, 45 Rue d'Ulm, Paris 14e, France
KATASEV (Dr L.A.), Commission on Comets and Meteors, Astronomical Council, USSR Academy of Sciences, Vavilov Str. 34, Moscow V-312, U.S.S.R.
* KATGERT-MERKELIJN (Mrs Dr J.K.), Sterrewacht, Rijksuniversiteit Leiden, Leiden, The Netherlands
KATO (Dr S.), Institute of Astrophysics, College of Science, University of Kyoto, Kyoto, Japan

KATSIS (Dr D.N.), 12, rue Varnis, Nea Smyrne, Athens, Greece

KAUFMANN (Prof. P.), Centro de Rádio-Astronomía e Astrofísica, Universidade Mackenzie, C.P. 8792, São Paulo, Brazil

KAULA (Dr W.M.), Institute of Geophysics and Planetary Physics, University of California, Los Angeles, California 90024, U.S.A.

KAWABATA (Dr Kin-aki), Tokyo Astronomical Observatory, University of Tokyo, Mitaka, Tokyo, Japan

* KAWABATA (Dr S.), Kyoto Gakuen University, Sogabe-cho, Kameoka, Kyoto, Japan

KAWAGUCHI (Dr I.), Kwasan Observatory, University of Kyoto, Yamashina, Kyoto, Japan

KAZES (Dr I.), Observatoire de Paris, Section d'Astrophysique, 92190 Meudon, France

KAZIMIRCHAK-POLONSKAYA (Mrs Dr H.I.), Institute of Theoretical Astronomy, USSR Academy of Sciences, Leningrad V-164, U.S.S.R.

KEAY (Dr C.S.L.), University of Newcastle, New South Wales, Australia

KEENAN (Dr P.C.), Perkins Observatory, Delaware, Ohio 43015, U.S.A.

* KEGEL (Dr W.H.), Institut für Theoretische Astrophysik der Universität, 6900 Heidelberg, Berliner Str. 19, F.R.G.

KELLER (Dr G.), Department of Astronomy, Ohio State University, 174 West 18th Avenue, Columbus, Ohio 43201, U.S.A.

KELLER (Dr H.U.), Sternwarte und Planetarium der Stadt Bochum, Castroper Strasse 67, D-4630 Bochum, F.R.G.

KELLERMAN (Dr K.I.), National Radio Astronomical Observatory, Box 2, Green Bank, West Virginia 24944, U.S.A.

KELLOG (Dr W.W.), National Center for Atmospheric Research, Boulder, Colorado 80302, U.S.A.

KENDERDINE (Dr S.), Mullard Radio Astronomy Observatory, Cavendish Laboratory, Free School Lane, Cambridge, U.K.

KENNEDY (Prof. J.E.), Department of Physics, University of Saskatchewan, Saskatoon, Canada

KERES (Prof. H.), University of Tartu, L. Koidula 6, Tartu, Estonia, U.S.S.R.

KERR (Dr F.J.), Astronomy Program, University of Maryland, College Park, Maryland 20742, U.S.A.

* KERR (Prof. R.P.), Department of Mathematics, University of Canterbury, Private Bag, Christchurch, New Zealand

KESSLER (Dr K.G.), Atomic Physics Division, National Bureau of Standards 13.00, Washington D.C. 20234, U.S.A.

* KESTEVEN (Prof. M.L.), Queen's University, Kingston, Ontario, Canada

KHACHIKIAN (Dr E.Y.), Byurakan Astrophysical Observatory, Byurakan, Armenia, U.S.S.R.

KHARADZE (Prof. Dr E.K.), Director, Astrophysical Observatory, Abastumani, Georgia, U.S.S.R.

* KHARIN (Dr A.S.), Main Astronomical Observatory of the Ukrainian Academy of Sciences, Kiev-252127, U.S.S.R.

* KHARITONOV (Dr A.V.), Astrophysical Institute of Kazak, Academy of Sciences, Alma-Ata 68, Kamenskoe plato, U.S.S.R.

KHAVTASI (Dr J.Sh.), Institute Cybernetics, Academy of Sciences, Chitadze str. 6, Tbilisi, Georgia, U.S.S.R.

KHETSURIANI (Dr T.S.), Abastumani Astrophysical Observatory, Abastumani, Georgia, U.S.S.R.

KHOKHLOVA (Mrs Dr V.L.), Astronomical Council, USSR Academy of Sciences, Vavilov Str. 34, Moscow, U.S.S.R.

KHOLOPOV (Dr P.N.), Astronomical Council of the USSR Academy of Sciences, Vavilov Str. 34, Moscow, U.S.S.R.

* KHOLSHEVNIKOV (Dr K.V.), Leningrad University, Faculty of Mathematics, 10 linia 33, Leningrad 199178, U.S.S.R.

KHROMOV (Dr G.S.), Sternberg Astronomical Institute, Moscow, U.S.S.R.

KIANG (Dr T.), Dunsink Observatory, Castleknock, Co. Dublin, Ireland

KIENLE (Prof. Dr H.), Ziegelhäuser Landstrasse 31, 6900 Heidelberg, F.R.G.
KIEPENHEUER (Prof. Dr K.O.), Direktor des Fraunhofer Instituts, Schöneckstrasse 6, 78 Freiburg i. Br., F.R.G.
KIKUCHI (Dr S.), Astronomical Institute, Tohoku University, Sendai, Japan.
KILADZE (Dr R.I.), Abastumani Astrophysical Observatory, Georgian Academy of Sciences, Abastumani, U.S.S.R.
* KILAR (Dr B.), Faculty of Architecture, Civil Engineering and Geodesy, Ljubljana University, Jamova 2, 51000 Ljubljana, Yugoslavia
KIM (Y.H.), Pyongyang Astronomical Observatory, Academy of Sciences, Pyongyang, Democratic People's Rep. of Korea
KING (Dr D.S.), Department of Astronomy, University of New Mexico, Albuquerque, New Mexico 87106, U.S.A.
KING (Dr H.C.), McLaughlin Planetarium, Royal Ontario Museum, University of Toronto, 100 Queen's Park, Toronto 5, Canada
KING (Dr I.R.), Berkeley Astronomy Department, University of California, Berkeley, California 94720, U.S.A.
KING (Dr R.B.), P.O. Box 725, Mendocino, California 95460, U.S.A.
* KING-HELE F.R.S. (D.G.), Royal Aircraft Establishment, Farnborough, Hants, U.K.
KINMAN (Dr T.D.), Kitt Peak National Observatory, P.O. Box 26732, 950 North Cherry Avenue, Tucson, Arizona 85726, U.S.A.
* KINOSHITA (Dr H.), Tokyo Astronomical Observatory, Mitaka, Tokyo, 181, Japan
KIPPENHAHN (Prof. R.), Universitäts-Sternwarte Göttingen, Geismarlandstrasse 11, 34 Göttingen, F.R.G.
KIPPER (Prof. A.), W. Struve Astrophysical Observatory, Tartu, Tôravere, Estonia, U.S.S.R.
* KIPPER (Dr T.A.), W. Struve Tartu Astrophysical Observatory, Tôravere, Tartu 202444, Estonia, U.S.S.R.
KIRAL (Dr A.), University Observatory, Beyazit, Istanbul, Turkey
KISELEV (Dr A.A.), Main Astronomical Observatory, USSR Academy of Sciences, Pulkovo, U.S.S.R.
KISLYAKOV (Dr A.G.), Radiophysical Research Institute, Gorkij, U.S.S.R.
KISSELL (Dr K.E.), Director, AF Avionics Lab/RSO, Wright-Patterson Air Force Base, Ohio 45433, U.S.A.
KITAMURA (Dr M.), Tokyo Astronomical Observatory, Mitaka, Tokyo, Japan
* KITAMURA (Dr S.), Science Education Institute of Osaka, Sumiyoshi-ku, Osaka, Japan
KIZILIRMAK (Prof. A.), Ege University Observatory, P.K. 21, Bornova, Izmir, Turkey
* KJAERGAARD (P.), Copenhagen University Observatory, Øster Voldgade 3, DK-1350 Copenhagen, Denmark
KJELDSETH MOE (Dr O.), Institute of Theoretical Astrophysics, University of Oslo, P.O. Box 1029, Blindern, Oslo 3, Norway
KLARE (Dr G.), Bad. Landessternwarte, 6900 Heidelberg-Königstuhl, F.R.G.
KLARMAN (Dr J.), Washington University, Department of Physics, Saint Louis, Missouri 63130, U.S.A.
KLECZEK (Dr J.), Astronomical Institute, Czechoslovak Academy of Sciences, Observatory Ondřejov, Czechoslovakia
* KLEIN (Dr M.J.), Jet Propulsion Laboratory, Pasadena, California 91103, U.S.A.
* KLEINMANN (Dr D.), Smithsonian Astrophysical Observatory, 60 Garden Street, Cambridge, Massachusetts 02138, U.S.A.
KLEMOLA (Dr A.R.), Lick Observatory, University of California, Santa Cruz, California 95060, U.S.A.
KLEPCZYNSKI (Dr W.J.), U.S. Naval Observatory, Washington, D.C. 20390, U.S.A.
* KLEMPERER (Dr W.K.), Space Environment Laboratory, National Oceanic and Atmospheric Agency, Boulder, Colorado 80302, U.S.A.

Klimishin (Dr I.A.), Astronomical Observatory, Lvov University, Lvov, U.S.S.R.
* Klinglesmith (Dr D.A.), Goddard Space Flight Center, Laboratory for Optical Astronomy, Greenbelt, Maryland 20771, U.S.A.
Klock (Dr B.L.), U.S. Naval Observatory, Department of Navy, Washington, D.C. 20390, U.S.A.
Knaflich (Mrs H.B.), The Boeing Company, Aerospace Group, M.S. 88-21, P.O. Box 3999, Seattle, Washington 98124, U.S.A.
* Knowles (S.B.), Naval Research Laboratory, Washington, D.C. 20390, U.S.A.
Ko (Dr Hsien-Ching), Department of Electrical Engineering, Ohio State University, Columbus, Ohio 43210, U.S.A.
Kobrin (Prof. Dr M.M.), Radiophysical Research Institute, Gorkij, U.S.S.R.
* Koch (Dr L.), Centre d'Etudes Nucléaires de Saclay, Service Electronique Physique, B.P. No. 2, 91 Gif-sur-Yvette, France
Koch (Dr R.H.), Flower and Cook Observatory, University of Pennsylvania, Philadelphia, Pennsylvania 19104, U.S.A.
Kodaira (Dr K.), Department of Astronomy, Faculty of Science, University of Tokyo, Bunkyo-ku, Tokyo, Japan
Koebcke (Dr F.), Astronomical Observatory, Sloneczna 36, Poznan, Poland
Koeckelenbergh (Dr A.), Observatoire Royal de Belgique, avenue Circulaire, 3, B-1180 Bruxelles, Belgium
Koehler (Prof. Dr J.A.R.), Department of Physics, University of Saskatchewan, Saskatoon, Canada
Koelbloed (Dr D.), Astronomical Institute, Roetersstraat 1a, Amsterdam, The Netherlands
Kogure (Dr T.), Department of Physics, Faculty of Science, Ibaraki University, Mito (310), Japan
Köhler (Prof. Dr H.), Sauerbruchstrasse 6, Heidenheim a.d. Brenz, F.R.G.
Kohoutek (Dr L.), Hamburger Sternwarte, 205 Hamburg-Bergedorf, F.R.G.
* Kolaczek (Dr B.), Al. Armii Ludowej 17/55, 00-632 Warszawa, Poland
Kolchinskij (Dr I.G.), Main Astronomical Observatory, Ukrainian Academy of Sciences, Kiev, U.S.S.R.
* Kolesnik (Mrs Dr L.N.), Main Astronomical Observatory, Academy of Sciences of the Ukrainian SSR, Kiev-252127, U.S.S.R.
Kolesov (Dr A.K.), Astronomical Observatory, Leningrad, U.S.S.R.
* Kollberg (Dr E.L.), Chalmers University of Technology, Research Laboratory of Electronics, Fack, S-402 20 Göteborg 5, Sweden
* Komarov (Dr N.S.), Odessa Astronomical Observatory, Shevchenko Park, Odessa 270014, U.S.S.R.
* Komesaroff (M.M.), CSIRO Division of Radiophysics, P.O. Box 76, Epping, N.S.W. 2121, Australia
* Kondo (Dr M.), Institute of Earth Science and Astronomy, College of General Education, University of Tokyo, Komaba 3–8–1, Meguroku, Tokyo, 153, Japan
Kondo (Dr Y.), Chief, Astrophysics Section (TN23), NASA-Manned Spacecraft Center, Houston, Texas 77058, U.S.A.
Konin (Dr V.V.), Nikolaev Observatory, Nikolaev, U.S.S.R.
Konopleva (Dr V.P.), Main Astronomical Observatory, Academy of Sciences, Ukrainian SSR, Kiev 127, U.S.S.R.
Konovovich (Dr E.V.), Sternberg Astronomical Institute, Moscow V-234, U.S.S.R.
Kopal (Prof. Z.), Department of Astronomy, The University, Manchester 13, U.K.
Kopecký (Dr M.), Astronomical Institute, Czechoslovak Academy of Sciences, Observatory Ondřejov, Czechoslovakia
Kopylov (Dr I.M.), Director, Special Astrophysical Observatory, USSR Academy of Sciences, St. Zelenchukskaya, Stavropolsky Kraj, U.S.S.R.

KORCHAK (Dr A.A.), Institute of Terrestrial Magnetism, Ionosphere and Propagation of Radio Waves, USSR Academy of Sciences, Moscow, U.S.S.R.
KORDYLEWSKI (Dr K.), Astronomical Observatory, Kopernika 27, Cracow, Poland
KOROL' (Dr A.K.), Main Astronomical Observatory, Ukrainian Academy of Sciences, Kiev, U.S.S.R.
KOROL'KOV (Dr D.V.), Special Astrophysical Observatory, USSR Academy of Sciences, Leningrad Branch of the SAO, Pulkovo 196140, U.S.S.R.
KOSIN (Dr G.S.), Main Astronomical Observatory, USSR Academy of Sciences, Pulkovo, U.S.S.R.
KOSTINA (Mrs Dr L.D.), Main Astronomical Observatory, USSR Academy of Sciences, Pulkovo, U.S.S.R.
KOSTYAKOVA (Mrs Dr E.B.), Sternberg Astronomical Institute, Moscow, U.S.S.R.
KOSTYLEV (Dr K.V.), Engelhardt Observatory, Kazan, U.S.S.R.
KOTEL'NIKOV (Acad. V.A.), Commission for Radioastronomy, USSR Academy of Sciences, Moscow, U.S.S.R.
KOTHARI (Dr D.S.), Department of Physics, University of Delhi, New Delhi, India
KOTSAKIS (Prof. Dr D.), University of Athens, Director of Astronomical Institute, National Observatory, 189, Hippocrates Str., Athens (708), Greece
KOURGANOFF (Prof. V.), Faculté des Sciences, Service d'Astronomie, 91-Orsay, France
* KOUTCHMY (Dr S.), Institut d'Astrophysique, 98 bis, Boulevard Arago, 75014 Paris, France
KOVAL' (Dr I.K.), Main Astronomical Observatory, Ukrainian Academy of Sciences, Kiev, U.S.S.R.
KOVALEVSKY (Dr J.), Groupe de Recherches de Géodesie Spatiale, Observatoire de Meudon, 92190 Meudon, France
KOVAR (Mrs Dr N.S.), Associate Professor, Physics Department, University of Houston, Houston, Texas 77004, U.S.A.
KOVAR (Dr R.P.), 1227 Saxony Lane, Nassau, Texas 77058, U.S.A.
KOVETZ (Dr A.), Department of Physics and Astronomy, Tel-Aviv University, Ramat-Aviv, Israel
KOWATCHEW (B.J.), Section d'Astronomie de l'Académie Bulgare des Sciences, Rue 7 Novembre No. 1, Sofia, Bulgaria
* KOYAMA (Dr S.), Kagawa University, Takamatsu, Kagawa, Japan
KOZAI (Dr Y.), Tokyo Astronomical Observatory, Mitaka, Tokyo, Japan
KOZHEVNIKOV (Dr N.I.), Sternberg Astronomical Institute, Moscow, U.S.S.R.
KOZIEŁ (Prof. Dr K.), Director, Jagellonian University Observatory, Kopernika 27/3, Cracow, Poland
KOZLOVSKY (Dr B.Z.), Department of Physics and Astronomy, Tel-Aviv, University, Ramat-Aviv, Israel
KOZYREV (Dr N.A.), Pulkovo Observatory, Leningrad, U.S.S.R.
KRAFT (Dr R.P.), Mount Wilson and Palomar Observatories, 813 Santa Barbara Street, Pasadena, California, U.S.A.
KRAMER (Dr K.N.), Astronomical Observatory, Odessa University, Odessa, U.S.S.R.
KRANJC (Dr A.), Osservatorio Astronomico di Bologna, Bologna, Italy
KRASOVSKIJ (Prof. Dr I.), Box 1269, Moscow K9, U.S.S.R.
KRAT (Prof. V.A.), Pulkovo Observatory, Leningrad M-140, U.S.S.R.
KRAUS (Dr J.D.), Radio Observatory, Ohio State University, Columbus, Ohio, U.S.A.
KRAUSE (Dr F.), Zentralinstitut für Astrophysik, DDR 15 Potsdam, Telegrafenberg, G.D.R.
KRAUSHAAR (Prof. W.L.), Department of Physics, University of Wisconsin, Sterling Hall, Madison, Wisconsin 53706, U.S.A.
* KREISEL (Dr E.), Zentralinstitut für Astrophysik, Telegrafenberg, DDR 15 Potsdam, G.D.R.
KRESÁK (Dr L.), Astronomical Institute, Slovak Academy of Sciences, 89930 Bratislava, Czechoslovakia

KRESÁKOVÁ (Mrs Dr M.), Astronomical Institute, Slovak Academy of Sciences, Dúbravská cesta A/11, Bratislava, Czechoslovakia

KRINOV (Dr E.L.), Committee on Meteorites of the USSR Academy of Sciences, ul. M. Ul'ianovoy 3, korpus 1, Moscow W-313, U.S.S.R.

KRISHNAN (Prof. T.), Director, Madras Institute of Technology, Chromepet, Madras 600044, India

KRISHNA SWAMY (Dr K.S.), Observational Astronomy Branch, Code 672, Laboratory for Optical Astronomy, Goddard Space Flight Center, Greenbelt, Maryland 20771, U.S.A.

KRISTENSON (Dr H.), Arkitektvägen 6, S-245 00 Staffanstorp, Sweden

KRISTIAN (Dr J.), Mount Wilson and Palomar Observatories, Carnegie Institution of Washington, California Institute of Technology, 813 Santa Barbara Street, Pasadena, California 91106, U.S.A.

* KRISTIANSSON (Prof. K.G.B.), Department of Physics, University of Lund, Sweden

KŘIVSKÝ (Dr L.), Astronomical Institute, Czechoslovak Academy of Sciences, Observatory Ondřejov, Czechoslovakia

KŘÍŽ (Dr S.), Astronomical Institute, Czechoslovak Academy of Sciences, Observatory Ondřejov, Czechoslovakia

KROGDAHL (Prof. W.S.), Department of Astronomy, University of Kentucky, Lexington, Kentucky, U.S.A.

KRON (Dr G.E.), Mt. Stromlo Observatory, Woden P.O., A.C.T. 2606, Australia

KRON (Mrs K.G.), Mt. Stromlo Observatory, Woden P.O., A.C.T. 2606, Australia

* KRONBERG (Dr P.P.), David Dunlap Observatory, University of Toronto, Richmond Hill, Ontario, Canada

KROOK (Dr M.), Harvard College Observatory, Cambridge, Massachusetts 02138, U.S.A.

KRÜGER (Dr A.), Heinrich-Hertz-Institut für Solar-Terrestrische Physik, Berlin-Adlershof, Rudower Chaussee, G.D.R.

KRÜGER (Prof. E.), Via Mauro Macchi 65, Milano, Italy

KRUSZEWSKI (Dr A.), Astronomical Observatory, Warsaw University, Al. Ujazdowskie 4, Warsaw, Poland.

KRZEMINSKI (Dr W.), Zakład Astronomii P.A.N., Aleje Ujazdowskie 4, Warszawa, Poland

* KSANFOMALITI (Dr L.V.), Space Research Institute, USSR Academy of Sciences, Profsojusnaja 88, Moscow 117485, U.S.S.R.

* KUBIAK (Dr M.), Astronomical Observatory, Warsaw University, Al. Ujazdowskie 4, 00–478 Warszawa, Poland

* KUBIČELA (A.), Astronomska Opservatorija, Volgina 7, 11050 Beograd, Yugoslavia

KUBOTA (Dr J.), Kwasan Observatory, Yamashina, Kyoto, Japan

KUHI (Dr L.V.), Berkeley Astronomy Department, University of California, Berkeley, California 94720, U.S.A.

* KÜHNE (Dr Ch.), Firma Carl Zeiss, 7082 Oberkochen, Postfach 35-36, F.R.G.

KUKARKIN (Prof. Dr B.V.), Sternberg Astronomical Institute, Moscow V-234, U.S.S.R.

KUKLIN (Dr G.V.), Post Office Meget, Sibizmiran, Irkutsk, U.S.S.R.

KULIKOV (Prof. Dr K.A.), Sternberg Astronomical Institute, Moscow V-234, U.S.S.R.

KULIKOVSKIJ (Dr P.G.), Sternberg State Astronomical Institute, Moscow V-234, U.S.S.R.

KULKARNI (Dr P.V.), Physical Research Laboratory, Nacrangpura, Ahmedabad-9, India

* KULSRUD (Dr R.M.), Princeton University, Princeton, New Jersey 08540, U.S.A.

KUMAR (Prof. S.S.), Leander McCormick Observatory, University of Virginia, Charlottesville, Virginia, U.S.A.

KUMSISHVILI (Dr J.I.), Abastumani Astrophysical Observatory, Georgian Academy of Sciences, Abastumani, U.S.S.R.

KUNDU (Dr M.R.), Astronomy Program, University of Maryland, College Park, Maryland 20742, U.S.A.

KUNKEL (Dr W.E.), Cerro Tololo Inter-American Observatory, Casilla 63-D, La Serena, Chile

KÜNZEL (H.), Heinrich-Hertz-Institut für Solar-Terrestrische Physik, DDR 15 Potsdam, Telegrafenberg, G.D.R.

KUPERUS (Dr M.), Astronomical Observatory 'Sonnenborgh', Servaas Bolwerk 13, Utrecht, The Netherlands

* KUPO (Mrs Dr I.D.), Astrophysical Institute, Kazakh SSR Academy of Sciences, Alma-Ata 480058, U.S.S.R.

KUPPERIAN JR (Dr J.E.), Astrophysics Branch, Goddard Space Flight Center, NASA, Greenbelt, Maryland, U.S.A.

KUPREVICH (Dr N.F.), Pulkovo Observatory, Leningrad, U.S.S.R.

KURIL'CHIK (Dr V.N.), Sternberg Astronomical Institute, Moscow, U.S.S.R.

* KUROCHA (Dr L.N.), Astronomical Observatory, Kiev University, street Observatornaya 3, Kiev 252053, U.S.S.R.

KURT (Dr V.G.), Sternberg Astronomical Institute, Moscow V-234, U.S.S.R.

KURTH (Dr R.), Professor of Mathematics, Georgia Institute of Technology, Atlanta, Ga., U.S.A.

KUSHWAHA (Prof. R.S.), Department of Mathematics, University of Jodhpur, Jodhpur (Raj), India

KUSTAANHEIMO (Prof. P.), Klaukkala, Finland.

* KUTTER (Dr G.S.), Evergreen State College, Olympia, Washington 98505, U.S.A.

KUTUZOV (Dr S.A.), The Astronomical Observatory, Leningrad University, Leningrad 199178, U.S.S.R.

KUZMENKO (Mrs Dr K.N.), Astronomical Observatory of the University, Odessa, U.S.S.R.

KUZ'MIN (Dr A.D.), Physical Institute, USSR Academy of Sciences, Moscow, U.S.S.R.

KUZMIN (Prof. Dr G.G.), W. Struve Astrophysical Observatory of Tartu, 202 444 Tôravere, Tartu, Estonia, U.S.S.R.

* KUZNETSOV (Dr D.A.), Special Astrophysical Observatory, USSR Academy of Sciences, ul. Berezhnogo 167, St. Zelenchukskaya, Stavropolsky Kraj 357140, U.S.S.R.

KWEE (Dr K.K.), Sterrewacht, Leiden, The Netherlands

* LABEYRIE (Dr A.), Observatoire de Paris, Section d'Astrophysique de Meudon, 92190 Meudon, France.

LABEYRIE (Dr L.), G.E.A., B.P. n° 2, 91 Gif s/Yvette, France.

LABORDE (Dr G.), Observatoire de Paris, Section d'Astrophysique, 92190 Meudon, France.

LABS (Dr D.), Landessternwarte, Königstuhl, 69 Heidelberg, F.R.G.

LACCHINI (Prof. G.B.), Borgo Durbecco, Via A. Cicognani No. 10, Faenza (Ravenna), Italy.

LACOMBE (Dr C.G.), Observatorio Nacional, Rua General Bruce, 586, São Cristovão (ZC-08), Rio de Janeiro, Gb., Brazil.

LACROUTE (Prof. P.), Directeur de l'Observatoire de Strasbourg, 11 Rue de l'Université, 67-Strasbourg (Bas-Rhin), France

LAFFINEUR (Dr M.), Institut d'Astrophysique, 98bis Boulevard Arago, 75014 Paris, France

LAGERQVIST (Prof. A.), Institute of Physics, Stockholms University, Vanadisvägen 9, S-113 46 Stockholm VA, Sweden

LAHIRI (Sir N.C.), Officer-in-charge, Nautical Almanac Division, Meteorological Department, Alipore, Calcutta, India

LALLEMAND (Dr A.), Observatoire de Paris, 61 Avenue de l'Observatoire, Paris-14e, France

* LAMBECK (Dr K.), Groupe de Recherches de Géodésie Spatiale, Observatoire de Paris, Section d'Astrophysique de Meudon, 92190 Meudon, France

LAMBERT (Dr D.), Department of Astrophysics, University Observatory, South Parks Rd, Oxford, U.K.

LAMBRECHT (Prof. Dr H.), Direktor der Universitäts-Sternwarte, Schillergässchen 2, 69 Jena, G.D.R.

LAMLA (Dr E.), Astronomische Institute der Universität, 53 Bonn, Poppelsdorfer Allee 49, F.R.G.

LANDINI (M.), Osservatorio Astrofisico di Arcetri, Firenze, Italy
LANDOLT (Dr A.), Department of Astronomy and Physics, Louisiana State University, Baton Rouge, Louisiana 70803, U.S.A.
* LANDSTREET (Prof. J.D.), Department of Astronomy, University of Western Ontario, London, Ontario, Canada
* LANG (Dr K.), California Institute of Technology, Pasadena, California 91109, U.S.A.
* LANTOS (Dr P.), Observatoire de Paris, Section d'Astrophysique de Meudon, 92190 Meudon, France
LA PAZ (Dr L.), Director Emeritus, Institute of Meteoritics, Campus P.O. Box 23, University of New Mexico, Albuquerque, New Mexico 87131, U.S.A.
LA POINTE (Dr S.M.), Department of Physics, University of Montreal, Montreal Quebec, Canada
LAPUSHKA (Dr K.K.), Astronomical Observatory, Latvian University, Riga, Latvian SSR, U.S.S.R.
LARGE (Dr M.I.), Chatterton Astronomy Department, University of Sydney, Sydney, Australia
* LARI (Dr C.), Laboratorio Nazionale di Radioastronomia, Via Irnerio, 46, Bologna, Italy
LARINK (Prof. Dr J.), Hamburger Sternwarte, 205 Hamburg-Bergedorf, F.R.G.
* LARSON (Dr R.B.), Yale University Observatory, New Haven, Connecticut 06520, U.S.A.
LARSSON-LEANDER (Dr G.), Lund Observatory, S-222 24 Lund, Sweden
* LASKARIDES (Dr P.G.), Laboratory of Astronomy, University of Athens, Panepistimiopolis, Athens (621), Greece
LASKER (Dr B.M.), Cerro Tololo Inter-American Observatory, Casilla 63-D, La Serena, Chile
* LATHAM (Dr D.W.), Smithsonian Institution Observatory, Cambridge, Mass 02138, U.S.A.
* LATOUR (Dr J.J.), Observatoire de Nice, Le Mont-Gros, 06000 Nice, France
LATYPOV (Dr A.A.), Astronomical Institute of the Usbek SSR, Academy of Sciences, Tashkent, U.S.S.R.
* LAURENT (Dr B.E.), Department of Physics, University of Stockholm, Vanadisvägen 9, S-113 46 Stockholm, Sweden
LAUSTSEN (Dr S.), European Southern Observatory, ETP Division, CERN, CH-1211 Geneva 23, Switzerland
* LAUTERBORN (Dr D.), Hamburger Sternwarte, 2050 Hamburg 80, Gojenbergsweg 112, F.R.G.
LAUTMAN (Dr D.A.), Smithsonian Astrophysical Observatory, 60 Garden Street, Cambridge, Mass. 02138, U.S.A.
* LAVAL (Dr A.), Observatoire de Marseille, 2, Place Le Verrier, 13004 Marseille, France
LAVDOVSKIJ (Dr V.V.), Pulkovo Observatory, Leningrad, U.S.S.R.
LAVROV (Dr M.I.), Engelhardt Astronomical Observatory, Kazan University, Kazan, U.S.S.R.
LAVRUKHINA (Mrs Prof. Dr A.K.), Institute of Geochemistry and Analytic Chemistry, USSR Academy of Sciences, Moscow, U.S.S.R.
LAWRENCE (Dr G.M.), LASP, Space Science Building, University of Colorado, Boulder, Colorado 80302, U.S.A.
LAYZER (Prof. D.), Harvard College Observatory, 60 Garden Street, Cambridge, Mass. 02138, U.S.A.
* LAZOVIĆ (Dr J.), Dept. of Astronomy, Faculty of Sciences, Belgrade University, Studentski trg 16, P.O. Box 550, 11001 Beograd, Yugoslavia
LEBEDINETS (Dr V.N.), Commission on Comets and Meteors, Astronomical Council, USSR Academy of Sciences, Moscow, U.S.S.R.
* LEBLANC (Dr Y.), Observatoire de Paris, Section d'Astrophysique de Meudon, 92190 Meudon, France
LEBOVITZ (Dr N.), Department of Mathematics, Eckhart Hall, Room 309, University of Chicago, 1118-32 East 58th Street, Chicago, Illinois 60637, U.S.A.
LECAR (Dr M.), Smithsonian Astrophysical Observatory ,60 Garden Street, Cambridge, Mass. 02138, U.S.A.

Le Contel (J.M.), Observatoire de Paris, 61 avenue de l'Observatoire, Paris 14e, France
Lederle (Dr T.), Astronomisches Rechen-Institut, Mönchhofstrasse 12–14, 69 Heidelberg, F.R.G.
Ledoux (Prof. Dr P.), Institut d'Astrophysique, Université de Liège, avenue de Cointe, 5, B-4200 Cointe-Ougrée, Belgium
* Lee (Dr P.D.), Louisiana State University, Baton Rouge, Louisiana 70803, U.S.A.
* Lefebvre (M.), C.N.E.S., DGP/PS, B.P. no. 4, 91 Bretigny s/Orge, France
Lefèvre (Dr J.), Observatoire de Nice, Le Mont-Gros, 06-Nice, France
Lehnert (Dr B.P.), Royal Institute of Technology, Department of Electronics, S-100 44 Stockholm 70, Sweden
Legg (Dr T.H.), Astrophysics Branch, National Research Council, Ottawa 7, Ontario, Canada
* Leibowitz (Dr E.M.), Tel-Aviv University, Ramat-Aviv, Tel-Aviv, Israel
Leighton (Prof. R.B.), California Institute of Technology, Pasadena, Calif. 91109, U.S.A.
Leikin (Dr G.A.), Astronomical Council of the USSR Academy of Sciences, Vavilov Str. 34, Moscow, V-312, U.S.S.R.
* Leinert (Dr Ch.), Max-Planck-Institut für Astronomie, 6900 Heidelberg 1, Königstuhl, F.R.G.
* Lemaire (Dr P.), C.N.R.S., Laboratoire de Physique Stellaire et Planétaire, B.P. No. 10, 91-Verrières-le-Buisson, France
* Lemke (Dr D.), Max-Planck-Institut für Astronomie, 6900 Heidelberg 1, Königstuhl, F.R.G.
Lena (Dr P.) Observatoire de Meudon, 92190 Meudon, France
Lengauer (Dr G.G.), Main Astronomical Observatory, USSR Academy of Sciences, Pulkovo, U.S.S.R.
Lequeux (J.), Observatoire de Paris, Section d'Astrophysique, 92190 Meudon, France
Leroy (Dr J.-L.), Observatoire du Pic-du-Midi, 65-Bagnères-de-Bigorre, Hautes Pyrénées, France
Le Squeren (Mrs A.M.), Observatoire de Paris, Section d'Astrophysique, 92190 Meudon, France
Letfus (Dr V.), Astronomical Institute, Czechoslovak Academy of Sciences, Observatory Ondřejov, Czechoslovakia
Leung (Dr Kam-Ching), Behlen Laboratory of Physics, University of Nebraska, Lincoln, Nebraska 68508, U.S.A.
Levallois (J.J.), Institut Geographique National, 2 Av. Pasteur, 94-Saint-Mandé, France
Levin (Dr B.J.), O. Schmidt Institute of Physics of the Earth, Academy of Sciences of the USSR, B. Gruzinskaya 10, Moscow D-242, U.S.S.R.
Lévy (Dr J.R.), Observatoire de Paris, 61 Avenue de l'Observatoire, Paris 14e, France
Li (J.Y.), Pyongyang Astronomical Observatory, Academy of Sciences, Pyongyang, Democratic People's Rep. of Korea
* Li (Dr N.C.), California University, 6531 Withworth Rd., Los Angeles, Calif. 90035, U.S.A.
Liddell (U.), National Aeronautics and Space Administration, Lunar and Planetary Program, Office of Space Science and Applications, Washington, D.C. 20546, U.S.A.
* Lieske (Dr J.H.), California Institute of Technology, Jet Propulsion Laboratory, 4800 Oak Drive, Pasadena, California 91103, U.S.A.
Liller (Dr M.), Harvard College Observatory, 60 Garden Street, Cambridge, Massachusetts 02138, U.S.A.
Liller (Prof. W.), Harvard College Observatory, 60 Garden Street, Cambridge, Massachusetts 02138, U.S.A.
Lilley (Prof. A.E.), Harvard College Observatory, Cambridge, Massachusetts 02138, U.S.A.
Limber (Dr D.N.), Yerkes Observatory, Williams Bay, Wisconsin, U.S.A.
Lin (Dr C.C.), Department of Mathematics, Massachusetts Institute of Technology, Cambridge, Massachusetts, U.S.A.
* Lincoln (Miss J.V.), Solar-Terrestrial Data Ser., National Geophysical & Solar Terrestrial Data Center, NOAA, Boulder, Colorado 80302, U.S.A.
Lindblad (Dr B.A.), Lund Observatory, S-222 24 Lund, Sweden
Lindblad (Dr P.O.), Stockholm Observatory, S-133 00 Saltsjöbaden, Sweden

LINDOFF (Dr U.I.G.S.), Institute of Astronomy, S-222 24 Lund, Sweden
LINDSAY (Dr E.M.), Director, Armagh Observatory, Armagh, U.K.
LINES (Dr A.W.), Director, Engineering Division, Science Research Council State House, High Holborn, London, WC1R 4TA, U.K.
LINFOOT (Dr E.H.), The Observatories, Madingley Road, Cambridge, U.K.
LING (Dr Chih-bing), Institute of Mathematics, Academia Sinica, P.O. Box No. 143, Taipei, Taiwan
LINK (Dr F.), Institut d'Astrophysique, 98bis, Boul. Arago, 75014 Paris, France
LINNELL (Dr A.P.), Department of Astronomy, Michigan State University, East Lansing, Michigan 48823, U.S.A.
LINNIK (Prof. Dr V.P.), Pulkovo Observatory, Leningrad, U.S.S.R.
* LINSKY (Dr J.), Joint Institute for Laboratory Astrophysics, Boulder, Colorado 80302, U.S.A.
LIPPINCOTT (S.L.), Sproul Observatory, Swarthmore College, Swarthmore, Pa., U.S.A.
LIPSKIJ (Dr Y.N.), Sternberg Astronomical Institute, Moscow, U.S.S.R.
LISZKA (Dr L.), Geophysical Observatory, S-981 00 Kiruna, Sweden
LITTLE (A.G.), Chatterton Astronomy Department, School of Physics, University of Sydney, Sydney, N.S.W., Australia
LITTLE (C.G.), Chief, Division 87, National Bureau of Standards, Room 3001, Radio Bldg., Boulder, Colorado, U.S.A.
* LITTLE (Dr L.T.), Electronics Laboratory, University of Kent at Canterbury, Kent, U.K.
LITTLEFIELD (Dr T.A.), Department of Physics, King's College, Newcastle upon Tyne, U.K.
* LITVAK (Dr M.), Smithsonian Astrophysical Observatory, Cambridge, Massachusetts 02138, U.S.A.
* LIU (Dr S.Y.), High Altitude Observatory, Boulder, Colorado, U.S.A.
LIVINGSTON (Dr W.C.), Kitt Peak Observatory, P.O. Box 26732, Tucson, Arizona 85726, U.S.A.
LIVSHITS (Dr M.A.), Institute of Terrestrial Magnetism, Ionosphere and Propagation of Radio Waves, USSR Academy of Sciences, Moscow, U.S.S.R.
* LLOYD EVANS (Dr T.H.H.), Radcliffe Observatory, P.O. Box 373, Pretoria, South Africa
LOCHTE-HOLTGREVEN (Prof. Dr W.), Institut für Experimentalphysik der Universität, Kiel, F.R.G.
LOCKE (Dr J.L.), Astrophysics Branch, National Research Council, Ottawa, Ontario, Canada
* LOCKWOOD (Dr G.W.), Lowell Observatory, Flagstaff, Arizona 86001, U.S.A.
LODÉN (Mrs Dr K.), Stockholm Observatory, S-133 00 Saltsjöbaden, Sweden
LODÉN (Dr L.O.), Astronomiska Observatoriet, Box 515, 751 20 Uppsala, Sweden
LOHMAN (Prof. Dr W.), Astronomisches Rechen-Institut, Mönchhofstrasse 12–14, 69 Heidelberg, F.R.G.
LONGAIR (Dr M.S.), Mullard Radio Astronomy Observatory, Cavendish Laboratory, Cambridge, U.K.
* LOPEZ (Dr J.A.), Pedro Echagüe 74 (oeste), San Juan, Argentina
* LOPEZ GARCIA (Dr Z.), Observatorio Astronómico, La Plata, Argentina
LORÓN (Dr M.M.), Director del Observatorio Astronómico, Alfonso XII, No. 3, Madrid, Spain
LORTET-ZUCKERMANN (Mrs M.C.), Institut d'Astrophysique, 98bis Boulevard Arago, 75014 Paris, France
LOTOVA (Mrs Dr N.A.), Physical Institute, USSR Academy of Sciences, Moscow, U.S.S.R.
LOUGHHEAD (Dr R.E.), CSIRO, Division of Physics, University Grounds, Chippendale, NSW 2008, Australia
LOUISE (Dr R.), L.A.S.-Traverse du Siphon, Les 3 Lucs, 13-Marseille 12e, France
LOURENS (J.v.B.), Royal Observatory, Observatory, Cape Province, South Africa
LOVAS (M.), Konkoly Observatory of the Hungarian Academy of Sciences, Konkoly Thege Miklós u. 13/17, Budapest XII, Hungary
LOVELL (Sir Bernard Prof. A.C.B.), Nuffield Radio Astronomy Laboratories, Jodrell Bank, Macclesfield, Cheshire SK11 9DN, U.K.

Low (Dr F.J.), Lunar and Planetary Laboratory, The University of Arizona, Tucson, Arizona 85721, U.S.A.

Lozinskij (Dr A.M.), Astronomical Council of the USSR Academy of Sciences, Vavilov Str. 34, Moscow, V-312, U.S.S.R.

Lübeck (K.), Hamburger Sternwarte, 205 Hamburg-Bergedorf, F.R.G.

* Lucy (Dr L.B.), Department of Physics, Faculty of Arts and Sciences, University of Pittsburgh, Pittsburgh, Pennsylvania 15213, U.S.A.

Lukatskaja (Mrs Dr F.I.), Main Astronomical Observatory, Ukrainian Academy of Sciences, Kiev, U.S.S.R.

* Lumme (Dr K.A.), Observatory and Astrophysics Laboratory, University of Helsinki, Tähtitornimäki, SF-00130 Helsinki 13, Finland

Lundquist (Dr C.A.), 218 Parker Street, Newton, Massachusetts, U.S.A.

Lunel (Mrs M.), Observatoire de Lyon, 69-Saint-Genis-Laval, France

Lüst (Prof. Dr R.), Max-Planck-Institut für extraterrestrische Physik, 8046 Garching b. München, F.R.G.

Lüst (Dr Rhea), Max-Planck-Institut für Physik und Astrophysik, Föhringer Ring 6, 8000 München 23, F.R.G.

* Lutz (Dr B.L.), Department of Earth & Space Sciences, State University of New York, Stony Brook, New York 11790, U.S.A.

* Lutz (Dr T.E.), Department of Pure and Applied Mathematics, Washington State University, Pullman, Washington 99163, U.S.A.

Luud (Dr L.S.), W. Struwe Astrophysical Observatory, Estonian Academy of Sciences, Tartu, Estonian SSR, U.S.S.R.

Luyten (Prof. W.J.), The Observatory, University of Minnesota, Minneapolis 14, Minnesota, U.S.A.

Lynden-bell (Dr D.), Royal Greenwich Observatory, Herstmonceux Castle, Hailsham, Sussex, U.K.

Lynds (Dr B.T.), Kitt Peak National Observatory, P.O. Box 26732, Tucson, Arizona 85726, U.S.A.

Lynds (Dr C.R.), Kitt Peak National Observatory, 950 North Cherry Avenue, P.O. Box 26732, Tucson, Arizona 85726, U.S.A.

* Lyne (Dr A.G.), Nuffield Radio Astronomy Laboratories, Jodrell Bank, Macclesfield, Cheshire, U.K.

Lyngå (Dr G.), Institutionen för astronomi, S-222 24 Lund, Sweden

Lyttkens (Dr E.), Skolgatan 33B, S-752 21 Uppsala, Sweden

Lyttleton (Dr R.A.), St. John's College, Cambridge, U.K.

* McCabe (Miss M.), Associate Astronomer, Institute for Astronomy, 2840 Kolowalu Street, Honolulu, Hawaii 96822, U.S.A.

McCarthy S.J. (Dr M.F.), Specola Vaticana, Castel Gandolfo, Vatican City State

McClain (E.F.), 225 Maple Road, Morningside, Maryland 20023, U.S.A.

McClure (Dr R.H.), Yale University, Department of Astronomy, New Haven, Connecticut 07520, U.S.A.

* McCluskey (Dr G.E.), LeHigh University, Bethlehem, Penn. 18015, U.S.A.

* McCord (Dr T.B.), Massachusetts Institute of Technology, 24–422, Cambridge, Mass. 02139, U.S.A.

McCracken (Prof. K.G.), CSIRO, Mineral Physics Station, P.O. Box 136, North Ryde, N.S.W. 2113, Australia

* McCray (Dr R.A.), Joint Institute for Laboratory Astrophysics, Boulder, Colorado, U.S.A.

McCrea (Prof. W.H.), The University of Sussex, School of Mathematical and Physical Sciences, Falmer, Sussex, U.K.

McCrosky (Dr R.E.), Smithsonian Astrophysical Observatory, 60 Garden Street, Cambridge, Massachusetts 02138, U.S.A.

McCuskey (Dr S.W.), Warner and Swasey Observatory, Case Institute of Technology, East Cleveland, Ohio 44112, U.S.A.

* McCutcheon (Dr W.H.), The University of British Columbia, Department of Physics, Vancouver 8, Canada

McDonald (Dr F.B.), NASA, Goddard Space Flight Center, Greenbelt, Maryland 20771, U.S.A.

McDonald (Mrs Dr J.K. Petrie), Dominion Astrophysical Observatory, Victoria, B.C., Canada

* McDonnell (Dr J.A.M.), Electronics Laboratories, University of Kent at Canterbury, Canterbury, Kent, U.K.

McElroy (Dr M.B.), Kitt Peak National Observatory, 950 North Cherry Ave., Tucson, Arizona, U.S.A.

McGee (Prof J.D.), Physics Department, Imperial College of Science and Technology, Prince Consort Road, London, S.W. 7, U.K.

McGee (Dr R.X.), CSIRO, Division of Radiophysics, P.O. Box 76, Epping, N.S.W., 2121, Australia

McIntosh (Dr B.A.), Astrophysics Branch, National Research Council, Ottawa 7, Ontario, Canada

* McKee (Dr Chr.), Harvard University, Cambridge, Massachusetts 02138, U.S.A.

McKenna-Lawlor (S.M.P.), Vishnu House, Blanchardstown, Dublin, Ireland

McKinley (Dr D.W.R.), Executive Director (Laboratories), National Research Council, Ottawa 7, Ontario, Canada

McLean (Dr D.J.), CSIRO, Division of Radiophysics, P.O. Box 76, Epping, N.S.W. 2121, Australia

McMullan (Dr D.), Royal Greenwich Observatory, Herstmonceux Castle, Hailsham, Sussex, U.K.

McNally (Dr D.), University of London Observatory, Mill Hill Park, London, N.W.7 ZQS, U.K.

McNamara (Dr D.H.), Department of Physics, Brigham Young University, Provo, Utah, U.S.A.

McNarry (L.R.), Astrophysics Branch, National Research Council, Ottawa, Ontario, Canada

McWhirter (Dr R.W.P.), Science Research Council, Astrophysics Research Unit, Culham Laboratory, Abingdon, Berkshire, U.K.

McVittie (Prof. G.C.), 74 Old Dover Road, Canterbury, Kent, U.K.

MacConnell (Dr D.J.), Department of Astronomy, Wesleyan University, Middletown, Connecticut 06457, U.S.A.

MacGarroll (Dr R.), Observatoire de Meudon, 92190 Meudon, France

MacLeod (Dr J.M.), Astrophysics Branch, National Research Council, Ottawa 7, Canada

MacQueen (Dr R.), High Altitude Observatory, Boulder, Colorado, U.S.A.

MacRae (Dr D.A.), Director of the David Dunlap Observatory, University of Toronto, Richmond Hill, Ontario, Canada

* Macchetto (Dr F.), Space Science Department, ESTEC, Noordwijk, The Netherlands

* Macdonald (Dr G.H.), The Electronics Laboratory, The University, Canterbury, Kent, U.K.

Machado (J.M.A. Braz), Serviços Radioelectricos dos C.T.T., Lisbon, Portugal

* Mackay (Dr C.D.), The Institute of Astronomy, Madingley Road, Cambridge, U.K.

Macris (Dr C.J.), Director, Research Center for Astronomy and Applied Mathematics, Academy of Athens, 14 Anagnostopoulou Street, Athens (136), Greece

Madeira (Dr J.A.), Observatorio Astronomico, Tapada, Lisbon, Portugal

Madwar (Prof. M.R.), 15 Rue Pahlewi, Dokki, Cairo, Arab Rep. of Egypt

* Maeder (Dr A.), Observatoire de Genève, Ch-1290 Sauverny, Switzerland

Maffei (Dr P.), Laboratorio di Astrofisica, Casella Postale 67, 00044 Frascati (Roma), Italy

Magelashvili (Dr N.L.), Abastumani Observatory, Abastumani, Georgia, U.S.S.R.

* Magnan (Dr Chr.), Institut d'Astrophysique, 98bis, Boulevard Arago, 75014 Paris, France

Magnaradze (Mrs Dr N.G.), State University, Tbilisi, Georgia, U.S.S.R.

MAÎTRE (Dr V.), Observatoire, 25-Besançon, France
MAKAROVA (Mrs Dr E.A.), Sternberg Astronomical Institute, Moscow, U.S.S.R.
MAKITA (Dr M.), Tokyo Astronomical Observatory, University of Tokyo, Mitaka, Tokyo, Japan
MALACARA (Dr D.), Instituto Nacional de Astrofísica, Optica y Electrónica, Tonantzintla, Pue., Apartados Postales Nos. 216 y 51, Puebla, Pue, Mexico
MALAISE (Dr D.J.), Institut d'Astrophysique, Université de Liège, avenue de Cointe, 5, B-4200 Cointe-Ougrée, Belgium
MALITSON (Mrs H.H.), Code 693, Laboratory for Extraterrestrial Physics, NASA-Goddard Space Flight Center, Greenbelt, Maryland 20771, U.S.A.
MALLIA (Dr E.A.), Department of Astrophysics, South Parks Road, Oxford OX1 3RQ, U.K.
MALMQUIST (Prof. K.G.), S:t Olofsgatan 10 A, S-753 21 Uppsala, Sweden
MALTBY (Dr P.), Institute of Theoretical Astrophysics, University of Oslo, Blindern, Oslo 3, Norway
MALVILLE (Dr J.M.), High Altitude Observatory, Boulder, Colorado, U.S.A.
MAMEDBEJLI (Dr G.D.), Shemakha Astrophysical Observatory, Azerbajdzhan Academy of Sciences, Baku, U.S.S.R.
MAMMAMO (A.), Osservatorio Astrofisico, Asiago (Vicenza), Italy
MANARA (Dr A.), Osservatorio Astronomico di Milano, via Brera 28, Milano, Italy
* MANCHESTER (Dr R.N.), University of Massachusetts, Amherst, Massachusetts 01002, U.S.A.
MANDEL'SHTAM (Dr S.L.), Laboratory of Spectroscopy, Lebedev Institute, Academy of Sciences of the USSR, Moscow, U.S.S.R.
MANGENEY (A.), Observatoire de Meudon, 5, place Janssen, 92190 Meudon, France
MANNINO (Prof. G.), Osservatorio Astronomico Universitario, Via Zamboni 33, Bologna, Italy
MARAN (Dr S.P.), Advanced Systems and Ground Observations Branch, Code 683, Goddard Space Flight Center, Greenbelt, Maryland 20771, U.S.A.
MARCUS (Prof. E.), Observatoire de Bucarest, 5 rue Cutitul de Argint, Bucarest, Romania
MARGONI (R.), Osservatorio Astrofisico, Asiago (Vicenza), Italy
MARIK (Dr M.), Astronomical Institute of the Eötvös Loránd University, Kun Béla tér 2, Budapest, Hungary
* MARK (Dr J.W.-K.), Massachusetts Institute of Technology, Department of Mathematics, Cambridge, Massachusetts 02139, U.S.A.
MARKARIAN (Dr B.E.), Byurakan Astrophysical Observatory, Byurakan, Armenia, U.S.S.R.
MARKOWITZ (Prof. Dr Wm.), Nova University, 8000 N. Ocean Drive, Dania, Florida 33004, U.S.A.
* MARLBOROUGH (Prof. J.M.), Department of Astronomy, University of Western Ontario, Ontario, Canada
* MAROCHNIK (Prof. Dr L.S.), Astrophysical Department, Rostov University, prospekt Stachki 192, Rostov-Don 344061, U.S.S.R.
* MARQUES (M.N.), Tapada da Ajuda, Lisboa 3, Portugal
MARSDEN (Dr B.G.), Smithsonian Astrophysical Observatory, 60 Garden Street, Cambridge, Massachusetts 02138, U.S.A.
* MARSH (M.Sc.J.C.D.), The Hatfield Polytechnic Observatory, Bayfordbury, Near Hertford, Herts, U.K.
MARTEL-CHOSSAT (Mrs Dr M.T.), Observatoire de Lyon, Saint-Genis-Laval, (Rhône), France
MARTIN (N.), Astronome Adjoint, Observatoire de Marseille, 2, place Le Verrier, 13004 Marseille, France
MARTIN JR (W.C.), Chief, Spectroscopy Section, National Bureau of Standards, Physics Building A-165, Washington D.C. 20234, U.S.A.
MARTINET (L.), Observatoire Cantonal de Genève, Genève, Sauverny 1290, Switzerland
* MARTINEZ (Dr M.), Instituto de Astronomía, UNAM, Apartado Postal 70-264, México D.F., Mexico
* MARTINI (Dr A.), Laboratorio di Astrofisica, Casella Postale 67, Frascati (Roma), Italy

MARTRES (Mrs M.J.), Observatoire de Meudon, Place Jules Janssen, 92190-Meudon, France
MARTYNOV (Prof. Dr D.Ya.), Director of the Sternberg Astronomical Institute, Moscow University, Leninskije Gory, Moscow-V234, U.S.S.R.
MARVIN (Dr U.B.), Smithsonian Astrophysical Observatory, Cambridge, Massachusetts 02138, U.S.A.
MARX (Prof. G.), Roland Eötvös University, Puskin u. 5–7, Budapest VIII, Hungary
MASANI (Prof. A.), Osservatorio Astronomico di Brera, Via Brera 28, Milano, Italy
* MASLOWSKI (Dr J.), Astronomical Observatory of the Jagellonian University, Cracow, Kopernika 27, Poland
* MASNOU (Dr F.), Observatoire de Paris, Section d'Astrophysique de Meudon, 92190 Meudon, France
MASSEVICH (Mrs Dr A.G.), Astronomical Council of the USSR Academy of Sciences, Vavilov Str. 34, Moscow V-312, U.S.S.R.
MATHEWS (Dr W.G.), Physics Department, University of California-San Diego, La Jolla, California 92038, U.S.A.
MATHEWSON (Dr D.S.), CSIRO, Division of Radiophysics, University Grounds, Chippendale, N.S.W., Australia
MATHIS (Dr J.S.), Astronomy Department, University of Wisconsin, Madison, Wisconsin 53706, U.S.A.
MATSUMANI (Dr N.), Tokyo Astronomical Observatory, Mitaka, Tokyo, Japan
MATSUOKA (Dr M.), Institute of Space and Aeronautical Science, University of Tokyo, Komaba, Meguro-Ku, Tokyo, Japan
MATSUSHIMA (Prof. Dr S.), Department of Astronomy, The Pennsylvania State University, University Park, Pennsylvania 16802, U.S.A.
MATTHEWS (Dr T.A.), Department of Physics and Astronomy, University of Maryland, College Park, Maryland 20740, U.S.A.
MATTIG (Dr W.), Fraunhofer Institut, Schöneckstrasse 6, 78 Freiburg i. Breisgau, F.R.G.
* MATTILA (Dr K.), University of Helsinki, Observatory and Astrophysics Laboratory, Tähtitornimäki, SF-00130 Helsinki 13, Finland
MATVEYENKO (Dr L.I.), Institute for Space Research, Profsojuznaja 88, Moscow V 485, U.S.S.R.
* MATZNER (Dr R.), The University of Texas (Austin), Austin, Texas, U.S.A.
MAUDER (Dr H.), Remeis-Sternwarte, Sternwartstrasse 7, 8600 Bamberg, F.R.G.
* MAURICE (Ing. E.), Observatoire de Marseille, 2, place Le Verrier, 13004 Marseille, France
MAVRIDIS (Prof. Dr L.N.), Department of Geodetic Astronomy, University of Thessaloniki, Thessaloniki, Greece
MAXWELL (Dr A.), Harvard Radio Astronomy Station, Fort Davis, Texas, U.S.A.
* MAY (Ing. J.), Observatorio Radioastronómico de Maipú, Universidad de Chile, Casilla 68, Maipú, Chile
MAYALL (Mrs M.W.), 187 Concord Ave., Cambridge, Massachusetts 02138, U.S.A.
MAYALL (Dr N.U.), Kitt Peak National Observatory, 950 North Cherry Ave., Tucson, P.O. Box 26732, Arizona 85726, U.S.A.
MAYER (C.H.), Code 7130, Naval Research Laboratory, Washington, D.C. 20375, U.S.A.
MAYER (Dr P.), Department of Astronomy and Astrophysics, Charles University, Švédská 8, Prague 5, Czechoslovakia
MAYER (Dr U.), Astronomisches Institut der Universität, Waldhäuser Str. 64, 74 Tübingen, F.R.G.
MAYFIELD (Dr E.B.), Solar Physics Department, The Aerospace Corporation, P.O. Box 92957, Los Angeles, California 90009, U.S.A.
* MAYOR (Dr M.), Observatoire de Genève, CH-1290 Sauverny, Switzerland
* MAZZUCCONI (Dr F.), Osservatorio Astrofisico di Arcetri, Largo Enrico Fermi, 5, Firenze, Italy
MEABURN (Dr J.), Department of Astronomy, The University, Manchester 13, U.K.
MEADOWS (Dr A.J.), Astronomy Department, University of Leicester, University Road, Leicester, U.K.

MEDD (W.J.), Astrophysics Branch, National Research Council, Ottawa, Ontario, Canada
MEEKS (Dr M.L.), M.I.T., Neroc Haystack Observatory, Westford, Massachusetts 01886, U.S.A.
MEFFROY (Dr J.), Chef de Travaux, Faculté des Sciences, 34-Montpellier, France
MEGRELISHVILI (Mrs Dr T.G.), Abastumani Astrophysical Observatory, Georgian Academy of Sciences, Abastumani, U.S.S.R.
MEHLTRETTER (Dr J.P.), Fraunhofer Institut, Schöneckstrasse 6, 78 Freiburg i. Breisgau, F.R.G.
MEIN (P.), Observatoire de Meudon, 5, place Janssen, 92190 Meudon, France
MEINEL (Dr A.B.), Kitt Peak National Observatory, 950 North Cherry Avenue, Tucson, P.O. Box 26732, Arizona 85726, U.S.A.
MEINIG (Dr M.), Zentralinstitut Physik der Erde, DDR 15 Potsdam, Telegrafenberg, G.D.R.
* MEISEL (Dr D.D.), Department of Physics and Astronomy, State University College of Arts and Sciences, Geneseo, N.Y. 14454, U.S.A.
MELCHIOR (Prof. Dr P.), Chef du Département d'Astronomie de Position et de Géodynamique, Observatoire Royal de Belgique, avenue Circulaire, 3, B-1180 Bruxelles, Belgium
MEL'NIKOV (Prof. Dr O.A.), Pulkovo Observatory, Leningrad, U.S.S.R.
MEN' (Dr A.V.), Institute of Radiophysics and Electronics, Academy of Sciences Ukrainian SSR, Kharkov, U.S.S.R.
MENDEZ (Dr M.), Instituto de Astronomía, UNAM, Apartado Postal 70–264, México, D.F., Mexico
MENDOZA V. (Dr E.E.), Instituto de Astronomía, UNAM, Apartado Postal 70–264, México, D.F., Mexico
* MENNESSIER (Dr M.-O.), Observatoire de Paris, 61, Avenue de l'Observatoire, 75014 Paris, France
MENON (Dr T.K.), Tata Institute of Fundamental Research, Homi Bhabha Road, Colaba, Bombay-5, India
MENZEL (Dr D.H.), Harvard College Observatory, 60 Garden Street, Cambridge, Massachusetts 02138, U.S.A.
MERGENTALER (Prof. Dr J.), Astronomical Observatory, Kopernika 11, Wrocław, Poland
MERMAN (Dr G.A.), Institute of Theoretical Astronomy, Leningrad, U.S.S.R.
MERMAN (Mrs Dr N.V.), Main Astronomical Observatory, USSR Academy of Sciences, Pulkovo, U.S.S.R.
MERRILL (Dr J.E.), University of Florida Observatories, Department of Physics and Astronomy, University of Florida, Gainesville, Florida 32601, U.S.A.
MERTON (Dr G.), 17 Holywell, Oxford, U.K.
* MERTZ (Dr L.), Harvard College Observatory, 60 Garden Street, Cambridge, Massachusetts 02138, U.S.A.
MESSAGE (Dr P.J.), Mathematics Department, University of Liverpool, Liverpool 3, U.K.
MESTEL (Prof. L.), Department of Mathematics, The University, Manchester 13, U.K.
MEURERS (Prof. Dr J.), Direktor der Universitätssternwarte, Türkenschanzstrasse 17, A-1180 Vienna, Austria
* MEWE (Dr R.), Laboratorium voor Ruimteonderzoek, Beneluxlaan 21, Utrecht, The Netherlands
MEYER (Dr F.), Max-Planck-Institut für Physik und Astrophysik, Föhringer Ring 6, 8000 München 23, F.R.G.
MEYER-HOFMEISTER (Mrs Dr E.), Max-Planck-Institut für Physik und Astrophysik, Föhringer Ring 6, 8000 München 23, F.R.G.
MEZGER (Dr P.G.), Max-Planck-Institut für Radioastronomie, 53 Bonn, Argelanderstrasse 3, F.R.G.
MIANES (P.), Observatoire de Lyon, 69-Saint-Genis-Laval, France
MICHARD (Prof. R.), Président de l'Observatoire de Paris, 61, avenue de l'Observatoire, 75-Paris 14e, France
* MICHAUD (Dr G.), 250 du Finistère, Saint-Lambert, Quebec, Canada

MICHEL (Prof. F.C.), Department of Space Science, Rice University, Houston, Texas 77001, U.S.A.

MICHEL (Ir.H.), rue Ten Bosch, 54, B-1050 Bruxelles, Belgium

MICHKOVITCH (Prof. V.V.), Serbian Academy of Sciences and Arts, Knez Mihailova, 35, Belgrade, Yugoslavia

MICZAIKA (Dr G.R.), TRW Systems, 1 Space Park, Redendo Beach, California, U.S.A.

MIDDLEHURST (Miss B.M.), The Lunar Science Institute, 3303 Nasa Road 1, Houston, Texas 77058, U.S.A.

* MIETELSKI (Dr J.), Astronomical Observatory of the Jagellonian University, Cracow, Kopernika 27, Poland

MIGEOTTE (Prof. Dr M.), Institut d'Astrophysique, Université de Liège, avenue de Cointe, 5, B-4200 Cointe-Ougrée, Belgium

MIHAILA (Dr I.), Observatorul Astronomic, Strada Cutitul de Argint 5, Bucuresti 28, Roumania

MIHALAS (Dr D.), High Altitude Observatory, P.O. Box 3000, Boulder, Colorado 80303, U.S.A.

MIKAIL (Dr J.S.), Helwan Observatory, Near Cairo, Arab Rep. of Egypt

MIKESELL (Dr A.H.), U.S. Naval Observatory, Washington, D.C. 20390, U.S.A.

MIKHAILOV (Prof. Dr A.A.), Director of the Pulkovo Observatory, Leningrad M 140, U.S.S.R.

MIKHEL'SON (Dr N.N.), Main Astronomical Observatory, Academy of Sciences, Pulkovo, U.S.S.R.

MILET (B.), Aide Astronome, Observatoire de Nice, Le Mont-Gros, 06-Nice, France

* MILEY (Dr G.K.), Sterrewacht, Leiden, The Netherlands

MILFORD (Dr S.N.), Department of Physics, University of Queensland, St. Lucia, Brisbane, Queensland 4067, Australia

MILLER (Dr F.D.), Department of Astronomy, The University of Michigan, Physics-Astronomy Building, Ann Arbor, Michigan 48104, U.S.A.

MILLER (Dr J.S.), Lick Observatory, University of California, Santa Cruz, California 95060, U.S.A.

MILLER (Dr R.H.), Institute for Computer Research, University of Chicago, 5640 Ellis Ave., Chicago, Illinois 60637, U.S.A.

MILLER S.J. (Rev. W.J.), Director, Fordham University Astronomical Laboratory, Box 88, Fordham University, New York, N.Y. 10458, U.S.A.

MILLIGAN (J.E.), Astrophysics Branch, Goddard Space Flight Center, Greenbelt, Maryland, U.S.A.

* MILLIS (Dr L.R.), Lowell Observatory, Flagstaff, Arizona 86001, U.S.A.

MILLMAN (Dr P.M.), Astrophysics Branch, National Research Council, Ottawa 7, Ontario, Canada

* MILLS (Dr A.A.), Department of Astronomy, The University, Leicester LE1 7RH, U.K.

MILLS (Dr B.Y.), School of Physics, University of Sydney, Sydney, N.S.W., Australia

MILOVANOVIĆ (Dr V.), Service de Variation de latitude et du service de l'heure à l'Observatoire, Astronomique de Belgrade, Belgrade, Yugoslavia

MININ (Dr I.N.), Leningrad University Observatory, Leningrad, U.S.S.R.

MINKOWSKI (Dr R.L.), Radio Astronomy Laboratory, University of California, Berkeley, California 94720, U.S.A.

MINNET (H.C.), CSIRO, Division of Radiophysics, P.O. Box 76, Epping, N.S.W. 2121, Australia

* MINTZ BLANCO (Ma.B.), Cerro Tololo Interamerican Observatory, Casilla 63-D, La Serena, Chile

MIRZABEKYAN (Dr E.G.), Astrophysical Observatory, Byurakan, Armenia, U.S.S.R.

MIRZOYAN (Prof. Dr L.V.), Byurakan Astrophysical Observatory, Byurakan, Armenia, U.S.S.R.

MISNER (Dr Ch.W.), Department of Astronomy, University of Maryland, College Park, Maryland 20742, U.S.A.

MISSANA (Prof. N.), Osservatorio Astronomico di Pino Torinese, (Torino), Italy

* MITALAS (Dr R.), University of Western Ontario, Faculty of Science, Department of Astronomy, London 72, Ontario, Canada

* MITCHELL (Prof. G.F.), Department of Physics, St. Mary's University, Halifax, Nova Scotia, Canada
MITCHELL (R.), c/o C.I.C.E.S.E. Av. Gastelum No. 898, Ensenada, Baja California, Mexico
MITCHELL JR (Dr W.E.), Perkins Observatory, Delaware, Ohio, U.S.A.
MITIĆ (L.A.), Astronomical Observatory, Volgina, 7, Belgrade, Yugoslavia
MITRA (Dr A.P.), Assistant Director, National Physical Laboratory, New Delhi, India
MITROFANOVA (Miss Dr L.A.), Main Astronomical Observatory, USSR Academy of Sciences, Pulkovo, U.S.S.R.
MIYADI (Dr M.), 3-6 Senkawa, Chôfu-shi, Tokyo, Japan
* MIYAMOTO (Dr M.), Tokyo Astronomical Observatory, Mitaka, Tokyo, Japan
MIYAMOTO (Prof. Dr S.), Director, Kwasan Observatory, Faculty of Science, University of Kyoto, Yamashina, Kyoto, Japan
* MIYAMOTO (Dr S.), Institute of Space and Aeronautical Science, University of Tokyo, Komaba, Meguro-ku, Tokyo, Japan
MOERDIJK (Dr W.G.), Sterrenkundig Institut, Rijksuniversiteit Gent, J. Plateaustraat, 22, B-9000 Gent, Belgium
MOFFET (Dr A.T.), Radio Astronomy Laboratory, California Institute of Technology, Pasadena, California 91109, U.S.A.
MOGILEVSKIJ (Prof. E.I.), Institute of Terrestrial Magnetism, Ionosphere and Propagation of Radiowaves, USSR Academy of Sciences, Moscow, U.S.S.R.
MOHLER (Prof. O.C.), McMath-Hulbert Observatory, 895 Lake Angelus Road North, Pontiac 4, Michigan, U.S.A.
MOHR (Prof. J.M.), Department of Astronomy and Astrophysics, Charles University, Švedská 8, Prague 5, Czechoslovakia
MOISEEV (Dr I.G.), Crimean Astrophysical Observatory, USSR Academy of Sciences, P/o Nauchny, Crimea, U.S.S.R.
MOLCHANOV (Prof. A.P.), Institute of Physics, Leningrad State University, V-164, Leningrad, U.S.S.R.
MØLLER (Dr O.), Ole Rømer Observatory, Arhus C, Denmark
MONAGHAN (Dr J.J.), Department of Mathematics, Monash University, Clayton, 3168, Australia
MONFILS (Prof. Dr A.), Institut d'Astrophysique, Université de Liège, avenue de Cointe, 5, B-4200 Cointe-Ougrée, Belgium
MONIN (Dr G.A.), Crimean Astrophysical Observatory of the Academy of Sciences of the USSR, Pochtovoe, Crimea, U.S.S.R.
MONNETT (Dr G.), Observatoire de Marseille, Place Le Verrier, 13004 Marseille, France
* MONSIGNORI FOSSI (Dr B.C.), Osservatorio Astrofisico di Arcetri, Largo Enrico Fermi, 5, Firenze, Italy
* MOOK (Dr D.E.), Darmouth College, Hanover, New Hampshire, U.S.A.
MOORE (P.O.B.E.), Farthings, 39 West Street, Selsey, Sussex, U.K.
MOORE-SITTERLY (Mrs Dr Ch.), National Bureau of Standards 151. 00, Washington, D.C. 20234, U.S.A.
* MORAN (Dr J.), Smithsonian Astrophysical Observatory, Cambridge, Massachusetts 02138, U.S.A.
MORANDO (Dr B.), Bureau des Longitudes, 3, rue Mazarine, 75 Paris, 6e, France
MOREAU (Dr F.), Astronome Honoraire, Observatoire Royal de Belgique, avenue Armand Huysmans, 219, B-1050 Bruxelles, Belgium
MORENO (Prof. H.), Observatorio Astronómico Nacional, Casilla 36-D, Santiago, Chile
MORETON (G.E.), 15-5 The Esplanade, Balmoral Beach, N.S.W., 2088, Australia
MORGAN (Dr W.W.), Yerkes Observatory, Williams Bay, Wisconsin 53191, U.S.A.
* MORGULEFF (Ing.N.), Institut d'Astrophysique, 98bis, Boulevard Arago, 75014 Paris, France
MORIMOTO (Dr M.), Tokyo Astronomical Observatory, University of Tokyo, Mitaka, Tokyo, Japan

MORIYAMA (Dr F.), Tokyo Astronomical Observatory, Mitaka, Tokyo, Japan
MOROZ (Dr V.I.), Sternberg Astronomical Institute, Moscow, U.S.S.R.
* MOROZHENKO (Dr A.V.), The Main Astronomical Observatory of the Ukrainian Academy of Sciences, Kiev-127, U.S.S.R.
* MOROZHENKO (Mrs Dr N.N.), Main Astronomical Observatory of Ukrainian Academy of Sciences, Observatory, Kiev 252127, U.S.S.R.
MORRIS (Dr S.C.), Dominion Astrophysical Observatory, R.R.7, Victoria, B.C. Canada
* MORRISON (Dr D.D.), Institute for Astronomy, University of Hawaii, Honolulu, Hawaii 96822, U.S.A.
MORRISON (L.V.), Royal Greenwich Observatory, Herstmonceux Castle, Hailsham, Sussex, U.K.
MORRISON (Dr P.), Department of Physics, Massachusetts Institute of Technology, Cambridge Massachusetts 02139, U.S
MORTON (Dr D.C.), Princeton University Observatory, Peyton Hall, Princeton, New Jersey 08540, U.S.A.
MORTON (Dr G.A.), R.C.A. Laboratories, Princeton, New Jersey, U.S.A.
* MOSS (Dr D.L.), Department of Mathematics, Manchester University, Manchester M13 9PL, U.K.
MOTZ (Dr L.), Rutherfure Observatory, Columbia University, Box 57 Pupin Hall, New York, New York 10027, U.S.A.
MOURADIAN (Z.), Observatoire de Meudon, 5, place Janssen, 92190 Meudon, France
MOUTSOULAS (Dr M.), Department of Astronomy, University of Athens, Athens 621, Greece
* MOVAHED-A (Dr R.), Tabriz University, Tabriz, Iran
MRKOS (A.), Department of Astronomy and Astrophysics, Charles University, Švédská 8, Prague 5-Smíchov, Czechoslovakia
MUGGLESTONE (Dr D.), University of Queensland, Brisbane, Queensland, Australia
MULDERS (Dr G.F.W.), c/o Federation of Amer. Soc. Experimental Biology, 9650 Rockville Pike, 20014 Bethesda (Maryland), U.S.A.
MULHOLLAND (Dr J.D.), Department of Astronomy, University of Texas, Austin, Texas 78712, U.S.A.
MULLALY (Dr R.F.), School of Electrical Engineering, University of Sydney, Sydney, Australia
MULLER (Dr A.B.), Kapteyn Astronomical Laboratory, Broerstraat 7, Groningen, The Netherlands
MULLER (Prof. Ir. C.A.), Odinksveld, 8, Delden, The Netherlands
MÜLLER (Prof. Edith A.), Observatoire de Genève, 1290 Sauverny (GE), Switzerland
* MÜLLER (Prof. Dr H.), Herzogenmühlestrasse 4, CH-8051 Zürich, Switzerland
MULLER (Dr P.), Observatoire de Paris, Section d'Astrophysique, 92190 Meudon, France
MÜLLER (Prof. Dr R.), Postfach 8, 8204 Degerndorf/Inn, F.R.G.
MUMFORD (G.S.), Department of Mathematics, Tufts University, Medford, Massachusetts 02155, U.S.A.
MÜNCH (Prof. G.), California Institute of Technology, 1201 E. California Street, Pasadena, California, U.S.A.
* MURDIN (Dr P.G.), Royal Greenwich Observatory, Herstmonceux Castle, Hailsham, Sussex, U.K.
* MURPHY (Dr R.E.), Maryland Academy of Sciences, 119 South Howard Street, Baltimore, Maryland 21201, U.S.A.
MURRAY (C.A.), Royal Greenwich Observatory, Herstmonceux Castle, Hailsham, Sussex, U.K.
MUSEN (Dr P.), National Aeronautics and Space Administration, Goddard Space Flight Center, Greenbelt, Maryland, U.S.A.
MUSMAN (Dr S.), Sacramento Peak Observatory, Sunspot, New Mexico 88349, U.S.A.
MUSTEL (Prof. Dr E.R.), Astronomical Council, USSR Academy of Sciences, Vavilov Str., 34, Moscow, V-312, U.S.S.R.
MUTSCHLECNER (Dr P.), Department of Astronomy, Indiana University, Bloomington, Indiana 47401, U.S.A.

MYACHIN (Dr V.F.), Institute of Theoretical Astronomy, USSR Academy of Sciences, Leningrad, U.S.S.R.
MYERSCOUGH (Mrs Dr V.P.), Department of Mathematics, Queen Mary College, Mile End Road, London E.1, U.K.

NACOZY (Dr P.E.), Department of Aerospace Engineering, The University of Texas, Austin, Texas 78712, U.S.A.
NADOLSCHI (Prof. Dr Doc.V.), Com. Ardeoani, of. Tescani, jud. Bacău, Roumania
* NADYOZHIN (Dr D.K.), Institute of Applied Mathematics, Miusskaja Ploszhad 4, Moscow 125047, U.S.S.R.
NAGASAWA (Dr S.), Tokyo Astronomical Observatory, Mitaka, Tokyo, Japan
* NAGIRNER (Dr D.I.), Astronomical Observatory of Leningrad University, 10 linia 33, Leningrad 199178, U.S.S.R.
NAHON (Prof. F.), Observatoire de Paris, 61, Avenue de l'Observatoire, 75-Paris-14e, France
NAKAGAWA (Dr Y.), High Altitude Observatory, Boulder, Colorado 80302, U.S.A.
NAKANO (S.), 47, Myogadani, Bunkyo-ku, Tokyo, Japan
* NAKANO (Dr T.), Department of Physics, Kyoto University, Kyoto, Japan
* NAKAYAMA (Dr S.), College of General Education, University of Tokyo, Komaba 3–8–1, Meguroku, Tokyo, Japan
NAMBA (Dr O.), Astronomical Observatory 'Sonnenborgh', Zonnenburg 2, Utrecht, The Netherlands
NANDY (Dr K.), Royal Observatory, Edinburgh 9, U.K.
* NAPIER (Dr W.M.), Royal Observatory, Blackford Hill, Edinburgh EH9 3HJ, U.K.
NAQVI (Dr A.M.), 6274 Del Rosa Ave., Apt. 11, San Bernardino, California 92404, U.S.A.
* NARANAN (Dr S.), Tata Institute of Fundamental Research, Bombay, India
NARAYANA (J.V.), Meteorologist, Astrophysical Observatory, Kodaikanal 3, India
NARIAI (Dr H.), Research Institute for Theoretical Physics, Hiroshima University, Takehara, Hiroshima-Ken, Japan
NARIAI (Dr K.), Tokyo Astronomical Observatory, Mitaka, Tokyo, Japan
NARLIKAR (Prof. J.V.), Tata Institute of Fundamental Research, Homi Bhabha Road, Bombay 5, India
* NATHER (Dr R.E.), Department of Astronomy, University of Texas at Austin, Austin, Texas 78712, U.S.A.
* NAUMOV (Dr V.A.), Pulkovo Observatory, Leningrad 196140, U.S.S.R.
NAZAROVA (Dr T.N.), Institute of Geochemistry and Analytic Chemistry, USSR Academy of Sciences, Moscow, U.S.S.R.
NECKEL (Dr H.), Hamburger Sternwarte, 205 Hamburg-Bergedorf, F.R.G.
NECKEL (Dr Th.), Max-Planck-Institut für Astronomie, 6900 Heidelberg-Königstuhl, F.R.G.
NE'EMAN (Dr Y.), Department of Physics and Astronomy, Tel-Aviv University, Ramat-Aviv, Israel
NEFED'EV (Dr A.A.), Director, Engelhardt Observatory, Kazan, U.S.S.R.
NEFED'EVA (Dr A.I.), Engelhardt Observatory, Kazan, U.S.S.R.
NEFF (Dr J.S.), Department of Physics and Astronomy, State University of Iowa, Iowa City, Iowa 52240, U.S.A.
NELSON (Prof. B.), Department of Astronomy, San Diego State College, San Diego 15, California, U.S.A.
NEMIRO (Dr A.A.), Pulkovo Observatory, Leningrad, U.S.S.R.
NESMYANOVICH (Dr A.T.), Kiev University, Kiev, Ukrainian SSR, U.S.S.R.
NESS (Dr N.F.), NASA, Goddard Space Flight Center, Greenbelt, Maryland 20771, U.S.A.
NEUGEBAUER (Dr G.), California Institute of Technology, Department of Astrophysics, Pasadena, California 91109, U.S.A.
NEUGEBAUER (Dr G.), Physics Department, California Institute of Technology, 1201 East California Street, Pasadena, California 91109, U.S.A.

NEUPERT (Dr W.M.), Solar Physics Branch, Goddard Space Flight Center, Greenbelt, Maryland, U.S.A.

NEUŽIL (Dr L.), Astronomical Institute, Czechoslovak Academy of Sciences, Observatory Ondřejov, Czechoslovakia

NEVEN (Dr L.), Chef du Département d'Astrophysique, Observatoire Royal de Belgique, avenue Circulaire, 3, B-1180 Bruxelles, Belgium

NEVIN (Prof. T.E.), University College, Dublin, Ireland

NEWKIRK JR (Dr G.A.), High Altitude Observatory of the University of Colorado, Boulder, Colorado 80302, U.S.A.

* NEWSOM (Dr G.H.), Department of Astronomy, 174 W. 18th Ave., Columbus, Ohio 43210, U.S.A.

NEWTON (H.W.), 'Alandale', Dean Road, Seaford, Sussex, U.K.

NEWTON (Dr R.R.), Applied Physics Laboratory, Johns Hopkins University, 8621 Georgia Avenue, Silver Spring, Maryland, U.S.A.

* NEY (Dr E.), University of Minnesota, School of Physics and Astronomy, Minneapolis, Minnesota 55455, U.S.A.

NEYMAN (Prof. J.), Statistical Laboratory, University of California, Berkeley 4, California, U.S.A.

NICHOLLS (Prof. R.W.), Department of Physics, York University, Keele Street, Toronto, Ontario, Canada

NICHOLSON (W.), Royal Greenwich Observatory, Herstmonceux Castle, Hailsham, Sussex, U.K.

NICOLET (Prof. Dr M.), Directeur de l'Institut d'Aéronomie Spatiale de Belgique, avenue Circulaire, 3, B-1180 Bruxelles, Belgium

NICOLINI (Prof. T.), Direttore, Osservatorio Astronomico di Capodimonte, Napoli, Italy

NICOLOV (N.S.), Université de Sofia, Faculté Physique, Sofia, Bulgaria

NICOLSON (G.D.), c/o National Institute for Telecommunications Research, P.O. Box 3718, Johannesburg, South Africa

* NIEUWENHUIJZEN (Dr H.), Sterrewacht, Servaas Bolwerk 13, Utrecht, The Netherlands

* NIIMI (Dr H.), Department of Aeronautical Engineering, Kyoto University, Kyoto, Japan

NIKITIN (Dr A.A.), Observatory, State University, Leningrad, U.S.S.R.

* NIKOLOFF (Dr I.), Perth Observatory, Bickley, W.A. 6076, Australia

NIKOLSKY (Dr G.M.), Institute of Terrestrial Magnetism, Ionosphere and Radio Wave Propagation, USSR Academy of Sciences, Moscow, U.S.S.R.

NIKONOV (Dr V.B.), Crimean Astrophysical Observatory, USSR Academy of Sciences, P/O Nauchny, Crimea 334413, U.S.S.R.

NILSSON (Dr C.), Smithsonian Astrophysics Observatory, 60 Garden Street, Cambridge, Massachusetts 02138, U.S.A.

NINGER-KOSIBOWA (Dr S.), Astronomical Observatory, Kopernika 11, Wrocław, Poland

NININGER (Dr H.H.), Sedona's Meteorite Museum, P.O. Box 146, Sedona, Arizona, U.S.A.

NISHI (Dr K.), Tokyo Astronomical Observatory, University of Tokyo, Mitaka, Tokyo, Japan

NISHIDA (Prof. M.), Department of Physics, Kyoto University, Kitashirakawa, Kyoto, Japan

NISHIMURA (Dr S.), Tokyo Astronomical Observatory, Mitaka, Tokyo, Japan

NISSEN (Prof. J.J.), Observatorio Astronómico, San Juan, Argentina

NISSEN (Dr P.E.), Ole Roemer Observatory, DK-8000 Aarhus C, Denmark

* NOBILI (Dr L.), Osservatorio Astronomico, Vicolo dell'Osservatorio, Padova, Italy

NOCI (Dr G.), Osservatorio Astrofisico di Arcetri, Firenze, Italy

* NOEL (F.), Departamento de Astronomía, Universidad de Chile, Casilla 36-D, Santiago de Chile, Chile

NOERDLINGER (Prof. P.D.), Department of Astronomy, Michigan State University, East Lansing, Michigan 48823, U.S.A.

* NOONAN (Dr Th.W.), State University College at Brockport, Brockport, New York 14420, U.S.A.

NORLIND (Dr W.), Lund Observatory, S-222 24 Lund, Sweden
NØRLUND (Prof. N.E.), Director, Geodetic Institute, Malmögade 6, Copenhagen, Denmark
* NORRIS (Dr J.), Yale University, New Haven, Connecticut, U.S.A.
NOTNI (Dr P.), Zentralinstitut für Astrophysik, Sternwarte Babelsberg, DDR 1502 Potsdam-Babelsberg, Rosa-Luxemburg-Strasse 17a, G.D.R.
NOTUKI (Dr N.), 1–23 Higashida, Suginami-ku, Tokyo, Japan
* NOVICK (Dr R.), Columbia University, Physics Department, 538 West 120th Street, New York, New York 10027, U.S.A.
NOVIKOV (Dr I.D.), Mathematical Institute, USSR Academy of Sciences, Moscow, U.S.S.R.
NOVOPASHENNYJ (Dr B.V.), Astronomical Observatory, Odessa University, Park Chevchenko, Odessa, U.S.S.R.
NOVOSELOV (Prof. Dr V.S.), Astronomical Observatory, Leningrad University, Leningrad, U.S.S.R.
NOYES (Dr R.W.), Smithsonian Institution, 60 Garden Street, Cambridge, Massachusetts 02138, U.S.A.
NUSSBAUMER (Dr H.), Atomic and Astrophysics Group ETH, Huttenstrasse 34, 8006 Zürich, Switzerland

OBASHEV (S.O.), Astrophysical Institute, Kazakh SSR Academy of Sciences, Alma-Ata 20, U.S.S.R.
OBI (Dr S.), Department of Earth Science, College of General Education, University of Tokyo, Komaba, Shibuya, Tokyo, Japan.
* OBREGON (Dr O.J.), Instituto de Astronomía, UNAM, Apartado Postal 70–264, México D.F., Mexico
OBRIDKO (Dr V.N.), Institute of Terrestrial Magnetism, Ionosphere, and Propagation of Radio-waves, USSR Academy of Sciences, Moscow, U.S.S.R.
OCCHIONERO (Dr F.), Laboratorio de Astrofisca, Casella Postale 67, 00044 Frascati (Roma), Italy
O'CONNEL S.J. (Dr D.J.K.), Borgo S. Spirito 5, 00193 Roma, Italy
O'CONNELL (Dr R.F.), Department of Physics and Astronomy, Louisiana State University, Baton Rouge, Louisiana 70803, U.S.A.
ODA (Dr M.), Institute of Space and Aeronautical Science, University of Tokyo, Komaba, Meguro-Ku, Tokyo, Japan
O'DELL (Dr C.R.), PD-LST, Marshall Space Flight Center, Huntsville, Alabama 35812, U.S.A.
ODGERS (Dr G.J.), Dominion Astrophysical Observatory, R.R. 7, Victoria, B.C., Canada
* OESTERWINTER (Dr C.), Naval Weapons Laboratory, Dahlgren, Virginia, U.S.A.
OETKEN (Dr. L.), Zentralinstitut für Astrophysik, DDR 15 Potsdam, Telegrafenberg, G.D.R.
OGORODNIKOV (Prof. Dr K.F.), Department of Astronomy, University, Leningrad, U.S.S.R.
O'HANDLEY (Dr D.), California Institute of Technology, Jet Propulsion Laboratory, 4800 Oak Grove Drive, Pasadena, California 91103, U.S.A.
ÖHMAN (Prof. Y.), Stockholm Observatory, S-133 00 Saltsjöbaden, Sweden
* O'HORA (N.P.J.), Royal Greenwich Observatory, Herstmonceux Castle, Hailsham, Sussex, U.K.
OHYAMA (Prof. N.), 1–22–26, Hirisawa, Hamamatsu, Japan
OJA (Dr T.), Kvistaberg Observatory, S-190 51 Bro, Sweden
* OKAMOTO (Dr I.), International Latitude Observatory, Mizusawa, Iwate, Japan
OKE (Dr J.B.), Mount Wilson and Palomar Observatories, 1201 E. California Street, Pasadena, California 91104, U.S.A.
O'KEEFE (J.A.), 3712 Thornapple Street, Chevy Chase 15, Maryland, U.S.A.
OKI (Prof. Dr T.), Department of Earth Science, Faculty of Education, Fukushima University, Hamada-cho 12–23, Fukushima, Japan
* OKOYE (Dr S.E.), Department of Physics, University of Nigeria, Nsukka, Nigeria.
OKUDA (Prof. Dr H.), Department of Physics, Kyoto University, Kitashirakawa, Sakyo-Ku, Japan

Okuda (Dr T.), Director of International Latitude Observatory of Mizusawa, Mizusawa-shi, Iwate-ken, Japan

Oleak (Dr H.), Zentralinstitut für Astrophysik, Sternwarte Babelsberg, DDR 1502 Potsdam-Babelsberg, Rosa-Luxemburg-Strasse 17a, G.D.R.

* O'Leary (Dr B.), Hampshire College, Amherst, Massachusetts 01002, U.S.A.

Olivier (Dr C.P.), American Meteor Society, 521 N. Wynnewood Avenue, Nerberth, Pennsylvania, U.S.A.

Ollongren (Dr A.), Centraal Rekeninstituut, Wassenaarseweg 80, Leiden, The Netherlands

Olsen (Dr K.H.), Los Alamos Scientific Laboratory, Box 1663 Group J-15, Los Alamos, New Mexico 87544, U.S.A.

Olson (Dr E.C.), University of Illinois Observatory, Urbana, Illinois 61801, U.S.A.

Omarov (Dr T.B.), Astrophysical Institute of the Kazakh Academy of Sciences, Alma-Ata, U.S.S.R.

Omer Jr (Dr G.C.), 1080 SW 11th Terrace, Gainesville, Florida 32601, U.S.A.

* Omont (Prof. A.), Université de Paris 7, 2 place Jussieu, 75005 Paris, France

Onderlička (Dr B.), Department of Astronomy, Purkyně University, Brno, Czechoslovakia

Onegina (Mrs Dr A.B.), Main Astronomical Observatory, Ukrainian Academy of Sciences, Kiev, U.S.S.R.

Ōno (Dr Yōrō), Department of Physics, Hokkaido University, Sapporo, Hokkaido, Japan

Oort (Prof. Dr J.H.), President Kennedylaan 169, Oegstgeest, The Netherlands

Oosterhoff (Prof. Dr P.Th.), Ridderhoflaan 107, Koudekerk aan de Rijn, The Netherlands

Opalski (Dr W.), Institut Astronomique de l'Ecole Polytechnique, Koszykowa 75, Warsaw, Poland

Öpik (Dr E.J.), Armagh Observatory, Armagh, U.K.

Opolski (Dr A.), Astronomical Observatory, Kopernika 11, Wrocław, Poland

Orlin (Dr H.), Coast & Geodetic Survey, Rockville, Maryland 20852, U.S.A.

Orlov (Dr A.A.), Sternberg Astronomical Institute, Moscow, U.S.S.R.

Orlova (Mrs Dr N.S.), The Pulkovo Observatory, Leningrad M-140, U.S.A.

Orrall (Prof. Dr F.Q.), Institute for Astronomy, 2840 Koluwalu St., University of Hawaii, Honolulu, Hawaii 96822, U.S.A.

Orte (D.A.), Observatorio de Marina, San Fernando, Cádiz, Spain

Orús (Pr. Dr J.J. de), Barcelona University, Barcelona, Spain

Ōsaki (Dr T.), Ryukoku University, Shimo-gyo-ku, Kyoto, Japan

Osaki (Dr Y.), Department of Astronomy, Faculty of Science, University of Tokyo,, Bunkyo-Ku, Tokyo, Japan

Osawa (Dr K.), Tokyo Astronomical Observatory, Mitaka, Tokyo, Japan

* Osborn (Dr W.), Instituto Venezolano de Astronomía, Apartado 264, Mérida, Venezuela

Oskanyan (V.), Astrophysical Observatory, Byurakan, Armenian SSR, U.S.S.R.

* Osmer (Dr P.), Cerro Tololo Inter-American Obervatory, Casilla 63-D, La Serena, Chile

Osório (Dr J.J.S.P.), Observatorio Astronomico, Universidade do Porto, Monte da Vírgem, Vila Nova de Gaia, Portugal

Oster (Prof. L.), Joint Institute for Laboratory Astrophysics, University of Colorado, Boulder, Colorado, U.S.A.

Osterbrock (Prof. D.E.), Lick Observatory, University of California, Santa Cruz, California 95064, U.S.A.

Ostriker (Dr J.P.), Princeton University Observatory, Peyton Hall, Princeton, New Jersey 08540, U.S.A.

Oterma (Dr L.), Sirkkalankatu 31, 20700-Turku 6, Finland

Ottelet (Dr I.J.G.J.), Institut d'Astrophysique, Université de Liège, avenue de Cointe, 5, B-4200 Cointe-Ougrée, Belgium

Ovenden (Prof. Dr M.W.), Professor of Astronomy, Department of Geophysics and Astronomy, The University of British Columbia, Vancouver 8, B.C., Canada

OWAKI (Dr N.), Department of Astronomy and Earth Science, Tokyo Gakugei University, 4–1–1 Nukui-kita, Koganei-shi, Tokyo, 184 Japan
OWEN (Dr T.), Department of Earth and Space Sciences, State University of New York, Stony Brook, New York 11790, U.S.A.
OWREN (Dr L.), Department of Physics, Div. B, University of Bergen, Allegaten 53–55, 5000 Bergen, Norway
* OXENIUS (Dr J.), Centre d'Etudes Nucléaires, 92 Fontenay-aux-Roses, France
ÖZEMRE (Dr K.), University Observatory, Beyazit, Istanbul, Turkey
OZERNOJ (L.M.), Lebedev Physical Institute, Moscow, U.S.S.R.
OZSVATH (Prof. I.), Earth & Planetary Sciences Laboratory, Graduate Research Center, P.O. Box 30365, Dallas, Texas 75230, U.S.A.

PACHNER (Prof. J.), Physics Department, University of Saskatchewan, Regina Campus, Regina, Sask., Canada
PACHOLCZYK (Prof. Dr A.G.), Steward Observatory, University of Arizona, Tucson, Arizona, U.S.A.
PACINI (Dr F.), Laboratorio di Astrofisica, Casella Postale 67, Frascati (Roma), Italy
PACZYŃSKI (Dr B.), Astronomical Observatory, Al. Ujazdowskie 4, Warszawa, Poland
PAGE (Prof. T.L.), Director, Van Vleck Observatory, Wesleyan University, Middletown, Conn. 06457, U.S.A.
PAGEL (Dr B.E.J.), Royal Greenwich Observatory, Herstmonceux Castle, Hailsham, Sussex, U.K.
PAJDUŠÁKOVÁ (Mrs Dr L.), Director, Skalnaté Pleso Observatory, Slovak Academy of Sciences, Tatranská Lomnica, Czechoslovakia
PAL (Dr A.), Université, Cluj, Roumania
PALMER (D.R.), Royal Greenwich Observatory, Herstmonceux Castle, Hailsham, Sussex, U.K.
PALMER (Dr H.P.), Nuffield Radio Astronomy Laboratories, Jodrell Bank, Macclesfield, Cheshire, U.K.
PALMER (Dr P.E.), Department of Astronomy and Astrophysics, Ryerson Laboratory 162, 1100–14 East 58th Street, Chicago, Illinois 60637, U.S.A.
* PANAGIA (Dr N.), Laboratorio di Astrofisica, Casella Postale 67, Frascati (Roma), Italy
PANAJOTOV (Dr L.A.), Pulkovo Observatory, Leningrad, U.S.S.R.
PANDE (M.C.), U.P. State Observatory, Nainital, India
PAPAGIANNIS (Dr M.), Department of Astronomy, Boston University, Boston, Massachusetts 02215, U.S.A.
* PAPALIOLIOS (Dr C.), Smithsonian Astrophysical Observatory, Cambridge, Massachusetts 02138, U.S.A.
PAQUET (Dr P.E.G.), Observatoire Royal de Belgique, avenue Circulaire, 3, B-1180 Bruxelles, Belgium
* PARCELIER (Dr P.), Observatoire de Paris, 61, Avenue de l'Observatoire, 75014 Paris, France
PARIJSKIJ (Dr N.N.), O. Schmidt Institute of Physics of the Earth, USSR Academy of Sciences, B. Gruzinskaya 10, Moscow, U.S.S.R.
PARIJSKIJ (Dr Yu.N.), Special Astrophysical Observatory, USSR Academy of Sciences, Leningrad Branch of the SAO, Pulkovo 196140, U.S.S.R.
* PARKER (Dr E.A.), Electronics Laboratories, University of Kent at Canterbury, Canterbury, Kent, U.K.
PARKER (Dr E.N.), Institute for Nuclear Studies, University of Chicago, Chicago, Illinois 60637, U.S.A.
PARKER (Dr R.A.R.), Washburn Observatory, University of Wisconsin, Madison, Wisconsin 53705, U.S.A.
* PARKINSON (Dr J.H.), X-Ray Astronomy Group, Physics Department, Leicester University, Leicester, U.K.

* PARKINSON (Dr T.), Kitt Peak National Observatory, 950 North Cherry Avenue, Tucson, Arizona 85726, U.S.A.

PARKINSON (Dr W.H.), Harvard College Observatory, 60 Garden Street, Cambridge, Massachusetts 02138, U.S.A.

* PARSONS (Dr S.B.), Department of Astronomy, University of Texas, Austin, Texas 78712, U.S.A.

* PASACHOFF (Dr J.M.), Williams College, Hopkins Observatory, Williamstown, Massachusetts 01267, U.S.A.

* PASCOAL (A.), Observatório da Universidade do Porto, Monte da Virgem, Villa Nova de Gaia, Portugal

PASCUAL MARTINEZ (M. de), Observatorio de Madrid, Madrid, Spain

PASINETTI (Prof. Dr L.E.), Osservatorio Astronomico di Milano-Merate, Via E. Bianchi 46, 22055 Merate, Italy

* PASTORIZA (Dr M.G.), Ayacucho 2142, Córdoba, Argentina

* PATERNÒ (Dr L.), Osservatorio Astrofisico, Città Universitaria, viale A. Doria, I-95125 Catania, Italy

* PATHRIA (Dr R.K.), Department of Physics, University of Waterloo, Waterloo, Ontario, Canada

PAULINY-TOTH (Dr I.I.K.), Max-Planck-Institut für Radioastronomie, Argelanderstrasse 3, 53 Bonn, F.R.G.

PAVLOV (Prof. Dr N.N.), Pulkovo Observatory, Leningrad, U.S.S.R.

PAVLOVSKAYA (Mrs Dr E.D.), Sternberg Astronomical Institute, Moscow, U.S.S.R.

PAYNE-GAPOSCHKIN (Mrs Dr C.H.), Harvard College Observatory, Cambridge, Massachusetts 02138, U.S.A.

PEACH (Dr G.), Department of Physics, University College, Gower Street, London W.C. 1, U.K.

PEACH (Dr J.V.), Department of Astrophysics, South Parks Road, Oxford, U.K.

PEALE (Dr S.J.), J.I.L.A., University of Colorado, Boulder, Colorado, U.S.A.

PEARSE (Prof. R.W.B.), Imperial College of Science and Technology, Prince Consort Road, London, S.W. 7, U.K.

PEAT (Dr D.W.), The Observatories, Madingley Road, Cambridge, U.K.

PECKER (Prof. J.-C.), Directeur, Institut d'Astrophysique, 98bis, boulevard Arago, 75014 Paris, France

* PEDERSEN (Prof. O.), History of Science Department, University of Aarhus, Ny Munkegade, DK-8000 Aarhus C, Copenhagen

PEDOUSSAUT (Dr A.), Observatoire de Toulouse, 31-Toulouse (Haute Garonne), France

PEEBLES (Dr P.J.E.), Department of Physics, Princeton University, Princeton, New Jersey 08540, U.S.A.

PEERY (Dr B.F.), Department of Astronomy, Indiana University, Bloomington, Indiana 47405, U.S.A.

PEIMBERT (Dr M.), Instituto de Astronomía, UNAM, Apartado Postal 70–264, México D.F., Mexico

PEKERIS (Prof. Ch.L.), The Weizmann Institute of Science, Department of Mathematics, Rehovoth, Israel

PELLAS (Dr P.), Muséum d'Histoire Naturelle-Minéralogie, 61, Rue de Buffon, Paris, France

PELSENEER (Prof. Dr J.), Université Libre de Bruxelles, avenue des Grenadiers, 76, B-1050 Bruxelles, Belgium

PELS-KLUYVER (Mrs Dr H.A.), Sterrewacht, Leiden, The Netherlands

PELTIER (L.C.), 327 S. Bredeick Street, Delphos, Ohio, U.S.A.

* PENNY (Miss C.J.A.), Royal Greenwich Observatory, Time Department, Herstmonceux Castle, Hailsham, Sussex, U.K.

PENSADO (J.), Observatoire, Alfonso XII, 3, Madrid, Spain

* PENSTON (Dr M.V.), Royal Greenwich Observatory, Herstmonceux Castle, Hailsham, Sussex, BN27 1RP, U.K.

PENZIAS (Dr A.), Bell Telephone Laboratories, Crawford Hill Laboratory, Box 400, Holmdel, New Jersey 07733, U.S.A.

* PERAIAH (Dr A.), Tata Institute of Fundamental Research, Bombay, India

PERCY (Dr J.R.), David Dunlap Observatory, University of Toronto, Richmond Hill, Ontario, Canada

PEREK (Dr L.), Astronomical Institute, Czechoslovak Academy of Sciences, Budečská 6, Prague 2, Czechoslovakia

PERINOTTO (Dr M.), Osservatorio Astrofisico, 50125 Arcetri, Largo E. Fermi 5, Firenze, Italy

* PERKINS (Dr F.W.), Princeton University, Plasma Physics Laboratory, P.O. Box 451, Princeton, New Jersey 08540, U.S.A.

PEROLA (Dr G.C.), Instituto di Scienze Fisiche, Via Celoria 16, 20133 Milano, Italy

PERRY (C.L.), Louisiana State University, Baton Rouge, Louisiana 70803, U.S.A.

* PERSIDES (Prof. Dr S.C.), Astronomical Department, University of Thessaloniki, Thessaloniki, Greece

PESCH (Prof. P.), Warner and Swasey Observatory, Case Institute of Technology, Taylor and Brunswick Roads, East Cleveland, Ohio 44112, U.S.A.

PETERSEN (J.O.), University Observatory, Øster Voldgade 3, DK 1350 Copenhagen, Denmark

PETERSON (Dr L.E.), Physics Department, Revelle College, University of California-San Diego, La Jolla, California 92038, U.S.A.

PETFORD (Dr A.D.), University Observatory, South Park Road, Oxford, U.K.

PETRI (Dr W.), Unterleiten 2, Post-Box 106, 8162 Schliersee (OBB), F.R.G.

* PETRINI (Dr D.), Observatoire de Nice, Le Mont-Gros, 06300 Nice, France

PETROSIAN (Dr V.), Institute of Plasma Physics, Stanford University, Stanford, California 94305, U.S.A.

PETROV (Prof. Dr G.I.), Institute of Cosmical Research, USSR Academy of Sciences, Moscow, U.S.S.R.

PETROV (Dr G.M.), Nikolaev Department of the Main Astronomical Observatory, USSR Academy of Sciences, Nikolaev, Ukrainian SSR, U.S.S.R.

PETROVSKAYA (Mrs Dr M.S.), Institute of Theoretical Astronomy, USSR Academy of Sciences, Leningrad, U.S.S.R.

PETTENGILL (Prof. G.H.), Department of Earth and Planetary Sciences, Massachusetts Institute of Technology, Cambridge, Massachusetts 02139, U.S.A.

PETTIT (Dr E.), Mount Wilson and Palomar Observatories, 813 Santa Barbara Street, Pasadena 4, California, U.S.A.

PEYTURAUX (Dr R.), Institut d'Astrophysique, 98bis, Boulevard Arago, Paris 14e, France

* PFENNIG (Dr H.), Max-Planck-Institut für Physik und Astrophysik, 8000 München, Föhringer Ring 6, F.R.G.

* PFLEIDERER (Dr J.), Institut für Astrophysik und Extraterrestrische Forschung, 5300 Bonn, Poppelsdorfer Allee 49, F.R.G.

PFLUG (Dr K.), Heinrich-Hertz-Institut für Solar-Terrestrische Physik, Berlin-Adlershof, Rudower Chaussee, G.D.R.

PHILIP (Dr A.G.D.), Dudley Observatory, 100 Fuller Road, Albany, New York 12208, U.S.A.

PHILLIPS (Prof. J.G.), Department of Astronomy, University of California, Berkeley, California, U.S.A.

PIASKOVSKY (Prof. D.V.), University Astronomical Observatory, Kiev, U.S.S.R.

PICK-GUTMAN (Mrs M.), Observatoire de Paris, Section d'Astrophysique, 92190-Meudon, France

PIDDINGTON (Dr J.H.), CSIRO Division of Physics, University Grounds, Chippendale, N.S.W. 2008, Australia

PIERCE (Dr A.K.), Kitt Peak National Observatory, 950 North Cherry Avenue, P.O. Box 26732, Tucson, Arizona 85726, U.S.A.

PIKEL'NER (Prof. S.B.), Sternberg Astronomical Institute, Moscow V-234, U.S.S.R.

PILOWSKI (Prof. Dr K.), Geodätisches Institut der Technischen Hochschule, Nienburger Strasse 1, Hannover, F.R.G.

* PINKAU (Prof. Dr K.), Max-Planck-Institut für Physik und Astrophysik, Institut für Extraterrestrische Physik, 8046 Garching b. München, F.R.G.

PINTO (Dr G.), Osservatorio Astronomico, Padova, Italy.

PIOTROWSKI (Dr S.), Astronomical Observatory, Ujazdowskie 4, Warsaw, Poland

PIŞMIŞ DE RECILLAS (Mrs Dr P.), Instituto de Astronomía, UNAM, Apartado Postal 70–264, México, D.F., Mexico

* PITZ (Dr E.), Max-Planck-Institut für Astronomie, 6900 Heidelberg 1, Köngstuhl, F.R.G.

* PIZZELLA (Dr G.), Istituto di Fisica dell'Università, Lecce, Italy

PLAKIDIS (Prof. Dr S.), 4B Eridanou Street, Athens (612), Greece

PLASKETT (Prof. H.H.), 48 Blenheim Drive, Oxford, U.K.

PLASSARD (Dr J.), Director, Ksara Observatory, Ksara, Lebanon

PLATZECK (Dr R.), Comision Nacional de la Energía Atómica, San Carlos de Bariloche, Argentina

PLAUT (Prof. Dr L.), Kapteyn Astronomical Laboratory, Postbus 800, Groningen, The Netherlands

PLAVCOVA (Mrs Dr Z.), Department of Astronomy, University of California, 405 Hilgard Avenue, Los Angeles, California 90024, U.S.A.

PLAVEC (Prof. Dr M.), Department of Astronomy, University of California, 405 Hilgard Avenue, Los Angeles, California 90024, U.S.A.

PNEUMAN (Dr G.), High Altitude Observatory, Boulder, Colorado 80302, U.S.A.

PODOBED (Dr V.V.), Sternberg Astronomical Institute, Moscow V-3234, U.S.S.R.

POGO (Dr A.), Mount Wilson and Palomar Observatories, 813 Santa Barbara Street, Pasadena 4, California, U.S.A.

POHL (Dr E.), Sternwarte Nürnberg, Lützowstrasse 10, 8500 Nürnberg, F.R.G.

POLETTO (Mrs Dr G.), Osservatorio Astrofisico Firenze, Largo E. Fermi 5, Firenze, Italy

POLOSKOV (Dr S.M.), Publishing House of Foreign Literature, Novo-Aleksejevskaja 52, Moscow, U.S.S.R.

POLOZHENTSEV (Dr D.D.), Pulkovo Observatory, Leningrad M-140, U.S.S.R.

POLUPAN (Dr P.N.), Astronomical Observatory, Kiev University, Kiev, U.S.S.R.

PONSONBY (Dr J.E.B.), Nuffield Radio Astronomy Laboratories, Jodrell Bank, Macclesfield, Cheshire, U.K.

* POOLEY (Dr G.G.), Mullard Radio Astronomy Observatory, Cavendish Laboratory, Cambridge, U.K.

* POPE (B.Sc.J.D.), Royal Greenwich Observatory, Herstmonceux Castle, Hailsham, Sussex, U.K.

POPOV (Dr N.A.), Gravimetrical Observatory, Ukrainian Academy of Sciences, Poltava, U.S.S.R.

POPOVA (Mrs Dr M.D.), Section d'Astronomie de l'Académie Bulgare des Sciences, Rue 7 Novembre No. 1, Sofia, Bulgaria

POPOVIĆ (Prof. Dr B.), Ognjena Price 80, Beograd, Yugoslavia

POPOVICI (Prof. Dr C.), Observatoire de Bucarest, 5 rue Cutitul de Argint, Bucarest, Roumania

* PÖPPEL (Dr W.), Instituto Argentino de Radioastronomía, Casilla No. 5, Villa Elisa, Prov. de Bs. As., Argentina

POPPER (Prof. D.M.), Department of Astronomy, University of California, Los Angeles 24, California, U.S.A.

PORFIR'EV (Dr V.V.), Main Astronomical Observatory, Ukrainian Academy of Sciences, Kiev, U.S.S.R.

PORTER (Dr J.G.), Whitestones, Hempstead Lane, Hailsham, Sussex, U.K.

PORTER (Prof. N.A.), Department of Physics, University College, Belfield, Dublin 4, Ireland

POSTOIEV (Dr A.), Instituto Astronômico e Geofísico, Universidade de São Paulo, Caixa Postal 30627, São Paulo, Brazil

POTTASCH (Prof. Dr S.R.), Kapteyn Astronomical Institute, Postbus 800, Groningen, The Netherlands

POTTER (Dr H.I.), Main Astronomical Observatory, Academy of Sciences, Pulkovo, U.S.S.R.
* POULAKOS (Dr C.), Research Center for Astronomy and Applied Mathematics, Academy of Athens, 14 Anagnostopoulou Street, Athens (136), Greece
POUNDS (Dr K.A.), University of Leicester, Department of Physics, University Road, Leicester, LE 17 RH, U.K.
POVEDA (Dr A.), Instituto de Astronomía, UNAM, Apartado Postal 70–264, México, D.F., Mexico
PRADERIE (Mrs Dr F.), Institut d'Astrophysique, 98 bis, Boulevard Arago, 75014 Paris, France
PRADHAN (Dr), Nizamiah Observatory, Hyderabad, India
PRENDERGAST (Dr K.H.), 1402 Pupin, Columbia University, New York, N.Y., U.S.A.
PRENTICE (J.P.M.), Star Ridge, Battisford, Needham Market, Suffolk, U.K.
PRESTON (Dr G.W. III), Lick Observatory, University of California, Santa Cruz, California 95060, U.S.A.
* PREVOT (Dr L.), Observatoire de Marseille, 2, Place Le Verrier, 13004 Marseille, France
PRICE (Dr M.J.), Astro Sciences Center, IIT Research Institute, 10 West 35th Street, Chicago, Illinois 60616, U.S.A.
PRICE (Dr R.M.), 26–463 M.I.T., Cambridge, Massachusetts 02139, U.S.A.
* PRIEST (Dr E.R.), Mathematical Institute, North Haugh, St Andrews Fife, U.K.
PRIESTER (Prof. Dr W.), Direktor, Institut für Astrophysik und Extraterrestrische Forschung, Universität Bonn, Auf dem Hügel 71, 53 Bonn, F.R.G.
PROBSTEIN (Dr R.F.), Department of Mechanical Engineering, Massachusetts Institute of Technology, Cambridge, Massachusetts 02139, U.S.A.
PROCHAZKA (Dr F.V.), Observatory of the University of Vienna, Türkenschanzstrasse 17, A-1180 Vienna, Austria.
* PRODAN (Dr Y.I.), Sternberg Astronomical Institute, Universitetskij prospect 13, Moscow 117234, U.S.S.R.
* PROISY (Dr P.E.), Observatoire de Lyon, 69230 Saint-Genis-Laval, France
* PROKAKIS (Dr Th.), Astronomical Institute, National Observatory of Athens, Athens (306), Greece
PROKOF'EV (Prof. V.K.), Crimean Astrophysical Observatory of the Academy of Sciences of the USSR, P/O Nauchny, Crimea 334413, U.S.S.R.
PROKOF'EVA (Mrs Dr I.A.), Main Astronomical Observatory, USSR Academy of Sciences, Pulkovo, U.S.S.R.
* PRONIK (Mrs Dr I.I.), Crimean Astrophysical Observatory of the USSR Academy of Sciences, P/O Nauchny, Crimea, U.S.S.R.
PRONIK (Dr V.I.), Crimean Astrophysical Observatory, USSR Academy of Sciences, Crimea, U.S.S.R.
PROTHEROE (Dr W.M.), Department of Astronomy, 174 W. Eighteenth Street, Ohio State University, Columbus, Ohio 43210, U.S.A.
PROTITCH (M.B.), Astronomical Observatory, Veliki Vracar, Belgrade, Yugoslavia
PROVERBIO (Prof. E.), Director, Institute of Astronomy, Via Ospedale 72, 09100 Cagliari, Italy
PRYCE (Dr M.H.L.), Institute of Astronomy and Space Research (Department of Physics), University of British Columbia, Vancouver 8, B.C., Canada
PRZYBYLSKI (Dr A.), Mount Stromlo Observatory, Canberra, A.C.T., Australia
PSKOVSKIJ (Dr Ju.P.), Sternberg Astronomical Institute, Moscow, U.S.S.R.
* PUNETHA (Dr L.M.), U.P. State Observatory, Manora Peak, Nainital, U.P., India
PURGATHOFER (Dr A.Th.), Universitäts-Sternwarte Wien, Türkenschanzstrasse 17, A-1180 Vienna, Austria
* PURTON (Prof. Ch.R.), York University, Toronto, Ontario, Canada
PYRER (Mrs Dr D.M.), Department of Physics and Astronomy, Tel-Aviv University, Ramat Aviv, Israel

* Querci (Dr F.), Observatoire de Paris, Section d'Astrophysique de Meudon, 92190 Meudon, France

Quijano (L.), Observatoire, San Fernando, Spain

* Quirk (Dr W.), Goddard Institute for Space Studies, 2880 Broadway, New York, New York 10025, U.S.A.

Quist (Dr B.), Åbo Akademi, Åbo, Finland

* Raadu (Dr M.A.), The Royal Institute of Technology, Department of Plasma Physics, Teknik, Ring 31–33, S-100 44 Stockholm 70, Sweden

Rabben (Dr H.H.), Max-Planck-Institut für Physik und Astrophysik, Abt. Extraterrestrische Physik, Garching b. München, F.R.G.

Rabe (Prof. Dr E.), Cincinnati Observatory, Observatory Place, Cincinnati, Ohio 45208, U.S.A.

* Rachkovsky (Dr D.N.), Crimean Astrophysical Observatory AN USSR, Crimea, Nauchny 334413, U.S.S.R.

* Racine (Dr R.), David Dunlap Observatory, Richmond Hill, Ontario, Canada

Rackham (Dr T.W.), Armagh Observatory and Planetarium, 46, Ashley Park, Armagh, U.K.

Radhakrishnan (Prof. V.), Director, Raman Research Institute, Bangalore-560006, India

* Rädler (Dr K.-H.), Zentralinstitut für Astrophysik, Telegrafenberg, DDR 15, Potsdam, G.D.R.

Radlova (Mrs Dr L.N.), Institute of Sciences and Technics Information, Department of Astronomy, USSR Academy of Sciences, Moscow, U.S.S.R.

Raghavan (N.), Department of Physics, Indian Institute of Technology, Kanpur, India

Rahe (Dr J.), Lehrstuhl für Astrophysik, T.U. Berlin, Strasse d. 17 Juni 135, 1000 Berlin, F.R.G.

Rahim (Dr M.H.A.), Helwan Observatory, Helwan (near Cairo), Arab Rep. of Egypt

Raimond (Dr E.), University Observatory, Sterrenwacht, Leiden, The Netherlands

Rajchl (Dr J.), Astronomical Institute, Czechoslovak Academy of Sciences, Observatory Ondřejov, Czechoslovakia

Raju (Dr P.K.), Lehrstuhl für Theoretische Astrophysik, Hausserstrasse 64, 7400 Tübingen, F.R.G.

Rakavy (Prof. G.), University of Jerusalem, Einstein Institute of Physics, Dept. of Theoretical Physics, The Hebrew University of Jerusalem, Jerusalem, Israel

Rakos (Dr K.D.), Universitätssternwarte, Türkenschanzstrasse 17, A-1180 Wien, Austria

Rakshit (Prof. H.), Bengal Engineering College, Sibpore, Hewrah, India

Ramaty (Dr R.), NASA, Goddard Space Flight Center, Greenbelt, Maryland, U.S.A.

Ramberg (Prof. Dr J.M.), Genvägen 4, S-133 00 Saltsjöbaden, Sweden

Randić (Dr L.), Technicki fakultet, Zagreb, p.p. 89, Yugoslavia

Rao (Dr U.R.), Physical Research Laboratory, Navrangpura, Ahmedabad-9, India

Rasmusen (Dr H.Q.), Vaerslevgaarden, 4400 Kalundborg, Denmark

Rayrole (J.), Observatoire de Meudon, 92190 Meudon, France

Razin (Dr V.A.), Radiophysical Research Institute, Gorkij, U.S.S.R.

Razmadze (Dr T.S.), Institute Geophysics, Academy of Sciences, Cosmic Ray Station, Tbilisi, U.S.S.R.

Reaves (Dr G.), Astronomy Department, University of Southern California, Los Angeles, California 90007, U.S.A.

* Reay (Dr N.K.), Astronomy Group, Physics Department, Imperial College, London SW 7, U.S.A.

Reber (Dr G.), CSIRO, Stowell Avenue, Hobart, Tasmania, Australia 7000

Reddish (Dr V.C.), Royal Observatory, Edinburgh 9, U.K.

Redman F.R.S. (Prof. R.O.), Director of the Observatories, Madingley Road, Cambridge, U.K.

Rees (Dr M.J.), Institute of Astronomy, Madingley Road, Cambridge CB3 OHA, U.K.

Reeves (Dr E.M.), Harvard College Observatory, 60 Garden Street, Cambridge, Massachusetts 02138, U.S.A.

Reeves (Dr H.), SEP-Saclay-BP No. 2, Gif-sur-Yvette (91), France

REFSDAL (Dr S.), Department of Physics, University of Nebraska, Lincoln, Nebraska, U.S.A.

* REGO FERNANDEZ (Dr M.), Departamento de Astronomía, Universidad Complutense, Ciudad Universitaria, Madrid 3, Spain

REID (Dr H.), Lockhead Electronics Company, 16811 El Camino Real, Houston, Texas 77058, U.S.A.

* REIMERS (Dr D.), Institut für Theoretische Physik und Sternwarte der Universität, 2300 Kiel, Olshausenstrasse, F.R.G.

REINMUTH (Dr K.), Häusserstrasse 61, Heidelberg, F.R.G.

REIZ (Prof. A.), University Observatory, Østervoldgade 3, DK-1350 Copenhagen-K, Denmark

REMY-BATTIAU (Mrs Dr L.), Conservateur, Institut d'Astrophysique, Université de Liège, avenue de Cointe, 5, B-4200 Cointe-Ougrée, Belgium

RENSE (W.A.), Physics Department, University of Colorado, Boulder, Colorado, U.S.A.

* RENSON (Dr P.F.M.L.), Institut d'Astrophysique de l'Université de Liège, Avenue de Cointe, 5, B-4200 Cointe-Ougrée, Belgium

RENZINI (Dr A.), Astronomical Observatory Bologna, via Zamboni 32, Bologna, Italy

REQUIÈME (Y.), Observatoire de Bordeaux, 33-Floirac, France

REUNING (Dr E.), Department of Physics and Astronomy, University of Georgia, Athens, Georgia 30601, U.S.A.

REYNOLDS (Dr J.H.), Department of Physics, University of California, Berkeley, California, U.S.A.

* RIBES (Dr J.-C.), Observatoire de Paris, Section d'Astrophysique de Meudon, 92190 Meudon, France

RICE (D.A.), Coast and Geodetic Survey, Rockville, Maryland 20852, U.S.A.

* RICE (Prof. J.B.), Department of Physics, Brandon University, Brandon, Manitoba, Canada

RICHARDSON (Dr E.H.), Dominion Astrophysical Observatory, R.R. 7, Victoria, B.C., Canada

RICHARDSON (R.S.), Griffith Observatory, P.O. Box 27787, Los Felix Station, Los Angeles 27, California, U.S.A.

RICHTER (Dr G.), Zentralinstitut für Astrophysik, Sternwarte Sonnenberg, DDR 64 Sonnenberg, G.D.R.

* RICHTER (Dr J.), Institut für Experimental-Physik, Olshausenstrasse, 40–60, Neue Universität, Gebäude D.6, 23 Kiel-1, F.R.G.

RICHTER (Dr N.), Zentralinstitut für Astrophysik, Karl-Schwarzschild-Observatorium, DDR 6901 Tautenburg, G.D.R.

RIEGEL (Dr K.W.), Astronomy-UCLA, Los Angeles, California 90024, U.S.A.

RIEU (Dr N.Q.), Observatoire de Meudon, 92190 Meudon, France

RIGHINI (Dr A.), Osservatorio Astrofisico di Catania, Città Universitaria, Catania, Sicilia, Italy

RIGHINI (Prof. G.), Direttore, Osservatorio Astrofisico di Arcetri, 'Giorgio Abetti', Largo Enrico Fermi, 5, 50125 Firenze, Italy

RIGUTTI (Prof. M.), Direttore, Osservatorio Astronomico di Capodimonte, via Moiariello, 16, 80131 Napoli, Italy

RIIHIMAA (Dr J.J.), Oulun yliopiston tähtitieteen laitos, Torikatu 7, Oulu, Finland

RIIVES (Dr V.), Tartu State University, Department of Theoretical Physics, Tartu, Estonia, U.S.S.R.

RINDLER (Prof. W.), Graduate Research Center, P.O. Box 30365, Dallas, Texas 75230, U.S.A.

RINEHART (Dr J.S.), Hyperdynamics, P.O. Box 392, Santa Fe, New Mexico 87501, U.S.A.

RING (Prof. J.), Department of Applied Physics, The University, Hull, U.K.

RINGES (Dr T.S.), Institute of Theoretical Astrophysics, University of Oslo, Blindern, Oslo 3, Norway

RINGUELET (Dr A.E.), Observatorio Astronómico, Universidad Nacional de la Plata, La Plata, Argentina

ROACH (Dr F.E.), 2987 Kalakaua Avenue (602), Honolulu, Hawaii 96815, U.S.A.

* ROARK (Dr T.), Department of Astronomy, 174 West 18th Ave., Ohio State University, Columbus, Ohio 43210, U.S.A.

* ROBBINS (Dr R.R.), Associate Professor, Department of Astronomy, University of Texas at Austin, Austin, Texas 78712, U.S.A.

ROBE (Dr H.A.Gh.), Institut d'Astrophysique, Université de Liège, avenue de Cointe, 5, B-4200 Cointe-Ougrée, Belgium

ROBERTS (Dr J.A.), CSIRO, Division of Radiophysics, Post Office Box 76, Epping, N.S.W., 2121, Australia

ROBERTS (Dr M.S.), National Radio Astronomy Observatory, Edgemont Road, Charlottesville, Virginia 22901, U.S.A.

ROBERTS (Dr W.O.), Director, National Center for Atmospheric Research, Boulder, Colorado, U.S.A.

* ROBERTS JR (Dr W.W.), University of Virginia, Department of Applied Mathematics and Computer Science, Thornton Hall, Charlottesville, Virginia 22901, U.S.A.

ROBERTSON (W.H.), Sydney Observatory, Sydney, N.S.W. 2000, Australia

ROBINSON (Dr B.J.), CSIRO, Division of Radiophysics, P.O. Box 76, Epping, N.S.W., 2121, Australia

ROBINSON (Prof. I.), Graduate Research Center, P.O. Box 30365, Dallas, Texas 75230, U.S.A.

* ROBINSON (Dr L.B.), Lick Observatory, University of California, Santa Cruz, U.S.A.

ROBLEY (Dr R.), Observatoire du Pic-du-Midi, Bagnères-de-Bigorre, Hautes-Pyrénées, France

* ROCHESTER (Prof. M.G.), Memorial University of Newfoundland, Department of Phsyics, St. John's, Newfoundland, Canada

RODDIER (F.), Faculté des Sciences de Nice, Avenue Valrose, 06-Nice, France

RODGERS (Dr A.W.), Mount Stromlo Observatory, Camberra, A.C.T., Australia

RODONO (Dr M.), Osservatorio Astrofisico, Città Universitaria, viale A. Doria, 95123 Catania, Sicilia, Italy

RODRIGUEZ (M.), Director del Instituto y Observatorio de Marina, San Fernando (Cadiz), Spain

ROEDER (Dr R.C.), Scarborough College, University of Toronto, 1265 Military Trail, West Hill, Ontario, Canada

ROEMER (Miss Dr E.), Lunar and Planetary Laboratory, University of Arizona, Tucson, Arizona 85721, U.S.A.

ROGER (Dr R.S.), Dominion Radio Astrophysical Observatory, Box 248, Penticton, B.C., Canada

ROGERS (Dr A.E.E.), Massachusetts Institute of Technology, Lincoln Laboratory, Lexington, Massachusetts 02173, U.S.A.

ROGERSON (Prof. J.B.), Princeton University Observatory, Peyton Hall, Princeton, New Jersey 08540, U.S.A.

ROGSTAD (Dr D.H.), Department of Radio Astronomy, California Institute of Technology Pasadena, California 91109, U.S.A.

ROHLFS (Prof. K.), Max-Planck-Institut für Radioastronomie, Argelanderstr. 3, 5300 Bonn, F.R.G.

ROLAND (Miss Dr C.), Institut d'Astrophysique, Université de Liège, avenue de Cointe 5, B-4200 Cointe-Ougrée, Belgium

ROMAN (Miss Dr N.G.), NASA Headquarters, Washington, D.C. 20546, U.S.A.

ROMANÁ (Dr A.), Director, Observatorio de Ebro, Roquetas, Tarragona, Spain

* ROMANCHUK (Dr P.R.), Astronomical Observatory, Observatornaya, 3, Kiev 252053, U.S.S.R.

ROMANO (Dr G.), Viale S. Francesco 7, Treviso, Italy

ROMPOLT (Dr B.), Astronomical Institute of Polish Academy of Sciences, Kopernika 11, Wrocław, Poland

RONAN (C.A.), 39 New Road, Barton, Cambridge CB3 7AY, U.K.

* RÖNNÄNG (Dr B.O.), Onsala Space Observatory, S-430 34 Onsala, Sweden

ROOD (Dr H.J.), Van Vleck Observatory, Wesleyan University, Middletown, Connecticut 06457, U.S.A.

* ROOD (Dr R.), Astronomy Department, University of Virginia, Charlottesville, Virginia 22903, U.S.A.

* Roosen (Dr R.G.), NASA, Goddard Space Flight Center, Laboratory for Solar Physics, New Mexico Station, 800 Yale Blvd, NE, Albuquerque, New Mexico 87131, U.S.A.
Rösch (Prof. J.), Directeur de l'Observatoire du Pic-du-Midi, 65 Bagnères-de-Bigorre, (Hautes-Pyrénées), France
Rose (Dr W.K.), Physics Department, Massachusetts Institute of Technology, Cambridge, Massachusetts 02139, U.S.A.
Rosen (Dr E.), The City College of the City University of New York, Department of History, New York, N.Y. 10031, U.S.A.
* Rosenberg (Dr J.), Sterrewacht, Servaas Bolwerk 13, Utrecht, The Netherlands
Rosino (Prof. L.), Director of the Astrophysical Observatory, Asiago (Vicenza), Italy
Roslund (Dr C.), Lund Observatory, S-222 24 Lund, Sweden
* Ross (Dr D.), Iowa State University, Erwin Fick Observatory, Ames, Iowa 50010, U.S.A.
Rosseland (Prof. S.), Institute of Theoretical Astrophysics, University of Oslo, Blindern, Oslo 3, Norway
Rossi (Dr B.), Department of Physics, Massachusetts Institute of Technology, Cambridge Massachusetts 02139, U.S.A.
Rostas (Dr F.), Observatoire de Meudon, 92190 Meudon, France
Rottenberg (Dr J.A.), 142 Chestnut Avenue, Pointe Claire 720, P.Q., Canada
Rountree Lesh (Mrs Dr J.), Department of Physics and Astronomy, Space Sciences Building, University of Denver, Denver, Colorado 80210, U.S.A.
Rouse (Dr C.A.), Intelcom Rad Tech, P.O. Box 80817, San Diego, California 92138, U.S.A.
* Rousseau (Dr J.), Observatoire de Lyon, 69230 Saint-Genis-Laval, France
* Rousseau (J.-M.), Observatoire de L'Université de Bordeaux, 33270 Floirac, France
* Routledge (Prof. D.), Department of Electrical Engineering, University of Alberta, Edmonton, Alberta, Canada
Routly (Dr P.M.), U.S. Naval Observatory, Washington, D.C. 20390, U.S.A.
* Rowan-Robinson (Dr G.M.), Department of Applied Mathematics, Queen Mary College, Mile End Road, London E1, U.K.
Roxburgh (Dr I.W.), Mathematics Department, Queen Mary College, Mile End Road, London E1, U.K.
Roy (Dr A.E.), Department of Astronomy, University of Glasgow, Glasgow, W. 2, U.K.
* Rozelot (Dr J.P.), Observatoire du Pic-du-Midi, 65200 Bagnères-de-Bigorre, France
Rozhkovski (Dr D.A.), Astrophysical Institute, Alma Ata, U.S.S.R.
Rubashev (Dr B.M.), Pulkovo Observatory, Leningrad, U.S.S.R.
Ruben (Dr G.), Zentralinstitut für Astrophysik, DDR 15 Potsdam, Telegrafenberg, G.D.R.
Rubin (Mrs Dr V.C.), Department of Terrestrial Magnetism, Carnegie Institution of Washington, 5241 Broad Branch Road N.W., Washington, D.C., U.S.A.
Rublev (Dr S.V.), Special Astrophysical Observatory, USSR Academy of Sciences, St. Zelenchukskaya, Stavropolsky Kraj, U.S.S.R.
* Rucinski (Dr S.), Astronomical Observatory of the Warsaw University, Al. Ujazdowskie 4, Warszawa, Poland
* Ruderman (Dr M.), Columbia University, Astronomy Department, New York, N.Y. 10027, U.S.A.
Rudkjøbing (Prof. M.), Director of the Ole Rømer Observatory, Aarhus C, Denmark
Rudnicki (Dr K.), Jagiellonian University Observatory, ul. Kopernika 27, Kraków, Poland
Rufener (Dr F.-G.), Observatoire de Genève, 1290 Sauverny (GE), Switzerland
Rule (B.R.), Mount Wilson and Palomar Observatories, 813 Santa Barbara Street, Pasadena, California 91106, U.S.A.
* Rumsey (N.J.), Physics and Engineering Laboratory, D.S.I.R., Lower Hutt, New Zealand
Runcorn (Prof. S.K.), School of Physics, The University, Newcastle-upon-Tyne, NE1 7RU, U.K.
Ruprecht (Dr J.), Astronomical Institute, Czechoslovak Academy of Sciences, Budečská 6, Prague 2, Czechoslovakia

RUSKOL (Mrs Dr E.L.), Institute of Physics of the Earth, USSR Academy of Sciences, Moscow, U.S.S.R.

RUSSELL (Dr J.A.), Astronomy Department, University of Southern California, University Park, Los Angeles, California 90007, U.S.A.

* RUSSEV (Dr R.), Bulgarian Academy of Sciences, Department of Astronomy, 7th November Street 1, Sofia, Bulgaria

* RUST (Dr D.M.), Sacramento Peak Observatory, Sunspot, New Mexico 88349, U.S.A.

* RUSU (Dr I.), Astronomical Observatory, Str. Cutitul de Argint 5, Bucarest 28, Roumania

RUSU (Mrs Dr L.), Observatorul Astronomic, Strada Cutitul de Argint 5, Bucuresti 28, Roumania

RŮŽIČKOVÁ-TOPOLOVÁ (Mrs Dr B.), Astronomical Institute, Czechoslovak Academy of Sciences, Observatory Ondřejov, Czechoslovakia

RYABOV (Dr Yu.A.), Sternberg Astronomical Institute, Moscow, U.S.S.R.

RYBICKI (Dr G.B.) Smithsonian Astrophysical Observatory, 60 Garden Street, Cambridge, Maschusetts 02138, U.S.A.

RYBKA (Psaof. Dr E.), Szopena 20/3, Kraków, Poland

RYBKA (Drr P.), Astronomical Institute of the Wrocław University, Dembowskiego 19, Wrocław 9, Poland

RYDBECK (Dr O.E.H.), Institute of Electronphysics I, Chalmers Technical University, Fack, S - 402 20 Göteborg 5, Sweden

RYLE F.R.S. (Prof. Sir Martin), Cavendish Laboratory, Free School Lane, Cambridge, U.K.

RYTER (DrCh.), CEN SACLAY-SEP, B.P. No 2, 91-Gif-sur-Yvette, France

* RYZHKOV (Dr N.F.), Leningrad Branch of Special Astrophysical Observatory, Academy of Sciences, Pulkovo, Leningrad, 196140, U.S.S.R.

RZHIGA (Dr O.N.), Institute of Radio and Electronics, USSR Academy of Sciences, Moscow, U.S.S.R.

SACK (Dr N.), Einstein Institute of Physics, Dept. of Theoretical Physics, The Hebrew University of Jerusalem, Jerusalem, Israel

* SACKMANN-CHRISTY (Mrs Dr I.J.), California Institute of Technology, Pasadena, California 91109, U.S.A.

SADEH (Dr D.), Department of Physics and Astronomy, Tel-Aviv University, Ramat-Aviv, Israel

SADLER (D.H.), 8 Collington Rise, Bexhill-on-Sea, Sussex, TN39 3RT, U.K.

SAFRONOV (Dr V.S.), O.J. Schmidt Institute of the Physics of the Earth, USSR Academy of Sciences, B. Gruzinskaya 10, Moscow 242, U.S.S.R.

SAGAN (Prof. C.), Center for Radiophysics and Space Research, Laboratory for Planetary Studies, Cornell University, Ithaca, New York 14850, U.S.A.

SAGGION (Prof. Dr A.), Istituto di Fisica 'G. Galilei', Via Marzolo 8, 35100 Padova, Italy

SAGITOV (Dr M.U.), Sternberg Astronomical Institute, Moscow 117234, U.S.S.R.

SAGNIER (Dr J.L.), Bureau des Longitudes, 3 rue Mazarine, Paris 8e, France

SAHADE (Dr J.), Director, Instituto de Astronomía y Física del Espacio, Casilla de Correo 67, Sucursal 28, Buenos Aires, Argentina

* SAHAL (Dr S.), Observatoire de Paris, Section d'Astrophysique de Meudon, 92190 Meudon, France

* SAÏSSAC (Dr J.), Observatoire du Pic-du-Midi, 65200 Bagnères de Bigorre, France

SAITO (Dr K.), Tokyo Astronomical Observatory, Mitaka, Tokyo, Japan

* SAITO (Dr M.), Tokyo Astronomical Observatory, Mitaka, Tokyo, Japan

SAITO (Dr S.), Kwasan Observatory, University of Kyoto, Yamashina, Kyoto, Japan

SAKASHITA (Dr S.), Department of Physics, Hokkaido University, Sapporo, Japan

SAKHAROV (Dr V.I.), Main Astronomical Observatory, USSR Academy of Sciences, Pulkovo, U.S.S.R.

SAKURAI (Dr K.), Code 693, Laboratory for Extraterrestrial Physics, NASA-Goddard Space Flight Center, Greenbelt, Maryland 20771, U.S.A.

SAKURAI (Prof. Dr T.), Department of Aeronautical Engineering, Faculty of Engineering, Kyoto University, Kyoto, Japan

* SALETIĆ (D.), Military Geographical Institute, Mije Kovacevica 5, Beograd 11000, Yugoslavia

SALIN (Dr A.), Observatoire de Meudon, 92190 Meudon, France

SALISBURY (Dr J.W.), Air Force Cambridge Research Laboratoires, Bldg. 1121, L.G. Hanscom Field, Bedford, Massachusetts 01730, U.S.A.

SALOMONOVICH (Dr A.E.), Physical Institute, USSR Academy of Sciences, Moscow, U.S.S.R.

SALPETER (Dr E.E.), Laboratory of Nuclear Studies, Cornell University, Ithaca, N.Y., U.S.A.

SALPETER S.J. (Dr E.W.), Specola Vaticana, Castel Gandolfo, Vatican City State

SALUKVADZE (Dr G.N.), Abastumani Astrophysical Observatory, Georgian Academy of Sciences, Abastumani, U.S.S.R.

SAMAHA (Prof. A.H.M.), 5 Wadi El-Nile Street, Maadi, Cairo, Arab Rep. of Egypt

SANAMIAN (Dr V.A.), Byurakan Astrophysical Observatory, Byurakan, Armenia, U.S.S.R.

* SANCHEZ MAGRO (Dr C.), Instituto Universitario de Astrofísica, Universidad de la Laguna, Tenerife, Spain

SANCHEZ-MARTINEZ (Dr F.), Instituto Universitario de Astrofísica, Universidad de La Laguna, Tenerife, Spain

SANCISI (R.), Kapteyn Laboratory, Postbus 800, Groningen, The Netherlands

SANDAGE (A.), Mount Wilson and Palomar Observatories, 813 Santa Barbara Street, Pasadena 4, California 91101, U.S.A.

SANDAKOVA (Dr E.V.), Kiev University Observatory, Kiev, U.S.S.R.

SANDERS (Dr P.), Observatoire Royal de Belgique, avenue Circulaire, 3, B-1180 Bruxelles, Belgium

* SANDERS (Dr R.), National Radio Astronomy Observatory, Edgemont Road, Charlottesville, Virginia 22901, U.S.A.

* SANDERS (Dr W.L.), New Mexico State University, Las Cruces, New Mexico, U.S.A.

SANDIG (Prof. H.-U.), Observatory of the Technical University, Lohrmann Institut, Dresden 8027, G.D.R.

* SANDQVIST (Dr Aa.), Stockholm Observatory, S-133 00 Saltsjöbaden, Sweden

SANDULEAK (Dr N.), Warner and Swasey Observatory, Taylor and Brunswick Roads, East Cleveland, Ohio 44112, U.S.A.

SANTOS (Dr A.J. Baptista dos), Observatorio Astronomico, Tapada, Lisbon, Portugal

SANWAL (Dr N.B.), Department of Astronomy, Osmani University, Hyderabad 7 (A.P.), India

SAPAR (Dr A.), W. Struve Astrophysical Observatory Tartu, Tôravere, Estonia, U.S.S.R.

SARGENT (Dr W.L.), California Institute of Technology, Pasadena, California 91109, U.S.A.

SARMA (M.B.K.), Department of Astronomy, Osmani University, Hyderabad, India

* SARTORI (Dr B.), University of Nebraska, Lincoln, Nebraska 68508, U.S.A.

SASLAW (Dr W.C.), Leander McCormick Observatory of the University of Virginia, P.O. Box 3818, University Station, Charlottesville, Virginia 22903, U.S.A.

SASTRY (Ch.V.), Astrophysical Observatory, Kodaikanal-3, India

* SATO (Dr F.), Chiba Prefecture Education Center, Katsuragi 2–10–1, Chiba, Japan

SATO (Dr H.), Department of Physics, Kyoto University, Kyoto, Japan

SATO (Dr Y.), Tokyo Astronomical Observatory, Mitaka, Tokyo, Japan

* SAUVAL (Dr A.J.P.), Observatoire Royal de Belgique, avenue Circulaire, 3, B-1180 Bruxelles, Belgium

SAVAGE (Dr B.), Washburn Observatory, University of Wisconsin, 475 North Charter Street, Madison, Wisconsin 53706, U.S.A.

SAVEDOFF (Prof. M.P.), Department of Physics and Astronomy, University of Rochester, River Campus Station, Rochester, N.Y. 14627, U.S.A.

SAWYER (Mrs Dr C.S.), Space Environment Laboratory, National Oceanic and Atmospheric Administration, Boulder, Colorado 80302, U.S.A.

SAWYER-HOGG (Mrs Prof. H.B.), David Dunlap Observatory, Richmond Hill, Ontario, Canada

SAXENA (Dr P.P.), Uttar Pradesh State Observatory, Manora Peak, Naini Tal, India
* SCALISE JR (E.), Centro de Rádio Astronomía e Astrofísica, Universidade Mackenzie, CP. 8792, São Paulo, SP, Brazil
SCAFRE (Dr C.D.), Department of Physics, University of Victoria, Victoria, B.C., Canada
SCARGLE (Dr J.D.), Lick Observatory, University of California, Santa Cruz, California 94060, U.S.A.
SCHADEE (Dr A.), Astronomical Observatory 'Sonnenborgh', Servaas Bolwerk 13, Utrecht, The Netherlands
SCHAIFERS (Dr K.), Landessternwarte, Heidelberg-Königstuhl, F.R.G.
SCHALÉN (Prof. C.), Lund Observatory, S-222 24 Lund, Sweden
* SCHANDA (Dr E.), Institut für angewandte Physik, Sidlerstrasse 5, CH-3012, Bern, Switzerland
SCHATZMAN (Prof. E.L.), Observatoire de Paris-Meudon, Département d'Astrophysique Fondamentale, 5, place Jansen, 92190 Meudon, France
SCHEEPMAKER (Dr A.C.), Boerhaavestraat 59, Voorhout, The Netherlands
SCHEFFLER (Dr H.), Landessternwarte, 69 Heidelberg-Königstuhl, F.R.G.
SCHEUER (Dr P.A.G.), Mullard Radio Astronomy Observatory, Cavendish Laboratory, Free School Lane, Cambridge, U.K.
SCHILD (Dr R.E.), Smithsonian Astrophysical Observatory, 60 Garden Street, Cambridge, Massachusetts 02138, U.S.A.
SCHILLER (Prof. Dr K.), Pirschweg 6, 6079 Buchschlag über Sprendlingen, F.R.G.
SCHILT (Prof. J.), 481 Greenville Avenue, Centerdale 11, Rhode Island, U.S.A.
SCHKODROV (Dr V.G.), Department of Astronomy, '7th November' Street 1, Sofia, Bulgaria
* SCHLESINGER (Dr B.), Indiana University, Bloomington, Indiana 47401, U.S.A.
* SCHLOSSER (Dr W.), Astronomisches Institut der Ruhr-Universität Bochum, 4630 Bochum-Querenburg, Postfach 2148, F.R.G.
SCHLUTER (Prof. Dr A.), Max-Planck-Institut für Physik und Astrophysik, Institut für Astrophysik, Aumeisterstrasse 6, München 23, F.R.G.
SCHLUTER (Prof. Dr D.), Institut für Theoretische Physik und Sternwarte der Universität, 23 Kiel, Olshausenstrasse, Neue Universität, Geb. 13, F.R.G.
SCHMAHL (Dr G.), Universitätssternwarte, Geismarlandstr. 11, 3400 Göttingen, F.R.G.
SCHMALBERGER (Dr D.C.), State University of New York at Albany, ES 314, Albany, New York 12203, U.S.A.
SCHMEIDLER (Prof. Dr F.), Universitäts-Sternwarte, Sternwartstrasse 23, München 27, F.R.G.
* SCHMID-BURGK (Dr J.), Institut für Theoretische Astrophysik der Universität, 6900 Heidelberg, Berliner Str. 19, F.R.G.
SCHMIDT (Prof. Dr H.), Direktor der Universitäts-Sternwarte, Auf dem Hügel 71, 53 Bonn, F.R.G.
SCHMIDT (Dr H.U.), Max-Planck-Institut für Physik und Astrophysik, Föhringer Ring 6, 8 München 23, F.R.G.
SCHMIDT (Dr K.-H.), Zentralinstitut für Astrophysik, Sternwarte Babelsberg, DDR 1502 Potsdam-Babelsberg, Rosa-Luxemburg-Strasse 17a, G.D.R.
SCHMIDT (Prof. M.), Mount Wilson and Palomar Observatories, 1201 East California Blvd., Pasadena, California 91109, U.S.A.
SCHMIDT (Dr Th.), Landessternwarte Heidelberg-Königstuhl, 69 Heidelberg, F.R.G.
SCHMIDT-KALER (Prof. Dr Th.), 581 Witten, Steinhügel 105, F.R.G.
SCHMITT (A.), Observatoire de Strasbourg, Strasbourg (Bas-Rhin), France
SCHMITTER Y MARTIN DEL CAMPO (Dr E.F.), Institute of Astronomy, UNAM, Apartado Postal 70–264, México, D.F., Mexico
SCHNEIDER (Dr M.), Observatoire de Nice, Le Mont-Gros, 06300 Nice, France
SCHNELL (Mrs Dr A.), Universitäts-Sternwarte, Türkenschanzstrasse 17, A-1180 Vienna, Austria
* SCHOLZ (Dr M.), Institut für Theoretische Astrophysik der Universität, 6900 Heidelberg, Berliner Str. 19, F.R.G.

SCHOMBERT (Dr J.L.), U.S. Naval Observatory, Washington, D.C. 20390, U.S.A.
SCHÖNEICH (Dr W.), Forschungsgruppenleiter am Zentralinstitut für Astrophysik (Bereich Sternphysik), Potsdam G.D.R.
* SCHRAMM (Dr D.N.), Departments of Astronomy and Physics, PMA 17.220, The University of Texas, Austin, Texas 78712, U.S.A.
SCHROEDER (Dr D.J.), Beloit College, Beloit, Wisconsin 53511, U.S.A.
SCHRÖTER (Dr E.H.), Universitäts-Sternwarte, Geismarlandstr. 11, 34 Göttingen, F.R.G.
SCHRUTKA-RECHTENSTAMM (Prof. Dr G.), Universitäts-Sternwarte, Türkenschanzstrasse 17, A-1180 Vienna, Austria
SCHUBART (Dr J.), Astronomisches Rechen-Institut, Mönchhofstrasse 12–14, 69 Heidelberg, F.R.G.
SCHUCKING (Dr E.L.), Department of Physics, New York University, New York, N.Y. 10012. U.S.A.
SCHULER (Dr W.), Sternwarte der Kantonsschule, 4500 Solothurn, Switzerland
SCHULTE (Dr D.H.), Itek Corporation, 10 Maguire Road, Lexington, Massachusetts, U.S.A.
SCHÜRER (Prof. Dr M.), Astronomisches Institut der Universität Bern, Sidlerstrasse 5, CH-Bern, Switzerland
SCHWARZ (Dr U.J.), Kapteyn Astronomical Laboratory, Postbus 800, Groningen, The Netherlands
SCHWARZSCHILD (Prof. M.), Princeton University Observatory, Peyton Hall, Princeton, New Jersey 08540, U.S.A.
SCIAMA (Prof. D.W.), Department of Astrophysics, South Parks Road, Oxford OX1 3RQ, U.K.
SCONZO (Dr P.), 29 Old Mystic Street, Arlington, Mass. 02174, U.S.A.
SCOTT (Prof. E.L.), Statistical Laboratory, University of California, Berkeley, California 94720, U.S.A.
SCOTT (F.P.), 531826 Road N., Arlington, Va. 22207, U.S.A.
SCOTT (Dr P.F.), Mullard Radio Astronomy Observatory, Cavendish Laboratory, Free School Lane, Cambridge, U.K.
SEAQUIST (Dr E.R.), David Dunlap Observatory, Richmond Hill, Ontario, Canada
SEARLE (Dr L.), Hale Observatories, 813 Santa Barbara Street, Pasadena, California 91106, U.S.A.
SEARS (Dr L.), Astronomy Department, University of Michigan, Ann Arbor, Michigan 48104, U.S.A.
SEATON (Prof. M.J.), Department of Physics, University College London, Gower Street, London WC1, U.K.
SEDDON (H.), The Royal Observatory, Edinburgh 9, U.K.
* SEDMAK (Dr G.), Osservatorio Astronomico, Via Tiepolo 11, Trieste, Italy
SEEGER (Prof. Ch.), Department of Astronomy, New Mexico State University, Box 4500, Las Cruces, N.M. 88001, U.S.A.
* SEEGER (Dr P.A.), Los Alamos Scientific Laboratory, University of California, Los Alamos, New Mexico 87544, U.S.A.
SEGGEWISS (Dr W.), Universitätssternwarte, Poppelsdorfer Allee 49, 5300 Bonn, F.R.G.
SEHNAL (Dr L.), Astronomical Institute, Czechoslovak Academy of Sciences, Observatory Ondřejov, Czechoslovakia
SEIDELMANN (Dr P.K.), U.S. Naval Observatory, Washington, D.C. 20390, U.S.A.
SEIELSTAD (Dr G.A.), California Institute of Technology, Pasadena, California 91109, U.S.A.
SEITTER (Dr W.C.), Universitäts-Sternwarte Bonn, Observatorium Hoher List, 5568 Daun/Eifel, F.R.G.
SEKANINA (Dr Z.), Smithsonian Astrophysical Observatory, 60 Garden Street, Cambridge, Mass. 02138, U.S.A.
SEKIGUCHI (Dr N.), Tokyo Astronomical Observatory, University of Tokyo, Mitaka, Tokyo, Japan

SEMEL (Dr M.), Observatoire de Meudon, 92190 Meudon, France

SEN (Dr H.K.), High Temperature Section, Thermal Radiation Laboratory, Air Force Cambridge Research Center, Bedford, Massachusetts, U.S.A.

* SEN (Dr S.N.), Indian Association for the Cultivation of Science, Jadavpur, India

SENGBUSCH (Dr K.), Max-Planck-Institut für Physik und Astrophysik, Föhringer Ring 6, 8000 München 23, F.R.G.

SERKOWSKI (Dr K.), Lunar and Planetary Laboratory, University of Arizona, Tucson, Arizona 85721, U.S.A.

SÉRSIC (Dr J.L.), Observatorio Astronómico, Córdoba, Argentina

SERVAJEAN (Dr R.), Observatoire de Paris, Section d'Astrophysique, 92190 Meudon, France

SETTI (G.), Laboratorio di Radioastronomia, Istituto di Fisica, Bologna, Italy

ŠEVARLIĆ (Prof. Dr B.M.), Directeur, Institut d'Astronomie, Faculté des sc. nat. et math., Université de Beograd, Volgina 7, 11050 Beograd, Yugoslavia

SEVERNYJ (Prof. Dr A.B.), Director, Crimean Astrophysical Observatory of the USSR Academy of Sciences, P/O Nauchny, Crimea, U.S.S.R.

SHAH (Dr G.A.), Indian Institute of Astrophysics, c/o Raman Research Institute, Hebbal, Bangalore-560006, India

SHAKESHAFT (Dr J.R.), Mullard Radio Astronomy Observatory, Cavendish Laboratory, Free School Lane, Cambridge, U.K.

* SHAKHBAZIAN (Dr Y.L.), Pulkovo Observatory, Leningrad 196140, U.S.S.R.

SHAKHOVSKOJ (Dr N.M.), Crimean Astrophysical Observatory, USSR Academy of Sciences, Crimea, U.S.S.R.

SHANE (Dr C.D.), P.O. Box 582, Santa Cruz, California 95060, U.S.A.

SHANE (W.W.), Sterrewacht, Leiden 2401, The Netherlands

SHAO (Mr Cheng-yuan), Harvard College Observatory, 60 Garden Street, Cambridge, Massachusetts 02138, U.S.A.

SHAPIRO (Dr I.I.), Building 54–622, Massachusetts Institute of Technology, Cambridge, Mass. 02139, U.S.A.

SHAPIRO (Dr M.M.), Laboratory for Cosmic Ray Physics, Code 7020, Naval Research Laboratory, Washington, D.C. 20390, U.S.A.

SHAPLEY (A.H.), CRPL, National Bureau of Standards, Boulder, Colorado, U.S.A.

SHAPLEY (Mrs M.B.), Sharon Cross Road, Peterboro, New Hampshire, U.S.A.

SHARAF (Mrs Dr Sh.G.), Institute of Theoretical Astronomy, USSR Academy of Sciences, Leningrad, U.S.S.R.

SHAROV (Dr A.S.), Sternberg Astronomical Institute, Moscow 117234, U.S.S.R.

SHARPLESS (Dr S.L.), Department of Physics and Astronomy, The University of Rochester, Rochester, New York 14627, U.S.A.

SHAVIV (Dr G.), Department of Physics and Astronomy, Tel-Aviv University, Ramat-Aviv, Israel

SHAW (Prof. J.H.), Department of Physics and Astronomy, Ohio State University, 174 W. 18th Ave., Columbus, Ohio 43210, U.S.A.

SHAW (Prof. R.W.), Clark Hall of Science, Cornell University, Ithaca, N.Y., U.S.A.

SHCHEGLOV (Dr P.V.), Sternberg Astronomical Institute, Moscow 117234, U.S.S.R.

SHCHEGLOV (Prof. Dr V.P.), Director of the Astronomical Observatory, Tashkent, U.S.S.R.

SHCHEGOLEV (Dr D.E.), Pulkovo Observatory, Leningrad 196140, U.S.S.R.

SHCHERBINA-SAMOJLOVA (Mrs Dr I.S.), Institute of Science and Technics Information, Department of Astronomy, USSR Academy of Sciences, Moscow, U.S.S.R.

SHCHIGOLEV (Prof. B.M.), Sternberg Astronomical Institute, Moscow 117234, U.S.S.R.

SHEELEY JR (Dr N.T.), Kitt Peak Observatory, 950 North Cherry Avenue, Tucson, P.O. Box 26732, Arizona 85726, U.S.A.

SHEN (Dr B.S.P.), Flower and Cook Observatory, University of Pennsylvania, Philadelphia, Pennsylvania 19104, U.S.A.

SHEN (Dr Zee), Department of Mathematics, National Taiwan University, Taipei, Taiwan
SHEPTUNOV (Dr G.S.), Main Astronomical Observatory, USSR Academy of Sciences, Pulkovo, U.S.S.R.
SHER (Dr D.), 2837 Minto Drive, Apt. 2, Cincinnati, Ohio 45208, U.S.A.
SHERIDAN (K.V.), CSIRO, Division of Radiophysics, P.O. Box 76, Epping, N.S.W. 2121, Australia
* SHIM (Dr Woon Taik), Department of Astrogeophysics, Sung Kyun Kwan University, Seoul, Republic of Korea
SHIMIZU (Prof. Dr M.), The Institute of Space and Aeronautical Science, The University of Tokyo, Tokyo, Japan
SHIMIZU (Dr T.), Institute of Astrophysics, College of Science, University of Kyoto, Kyoto, Japan
* SHIMMINS (A.J.), Australian National Radio Astronomy Observatory, P.O. Box 276, Parkes, N.S.W. 2870, Australia
SHIRYAEV (Dr A.V.), Astronomical Observatory, Leningrad University, Leningrad, U.S.S.R.
SHKLOVSKY (Dr I.S.), Sternberg Astronomical Institute, Universit. prospekt 13, Leningrad 117234, U.S.S.R.
* SHOBBROOK (Dr R.R.), The Chatterton Astronomy Department, School of Physics, University of Sydney, Sydney, N.S.W. 2006, Australia
SHOEMAKER (E.), U.S. Geological Survey, Menlo Park, California, U.S.A.
* SHOLOMITSKY (Dr G.B.), Institute for Space Research, Profsoyuznaja ul. 88, Moscow 117485, U.S.S.R.
SHORE (Dr B.W.), T Division, Lawrence Livermore Laboratory, P.O. Box 808, Livermore, California 94550, U.S.A.
SHTEINS (Dr K.A.), Latvian University, Riga, U.S.S.R.
SHU (Dr F.), Department of Earth and Space Sciences, State University of New York, Stony Brook, N.Y. 11790, U.S.A.
SHUKLA (K.S.), Reader, Department of Mathematics and Astronomy, Lucknow University, Lucknow, U.P., India
SHUL'BERG (Dr A.R.), Astronomical Observatory, Odessa University, Odessa, U.S.S.R.
* SHUL'MAN (Dr L.M.), Main Astronomical Observatory of the Ukrainian Academy of Sciences, Kiev 252127, U.S.S.R.
* SHULOV (Dr O.S.), Astronomical Observatory of Leningrad University, 10 linia 33, Leningrad 199178, U.S.S.R.
SHUTER (Dr W.L.H.), Department of Physics, University of British Columbia, Vancouver 8, B.C., Canada
SIDA (Prof. D.W.), Carleton University, Ottawa, Ontario, Canada
SIEGEL (C.L.), Göttingen University, Bunsenstrasse 3–5, Göttingen, F.R.G.
* SILK (Dr J.), University of California, Astronomy Department, Berkeley, California 94720, U.S.A.
* SILVERBERG (Dr E.), University of Texas, McDonald Observatory, Austin, Texas 78712, U.S.A.
ŠIMEK (Dr M.), Astronomical Institute, Czechoslovak Academy of Sciences, Observatory Ondřejov, Czechoslovakia
* SIMKIN (Dr S.), Research Associate & Lecturer, Columbia University, New York, N.Y. 11027, U.S.A.
SIMODA (Dr M.), Department of Astronomy and Earth Science, Tokyo Gakugei University, Koganei-shi, Tokyo, 184 Japan
SIMON (Dr G.W.), Sacramento Peak Observatory, Sunspot, New Mexico 88349, U.S.A.
SIMON (Dr M.), State University of New York, Stony Brook, New York 11790, U.S.A.
* SIMON (Dr N.R.) Behlen Observatory, University of Nebraska, Lincoln, Nebraska 68508, U.S.A.
SIMON (Dr P.), Observatoire de Paris, Section d'Astrophysique, 92190 Meudon, France
SIMON (Prof. Dr R.L.E.), Institut d'Astrophysique, Université de Liège, avenue de Cointe, 5, B-4200 Cointe-Ougrée, Belgium

* SIMONENKO (Mrs Dr A.N.), Committee on Meteorites, USSR Academy of Sciences, ul. M. Ulianovoj 3, Korp. 1, Moscow 117313, U.S.S.R.
* SIMONSON (Dr S.C.), University of Maryland, College Park, Maryland 20742, U.S.A.

SIMOVLJEVIĆ (Dr J.), Prirodno matem. fakultet, katedra za astronomiju Beograd, Studentski trg 16, Beograd, Yugoslavia

* SINCLAIR (Dr A.Th.), Royal Greenwich Observatory, Herstmonceux Castle, Hailsham, Sussex, U.K.

SINNERSTAD (Dr U.), Stockholm Observatory, S-133 00 Saltsjöbaden, Sweden

SINTON (Dr W.M.), Institute for Astronomy, 2840 Koluwalu St., University of Hawaii, Honolulu, Hawaii 96822, U.S.A.

SINVHAL (Dr S.D.), Director, Uttar Pradesh State Observatory, Naini Tal, India

SINZI (Dr A.M.), Hydrographic Office, 5 chome, Tsukuji, Chuo-ku, Tokyo, Japan

SISSON (G.M.), Planetrees, Wall, Hexham, Northumberland, U.K.

* SISTERO (Dr R.F.), Sucre 1275, 2° E Córdoba, Argentina

SITARSKI (Dr G.), Astronomical Institute of Polish Academy of Sciences, Al. Ujazdowskie 4, Warszawa, Poland

SITNIK (Dr G.F.), Sternberg Astronomical Institute, Moscow 117234, U.S.S.R.

SIVARAMAN (K.R.), Astrophysical Observatory, Kodaikanal 3, India

SKALAFURIS (Dr A.), Department of Physics, Bartol Research Foundation of the Franklin Institute, Swarthmore, Pennsylvania 19081, U.S.A.

SKUMANICH (Dr A.), High Altitude Observatory, Boulder, Colorado 80302, U.S.A.

SLAUCITAJS (Prof. Dr S.), Observatorio Astronómico, La Plata, Argentina

SLAVENAS (Prof. Dr P.V.), Kestucio gatve 17–3, Vilnius, Lithuanian SSR, U.S.S.R.

SLEBARSKI (T.B.), University Observatory, St. Andrews, Scotland, U.K.

SLEE (Dr O.B.), CSIRO, Division of Radiophysics, P.O. Box 76, Epping, N.S.W. 2121, Australia

SLETTEBAK (Prof. Dr A.), Perkins Observatory, Delaware, Ohio, U.S.A.

SLONIM (Mrs Dr E.M.), Astronomical Observatory, Tashkent, U.S.S.R.

SLYSH (Dr V.I.), Institute of Cosmical Research, USSR Academy of Sciences, Moscow, U.S.S.R.

SMAK (Dr J.), Astronomical Observatory of Warsaw University, Al. Ujazdowskie 4, Warsaw, Poland

SMART (Prof. W.M.), Westbourne House, Westbourne Road, Lancaster, U.K.

SMERD (Dr S.F.), CSIRO, Division of Radiophysics, P.O. Box 76, Epping, N.S.W. 2121, Australia

SMEYERS (Dr P., Abbé), Astronomisch Instituut, Katholieke Universiteit Leuven, Naamsestraat 61, B-3000 Leuven, Belgium

SMILEY (Dr C.H.), Ladd Observatory, Brown University, Providence, R.I., U.S.A.

SMIT (Prof. J.A.), Physical Laboratory of the University, Bijlhouwerstraat 6, Utrecht, The Netherlands

SMITH (Dr A.G.), Department of Physics, University of Florida, Gainesville, Florida, U.S.A.

* SMITH (Dr A.M.), Observational Astronomy Branch, Goddard Space Flight Center, Greenbelt, Maryland 20771, U.S.A.

SMITH (B.A.), The Observatory, New Mexico State University, Las Cruces, New Mexico 88001, U.S.A.

* SMITH (Dr D.), High Altitude Observatory, Boulder, Colorado, U.S.A.

SMITH (Mrs Dr E.v.P.), Department of Physics and Astronomy, University of Maryland, College Park, Maryland 20742, U.S.A.

SMITH (Prof. F. Graham), Nuffield Radio Astronomy Laboratories, Jodrell Bank, Macclesfield, Cheshire, U.K.

SMITH (Dr G.), Department of Astrophysics, University Observatory, South Parks, Road, Oxford, U.K.

SMITH (Prof. Harlan J.), Director, McDonald Observatory, Astronomy Department, Room 407 Physics Bld., The University of Texas, Austin, Texas 78712, U.S.A.

SMITH (Dr Henry J.), National Aeronautics and Space Administration, Code SG, Washington 25, D.C., U.S.A.
SMITH (H.M.), Royal Greenwich Observatory, Herstmonceux Castle, Hailsham, Sussex, U.K.
* SMITH (Dr L.F.), Max-Planck-Institut für Radioastronomie, 5300 Bonn, Argelanderstrasse 3, F.R.G.
SMITH (Dr M.G.), Observatorio Interamericano de Cerro Tololo, Casilla 63-D, La Serena, Chile
* SMITH (Dr R. Connon), Astronomy Centre, University of Sussex, Falmer, Brighton, Sussex, BN1 9QH, U.K.
* SMITH (Dr W.H.), Princeton University, Princeton, New Jersey 08540, U.S.A.
* SMOLINSKI (Dr J.), Krasinkskiego 63a m 12, 87–100 Toruń, Poland
SMOLUCHOWSKI (Prof. R.), Solid State Sciences, Princeton University, Princeton, New Jersey 08540, U.S.A.
* SMRIGLIO (Dr F.), Osservatorio Astronomico, Viale del Parco Mellini 84, Roma, Italy
SMYTH (Dr M.J.), University of Edinburgh, Department of Astronomy, Royal Observatory, Edinburgh EH9 3HJ, U.K.
* SNYDER (Dr L.E.), Astronomy Department, University of Virginia, Charlottesville, Virginia 22903, U.S.A.
SOBERMAN (Dr R.K.), General Electric Co., Valley Forge Space Technology Center, Box 8555, Philadelphia, Pa. 19101, U.S.A.
SOBIESKI (Dr S.), Astrophysics Branch, Code 613, NASA, Goddard Space Flight Center, Greenbelt, Maryland 20771, U.S.A.
SOBOLEV (Dr V.M.), Main Astronomical Observatory, USSR Academy of Sciences, Pulkovo, U.S.S.R.
SOBOLEV (Prof. Dr V.V.), The University, Leningrad, U.S.S.R.
SOBOLEVA (Mrs Dr N.S.), Special Astrophysical Observatory, USSR Academy of Sciences, Leningrad Branch of the SAO, Pulkovo 196140, U.S.S.R.
SOCHER (Dr H.), Universitäts-Sternwarte Wien, Türkenschanzstrasse 17, A-1180 Vienna, Austria
SOFIA (Dr S.), University of South Florida, Department of Astronomy, Tampa, Florida 33620, U.S.A.
* SOLHEIM (Dr J.-E.), Nordlysobservatoriet, Universitetet i Tromsö, 9001 Tromsö, Norway
SOLOMON (Dr P.M.), Astronomy Department, Columbia University, Box 110 Pupin Hall, New York, N.Y. 10027, U.S.A.
SOMERVILLE (Dr W.B.), University College, Gower Street, London WC1, U.K.
SONETT (Dr C.P.), Department of Planetary Sciences and Lunar and Planetary Laboratory, University of Arizona, Tucson, Arizona 85721, U.S.A.
* SØRENSEN (N.G.), Department of Physics, University of Aarhus, Aarhus, Denmark
SOROCHENKO (Dr R.L.), Physical Institute, USSR Academy of Sciences, Moscow, U.S.S.R.
SOUFFRIN (P.), Observatoire de Nice, Le Mont-Gros, 06300 Nice, France
SOUFFRIN (Mrs Dr S.), Institut d'Astrophysique, 98bis boulevard Arago, 75014 Paris, France
* SOULIE (G.), Observatoire de l'Université de Bordeaux, 33270 Floirac, France
SOUTHWORTH (Dr R.B.), Harvard College Observatory, 60 Garden Street, Cambridge, Massachusetts 02138, U.S.A.
* SPADA (G.), Laboratorio di Astrofisica, Frascati, Roma, Casella Postale 67, Italy
* SPARKS (Dr W.M.), Laboratory for Optical Astronomy, Goddard Space Flight Center, Greenbelt, Maryland 20771, U.S.A.
SPEER (Dr R.J.), Physics Department, Imperial College, Prince Consort Road, London SW 7, U.K.
SPIEGEL (Dr E.), Courant Institute of Mathematical Sciences, 251 Mercer Street, New York, N.Y. 10003, U.S.A.
SPINRAD (Dr H.), Berkeley Astronomical Department, University of California, Berkeley, California, U.S.A.

SPITE (F.), Observatoire de Meudon, 5, place Janssen, 92190 Meudon, France

SPITE (Mrs M.), Observatoire de Meudon, 92190 Meudon, France

SPITZER JR (Dr L.), Director of the Princeton University Observatory, Peyton Hall, Princeton, New Jersey 08540, U.S.A.

* SREEKANTAN (Dr B.V.), Tata Institute of Fundamental Research, Homi Bhabha Road, Bombay 400 005, India

SREENIVASAN (Dr S.R.), Department of Physics, University of Calgary, Calgary, Alberta, Canada

STABELL (R.), Institute of Theoretical Astrophysics, University of Oslo, P.O. Box 1029, Blindern, Oslo 3, Norway

STAHR-CARPENTER (Mrs Dr M.), Leander McCormick Observatory, Box 3818, University Station, Charlottesville, Virginia 22903, U.S.A.

* STANDISH Jr. (Dr E.M.), Jet Propulsion Laboratory, California Institute of Technology, 4800 Oak Grove Drive, Mail Stop CPB-104, Pasadena, California 91103, U.S.A.

STANILA (Dr G.), Observatoire Astronomique, str. Cutitul de Argint 5, Bucarest 28, Roumania

STANLEY (G.J.), California Institute of Technology, Pasadena 4, California, U.S.A.

* STARRFIELD (Dr S.G.), Arizona State University, Tempe, Arizona 85281, U.S.A.

STARYTSIN (Dr G.V.), Main Astronomical Observatory, USSR Academy of Sciences, Pulkovo 196140, U.S.S.R.

STAWIKOWSKI (Dr A.), Astronomical Observatory of Toruń University, Sienkiewicza 30, 87–100 Toruń, Poland

STEAD (Dr Elizabeth A.), University of Wales, Department of Applied Mathematics, University College of Swansea, Singleton Park, Swansea, GLAM SA2 8PP, U.K.

STEARNS (Dr C.L.), Van Vleck Observatory, Middletown, Connecticut, U.S.A.

STECHER (T.P.), Code 672, Goddard Space Flight Center, Greenbelt, Maryland 20771, U.S.A.

STEIGER (Prof. W.R.), Hawaii Institute of Geophysics, University of Hawaii, 2525 Correa Rd., Honolulu, Hawaii 96822, U.S.A.

* STEIGMAN (Dr G.), Yale University Observatory, New Haven, Conn. 06520, U.S.A.

STEIN (Dr R.F.), Brandeis University, Waltham, Massachusetts, U.S.A.

STEIN (Dr W.A.), University of California, Physics Department, La Jolla, California, U.S.A.

STEINBERG (Dr J.L.), Service de Radioastronomie, Observatoire de Paris, Section d'Astrophysique, 92190 Meudon, France

STEINITZ (Dr R.), Institute of Planetary and Space Physics, Tel-Aviv University, Ramat-Aviv, Israel

STEINLIN (Dr U.W.), Astronomisches Institut der Universität Basel, Venusstrasse 7, CH-4102 Binningen, Switzerland

* STELLMACHER (Dr G.), Institut d'Astrophysique, 98bis, boulevard Arago, 75014 Paris, France

STENFLO (Dr J.O.), Institute of Astronomy, S-222 24 Lund, Sweden

* STEPANIAN (Dr A.A.), Crimean Astrophysical Observatory, p. Nauchny, Crimea 334413, U.S.S.R.

* STEPANIAN (Mrs Dr N.N.), Crimean Astrophysical Observatory, p. Nauchny, Crimea 334413, U.S.S.R.

STEPANOV (Dr W.E.), Post Office 65, Sibizmiran, Irkutsk, U.S.S.R.

STEPHENSON (Prof. C.B.), Warner and Swasey Observatory, Taylor and Brunswick Roads, East Cleveland, Ohio 44112, U.S.A.

* STEPIEN (Dr K.), Astronomical Observatory of the Warsaw University, Al. Ujazdowskie 4, Warszawa 00–478, Poland

ŠTERNBERK (Dr B.), Astronomical Institute, Czechoslovak Academy of Sciences, Budečska 6, Prague 2, Czechoslovakia

STESHENKO (Dr N.V.), Crimean Astrophysical Observatory, USSR Academy of Sciences, P/O Nauchny, Crimea 334413, U.S.S.R.

* STEVENS (Dr G.A.), Laboratorium voor Ruimteonderzoek, Beneluxlaan 21, Utrecht, The Netherlands

STEWART (Dr P.), Department of Mathematics, The University, Manchester M13 9PL, U.K.

STEYAERT (Dr H.L.C.), Sterrenkundig Instituut, Rijksuniversiteit Gent, J. Plateaustraat, 22, B-9000 Gent, Belgium

STIBBS (Prof. D.W.N.), University Observatory, Buchanan Gardens, St. Andrews, Five, U.K.

STICKER (Prof. Dr B.), Direktor des Instituts für Geschichte der Naturwissenschaften, Hartungstrasse 5, Hamburg 13, F.R.G.

STIEFEL (E.), Mathematische Abteilung, Eidgenössische Technische Hochschule, Zürich, Switzerland

* STIX (Dr M.), Universitäts-Sternwarte, 3400 Göttingen, Geismarlandstr. 11., F.R.G.

STOBIE (Dr R.S.), The Royal Observatory, Blackford Hill, Edinburgh EH9 3HJ, U.K.

STOCK (Dr J.), CONICIT, Apartado 264, Mérida, Caracas, Venezuela

* STOCKTON (Dr A.), Institute for Astronomy, University of Hawaii, Honolulu, Hawaii 96822, U.S.A.

STODÓŁKIEWICZ (Dr J.S.), Astronomical Observatory of Warsaw University, Warszawa, Al. Ujazdowskie 4, Poland

ŠTOHL (Dr J.), Astronomical Institute, Slovak Academy of Sciences, Dúbravska cesta, Bratislava, Czechoslovakia

STONE (Dr R.G.), Code 693, Laboratory for Extraterrestrial Physics, NASA-Goddard Space Flight Center, Greenbelt, Maryland 20771, U.S.A.

STOY (Dr R.H.), Royal Observatory, Edinburgh 9, U.K.

STOYKO (Mrs A.), Observatoire de Paris, 61, avenue de l'Observatoire, 75014 Paris, France

STOYKO (Dr N.), Observatoire de Paris, 61, avenue de l'Observatoire, 75014 Paris, France

STRAIŽYS (Dr V.), Astronomijos observatorija, Vilnius 31, Čiurlionio 29, Lithuania, U.S.S.R.

STRAND (Dr K.Aa.), U.S. Naval Observatory, Washington, D.C. 20390, U.S.A.

STRASSL (Prof. Dr H.), Direktor des Astronomischen Instituts der Universität, Steinfurterstrasse 107, Münster (Westfalen), F.R.G.

* STRAUSS (Dr F.M.), Instituto de Física, UFRGS, Depto. Astronomía, R. Luiz Englert S/N, 90000 Porto-Alegre, RGS., Brazil

STRITTMATTER (P.A.), Peterhouse, Cambridge, U.K.

STROBEL (Dr W.), Astronomisches Rechen-Institut, Mönchhofstrasse 12–14, 69 Heidelberg, F.R.G.

STROHMEIER (Prof. Dr W.), Direktor der Remeis-Sternwarte, Sternwartstrasse 7, Bamberg, F.R.G.

STROM (Dr R.G.), Lunar and Planetary Laboratory, University of Arizona, Tucson, Arizona 75721, U.S.A.

* STROM (Dr R.G.), Sterrewacht, 2401, Leiden, The Netherlands

STROM (Dr S.E.), Kitt Peak National Observatory, P.O. Box 26732, Tucson, Arizona 85726, U.S.A.

STRÖMGREN (Prof. B.), Observatoriet, Østervoldgade 3, DK-1350 Copenhagen, Denmark

STRONG (Prof. J.D.), Astronomy Research Facility, University of Massachusetts, Amherst, Mass. 01002, U.S.A.

STUMPFF (Dr P.), Max-Planck-Institut für Radioastronomie, Argelanderstrasse 3, 53 Bonn, F.R.G.

STURCH (Dr C.), Department of Astronomy, University of Western Ontario, London 72, Ontario, Canada

STURROCK (Dr P.A.), Institute for Plasma Research, Stanford University, Stanford, California, U.S.A.

SUDA (Dr K.), Astronomical Institute, Tohoku University, Sendai, Japan

SUEMOTO (Dr Z.), Department of Astronomy, Faculty of Science, University of Tokyo, Bunkyo-ku, Tokyo, Japan

SUGAWA (Dr C.), International Latitude Observatory of Mizusawa, Mizusawa-shi, Iwate-ken, Japan

SUGIMOTO (Dr D.), Institute of Earth Science and Astronomy, College of General Education, University of Tokyo, 3–8–1 Komaba, Meguro-ku, Tokyo, 153 Japan

SUKHAREV (Dr L.A.), Pulkovo Observatory, Leningrad 196140, U.S.S.R.
* SULLIVAN (Dr W.T. III), Department of Astronomy, University of Washington, Seattle, Washington 98195, U.S.A.
SULTANOV (Dr G.F.), Shemakha Astrophysical Observatory, Azerbajdzhan Academy of Sciences, Baku, U.S.S.R.
SUSLOV (Dr A.K.), Saltykov-Shtshedrin st.h. 12 fl. 28, Leningrad D-28, U.S.S.R.
* SUZUKI (Dr Y.), 23–1 Nakajima Hironocho Uji-si, Kyoto, Japan
SVATOŠ (Dr J.), Department of Astronomy and Astrophysics, Charles University, Švédská 8, Prague 5-Smíchov, Czechoslovakia
SVECHNIKOVA (Mrs Dr M.A.), Astronomical Observatory, Ural University, Sverdlovsk, U.S.S.R.
ŠVESTKA (Dr Z.), Fraunhofer Institut, Schöneckstrasse 6, 78 Freiburg im Breisgau, F.R.G.
SVOLOPOULOS (Dr S.N.), Professor of Astrophysics, University of Athens, Panepistimiopolis, Athens (621), Greece
* SWANENBURG (Dr B.N.), Cosmic Ray Working Group, Kamerlingh Onnes Laboratorium, Nieuwsteeg 18, Leiden, The Netherlands
SWARUP (Dr G.), Tata Institute of Fundamental Research, Bombay, India
SWEET (Prof. P.A.), The University Observatory, Glasgow W2, U.K.
* SWEIGART (Dr A.V.), Yale University, New Haven, Conn. 06520, U.S.A.
SWENSON JR (Dr G.W.), Vermillion River Observatory, University of Illinois, Urbana, Ill. 61801, U.S.A.
SWENSSON (Dr J.W.), Institute of Physics, Sölvegatan 14, S-223 62 Lund, Sweden
SWIHART (Dr T.L.), Steward Observatory, University of Arizona, Tucson, Arizona 85721, U.S.A.
SWINGS (Dr J.P.), 23, avenue Léon Souguenet, B-4050 Esneux, Belgium
SWINGS (Prof. Dr P.), 23, avenue Léon Souguenet, B-4050 Esneux, Belgium
SWOPE (Miss H.H.) Mount Wilson and Palomar Observatories, 813 Santa Barbara Street, Pasadena, California 91106, U.S.A.
SYKES (Dr J.B.), 20 Milton Lane, Steventon, Abingdon, Berkshire OX13 6SA, U.K.
SYKES-HART (Mrs A.B.), University Observatory, Oxford, U.K.
SYMMS (L.S.T.), 9, The Grove, Bexhill-on-Sea, Sussex, U.K.
SYROVATSKIJ (S.I.), Lebedev Physical Institute, Moscow, U.S.S.R.
SYTINSKAYA (Mrs Dr N.N.), University Observatory, Leningrad, U.S.S.R.
SYUNYAEV (Dr R.A.), Institute of Applied Mathematics, USSR Academy of Sciences, Moscow, U.S.S.R.
SZAFRANIEC (Miss Dr R.), Astronomical Observatory, Kopernika 27, Cracow, Poland
SZEBEHELY (Dr V.), Dept. of Aerospace Engineering, The University of Texas, 2315 Speedway, Austin, Texas 78712, U.S.A.
SZEIDL (Dr B.), Konkoly Observatory, P.O. 114, Box 67, Budapest XII, Hungary

* TABARA (Dr H.), Tokyo Astronomical Observatory, Mitaka, Tokyo, Japan
* TADEMARU (Dr E.), University of Massachusetts, Amherst, Mass. 01002, U.S.A.
TAFFARA (Prof. S.), Osservatorio Astronomico, Padova, Italy
TAGLIAFERRI (Prof. Dr G.), Osservatorio Astrofisico di Arcetri, Largo Enrico Fermi 5, Florence, Italy
TAI (Dr Yuin-Kwei), Director, Institute of Geophysics, Central University, Taipei, Taiwan
TAKAGI (Dr S.), International Latitude Observatory of Mizusawa, Mizusawa-shi, Iwate-ken, Japan
TAKAKUBO (Dr K.), Astronomical Institute, Faculty of Science, Tohoku University, Sendai, Japan
TAKAKURA (Dr T.), Department of Astronomy, Faculty of Science, University of Tokyo, Bunkyo-ku, Tokyo, Japan
* TAKARADA (Dr K.), Kyoto University of Industrial Arts and Textile Fibers, Matsugasaki, Sa kyoku, Kyoto, Japan

TAKASE (Dr B.), Tokyo Astronomical Observatory, Mitaka, Tokyo, Japan
TAKAYANAGI (K.), Institute of Space and Aeronautical Science, University of Tokyo, Komaba, Meguro-ku, Tokyo, Japan
TAKENOUCHI (Dr T.), Tokyo Astronomical Observatory, Mitaka, Tokyo, Japan
* TAKEUTI (Dr M.), Astronomical Institute, Tohoku University, Sendai, Japan
TALWAR (S.P.), Department of Physics, University of Delhi, Delhi, India
TAMMANN (Dr G.A.), Astronomisches Institut der Universität Basel, Venusstrasse 7, CH-4102 Binningen, Switzerland
* TAMURA (Dr S.), Astronomical Institute, Tohoku University, Sendai, Japan
TANABE (Dr H.), Tokyo Astronomical Observatory, University of Tokyo, Mitaka, Tokyo, Japan
TANAKA (Dr H.), Research Institute of Atmospherics, Nagoya University, Ichida-machi, Toyokawa-shi, Aichi, Japan
* TANAKA (Dr K.), Tokyo Astronomical Observatory, Mitaka, Tokyo, Japan
* TANAKA (Dr R.), University of Niigata, Asahimachi, Niigata, Japan
TANAKA (Dr Y.), Werkgroep Kosmische Straling, Kamerlingh Onnes Laboratorium, Nieuwsteeg 18, Leiden, The Netherlands
* TANAKA (Dr Y.), Faculty of Education, Ibaraki University, Mito, Japan
TANDBERG-HANSSEN (Dr E.), High Altitude Observatory of the University of Colorado, Boulder, Colorado, U.S.A.
* TANDON (Dr J.N.), Department of Physics & Astrophysics, University of Delhi, Delhi-110007, India
TANNAHILL (Dr T.R.), University Observatory, Glasgow W2, U.K.
TANNER (R.W.), Seismology Division, Earth Physics Branch, Energy Mines and Resources, Ottawa, Canada
* TAPLEY (Dr B.D.), Dept. of Aerospace Engineering, University of Texas at Austin, Texas 78712, U.S.A.
TASSOUL (Dr J.L.), Département de Physique, Université de Montréal, P.O. Box 6128, Montréal 3, P.Q., Canada
TATEVYAN (Dr S.K.), Astronomical Council, USSR Academy of Sciences, Vavilov Str. 34, Moscow V-312, U.S.S.R.
* TATUM (Prof. J.B.), Department of Physics, University of Victoria, Victoria, British Columbia, Canada
TAUBER (Prof. G.E.), Department of Physics and Astronomy, Tel-Aviv University, Ramat-Aviv, Israel
TAVARES (Dr J.T.L.), Avenida Dias da Silva, 173, R/C Esq., Coimbra, Portugal
TAVASTSHERNA (Dr K.N.), Main Astronomical Observatory, USSR Academy of Sciences, Pulkovo, Leningrad M-140, U.S.S.R.
TAYLER (Dr R.J.), Astronomy Centre, University of Sussex, School of Mathematical and Physical Sciences, Falmer, Sussex, U.K.
TAYLOR (Dr D.J.), Steward Observatory, University of Arizona, Tucson, Arizona 85721, U.S.A.
TAYLOR (G.E.), H.M. Nautical Almanac Office, Royal Greenwich Observatory, Herstmonceux Castle, Hailsham, Sussex, U.K.
* TAYLOR JR (Dr J.H.), University of Massachusetts, Amherst, Mass. 01002, U.S.A.
TECH (Dr J.), U.S. Department of Commerce, National Bureau of Standards, Spectroscopy Section, Washington, D.C. 20234, U.S.A.
TÉHÉRANY (D.), 83, avenue Rey, Teheran, Iran
TEJFEL' (Dr V.G.), Astrophysical Institute of the Kazakh Academy of Sciences, Alma-Ata, U.S.S.R.
TELEKI (Dr G.), Astronomical Observatory, Volgina Str. 7, 11050 Belgrade, Yugoslavia
TEMESVÁRY (Prof. Dr S.), Department of Astronomy and Space Science, State University of New York, Albany, N.Y. 12203, U.S.A.
TEMPESTI (Prof. P.), Osservatorio Astronomico di Collurania, Teramo, Italy

Teplitskaya (Mrs Dr R.B.), SibIZMIR, USSR Academy of Sciences, Irkutsk, U.S.S.R.

* Terashita (Dr Y.), Technical University of Kanazawa, Nonoichi-cho Minami-Kyoku, Kanazawa, Japan

Ter Haar (Dr D.), Magdalen College, Oxford, U.K.

Terrazas (L.R.), Instituto Nacional de Astrofísica, Optica y Electrónica, Tonantzintla, Apartados Postales Nos. 216 y 51, Puebla, Mexico

Terien (Dr J.), Directeur, Bureau International des Poids et Mesures, Pavillon de Breteuil, F 92-Sèvres (Seine-et-Oise), France

Terzan (Dr A.), Observatoire de Lyon, 69-St-Genis-Laval, France

Terzian (Dr Y.), Center for Radiophysics and Space Research, Space Sciences Building, Cornell University, Ithaca, N.Y. 14850, U.S.A.

Teske (Dr R.G.), Astronomy Department, University of Michigan, Ann Arbor, Michigan, U.S.A.

Texereau (J.), Observatoire de Paris, 61, avenue de l'Observatoire, 75014 Paris, France

Thackeray (Prof. Dr A.D.), c/o Department of Astronomy, University of Cape Town, Rondebosch, 7700, South Africa

* Thaddeus (Dr P.), Institute for Space Studies, 2880 Broadway, New York, N.Y. 10025, U.S.A.

Thé (Dr Pik Sin), Astronomical Institute, University of Amsterdam, Roetersstraat 15, Amsterdam C, The Netherlands

Thernöe (K.A.), University Observatory, Østervoldgade 3, DK-1350 Copenhagen, Denmark

Thiry (Prof. Y.), Professeur à la Faculté des Sciences de l'Université de Paris, 7–9 Quai St. Bernard, Paris 5, France

Thomas (Dr D.V.), Royal Greenwich Observatory, Herstmonceux Castle, Hailsham, Sussex, U.K.

Thomas (Dr H.C.), Max-Planck-Institut für Physik und Astrophysik, Föhringer Ring 6, 8000 München 23, F.R.G.

* Thomas (Dr J.H.), Dept. of Mechanical & Aerospace Sciences, The University of Rochester, Rochester, New York 14627, U.S.A.

* Thomas (Dr R.J.), Code 682, NASA-Goddard Space Flight Center, Greenbelt, Maryland 20771, U.S.A.

Thomas (Dr R.N.), Institut d'Astrophysique, 98bis, boulevard Arago, 75014 Paris, France

Thompson (Dr A.R.), National Radio Astronomy Observatory, Edgemont Road, Charlottesville, Virginia 22901, U.S.A.

Thompson (Dr G.I.), The Royal Observatory, Edinburgh 9, U.K.

Thomson (M.M.), 209 Dovercourt Avenue, Ottawa, Ontario K1Z 7H3, Canada

Thorne (Dr K.), Physics Department, California Institute of Technology, Pasadena, California 91109, U.S.A.

Tifft (Dr W.G.), Steward Observatory, University of Arizona, Tucson, Arizona 85721, U.S.A.

Tifrea (Dr Emilia), Observatoire Astronomique, str. Cutitul de Argint 5, Bucarest 28, Roumania

* Tinbergen (Dr J.), Sterrewacht, Leiden, The Netherlands

Ting (Yeou-Tswen), Chief of Astronomical Department, Taiwan Provincial Weather Bureau, 64, Kung Yuen Road, Taipei, Taiwan

* Tinsley (Dr Beatrice), Institute for Physical Science, The University of Texas at Dallas, P.O. Box 30365, Dallas, Texas 75230, U.S.A.

Tiuri (Prof. Dr M.), Technical University, Department of Electrotechnics, Albert Str. 40–42, Helsinki, Finland

* Tjin a Djie (Dr H.R.E.), Sterrekundig Instituut, Universiteit van Amsterdam, Roetersstraat 15, Amsterdam, The Netherlands

Tlamicha (Dr A.), Astronomical Institute, Czechoslovak Academy of Sciences, Observatory Ondřejov, Czechoslovakia

Todoran (Dr I.), Observatoire, Université, Cluj, Roumania

* Tofani (Ing. G.), Osservatorio Astrofisico di Arcetri, Largo Enrico Fermi 5, Firenze, Italy

TOLBERT (Dr C.R.), Leander McCormick Observatory, University of Virginia, Charlottesville, Virginia 22903, U.S.A.
* TOMASKO (Dr M.G.), Lunar & Planetary Laboratory, University of Arizona, Tucson, Arizona, U.S.A.
TOMBAUGH (C.W.), New Mexico State University, Research Center, University Park, New Mexico, U.S.A.
* TOMITA (Dr K.), Research Institute for Theoretical Physics, Hiroshima University, Takehara, Hiroshima-ken, Japan
* TOMITA (K.), Tokyo Astronomical Observatory, Mitaka, Tokyo, Japan
TOOMRE (Dr A.), Department of Mathematics, Massachusetts Institute of Technology, Cambridge, Mass. 02139, U.S.A.
* TOOMRE (Dr J.), University of Colorado, Boulder, Colorado 80302, U.S.A.
TORAO (Prof. Dr M.) Yoyogi 5–7–9–401, Shibuya, Tokyo, Japan
* TORELLI (Dr M.), Osservatorio Astronomico, Viale del Parco Mellini, 84, Roma, Italy
TORGÅRD (Miss Dr I.H.M.), Nils Ericssonsgatan 28, 521 00 Falköping, Sweden
TORRES-PEIMBERT (Dr S.), Instituto de Astronomía, UNAM, Apartado Postal 70–264, México D.F., Mexico
TORROJA (Prof. J.M.), University of Madrid, Fac. de Ciencias, Madrid 3, Spain
TOUSEY (Dr R.), U.S. Naval Research Laboratory, Washington 25, D.C., U.S.A.
TOVMASSIAN (Dr H.M.), Byurakan Astrophysical Observatory, Byurakan, Armenia, U.S.S.R.
TOWNES (Dr C.H.), Physics Department, University of California, Berkeley, California 94720, U.S.A.
TRAFTON (Dr L.M.), Department of Astronomy, University of Texas, Austin, Texas 78712, U.S.A.
TRAVING (Prof. Dr G.), Lehrstuhl für Theoretische Astrophysik der Universität Heidelberg, 69 Heidelberg, Berliner Str. 19, F.R.G.
TREANOR S.J. (Dr P.J.), Specola Vaticana, Castel Gandolfo, Vatican City State
TREDER (Prof. Dr H.-J.), Zentralinstitut für Astrophysik, Sternwarte Babelsberg, DDR 1502 Potsdam-Babelsberg, Rosa-Luxemburg-Strasse 17a, G.D.R.
TREFFTZ (Miss Dr E.), Max-Planck-Institut für Physik und Astrophysik, Aumeisterstrasse 6, München 23, F.R.G.
TREHAN (Dr S.K.), Department of Mathematics, Panjab University, Chandigarh 14, India
TRELLIS (M.), Observatoire de Nice, Le Mont-Gros, 06300 Nice, France
TREMKO (Dr J.), Skalnaté Pleso Observatory, Slovak Academy of Sciences, Tatranská Lomnica, Czechoslovakia
TREXLER (J.H.), U.S. Naval Research Laboratory, Washington 25, D.C., U.S.A.
* TRIMBLE (Dr V.L.), Department of Physics, University of California, Irvin, CA 92664, U.S.A.
TROITSKY (Prof. Dr V.S.), Radiophysical Research Institute, Gorkij, U.S.S.R.
* TRÜMPER (Prof. Dr J.), Astronomisches Institut der Universität Tübingen, 7400 Tübingen, Waldhäuserstr. 64, F.R.G.
TRURAN (Dr J.W.), Belfer Graduate School of Science, Yeshiva University, Amsterdam Ave. and 186th Str., New York, 10033, U.S.A.
TRUTSE (Dr Yu.L.), Institute of Physics of Atmosphere, USSR Academy of Sciences, Moscow, U.S.S.R.
TS'AI (Chang-hsien), Taipei Observatory, Yuan Shan, Taipei, Taiwan
TSAO (Prof. Mo), 47, 3rd Section, Hsin-I Road, Taipei, Taiwan
TSAP (Dr T.T.), Crimean Astrophysical Observatory, USSR Academy of Sciences, Crimea, U.S.S.R.
TSCHARNUTER (Dr W.M.), Institut für Theoretische Astronomie, University of Vienna, Türkenschanzstrasse 17, A-1180 Vienna, Austria
TSESEVICH (Prof. Dr V.P.), Director of the Astronomical Observatory, Odessa, U.S.S.R.
TSUBAKI (Dr T.), Institute of Earth Science and Astrophysics, University of Shiga, Ohtsu, Japan

TSUBOKAWA (Dr I.), Professor of Tokyo University, Seismological Observatory Institute, No. 1–1 gou-1 chome, Yayoimachi, Bunkyoku, Tokyo, Japan

TSUCHIYA (A.), Tokyo Astronomical Observatory, University of Tokyo, Mitaka, Tokyo, Japan

TSUJI (Dr T.), Department of Astronomy, Faculty of Science, University of Tokyo, Bunkyo-ku, Tokyo 113, Japan

TSURUTA (Dr S.), Theoretical Studies Branch, Code 641, NASA, Goddard Space Flight Center, Greenbelt, Maryland 20771, U.S.A.

TUCKER (R.H.), Royal Greenwich Observatory, Herstmonceux Castle, Hailsham, Sussex, U.K.

TÜFEKÇIOĞLU (Dr Z.), Astronomy Department, University of Ankara, Fen Fakultesi, Ankara, Turkey

TULL (Dr R.C.), Department of Astronomy, University of Texas, Austin, Texas 78712, U.S.A.

* TUOMINEN (Dr I.), Observatory and Astrophysics Laboratory, University of Helsinki, Tähtitorninmäki, SF-00130 Helsinki 13, Finland

TUOMINEN (Prof. J.V.), University of Helsinki, Observatory and Astrophysics Laboratory, Tähtitorninmäki, SF-00130 Helsinki 13, Finland

TURŁO (Dr Z.), Astronomical Observatory, Toruń-Piwnice, Poland

* TURNER (Dr B.E.), National Radio Astronomy Observatory, Charlottesville, Virginia 22901, U.S.A.

TURNER (Dr K.C.), Department of Terrestrial Magnetism, Carnegie Institute of Washington, 5241 Broad Branch Road, N.W., U.S.A.

TURTLE (Dr A.J.), The Chatterton Astrophysics Department, School of Physics, University of Sydney, 2006 Australia

TUTUKOV (Dr A.V.), Astronomical Council, USSR Academy of Sciences, Vavilov Str. 34, Moscow V-312, U.S.S.R.

TUVE (Dr M.A.), Carnegie Institution of Washington, Department of Terrestrial Magnetism, 5241 Broad Branch Road, N.W., Washington 15, D.C., U.S.A.

TUZI (Dr K.), 2–2–20 Ôsawa, Mitaki-shi, Tokyo, Japan

TWISS (Dr R.Q.), National Physical Laboratory, Teddington, Middlesex, U.K.

TYLER JR (Dr G.L.), Center for Radar Astronomy, Stanford University, Stanford, California 94305, U.S.A.

UCHIDA (Dr J.), Department of Technology, Tohoku-Gakuin University, Tagajo-machi, Miyagi-Ken, Japan

UCHIDA (Dr Y.), Tokyo Astronomical Observatory, University of Tokyo, Mitaka, Tokyo, Japan

UDAL'TSOV (Dr V.A.), Physical Institute, USSR Academy of Sciences, Moscow, U.S.S.R.

UENO (Dr S.), Department of Astronomy, University of Southern California, University Park, Los Angeles, California 90007, U.S.A.

UESUGI (Dr A.), Department of Astronomy, University of Kyoto, Kyoto, Japan

UETA (Dr J.), Department of Astronomy, Faculty of Science, Kyoto University, Kyoto, Japan

* ULMSCHNEIDER (Dr P.), Astronomisches Institut und Sternwarte der Universität, 8700 Würzberg, Büttnerstrasse 72, F.R.G.

ULRICH (Dr B.), Astronomy Department, University of Texas, Austin, Texas 78712, U.S.A.

ULRICH (Mrs Dr M.H.), Institut d'Astrophysique, 98bis, Boulevard Arago, Paris 75014, France

* ULRICH (Dr R.K.), Department of Astronomy, University of California, LA, Los Angeles, California 90024, U.S.A.

UNDERHILL (Dr A.B.), Code 670, Goddard Space Flight Center, Greenbelt, MD. 20771, U.S.A.

UNDERWOOD (J.H.), Aerospace Corporation, P.O. Box 95085, Los Angeles, California 90045, Mail Station 120/1909, U.S.A.

UNNO (Dr W.), Department of Astronomy, Faculty of Science, University of Tokyo, Bunkyo-ku, Tokyo, Japan

UNSÖLD (Prof. Dr A.), Direktor des Instituts für Theoretische Physik, Olshausenstrasse, Neue Universität, Bau 13/I, Kiel, F.R.G.

UPGREN (Dr A.), Department of Astronomy, Wesleyan University, Middletown, Connecticut 06457, U.S.A.

UPTON (Dr E.K.L.), Astronomy Department, University of California, Los Angeles, California 90024, U.S.A.

URBARZ (Dr H.), Astronomisches Institut der Universität, Waldhäuser-Strasse 64, 74 Tübingen, F.R.G.

URECHE (Dr V.), University of Cluj, Cluj, Roumania

UREY (Prof. H.C.), University of California, La Jolla, California, U.S.A.

USHER (Dr P.D.), 507 Davey Laboratory, Department of Astronomy, Pennsylvania State University, University Park, Pennsylvania 16802, U.S.A.

* UTSUMI (Dr K.), Department of Astronomy, Faculty of General Education, Hiroshima University, Hiroshima, Japan

* UUS (Dr H.U.), W. Struve Tartu Astrophysical Observatory, Tôravere, Tartu 202444 Estonia, U.S.S.R.

* VAIANA (Dr G.), American Science & Engineering Inc., 955 Massachusetts Avenue, Cambridge, Massachusetts 02139, U.S.A.

VAINSTEIN (Dr L.A.), Physical Institute, USSR Academy of Sciences, Moscow, U.S.S.R.

VALNÍČEK (Dr B.), Astronomical Institute, Czechoslovak Academy of Sciences, Observatory Ondřejov, Czechoslovakia

VAN AGT (Dr S.L.Th.J.), Sterrekundig Instituut, Toernooiveld, 6805 Nijmegen, The Netherlands

VAN ALBADA (Dr T.S.), Kapteyn Astronomical Laboratory, Hoogbouw W.S.N.-Postbus 800, Groningen 8002, The Netherlands

VAN ALLEN (Dr J.A.), Department of Physics and Astronomy, University of Iowa, Iowa City, Iowa 52240, U.S.A.

VAN ALTENA (Dr W.F.), Yerkes Observatory, Astrometric Program, Williams Bay, Wisconsin 53191, U.S.A.

* VAN BLERKOM (Dr D.J.), University of Massachusetts, Amherst, Massachusetts 01002, U.S.A.

VAN BREDA (Dr I.G.), University Observatory, Buchanan Gardens, St Andrews, U.K.

VAN BUEREN (Prof. H.G.), Astronomical Observatory 'Sonnenborgh', Zonnenburg 2, Utrecht, The Netherlands

VAN DE HULST (Prof. Dr H.C.), Sterrewacht, Leiden, Kaiserstraat, The Netherlands

VAN DE KAMP (Prof. Dr P.), Director of the Sproul Observatory, Swarthmore College, Swarthmore, Pennsylvania 19081, U.S.A.

VANDEKERKHOVE (Dr E.), Astronome Honoraire, Observatoire Royal de Belgique, avenue de Sumatra, 7, B-1180 Bruxelles, Belgium

VAN DEN BERGH (Prof. S.), David Dunlap Observatory, Richmond Hill, Ontario, Canada

* VANDEN BOUT (Dr P.), Department of Astronomy, College of Natural Sciences, University of Texas at Austin, Austin, Texas 78712, U.S.A.

VAN DEN HEUVEL (E.P.J.), Astronomical Institute, Zonnenburg 2, Utrecht, The Netherlands

VAN DER BORGHT (Prof. R.), Department of Mathematics, Monash University, Melbourne, Australia

* VAN DER KRUIT (Dr P.C.), Hale Observatories, 813 Santa Barbara Street, Pasadena, California 91101, U.S.A.

VAN DER LAAN (Prof. Dr H.), Sterrewacht, Kaiserstraat 57, Leiden, The Netherlands

VANDERLINDEN (Prof. ém. Dr H.L.), University Ghent, Hertogsdreef, 15, B-1170 Brussels, Belgium

VANDERVOORT (Prof. P.O.), Ryerson Physical Laboratory, 1100–14 East 58th Street, Chicago, Illinois 60637, U.S.A.

VAN DIGGELEN (Dr J.), Sterrewacht 'Sonnenborgh', Zonnenburg 2, Utrecht, The Netherlands

VAN DUINEN (R.J.), University of Groningen, Astronomy Department, Space Research Division, Groningen, The Netherlands

VAN FLANDERN (Dr T.C.), U.S. Naval Observatory, Washington D.C. 20390, U.S.A.

VAN GENDEREN (Dr A.M.), Leiden Southern Station, P.O. Box 13, Broederstroom, Transvaal, Republic of South Africa

VAN HERK (Dr G.), Sterrewacht, Leiden, The Netherlands

VAN HOOF (Prof. Dr A.), Astronomisch Instituut, Katholieke Universiteit Leuven, Naamsestraat, 61, B-3000 Leuven, Belgium

VAN HORN (Dr H.M.), University of Rochester, River Campus Station, Rochester, New York 14627, U.S.A.

VAN HOUTEN (Dr C.J.), Sterrewacht, Leiden, The Netherlands

VAN HOUTEN-GROENEVELD (Mrs Dr I.), Sterrewacht, Leiden, The Netherlands

VAN LEER (Dr B.), The Observatory, Leiden, The Netherlands

* VAN NIEUWKOOP (Dr Ir. J.), Astronomical Institute, Servaas Bolwerk 13, Utrecht, The Netherlands

VAN REGEMORTER (Dr H.), Observatoire de Paris, Section d'Astrophysique, 92190 Meudon, France

* VAN RENSBERGEN (Dr W.), Dorpsstraat, 36, B-2958 Weerde, Belgium

VAN T'VEER (Mrs Dr C.), Institut d'Astrophysique, 98bis, Boulevard Arago, 75014 Paris, France

VAN T'VEER (Dr F.), Institut d'Astrophysique, 98bis, Boulevard Arago, 75014 Paris, France

VAN WOERDEN (Dr H.), Kapteyn Astronomical Laboratory, Postbus 800, Groningen, The Netherlands

VANÝSEK (Prof. V.), Department of Astronomy and Astrophysics, Charles University, Švédská 8, 15000 Praha 5-Smíchov, Czechoslovakia

* VARDANIAN (Dr R.A.), Byurakan Astrophysical Observatory, Byurakan, Armenia, U.S.S.R.

VARDYA (Dr M.S.), Astrophysics Division, Tata Institute of Fundamental Research, Colaba, Bombay 5, India

VARSAVSKY (Dr C.M.), Las Heras 1975 (6°A), Buenos Aires, Argentina

VASIL'EV (Dr V.M.), Main Astronomical Observatory, USSR Academy of Sciences, Pulkovo, U.S.S.R.

VASIL'EVA (Mrs Dr G.J.), Main Astronomical Observatory, USSR Academy of Sciences, Pulkovo, U.S.S.R.

VASILEVSKIS (Dr S.), Lick Observatory, University of California, Santa Cruz, California 95064, U.S.A.

VASILYEV (Dr O.B.), Observatory of the University, Pulkovo h. 13 fl. 24, Leningrad M-140, U.S.S.R.

* VAUCLAIR (Dr G.), Observatoire de Paris, Section d'Astrophysique de Meudon, 92190 Meudon, France

VAUGHAN JR (Dr A.H.), Mount Wilson and Palomar Observatory, 813 Santa Barbara Street, Pasadena, California 91106, U.S.A.

* VEISMANN (Dr U.K.), Tartu Astrophysical Observatory, Tôravere, 202444 Estonia, U.S.S.R.

VELGHE (Prof. Dr A.G.), Directeur de l'Observatoire Royal de Belgique, avenue Circulaire, 3, B-1180 Bruxelles, Belgium

* VELTMANN (Dr Ü.I.K.), W. Struve Astrophysical Observatory of Tartu, Tôravere 202444, Estonia, U.S.S.R.

* VENUGOPAL (Dr V.R.), Radio Astronomy Centre, T.I.F.R., P. Box 8, Ootacamund, Tamil Nadu, India

VERBAANDERT (Prof. Dr A.G.), Astronome Honoraire, Observatoire Royal de Belgique, avenue de l'Observatoire, 90, B-1180 Bruxelles, Belgium

VERGNANO (Prof. A.), Osservatorio Astronomico, Pino Torinese, Italy

VERNIANI (Prof. F.), Instituto di Fisica, Via Irnerio, 46, Bologna 40126, Italy

VERON (Dr P.), Observatoire de Meudon, 92190 Meudon, France

VERSCHUUR (Dr G.L.), Fiske Planetarium, Department of Astro-geophysics, University of Colorado, Boulder, Colorado 80302, U.S.A.

* VESECKY (Dr J.F.), Department of Astronomy, University of Leicester, Leicester, LE1 7RH, U.K.

VETEŠNÍK (Dr M.), Astronomical Institute, Purkyně University, Kotlářská 2, Brno, Czechoslovakia

* VEVERKA (Dr J.), Center for Radiophysics and Space Research, Cornell University, Ithaca, New York 14850, U.S.A.

VICENTE (Prof. Dr R.O.), R. Mestre Aviz, 30, R/c, Algés, Portugal

VIDAL (Dr J.M.C.), Avenida José Antonio 677, Barcelona, Spain

* VIDAL-MADJAR (A.), Laboratoire de Physique Stellaire et Planétaire, B.P. n° 10, 91370 Verrières-le-Buisson, France

VILA (Dr S.C.), Department of Astronomy, University of Pennsylvania, Philadelphia, Pennsylvania 19104, U.S.A.

* VILHU (Dr O.), Observatory and Astrophysics Laboratory, University of Helsinki, Tähtitornimäki, SF-00130 Helsinki 13, Finland

VINTI (Dr P.J.), Rm. 37–340, M.I.T. Measurement Systems Laboratory, 70 Vassar Street, Cambridge, Massachusetts 02139, U.S.A.

* VIOTTI (Dr R.), Laboratorio di Astrofisica, C.P. 67, 00044 Frascati (Roma), Italy

VIRGOPIA (Dr N.), Osservatorio Astronomico, Via Trionfale 204, 00136, Rome, Italy

VISVANATHAN (Dr N.), 787 Sawmill Brook Parkway, Newton Center, Massachusetts 02159, U.S.A.

VISWANATHAN (N.), Mount Wilson and Palomer Observatories, 813 Santa Barbara Street, Pasadena, California, U.S.A.

VITINSKIJ (Dr Yu. I.), Main Astronomical Observatory, USSR Academy of Sciences, Pulkovo, U.S.S.R.

* VLACHOS (Dr D.G.), Laboratory of Astronomy, Technical University of Athens, 5 K. Zografou Street, Zografou, Athens (624), Greece

VOIGT (Prof. Dr H.H.), Direktor der Universitäts-Sternwarte, Geismarlandstrasse 11, 34 Göttingen, F.R.G.

VOLLAND (Dr H.), Astronomisches Institut der Universität, Poppelsdorfer Allee 49, 53 Bonn, F.R.G.

VON DER HEIDE (Dr J.E.B.), Hamburger Sternwarte, Hamburg-Bergedorf, F.R.G.

VON HOERNER (Dr S.), National Radio Astronomy Observatory, P.O. Box 2, Green Bank, West Virginia, U.S.A.

VON KLÜBER (Dr H.), The Observatories, Madingley Road, Cambridge, U.K.

VON WEIZSÄCKER (Prof. C.F.), Max-Planck-Institut, Riemerschmidstrasse 7, 813 Starnberg, F.R.G.

VORONTSOV-VEL'YAMINOV (Prof. Dr B.A.), Sternberg Astronomical Institute, Moscow V-234, U.S.S.R.

VOROSHILOV (Dr V.I.), Main Astronomical Observatory, Ukrainian Academy of Sciences, Kiev, U.S.S.R.

VSEKHSVYATSKIJ (Prof. Dr S.K.), Astronomical Observatory, Kiev 53, U.S.S.R.

* VUJNOVIĆ (Dr V.), Institute of Physics of the University, P. Box 304, 41001 Zagreb, Yugoslavia

* VUKICEVIĆ-KARABIN (Dr M.), Department of Astronomy, Faculty of Sciences, Belgrade University, Studentski trg 16, P.O. Box 550, 11001 Beograd, Yugoslavia

WACHMANN (Prof. Dr A.A.), Augustastrasse 2, 205 Hamburg 80, F.R.G.

WACKERNAGEL (Dr H.B.), 2939 Country Club Drive, Colorado Springs, Colorado 80909, U.S.A.

* WADDINGTON (Dr C.J.), University of Minnesota, Minneapolis, Minn. 55455, U.S.A.

WADE (Dr C.M.), National Radio Astronomy Observatory, P.O. Box 2, Green Bank, West Virginia, U.S.A.

WAGMAN (Dr N.E.), Allegheny Observatory, Riverview Park, Pittsburgh, Pennsylvania 15214, U.S.A.

* WAGNER (Dr W.J.), Sacramento Peak Observatory, Sunspot, New Mexico 88349, U.S.A.
* WAKO (Dr Y.), International Latitude Observatory, Mizusawa, Iwate, Japan
* WALBORN (Dr N.R.), Cerro Tololo Inter-American Observatory, Casilla 63-D, La Serena, Chile

WALDMEIER (Prof. Dr M.), Director, Swiss Federal Observatory, Schmelzbergstrasse 25, 8006 Zürich, Switzerland

* WALKER Jr. (Prof. A.B.C.), Institute for Plasma Research, Stanford University, Via Crespi, Stanford, California 95305, U.S.A.

WALKER (Dr G.A.H.), Geophysics and Astronomy Department, University of British Columbia, Vancouver 8, B.C., Canada

WALKER (Dr M.F.), Lick Observatory, University of California, Santa Cruz, California 95060, U.S.A.

* WALKER (Dr R.L.), U.S. Naval Observatory, Flagstaff Station, Flagstaff, Arizona 86001, U.S.A.

WALKER (Dr R.), Washington University, St Louis, Missouri 63130, U.S.A.

WALKER (W.S.G.), Variable Star Section, RAS of New Zealand, 18 Pooles Road, Greerton, Tauranga, New Zealand

WALLACE (Dr L.V.), Kitt Peak National Observatory, 950 North Cherry Avenue, Tucson, P.O. Box 26732, Arizona 85726, U.S.A.

WALLENQUIST (Prof. Dr Å.A.E.), Norrlandsgatan 34D, S-752 29 Uppsala, Sweden

WALLERSTEIN (Dr G.N.), Department of Astronomy, Seattle, Washington, U.S.A.

* WALLIS (Dr M.K.), Mathematical Institute, 24 St Giles, Oxford, U.K.

WALRAVEN (Dr Th.), Leiden Southern Station, P.O. Box 13, Broederstroom, Transvaal, Republic of South Africa

WALSH (Prof. D.), Nuffield Radio Astronomy Laboratories, Jodrell Bank, Macclesfield, U.K.

* WALTER (Dr H.G.), Mecklenburger Str. 103, 6100 Darmstadt - Eberstadt, F.R.G.

WALTER (Prof. Dr K.), Astronomisches Institut der Universität Waldhäuserstrasse 64, 74 Tübingen, F.R.G.

WAMPLER (Dr E.J.), Lick Observatory, Mount Hamilton, California 95140, U.S.A.

WAPSTRA (Prof. A.), Van Woensel Kooylaan 48, Naarden, The Netherlands

WARES (Dr G.W.), CRF Astrophysics, Space Physics Laboratory, Air Force Cambridge Research Laboratory, Laurence G. Hanscom Field, Bedford, Massachusetts 01730, U.S.A.

* WARMAN (Dr J.), Instituto di Astronomía, UNAM, Apartado Postal 70–264, México D.F., Mexico

WARNER (Dr B.), Department of Astronomy, University of Cape Town, Rondebosch, C.P., South Africa

WARWICK (Dr J.W.), High Altitude Observatory of the University of Colorado, Boulder, Colorado, U.S.A.

WARZEE (Dr J.), Astronome Honoraire, Observatoire Royal de Belgique, rue Mattot, 115, B-1410 Waterloo, Belgium

WATERFIELD (Dr R.L.), The Observatory, Woolston, Yeovil, Somerset, U.K.

* WATERWORTH (Dr M.D.), University of Tasmania, Department of Physics, Box 252C, G.P.O., Hobart, Tasmania 7001, Australia
* WATSON (Dr W.), Department of Physics and Department of Astronomy, University of Illinois, Urbana, Illinois 61801, U.S.A.

WATTENBERG (Prof. D.), Direktor der Archenhold-Sternwarte, Berlin-Treptow, G.D.R.

WAYMAN (Dr P.A.), Director, Dunsink Observatory, Castleknock, Co. Dublin, Ireland

WEAVER (Dr H.F.), Radio Astronomy Laboratory, University of California, Berkeley 4, California, U.S.A.

* WEBBER (Dr J.), University of Illinois, Urbana, Illinois 61801, U.S.A.

WEBROVÁ (Mrs Dr L.), Astronomical Institute, Czechoslovak Academy of Sciences, Budečská 6, Prague 2, Czechoslovakia

* WEBSTER (Mrs Dr B.L.), Radcliffe Observatory, P.O. Box 373, Pretoria, South Africa
* WEEDMAN (Dr D.), Dyer Observatory, Vanderbilt University, Box 1803 Station B, Nashville, Tenn 37203, U.S.A.
* WEEKES (Dr T.C.), Mount Hopkins Observatory, Box 97, Amado, Arizona 85640, U.S.A.
* WEHINGER (Dr P.A.), Tel-Aviv University, Ramat-Aviv, Tel-Aviv, Israel
WEHLAU (Dr A.), Department of Astronomy, University of Western Ontario, London, Ontario, Canada
WEHLAU (Prof. Dr W.H.), Department of Astronomy, University of Western Ontario, London, Ontario, Canada
WEIDEMANN (Prof. Dr V.), Institut für Theoretische Physik und Sternwarte der Universität, 23 Kiel, Olshausenstrasse, Neue Universität, Haus C 4/1, F.R.G.
WEIGERT (Dr A.), Director, Hamburg Observatory, 205 Hamburg 80, F.R.G.
* WEILER (Dr K.W.), Kapteyn Astronomy Laboratory, University of Groningen, P.O. Box 800, Groningen, The Netherlands
WEILL (G.), Institut d'Astrophysique, 98bis, Boulevard Arago, 75014 Paris, France
WEIMER (Dr Th.), Observatoire de Paris, 61, avenue de l'Observatoire, Paris-14e, France
WEINBERG (Dr J.L.), Space Astronomy Laboratory, State University of New York at Albany, Executive Park East, Stuyvesant Plaza, Albany, New York 12203, U.S.A.
WEISS (Dr N.O.), Department of Applied Mathematics and Theoretical Physics, Silver Street, Cambridge, CB3 9EW, U.K.
* WEISS (Dr W.W.), Universitätssternwarte Wien, Figl-Observatorium für Astrophysik, Türkenschanzstrasse 17, A-1180 Wien, Austria
* WELCH (Dr W.J.), Radio Astronomy Laboratory, University of California, Berkeley, California 94720, U.S.A.
WELIACHEW (Dr L.), Observatoire de Meudon, 92190-Meudon, France
WELLGATE (G.B.), Royal Greenwich Observatory, Herstmonceux Castle, Hailsham, Sussex, U.K.
WELLMANN (Prof. P.), Universitäts-Sternwarte, Sternwartstrasse 23, München 27, F.R.G.
WEMPE (Prof. Dr J.), Zentralinstitut für Astrophysik, DDR 15 Potsdam, Telegrafenberg, G.D.R.
WENDKER (Dr H.J.), Hamburger Sternwarte, 205 Hamburg 80, Gojenbergsweg 112, F.R.G.
* WENIGER (Dr S.), Observatoire de Paris, Section d'Astrophysique de Meudon, 92190 Meudon, France
WENTZEL (Dr D.G.), Astronomy Program, University of Maryland, College Park, Maryland 20740, U.S.A.
WENZEL (Dr W.), Zentralinstitut für Astrophysik, Sternwarte Sonneberg, DDR 64 Sonneberg, G.D.R.
WESSELINK (Dr A.J.), Yale University Observatory, Box 2023 Yale Station, New Haven, Connecticut 06520, U.S.A.
* WESSELIUS (Dr P.R.), Department of Space Research, University of Groningen, P. O. Box 800, Groningen, The Netherlands
WEST (Dr R.M.), ESO Sky Atlas Laboratory, ESO TP Division, CERN, CH-1211 Genève 23, Switzerland
WESTERHOUT (Dr G.), Astronomy Program, University of Maryland, College Park, Maryland 20740, U.S.A.
WESTERLUND (Dr B.E.), European Southern Observatory, Bergedorfer Strasse 131, 205 Hamburg 80, F.R.G.
WESTFOLD (Prof. K.C.), Monash University, Clayton 3168, Australia
* WETHERILL (Dr G.W.), University of California, LA, Los Angeles, California, U.S.A.
WEYMANN (Prof. R.J.), Steward Observatory, University of Arizona, Tucson, Arizona, U.S.A.
WHEELER (Dr J.A.), Joseph Henry Laboratory, Princeton University, Princeton, New Jersey 08540. U.S.A.

* WHEELER (Dr J.C.), Harvard College Observatory, 60 Garden Street, Cambridge, Massachusetts 02138, U.S.A.
* WHELAN (Dr J.A.J.), Institute of Astronomy, University of Cambridge, Cambridge, U.K.

WHIPPLE (Dr F.L.), Director, Smithsonian Astrophysical Observatory, 60 Garden Street, Cambridge, Massachusetts 02138, U.S.A.

WHITAKER (E.A.), Lunar and Planetary Laboratory, University of Arizona, Tucson, Arizona, U.S.A.

WHITE (Dr O.R.), High Altitude Observatory, P.O. Box 1558, Boulder, Colorado 80302, U.S.A.

WHITE (Dr R.E.), Steward Observatory, The University of Arizona, Tucson, Arizona 85721, U.S.A.

WHITEOAK (Dr J.B.), CSIRO, Division of Radiophysics, P.O. Box 76, Epping, N.S.W. 2121, Australia

WHITFORD (Dr A.E.), Director of the Lick Observatory, University of California, Santa Cruz, California 95060, U.S.A.

WHITNEY (Prof. B.S.), University of Oklahoma Observatory, Norman, Oklahoma, U.S.A.

WHITNEY (Dr C.A.), Smithsonian Astrophysical Observatory, 60 Garden Street, Cambridge, Massachusetts 02138, U.S.A.

WHITROW (Dr G.J.), Department of Mathematics, Imperial College of Science and Technology, Prince Consort Road, London, S.W. 7, U.K.

WICKRAMASINGHE (Sc.D. Prof. N.C.), University College, Department of Applied Mathematics and Mathematical Physics, P.O. Box 78, Cardiff CFL XL, U.K.

WIDING (Dr K.D.), Code 7144, U.S. Naval Research Laboratory, Washington D.C. 20390, U.S.A.

WIDORN (Dr T.), Universitäts-Sternwarte, Türkenschanzstrasse 17, A-1180 Vienna, Austria

WIEDLING (Dr T.), Villavägen 15, S-611 00 Nyköping, Sweden

* WIEHR (Dr E.), Universitäts-Sternwarte, 3400 Göttingen, Geismarlandstrasse 11, F.R.G.

WIELEBINSKI (Dr R.), Max-Planck-Institut für Radioastronomie, Argelanderstrasse 3, 5300 Bonn, F.R.G.

WIELEN (Dr R.), Astronomisches Recheninstitut, Mönchhofstrasse 12–14, 6900 Heidelberg, F.R.G.

WIERZBIŃSKI (Prof. Dr St.), Astronomical Observatory, Kopernika 11, Wrocław, Poland

WIESE (W.L.), Chief, Plasma Spectroscopy Section, National Bureau of Standards, Physics Building A-149, Washington, D.C. 20234, U.S.A.

WIETH-KNUDSEN (Dr N.P.), Svend Trostsvej 12^{III}, Copenhagen-V, Denmark

WILCOCK (Dr W.L.), Department of Physics, University College of North Wales, Bangor, Caernarvonshire, U.K.

WILCOX (Dr J.M.), Institute for Plasma Research, Stanford University, Via Crespi, Stanford, California 94305, U.S.A.

WILD (Dr J.P.), CSIRO, Division of Radiophysics, P.O. Box 76, Epping, N.S.W., 2121, Australia

WILD (P.), Astronomisches Institut der Universität Bern, Sidlerstrasse 5, Bern, Switzerland

WILD (Dr P.A.T.), Astronomy Department, University of Cape Town, Rondebosch, Cape, South Africa

WILDEY (Dr R.L.), Center of Astrogeology, U.S. Geological Survey, 601 E.-Cedar Avenue, Flagstaff, Arizona 86001, U.S.A.

WILDT (Prof. R.), Yale University Observatory, Box 2023, Yale Station, New Haven, Connecticut 06520, U.S.A.

WILKINS (Dr G.A.), H.M. Nautical Almanac Office, Royal Greenwich Observatory, Herstmonceux Castle, Hailsham, Sussex, U.K.

WILLIAMS (Mrs Dr C.A.), Department of Astronomy, University of South Florida, Tampa, Florida 33620, U.S.A.

WILLIAMS (Dr D.), Radio Astronomy Laboratory, University of California, Berkeley, California, U.S.A.

WILLIAMS (Dr D.A.), Department of Mathematics, U.M.I.S.T., Manchester 1, U.K.
WILLIAMS (Dr I.P.), Department of Mathematics, Queen Mary College, Mile End Road, London E 1, U.K.
WILLIAMS (Dr J.A.), Physics Department, Albion College, Albion, Michigan 49224, U.S.A.
* WILLIAMS (Dr J.G.), Jet Propulsion Laboratory, 4800 Oak Grove Drive, Pasadena, California 91103, U.S.A.
* WILLIAMS (Dr P.M.), Institute of Astronomy, Madingley Road, Cambridge, U.K.
WILLIAMS (Dr R.E.), Steward Observatory, University of Arizona, Tucson 85721, U.S.A.
* WILLS (Dr B.J.), Department of Astronomy, University of Texas, Austin, Texas 78712, U.S.A.
WILLS (Dr D.), Department of Astronomy, University of Texas, Austin, Texas 78712, U.S.A.
WILLSTROP (Dr R.V.), University Observatories, Madingley Road, Cambridge, U.K.
WILSON (Dr A.G.), P.O. Box 113, Topanga, California, U.S.A.
WILSON (Prof. B.G.), Deputy President, Simon Fraser University, Burnaby, B.C., Canada
* WILSON (Dr L.), Department of Environmental Sciences, University of Lancaster, Bailrigg, Lancaster, U.K.
WILSON (Dr O.C.), Mount Wilson and Palomar Observatories, 813 Santa Barbara Street, Pasadena 4, California, U.S.A.
WILSON (Dr P.R.), Department of Applied Mathematics, University of Sydney, Sydney 2006, N.S.W., Australia
WILSON (Prof. R.), Department of Physics & Astronomy, University College London, Gower Street, London WC1E 6BT, U.K.
WILSON (Dr R.E.), Department of Astronomy, University of South Florida, Tampa, Florida 33620, U.S.A.
WILSON (Dr R.W.), Bell Laboratories, Crawford Hill Laboratory, Holmdel, New Jersey 07733, U.S.A.
* WILSON (Dr Th.L.), Max-Planck-Institut für Radioastronomie, 5300 Bonn, Argelanderstrasse 3, F.R.G.
WINCKLER (Dr J.R.), School of Physics and Astronomy, University of Minnesota, Minneapolis, Minnesota 55455, U.S.A.
* WING (Dr R.), The Ohio State University, Columbus, Ohio 43210, U.S.A.
WINKLER (Dr G.M.R.), U.S. Naval Observatory, Department of the Navy, Washington D.C. 20390, U.S.A.
* WINNBERG (Dr A.), Max-Planck-Institut für Radioastronomie, D-53 Bonn, Argelanderstrasse 3, F.R.G.
WISNIEWSKI (Dr W.), The University of Arizona, Lunar and Planetary Laboratory, Tucson, Arizona, U.S.A.
WITHBROE (Dr G.), Harvard College Observatory, 60 Garden Street, Cambridge, Massachusetts 02138, U.S.A.
WITKOWSKI (Prof. Dr J.), Director of the Observatory of Poznań, Ul. Sloneczna 36, Poznań, Poland
WITT (Dr A.N.), Department of Physics and Astronomy, University of Toledo, 2801 W. Bancroft Street, Toledo, Ohio 43606, U.S.A.
WLÉRICK (Dr G.), Observatoire de Paris, Section d'Astrophysique, 92190 Meudon, France
WOLFF (Dr S.C.), Institute of Astronomy, University of Hawaii, 2525 Correa Road, Honolulu, Hawaii 96822, U.S.A.
WOLSTENCROFT (Dr R.D.), Institute for Astronomy, University of Hawaii, Honolulu, Hawaii 96822, U.S.A.
WOLTJER (Prof. L.), Department of Astronomy, Columbia University, New York 10027, N.Y., U.S.A.
* WOOD (Dr D.B.), Goddard Space Flight Center, Greenbelt, Maryland, U.S.A.
WOOD (Prof. F.B.), University of Florida, Department of Physics and Astronomy, Gainesville, Florida 32601, U.S.A.

WOOD (Dr H.J. III), Department of Astronomy, University of Virginia, Charlottesville, Va. 22903, U.S.A.

WOOD (Dr H.W.), Government Astronomer, The Observatory, Sydney, N.S.W. 2000, Australia

WOOD (Dr J.A.), Smithsonian Astrophysical Observatory, 60 Garden Street, Cambridge, Massachusetts 02138, U.S.A.

WOOLF (Dr N.J.), School of Physics and Astronomy, University of Minnesota, Minneapolis, Minnesota 55455, U.S.A.

* WOOLFSON (Dr M.M.), Physics Department, University of York, Heslington, York Y01 5DD, U.K.

WOOLLEY (Sir Richard), Astronomer Royal, Royal Greenwich Observatory, Herstmonceux Castle, Hailsham, Sussex, U.K. – Until December 1976 South African Astronomical Observatory, P.O. Box 9, Observatory, Cape, South Africa

WOOLSEY (E.G.), Earth Physics Branch, Department of Energy, Mines & Resources, Ottawa, Ontario, Canada

WORLEY (C.E.), U.S. Naval Observatory, Washington, D.C. 20390, U.S.A.

WORRALL (Dr G.), The Observatories, Madingley Road, Cambridge, U.K.

WOSZCZYK (Dr A.), Astronomical Observatory, Toruń University, Sienkiewicza 30, Toruń, Poland

WRAY (Dr J.D.), Astronomy Department, Northwestern University, Evanston, Illinois 60201, U.S.A.

WRIGHT (Miss Dr F.W.), Harvard College Observatory, Cambridge, Massachusetts 02138, U.S.A.

WRIGHT (Miss Dr H.), 6040 Boulevard E., West New York, New Jersey 07093, U.S.A.

WRIGHT (Dr J.P.), Smithsonian Astrophysical Observatory, 60 Garden Street, Cambridge, Massachusetts 02138, U.S.A.

WRIGHT (Dr K.O.), Director, Dominion Astrophysical Observatory, R.R. 7, Victoria B.C., Canada

* WRIGHT (Dr M.C.H.), National Radio Astronomy Observatory, Edgemont Road, Charlottesville, Virginia 22901, U.S.A.

* WRIXON (Dr G.T.), Bell Telephone Labs, Crawford Hill Laboratory, Holmdel, N.J. 07733, U.S.A.

WURM (Prof. K.), Gartenweg 3, 8204 Brannenburg/Inn, F.R.G.

WYATT JR (Dr S.P.), University of Illinois Observatory, Urbana, Illinois, U.S.A.

* WYCKOFF (Dr S.), Tel-Aviv University, Ramat-Aviv, Tel-Aviv, Israel

WYLLER (Prof. A.A.), Stockholm Observatory, S-133 00 Saltsjöbaden, Sweden

WYNNE (Prof. Ch.G.), Applied Optics Section, Department of Physics, Imperial College of Science and Technology, Prince Consort Road, London S.W. 7, U.K.

XANTHAKIS (Prof. J.), Research and Computing Center, Academy of Sciences of Athens, Athens, Greece

YABUSHITA (Dr S.), Department of Applied Mathematics and Physics, Kyoto University, Kyoto, Japan

YABUUTI (Dr K.), Research Institute of Humanistic Science, Kyoto University, Kyoto, Japan

YAKHONTOVA (Mrs Dr N. Samojlova), Institute of Theoretical Astronomy, Leningrad, U.S.S.R.

YAKOVKIN (Prof. Dr A.A.), Main Astronomical Observatory of the Academy of Sciences of the Ukrainian SSR, Kiev 41, U.S.S.R.

YAKOVKIN (Dr N.A.), Kiev University Observatory, Kiev, U.S.S.R.

* YAMASHITA (Dr K.), Department of Physics, Nagoya University, Chikusa-ku, Nagoya, Japan

YAMASHITA (Dr Y.), Department of Astronomy, University of Tokyo, Yayoi, Bunkyo-ku, Tokyo, Japan

* YAMAZAKI (Dr A.), Hydrographic Department, 1st Regional Maritime Safety Headquarters, Otaru, Japan

YAPLEE (B.S.), Code 7134, U.S. Naval Research Laboratory, Washington D.C. 20390, U.S.A.
YAROV-YAROVOJ (Dr M.S.), Sternberg Astronomical Institute, Moscow, U.S.S.R.
YASUDA (Dr H.), Tokyo Astronomical Observatory, Mitaka, Tokyo, Japan
* YATSKIV (Dr Y.S.), Main Astronomical Observatory of the Ukrainian Academy of Sciences, Kiev-252127, U.S.S.R.
YAVNEL' (Dr A.A.), Meteorite Committee, USSR Academy of Sciences, Moscow, U.S.S.R.
* YEIVIN (Prof. Y.), Tel-Aviv University, Ramat-Aviv, Israel
YEN (Prof. J.L.), Department of Electrical Engineering, University of Toronto, Toronto 5, Ontario, Canada
YILMAZ (Dr F.), Observatoire de l'Université, Istanbul, Turkey
YILMAZ (Dr N.), Faculty of Science, Department of Astronomy, Ankara, Turkey
YOSS (Dr K.M.), University of Illinois Observatory, Urbana, Illinois, U.S.A.
YOUNG (Dr A.), School of Physical Sciences, The New University of Ulster, Coleraine, Co. Londonderry, U.K.
YOUNG (Dr A.T.), Department of Physics, Texas A+M University, College Station, Texas 77843, U.S.A.
* YOUSEF (Miss Dr M.A.M.), Astronomy Department, Faculty of Science, Cairo University, Cairo, Arab Republic of Egypt
YÜ (Dr Ching-Sung), Williams Observatory, Hood College, Frederick, Maryland, U.S.A.
* YU (Prof. Kyung Loh), Department of Earth Sciences, Teachers College, Seoul National University, Seoul, Republic of Korea
* YUAN (Dr Chi), Physics Department, City College of New York, New York, N.Y. 10031, U.S.A.
YUMI (Dr S.), International Latitude Observatory of Mizusawa, Mizusawa-shi, Iwate-ken, Japan
* YUN (Dr H.S.), Bartol Research Foundation, Pa 19081, U.S.A.

ZABRISKIE (Prof. Dr F.R.), Department of Astronomy, 102 Whitmore Laboratory, Pennsylvania State University, University Park, Pennsylvania 16802, U.S.A.
ZACHAROV (Dr I.), Astronomical Institute, Czechoslovak Academy of Sciences, Observatory Ondřejov, Czechoslovakia
ZADUNAISKY (Prof. P.E.), Observatorio Nacional de Física Cósmica, Depto. de Matemática Aplicada, A. Mitre 3.100-SAN MIGUEL-F.C.G.S.M. Pcia. de Buenos Aires, Argentina
ZAGAR (Prof. F.), Direttore dell' Osservatorio Astronomico, Via Brera 28, Milano, Italy
ZAHN (J.-P.), Observatoire de Nice, Le Mont-Gros, 06-Nice, France
* ZANDER (Dr R.J.), Institut d'Astrophysique de l'Université de Liège, B-4200 Cointe-Ougrée, Belgium
ZEL'DOVICH (Acad. Ya.B.), Institute of Applied Mathematics, Miusscaya pl. 4, Moscow A-47, U.S.S.R.
ZEL'MANOV (Dr A.L.), Sternberg Astronomical Institute, Moscow, U.S.S.R.
ZHELEZNIAKOV (Prof. Dr V.V.), Radiophysical Research Institute, Gorkij, U.S.S.R.
ZHEVAKIN (Prof. Dr S.A.), Radiophysical Research Institute, Gorkii, U.S.S.R.
ZHONGOLOVICH (Prof. Dr I.D.), Institute of Theoretical Astronomy, Leningrad, U.S.S.R.
* ZIEBA (Prof. A.), Astronomical Observatory, Jagellonian Observatory, Cracow 31-501 Krakow, Kopernika 27, Poland
ZIKIDES (Dr M.C.), Laboratory of Astronomy, University of Athens, Panepistimiopolis, Athens (621), Greece
ZIMMERMANN (Prof. Dr G.K.), Nikolaiev Observatory, Nikolaiev, U.S.S.R.
ZIMMERMANN (Dr H.) University Observatory, Jena, Schillergässchen 2, G.D.R.
ZIÓŁKOWSKI (Dr J.), Polish Academy of Sciences, Institute of Astronomy, Al. Ujazdowskie 4, Warsaw, Poland
ZIRIN (Dr H.), Hale Observatories, California Institute of Technology, 1201 E. California Boulevard, Pasadena, California 91109, U.S.A.

ZIRKER (Dr J.B.), Institute for Astronomy, 2840 Koluwalu St., University of Hawaii, Honolulu, Hawaii 96822, U.S.A.

ZONN (Prof. W.), Astronomical Observatory, Ujazdowskie 4, Warsaw, Poland

ZUCKERMANN (Dr B.M.), Department of Physics and Astronomy, University of Maryland, College Park, Maryland, U.S.A.

ZVEREV (Prof. Dr M.S.), Pulkovo Observatory, Leningrad-M 140, U.S.S.R.

ZWAAN (Dr C.), Astronomical Observatory 'Sonneborgh', Zonnenburg 2, Utrecht, The Netherlands

GPSR Compliance

The European Union's (EU) General Product Safety Regulation (GPSR) is a set of rules that requires consumer products to be safe and our obligations to ensure this.

If you have any concerns about our products, you can contact us on ProductSafety@springernature.com

In case Publisher is established outside the EU, the EU authorized representative is:

Springer Nature Customer Service Center GmbH
Europaplatz 3
69115 Heidelberg, Germany

Batch number: 10800581

Printed by Printforce, the Netherlands